Duxbury

Test Bank

for

Johnson and Kuby's

Elementary Statistics

Tenth Edition

Mohammed A. El-Saidi
Ferris State University

THOMSON
BROOKS/COLE

Australia • Brazil • Canada • Mexico • Singapore • Spain • United Kingdom • United States

Printed in the United States of America
1 2 3 4 5 6 7 09 08 07 06 05

Printer: Thomson/West

0-495-10530-9

Thomson Higher Education
10 Davis Drive
Belmont, CA 94002-3098
USA

Contents

Dedication

To my very special friend

Dr. Asmahan A. El Ezzi
Associate Professor of Biochemistry

Lebanese University, Beirut, Lebanon

Preface

This *Test Bank* is designed to accompany the tenth edition of *Elementary Statistics* by Robert Johnson and Patricia Kuby. Throughout this *Test Bank*, I have attempted to provide a good mix of difficulty level versus question type. In its entirety, this *Test Bank* contains 3925 true-false questions, multiple-choice questions, short-answer questions, and applied and computational questions. If you have any comments, corrections, or suggestions, please communicate with me via e-mail at *elsaidim@ferris.edu.*

In producing statistics exams, I try to satisfy several goals. These are (1) Test different aspects of the material covered, (2) Clearly differentiate between students of different ability, and (3) Require no memorization and make it easy to mark.

With these goals in mind, this *Test Bank* is organized by section within each chapter, and to some extent by the type of technique or concept. For each chapter, if possible, there are numerous true-false questions, multiple-choice questions, short-answer questions, and applied and computational questions followed immediately by the answers. Some of the questions are fairly straightforward, primarily requiring students to demonstrate their ability to perform various types of computations. Other questions describe an experiment and require students to recognize the appropriate statistical technique to use. Other questions require students to interpret Excel and Minitab output, perform statistical tests, and produce graphs using available statistical software.

I usually assume that students will be supplied with statistical tables and are allowed to bring in a formula sheet containing any information they choose to include. I usually allow one eight-and-a-half by eleven inches two-sided formula sheet for each exam.

My daughters Abby, Lily, Heidi, and Jessica deserve more than thanks for their patience during the preparation of this *Test Bank.* I extend my deepest regards to my parents Ahmed and Zenuib, my brothers Naguib and Emad, and my sister Laila in Egypt, and to my brother Gamal in U.S.A., for their continual love, encouragement and support. I am grateful to Ms. Ann Day, Senior Assistant Editor of Duxbury, for her patience, assistance, and great help and cooperation in the preparation of this *Test Bank.* Last, but certainly not least, I am thankful for the support of my friend Margaret T. Morrow for her great help in typing and editing this edition of the *Test Bank*

Mohammed A. El-Saidi, Ph.D.
Professor and Assistant Dean
Ferris State University

Chapter 1

Statistics

Sections1.1 and 1.2

True-False Questions

1. The field of statistics can be roughly subdivided into two areas: descriptive statistics and probability.
   ANSWER: F

2. Descriptive statistics includes the collection, presentation, and description of sample data.
   ANSWER: T

3. In the field of statistics, descriptive statistics includes collecting and describing data while inferential statistics involves interpreting results from the data.
   ANSWER: T

4. Eye color would be an example of qualitative data.
   ANSWER: T

5. Heights of professional basketball players would be an example of qualitative data.
   ANSWER: F

6. A company employs 750 individuals. To ascertain how the employees feel regarding a pension plan, 75 of the employees are surveyed. The proportion of the 75 employees who favor the pension plan is a parameter.
   ANSWER: F

7. A balance is used to measure weights to three decimal places. The data that result from this process would be classified as qualitative data.
   ANSWER: F

8. Attribute data and qualitative data are the same.
ANSWER: T

9. A population is the complete collection of individuals or objects or events whose properties are to be analyzed.
ANSWER: T

10. A variable is a characteristic of interest about each individual element of a population or sample.
ANSWER: T

11. A quantitative variable that can assume a countable number of values is referred to as continuous variable.
ANSWER: F

12. A quantitative variable that can assume an uncountable number of values is referred to as discrete variable.
ANSWER: F

13. A qualitative variable that categorizes or describes or names an element of a population is referred to as a nominal variable.
ANSWER: T

14. Inferential statistics is the study and description of data that result from an experiment.
ANSWER: F

15. Descriptive statistics is the study of a sample that enables us to make projections or estimates about the population from which the sample is drawn.
ANSWER: F

16. A population is typically a very large collection of individuals or objects about which we desire information.
ANSWER: T

17. A statistic is the calculated measure of some characteristics of a population.
ANSWER: F

18. A parameter is the measure of some characteristic of a sample.
ANSWER: F

19. As a result of surveying 50 freshmen, it was found that 16 had participated in interscholastic sports, 23 had served as officers of classes and clubs, and 18 had been in school plays during their high school years. This is an example of numerical data.
ANSWER: F

20. The "number of rotten oranges per shipping crate" is an example of a quantitative variable.
ANSWER: T

21. The "thickness of a sheet of sheet metal" used in a manufacturing process is an example of a quantitative variable.
ANSWER: T

22. The basic objective of statistics is that of obtaining a sample, inspecting this sample, and then making inferences about the unknown characteristics of the population from which the sample was drawn.
ANSWER: T

Multiple-Choice Questions

23. Which of the following best describes the data: zip codes for students attending college in the state of Michigan?

A) Attribute data
B) Numerical data
C) Quantitative data
D) Sample data
ANSWER: A

24. Which of the following best describes the data: grade point averages for athletes?

A) Attribute data
B) Quantitative data
C) Qualitative data
D) Sample data
ANSWER: B

25. Which of the following best describes the data: classifications of unlikely, likely, or very likely to describe possible buying of a product?

A) Attribute data
B) Numerical data
C) Quantitative data
D) Sample data
ANSWER: A

26. Consider the following data: the height in centimeters of children in a third-grade class. Which of the following best describes these data?

A) Attribute data
B) Qualitative data
C) Quantitative data
D) Sample data
ANSWER: C

27. Consider the following data: the 18-hole score for all rounds of golf played at Oak Hill Country Club last year. Which of the following best describes these data?

A) Attribute data
B) Quantitative data
C) Qualitative data
D) Statistic
ANSWER: B

28. Consider the following data: like no preference or dislike. Which of the following best describes these data?

A) Qualitative data
B) Numerical data
C) Quantitative data
D) Statistic
ANSWER: A

29. Consider the following data: the weights of babies born in a given hospital. Which of the following best describes these data?

A) Attribute data
B) Qualitative data
C) Quantitative data
D) Statistic
ANSWER: C

30. Which of the following data types would *not* be considered quantitative data?

A) Heights of basketball players
B) Weight of newborn babies
C) Grade point averages of college sophomores
D) Zip codes within the state of Ohio
ANSWER: D

31. A company has developed a new battery, but the average lifetime is unknown. In order to estimate this average, a sample of 100 batteries is tested and the average lifetime of this sample is found to be 250 hours. The 250 hours is the value of a:

A) parameter.
B) statistic.
C) sampling frame.
D) population.
ANSWER: B

32. Which of the following data types would be considered attribute data?

A) Hair color
B) Ages of college freshmen
C) 18 hole score of golfers
D) Shoe size of 3rd grade students
ANSWER: A

Short-Answer Questions

33. The Nielsen Company reports that 30% of the television audience watched a world-premiere movie. Is this an example of descriptive or inferential statistics?

ANSWER:
Inferential statistics

34. As part of the graduation paperwork, seniors at a particular college were asked to indicate their post-graduation plans. Results showed that 15% planned to start graduate school right after college graduation. Is this an example of descriptive or inferential statistics?

ANSWER:
Descriptive statistics

35. In statistics, what name do we give to a numerical characteristic of a sample?

ANSWER:
Statistic

36. In statistics, what name do we give to a numerical characteristic of a population?

ANSWER:
Parameter

37. In statistics, what name do we give to a set of all individuals whose properties are to be analyzed?

ANSWER:
Population

38. In statistics, what name do we give to a subset of a population?

ANSWER:
Sample

39. What is the difference between descriptive statistics and inferential statistics?

ANSWER:
Descriptive statistics: collect, present, describe sample data. Inferential statistics: interpret based on descriptive statistics, make decisions and draw conclusions about the population from which the sample was drawn.

40. What is the difference between a finite population and an infinite population?

ANSWER:
A population is finite when the membership could be physically listed. When the membership is unlimited, the population is infinite.

41. Discuss the difference between a variable and a parameter. Include an illustration.

ANSWER:
Variable: characteristic of interest about each element of a population. Parameter: numerical value summarizing all data of a population, therefore characteristic of the population. For example, weight of one baby is a variable, while average weight of all babies is a parameter.

42. In completing a survey, respondents use the following numbers to indicate marital status.
1 = Single (never married), 2 = Married, 3 = Divorced, 4 = Widowed
Is this data qualitative or quantitative? Explain.

ANSWER:
Even though marital status is coded by number, the data is qualitative as it categorizes each individual respondent. Also, the average of single and divorced is meaningless.

43. In completing a survey, respondents use the following numbers to indicate ages.
1 = age 19 years and under, 2 = 20 to 29 years of age
3 = 30 to 39 years of age, 4 = age 40 years and older
Is this data qualitative or quantitative? Explain.

ANSWER:
This is quantitative data; an average age.

44. Explain the difference between the terms "variable" and "data." Include an illustration that demonstrates this difference.

ANSWER:
Variable: a characteristic of interest about each individual element of a population or sample whereas data refers to the value or values of the variable (e.g., age of a person when first attends professional sporting event would be characteristic of interest about each person and is a variable. Jim was 17 when he first attended a professional sporting event; 17 is the value of the variable for Jim and is data).

Applied and Computational Questions

QUESTIONS 45 THROUGH 53 ARE BASED ON THE FOLLOWING INFORMATION:
An office supply warehouse has boxes of pencils, 100 pencils to the box. Information about the entire warehouse as well as a sample of the boxes is shown below:

Number of defectives per box	Number of boxes (in warehouse)	Number of boxes (in sample)
0	1500	50
1	250	20
2	75	3
3	40	3
4	10	1

45. Describe the population.

ANSWER:
All boxes of pencils in the warehouse

46. What is the population size?

ANSWER:
1875 boxes

47. Is the population finite or infinite? Why?

ANSWER:
Finite; since the number of boxes in the population can be (or could be) physically listed.

48. Describe the sample.

ANSWER:
The boxes of pencils sampled.

49. What is the sample size?

ANSWER:
77 boxes

50. A quality control technician is interested in the number of boxes with more than two defectives. What is the value of the parameter?

ANSWER:
50

51. A quality control technician is interested in the number of boxes with more than two defectives. What is the value of the sample?

ANSWER:
4

52. A quality control technician is interested in the proportion of boxes with no more than one defective pencil. What is the value of the parameter?

ANSWER:
1750/1875 = 0.933

53. A quality control technician is interested in the proportion of boxes with no more than one defective pencil. What is the value of the statistic?

ANSWER:
70/77 = 0.909

<u>**QUESTIONS 54 AND 55 ARE BASED ON THE FOLLOWING INFORMATION:**</u>
A paper company is interested in estimating the proportion of trees in a 500-acre forest with diameters exceeding 2 feet. The company selects 25 plots (100 feet by 100 feet) from the forest and utilizes the information from the 25 plots to help estimate the proportion for the whole forest.

54. What statistical term describes the 500-acre forest?

ANSWER:
Population

55. What statistical term describes the 25 plots?

ANSWER:
Sample

56. At a large community college 120 students are randomly selected and asked the distance of their commute to campus. From this group a mean of 9.8 miles is computed. Match the items in Column II with the statistical term in Column I.

Column I	Column II
1. Data (one)	a. The process used to select the 120 students and determine their distance
2. Data (set)	b. The computed 9.8 miles
3. Experiment	c. All students enrolled at the college
4. Parameter	d. The 120 commute distances
5. Population	e. The 120 students
6. Sample	f. The commute distance for one student
7. Statistic	g. 8 miles distance for one student
8. Variable	h. The mean commute distance for all students

ANSWER:
(1, g), (2, d), (3, a), (4, h), (5, c), (6, e), (7, b), (8, f)

57. In a community of 10,987, 100 homeowners were randomly selected and asked the amount of their January heating bill. From this group a mean of $76.98 is computed. Match the items in Column II with the statistical term in Column I.

Column I	Column II
1. Data (one)	a. The computed $76.98
2. Data (set)	b. The community of 10,987
3. Experiment	c. The 100 homeowners
4. Parameter	d. The 100 heating bills
5. Population	e. The heating bill for one home
6. Sample	f. The mean bill for all homes
7. Statistic	g. $88.76 bill for one home
8. Variable	h. The process used to select the 100 homeowners and determine their heating bill

ANSWER:
(1, f), (2, c), (3, h), (4, g), (5, b), (6, d), (7, a), (8, e)

QUESTIONS 58 THROUGH 61 ARE BASED ON THE FOLLOWING INFORMATION:
A quality-control inspector selects assembled parts from an assembly line and records the information concerning each part as: A: defective or nondefective, B: the employee number of the individual who assembled the part, and C: the weight of the part.

58. What is the population?

ANSWER:
All assembled parts from the assembly line

59. Is the population finite or infinite? Why?

ANSWER:
Infinite; since all assembled parts from the assembly line can't be (or couldn't be) physically listed.

60. What is the sample?

ANSWER:
The parts checked

61. Classify the three variables as either attribute or quantitative.

ANSWER:
A: attribute, B: attribute (it identifies the assembler), C: quantitative

QUESTIONS 62 THROUGH 65 ARE BASED ON THE FOLLOWING INFORMATION:
Select ten students currently enrolled at your college and collect data for these three variables: X: number of courses enrolled in, Y: total cost of textbooks and supplies for courses, and Z: method of payment used for textbooks and supplies

62. What is the population?

ANSWER:
All students currently enrolled at the college

63. Is the population finite or infinite?

ANSWER:
Finite

64. What is the sample?

ANSWER:
The 10 students selected

65. Classify the three variables as nominal, ordinal, discrete, or continuous.

ANSWER:
X: discrete, Y: continuous (cost rounded to nearest cent), Z: nominal

66. Identify the statement "A poll of registered voters asking which candidate they support" as an example of (1) nominal, (2) ordinal, (3) discrete, or (4) continuous variables.

ANSWER:
Nominal

67. Identify the statement "The length of time required for a wound to heal when using a new medicine." as an example of (1) nominal, (2) ordinal, (3) discrete, or (4) continuous variables.

ANSWER:
Continuous

68. Identify the statement "The number of telephone calls arriving at a switchboard per ten-minute period." as an example of (1) nominal, (2) ordinal, (3) discrete, or (4) continuous variables.

ANSWER:
Discrete

69. Identify the statement "The distance first-year college women can kick a football." as an example of (1) nominal, (2) ordinal, (3) discrete, or (4) continuous variables.

ANSWER:
Continuous

70. Identify the statement "The number of pages per job coming off a computer printer." as an example of (1) nominal, (2) ordinal, (3) discrete, or (4) continuous variables.

ANSWER:
Discrete

71. Identify the statement "The kind of tree used as a Christmas tree." as an example of (1) nominal, (2) ordinal, (3) discrete, or (4) continuous variables.

ANSWER:
Nominal

QUESTIONS 72 THROUGH 75 ARE BASED ON THE FOLLOWING INFORMATION:
A study by health economists at Midwestern University indicated that Alzheimer's disease cost the nation $90 billion a year in medical expenses and lost productivity. Patients' earning loss was $25 billion, the value of time of unpaid caregivers was $40 billion, and the cost of paid care was $30 billion.

72. What is the population?

ANSWER:
The population consists of all the Alzheimer patients in the U.S.

73. What is the response variable?

ANSWER:
The response variable is the cost in medical expenses and lost productivity per patient per year.

74. What is the parameter?

ANSWER:
The total cost per year for all Alzheimer patients in the U.S.

75. What is the statistic?

ANSWER:
The total cost per year for the Alzheimer patients used as a sample by the health economists at the Midwestern University, the basis for the estimation.

QUESTIONS 76 THROUGH 79 ARE BASED ON THE FOLLOWING INFORMATON:
A health magazine presented results of a recent study that analyzed data collected by the U.S. Census Bureau in 2000. Results reveal that for both men and women in the United States, heart disease remains the number one killer, victimizing 500,000 people annually. Age, obesity, and inactivity all contribute to heart disease, and all three of these factors vary considerably from one location to the next. The highest mortality rates (deaths per 100,000 people) were in New York, Florida, Oklahoma, and Arkansas, whereas the lowest were reported in Alaska, Utah, Colorado, and New Mexico.

76. What is the population?

ANSWER:
All people in the US who died in 2000

77. What are the characteristics of interest?

ANSWER:
Death from heart disease, state of residence, age, obesity, inactivity

78. What is the parameter?

ANSWER:
Mortality rate per 100,000 people in the US

79. Classify all the variables of the study as either attribute or numerical.

ANSWER:
Death from heart disease – attribute,
State of residence – attribute,
Age at death – numerical,
Obesity – attribute, Inactivity – attribute.

QUESTIONS 80 THROUGH 83 ARE BASED ON THE FOLLOWING INFORMATION:
Select twenty employees currently working at a local supermarket and collect data for the following four variables:
W: Marital status
X: number of children they have
Y: total cost of cloth and toys they spend every year for their children
Z: method of payment used for purchasing cloth and toys

80. What is the population?

ANSWER:
The population consists of all employees currently working at the local supermarket.

81. Is the population finite or infinite?

ANSWER:
Finite

82. What is the sample?

ANSWER:
The 20 employees selected

83. Classify the three variables as nominal, ordinal, discrete, or continuous.

ANSWER:
W – nominal, X - discrete, Y - continuous (cost rounded to nearest cent), Z - nominal

84. Identify the following statement as an example of nominal, ordinal, discrete, or continuous variables: "A poll of registered voters in Michigan asking which candidate they support."

ANSWER:
Nominal

85. Identify the following statement as an example of nominal, ordinal, discrete, or continuous variables: "The length of time required for a wound to heal when using a new medicine."

ANSWER:
Continuous

86. Identify the following statement as an example of nominal, ordinal, discrete, or continuous variables: "The number of telephone calls arriving at a switchboard per five-minute period."

ANSWER:
Discrete

87. Identify the following statement as an example of nominal, ordinal, discrete, or continuous variables: "The distance first-year college football players can kick a ball."

ANSWER:
Continuous

88. Identify the following statement as an example of nominal, ordinal, discrete, or continuous variables: "The number of pages in your statistics textbook."

ANSWER:
Discrete

QUESTIONS 89 THROUGH 92 ARE BASED ON THE FOLLOWING INFORMATION:
A study by health economists at a southern university indicated that Parkinson's disease cost the nation $85 billion a year in medical expenses and lost productivity. Patients' earning loss was $25 billion, the value of time of unpaid caregivers was $35 billion, and the cost of paid care was $22 billion.

89. What is the population?

ANSWER:
The population consists of all the Parkinson patients in the U.S.

90. What is the response variable?

ANSWER:
The response variable is the cost in medical expenses and lost productivity per patient per year.

91. What is the parameter?

ANSWER:
The total cost per year for all Parkinson patients in the U.S.

92. What is the statistic?

ANSWER:
The total cost per year for the Parkinson patients used as a sample by the health economists at the southern university; the basis for the estimation.

QUESTIONS 93 THROUGH 97 ARE BASED ON THE FOLLOWING INFORMATION:
At Central Michigan University, 500 students are randomly selected and asked the distance of their commute to campus. From this group a mean of 15.6 miles is computed.

93. What is the statistic?

ANSWER:
The computed 15.6 miles

94. What is the variable of interest?

ANSWER:
The commute distance for students to campus

95. What is the parameter?

ANSWER:
The mean commute distance for all students at the college to the campus

96. What is the sample?

ANSWER:
The 150 randomly selected students

97. Describe in detail how you would select a 5% systematic sample of the adults in Detroit in order to complete a survey about a political issue.

ANSWER:
Randomly select an integer value between 1 and 20. This integer represents the first item in the sample. Then, select every 20th data thereafter until you have the desired number of data for the sample.

98. Identify the following statement as descriptive in nature, or as inferential: "The average age of 500 surveyed students in your institution is 23 years."

ANSWER:
Descriptive

99. Identify the following statement as descriptive in nature, or as inferential: "Based on a sample of 10,000 Americans, it is fair to say that 70% of all American are overweight."

ANSWER:
Inferential

100. Identify the following statement as descriptive in nature, or as inferential: "Based on a sample of 20,000 college students, we may conclude that 80% of all college students dislike true-false questions."

ANSWER:
Inferential

101. Identify the following statement as descriptive in nature, or as inferential: "80% of students in your statistics class dislike true-false questions."

ANSWER:
Descriptive

102. Explain why the polls that are so frequently quoted during early returns on Election Day TV coverage are an example of cluster sampling.

ANSWER:
Each precinct is considered a cluster, and not all precincts are sampled.

Sections 1.3 and 1.4

True-False Questions

103. Within a set of experimental data, we always expect variation.
ANSWER: T

104. Statistical process control uses statistical methodology to control (or reduce) variability in a manufacturing process.
ANSWER: T

105. In statistics, a random sample means a sample that is selected haphazardly (without pattern).
ANSWER: F

106. To say that a sample is selected in such a way that every element in the population has an equal chance of being selected is equivalent to saying that all samples of size *n* have an equal chance of being selected.
ANSWER: T

107. A list of elements belonging to the population from which the sample will be drawn is referred to as the sampling frame.
ANSWER: T

108. When a judgment sample is drawn, the person selecting the sample chooses items in such a way that every element in the population has an equal probability of being chosen.
ANSWER: F

109. If we desire to select a 4% systematic sample, first we will randomly select one element from the first 40 elements, and then proceed to select every 4th item thereafter until we have the desired number of data for our sample.
ANSWER: F

110. In general, probability samples are samples in which the elements to be selected are drawn on the basis of probability in such a way that each element in the population has the same probability of being selected as part of the sample.
ANSWER: F

111. Cluster sample is a sample obtained by selecting some of, but not all of, the possible subdivisions within a population. These subdivisions, called clusters, often occur naturally within the population.
ANSWER: T

112. When a proportional random sample is drawn, the sampling frame is subdivided into various strata, and then a subsample is drawn from each stratum.
ANSWER: T

113. A stratified random sample is obtained by stratifying the sampling frame, and then selecting a fixed number of items from some of, but not all of, the strata by means of a simple random sampling technique.
ANSWER: F

114. A representative sample is a sample obtained in such a way that all individuals had an equal chance to be selected.
ANSWER: F

Multiple-Choice Questions

115. In the 1936 presidential election, Alfred Landon was predicted (incorrectly) to beat Franklin D. Roosevelt based on the results of a telephone survey. Because telephones were considered a luxury item during this period, the survey was biased because it related only to the opinion of those who could be reached by telephone. This incident represents which of the following?

A) An improperly defined parameter
B) An improperly defined sampling frame
C) A poorly defined population
D) A sample with no statistic defined for the sample
ANSWER: B

116. Choose the item that best completes the following statement: No matter what the variable is, if the tool of measurement is precise enough, there will be____________.

A) uncertainty
B) variability
C) probability
D) measurability
ANSWER: B

Short-Answer Questions

117. In statistics, what name do we give to a list of elements belonging to a population from which a sample will be drawn?

ANSWER:
Sampling frame

118. In statistics, what name do we give to a list of every element in a population?

ANSWER:
Census

119. Explain the relationship between a census and a sampling frame.

ANSWER:
Census is a listing of every element of the population.
Sampling frame is a subset of the population (or census) from which the sample is selected.

120. Discuss what the lack of variability in a quantitative response variable would tend to indicate. Include an illustration.

ANSWER:
Lack of variability in a quantitative response variable tends to indicate lack of precision in the tool of measurement. For example, if the heights of chair seats in classroom are reported to be 17", when, in fact, the heights vary from 16 ½ to 17 ¾, then lack of precision in the measuring instrument is present.

121. Discuss the difference between the following two methods of data collection: experiment and survey. Include an illustration of each.

ANSWER:
Experiment: investigator controls or modifies the environment and observes the effect of the variable under study
An illustration of an experiment: A doctor prescribes different drug dosages to different people to determine effectiveness.

Survey: investigator collects data by sampling a population but not modify the environment.
An illustration of a survey: A researcher stops people in a mall and asks them about the medicines they take and its effectiveness.

Applied and Computational Questions

122. Describe in detail how you would select a 4% systematic sample of the adults in a nearby large city in order to complete a survey about a political issue.

ANSWER:
Randomly select an integer between 1 and 25 (100/x = 100/4 = 25). This integer represents the first item in the sample. Then, select every 25th data item thereafter until you have the desired number of data for the sample.

Sections 1.5 and 1.6

True-False Questions

123. If it were *not* for the laws of probability, the theory of statistics would *not* be possible.
ANSWER: T

Multiple-Choice Questions

124. Suppose you are interested in determining the preferred candidate for governor of Michigan among registered voters in Mecosta County. Which of the following best describes this problem?

A) This is a problem in probability.
B) This is a problem in statistics.
C) Neither A nor B
D) Both A and B
ANSWER: B

125. Suppose you are interested in determining the likelihood of winning a state lottery by purchasing one ticket. Which of the following best describes this problem?

A) This is a problem in probability.
B) This is a problem in statistics.
C) Neither A nor B
D) Both A and B
ANSWER: A

126. Suppose you are interested in determining the mean age of all students attending community colleges in the state of Texas. Which of the following best describes this problem?

A) This is a problem in probability.
B) This is a problem in statistics.
C) Neither A nor B
D) Both A and B
ANSWER: B

Short-Answer Questions

127. Discuss the validity of the following statement: "Computers can analyze any and all sets of data and give statistically correct results."

ANSWER:
Standard statistical packages are good at performing tedious operations; however the user must insure that appropriate methods are correctly applied and that accurate conclusions are drawn.

128. Explain the difference between probability and statistics. Include an illustration.

ANSWER:
In probability we know the population and are interested in the likelihood of a particular sample (e.g., rolling a die we know likelihood that the number will be even). In statistics, draw a sample and then make inference about the population (e.g., roll a die 100 times and keep record).

Applied and Computational Questions

129. Classify the following statement as a probability or a statistics problem: "Determining whether a new drug shortens the recovery time from a certain illness."

ANSWER:
Statistics

130. Classify the following statement as a probability or a statistics problem: "Determining the chance that heads will result when a coin is flipped."

ANSWER:
Probability

131. Classify the following statement as a probability or a statistics problem: "Determining the amount of waiting time required to check out at a certain grocery store".

ANSWER:
Statistics

132. Classify the following statement as a probability or a statistics problem: "Determining the chance that you will be dealt a "blackjack"

ANSWER:
Probability

133. Classify the following statement as a probability or a statistics problem: "Determining how long it takes to handle a typical telephone inquiry at a real estate office."

ANSWER:
Statistics

134. Classify the following statement as a probability or a statistics problem: "Determining the length of life for the 100-meg zip disk produced by Fuji."

ANSWER:
Statistics

135. Classify the following statement as a probability or a statistics problem: "Determining the chance that a blue ball will be drawn from a bowl that contains 15 balls, of which 5 are blue."

ANSWER:
Probability

136. Classify the following statement as a probability or a statistics problem: "Determining the average price of the new computers that your company just purchased."

ANSWER:
Statistics

137. Classify the following statement as a probability or a statistics problem: "Chance of getting "doubles" when you roll a pair of dice."

ANSWER:
Probability

138. Classify the following statement as a probability or a statistics problem: "Determining whether a new drug shortens the recovery time from a certain illness."

ANSWER:
Statistics

139. Classify the following statement as a probability or a statistics problem: "Determining the chance that tails will result when a coin is tossed twice."

ANSWER:
Probability

140. Classify the following statement as a probability or a statistics problem: "Determining the amount of waiting time required to check out at a grocery store."

ANSWER:
Statistics

141. Classify the following statement as a probability or a statistics problem: "Determining the chance that you will receive an "A" grade in your statistics class."

ANSWER:
Probability

142. Classify the following statement as a probability or a statistics problem: "Chance of getting "doubles" when you roll a pair of dice.

ANSWER:
Probability

143. Classify the following statement as a probability or a statistics problem: "Determining the length of life for the 100-watt light bulbs a company produces".

ANSWER:
Statistics

144. Classify the following statement as a probability or a statistics problem: "Determining the shearing strength of the rivets that your company just purchased for building airplanes."

ANSWER:
Statistics

Chapter 2

Descriptive Analysis and Presentation of Single-Variable Data

Sections 2.1 and 2.2

True-False Questions

1. Circle graphs and bar graphs are graphs that are used to summarize qualitative, or attribute, or categorical data.
ANSWER: T

2. All graphic representations of sets of data need to be completely self-explanatory. That includes a descriptive meaningful title, and identification of the vertical and horizontal scales.
ANSWER: T

3. The *stem-and-leaf display* for summarizing numerical data is a combination of a graphic technique and a sorting technique.
ANSWER: T

4. There is no single correct answer when constructing a graphic display. The analyst's judgment and the circumstances surrounding the problem play a major role in the development of the graphic.
ANSWER: T

5. Circle graphs and bar graphs are graphs that are used to summarize quantitative data.
ANSWER: F

6. Circle graphs (pie diagrams) show the amount of data that belong to each category as a proportional part of a circle.
ANSWER: T

7. Circle graphs show the amount of data that belong to each category as a frequency.
ANSWER: F

8. Bar graphs show the amount of data that belong to each category as a proportionally sized rectangular area.
ANSWER: T

9. Bar graphs of attribute data should be drawn with connected bars of equal width.
ANSWER: F

10. One major reason for constructing a graph of quantitative data is to display its distribution.
ANSWER: T

Multiple-Choice Questions

11. Which of the following statements is false?

A) Pareto diagram is a bar graph with the bars arranged from the most numerous categories to the least numerous categories.
B) Pareto diagram includes a line graph displaying the cumulative percentages and counts for the bars.
C) A Pareto diagram of types of defects will show the ones that have the greatest effect on the defective rate in order of effect. It is then easy to see which defects should be targeted in order to most effectively lower the defective rate.
D) None of the above.
ANSWER: D

12. Which of the following statements is false?

A) Dotplot displays the data of a sample by representing each data with a dot positioned along a scale. This scale can be either horizontal or vertical. The frequency of the values is represented along the other scale.
B) Pareto diagram includes a line graph displaying the frequency (counts) for the bars.
C) Dotplot display is a convenient technique to use as you first begin to analyze the data. It results in a picture of the data as well as sorts the data into numerical order.
D) The stem-and-leaf display is a combination of a graphic technique and a sorting technique. This display is simple to create and use, and it is well suited to computer applications.
ANSWER: B

Short-Answer Questions

13. Complete the following statement: A stem-and-leaf display is a combination of a sorting technique and a __________ technique.

ANSWER:
graphing

14. Complete the following statement: Circle graphs and bar graphs are often used to summarize ____________ data.

ANSWER:
attribute

15. Data for the distribution of land in a particular county is given in percentages. Name two types of graphs that would be most appropriate to display these results.

ANSWER:
Bar graph or circle graph

Applied and Computational Questions

16. Construct a stem-and-leaf display for the data below.

219	225	222	243	234	241	231	235	234
231	240	231	246	232	229	233	233	226
225	227	230	229	227	218	216	234	240

ANSWER:

21	6 8 9
22	2 5 5 6 7 7 9 9
23	0 1 1 1 2 3 3 4 4 4 5
24	0 0 1 3 6

17. The number of vehicles passing a tollgate between 7 a.m. and 8 a.m. were recorded for twenty different days. Construct a stem-and-leaf display for these data.

10 26 32 15 16 22 31 46 27 33 27 15 16 19 20 16 12 22 30 41

ANSWER:

1	0 2 5 5 6 6 6 9
2	0 2 2 6 7 7
3	0 1 2 3
4	1 6

18. A group of hypertensive patients (with diastolic blood pressure between 110 and 130) were given a medication for reducing elevated blood pressure. The decreases in blood pressure produced by the medication were categorized into four categories as follows:

Category	*Decrease in Pressure*
A--Marked decrease in blood pressure	15 or more units
B--Moderate decrease in blood pressure	10 to less than 15 units
C--Slight decrease in blood pressure	5 to less than 10 units
D--Stationary blood pressure	0 to less than 5 units

Thirty patients who used the medication experienced the following blood pressure reductions. Give the height of each at the four bars of a bar graph for these results.

12	15	6	4	20	17	25	4	5	18
10	12	18	13	14	20	30	12	14	17
30	18	10	8	16	32	27	13	8	4

ANSWER:

Category	*Height of bar*
A	14
B	9
C	4
D	3

19. A random sample of test scores was taken from two different sections of an introductory statistics course. Construct a back-to-back stem-and-leaf display for this set of data.

Section A: 46 97 99 64 78 76 45 73 81 51 68 81 81 79 100
Section B: 80 69 92 75 88 47 98 92 90 81 42 50 59 66 67 66

ANSWER:

Sec. A		***Sec. B***
56	4	27
1	5	09
48	6	6679
3689	7	5
111	8	018
79	9	0228
0	10	

20. The total amount spent for textbooks (to the nearest dollar) was recorded for several students. Some of the information was collected for the summer session (denoted by S), and some was collected for the fall semester (denoted by F). Construct a back-to-back stem-and-leaf display for this set of data.

Semester:	S	F	F	S	F	F	F	F	S	F	S
Amount:	25	90	115	40	80	75	95	60	29	120	46
Semester:	S	F	F	S	F	F	F	S	F	F	S
Amount:	35	75	80	50	122	95	79	20	95	65	42
Semester:	F	F	F	F	F	S	F	F			
Amount:	80	69	112	105	108	37	98	92			

ANSWER:

Summer		*Fall*
059	02	
57	03	
026	04	
0	05	
	06	059
	07	559
	08	000
	09	025558
	10	58
	11	25
	12	02

21. A department of mathematical sciences has majors in four areas.

Major	*Number of Majors*
Mathematics	50
Computer Science	22
Actuarial Science	15
Statistics	10

If a circle graph is constructed for these data, what would be the percentage of the graph for each major?

ANSWER:

Major	*% of Majors*
Mathematics	51.5
Computer Science	22.7
Actuarial Science	15.5
Statistics	10.3

<u>QUESTIONS 22 THROUGH 25 ARE BASED ON THE FOLLOWING INFORMATION:</u>
The final-inspection defect report for an assembly line is reported on the table and Pareto diagram as shown below:

Defect	Blemish	Scratch	Chip	Bend	Dent	Others
Count	61	50	28	17	13	11

Pareto Chart for Product Defects

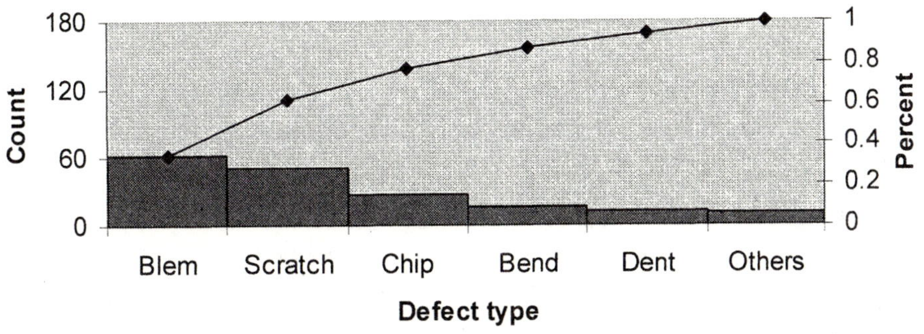

22. What is the total defect count in the report?

ANSWER:
180 defects

23. Find the percentage for "chip" defect items.

ANSWER:
Percent of chip = (50/180) · 100% = 15.56%

24. Find the "cum % for bend", and explain what that value means.

ANSWER:
[(61+50+28+17) /180] · 100% = (156/180) · 100% = 86.67%. The value 86.67% is the sum of the percentages for all defects that occurred more often than Bend, including Bend.

25. Management has given the production line the goal of reducing their defects by 50%. What two defects would you suggest they give special attention to in working toward this goal? Explain.

ANSWER:
The two defects, Blemish and Scratch, total 61.67%. If they can control these two defects, the goal should be within reach.

<u>QUESTIONS 26 THROUGH 29 ARE BASED ON THE FOLLOWING INFORMATION:</u>
The points scored by the winning teams on opening night of a recent NBA season are shown in the table below:

Team	Detroit	Dallas	Chicago
Score	90	110	92

26. Draw a bar graph of these scores using a vertical scale ranging from 80 to 120.

ANSWER:

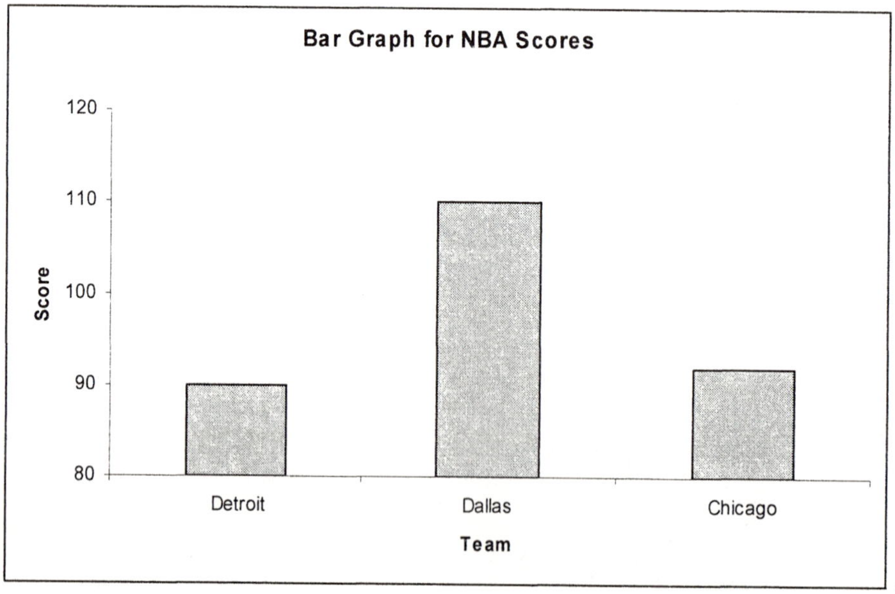

27. Draw a bar graph of the scores using a vertical scale ranging from 50 to 120.

ANSWER:

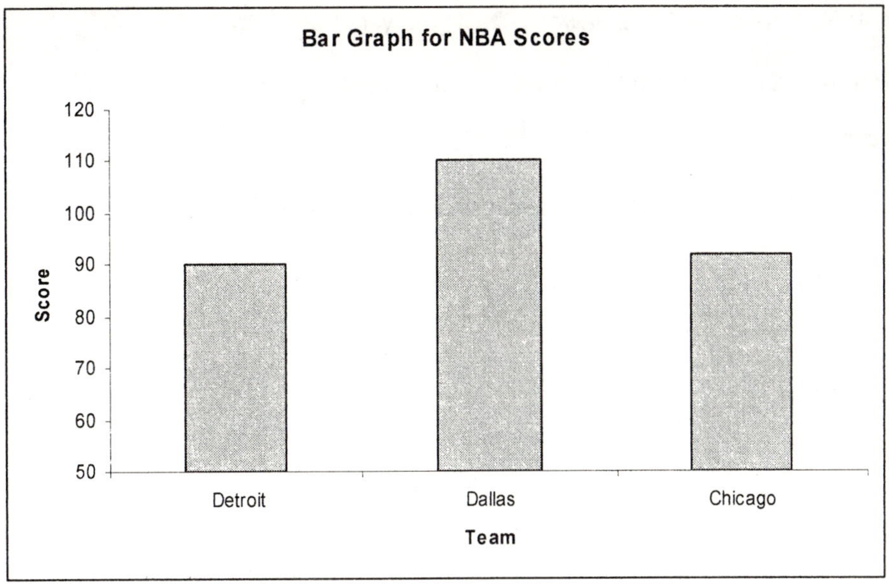

28. In which bar graph does it appear that the NBA scores vary more? Why?

ANSWER:
Bar graph in question 27 emphasizes the variation in the scores as it focuses only on the variation and not the relative size of the scores.

29. How could you create an accurate representation of the relative size and variation between these scores? Draw this new bar graph.

ANSWER:
An accurate representation of both the size and variation of the values would be best served by starting the vertical scale at zero.

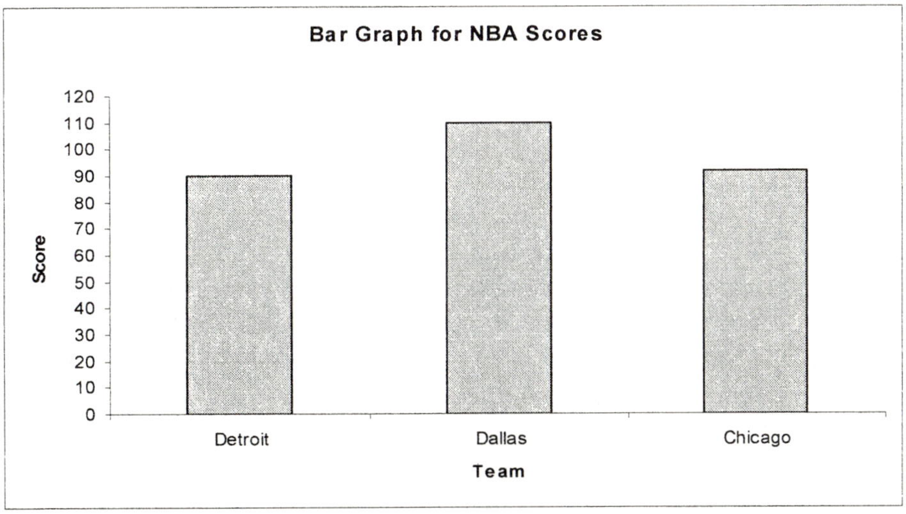

Chapter 2

<u>QUESTIONS 30 THROUGH 33 ARE BASED ON THE FOLLOWING INFORMATION:</u>
What not to get them on Valentines Day! A recent study among adults in USA shows that adults prefer not to receive certain items as gifts on Valentine's Day as shown below:

Teddy bears: 45%; Chocolate: 25%; Jewelry: 15%; Flowers: 12%; Don't Know: 3%.

30. Draw a bar graph picturing the percentages of "Presents not wanted".

ANSWER:

31. Draw a Pareto diagram picturing the "Presents not wanted".

ANSWER:

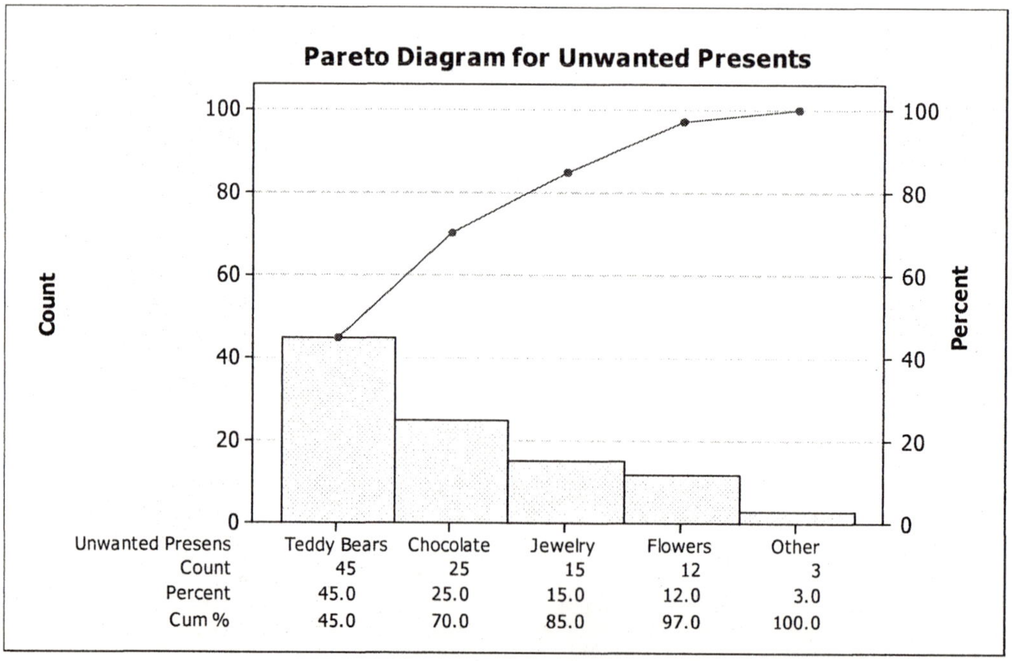

32. If you want to be 80% sure you did not get your valentine something unwanted, what should you avoid buying? How does the Pareto diagram show this?

ANSWER:
Teddy bears, chocolates, jewelry; these are listed first in the Pareto diagram.

33. 400 adults are to be surveyed, what frequencies would you expect to occur for each unwanted item listed on the snapshot?

ANSWER:
The frequencies are 180, 100, 60, 48, and 12 for teddy bears, chocolates, jewelry, flowers, and don't know, respectively.

QUESTIONS 34 THROUGH 36 ARE BASED ON THE FOLLOWING INFORMATION:
The points scored during each game by the Big Rapids High School basketball team last season were: 60, 58, 65, 75, 50, 65, 60, 72, 64, 70, 58, 65, 56, 40, 68, and 55.

34. Construct a dotplot of these data.

ANSWER:

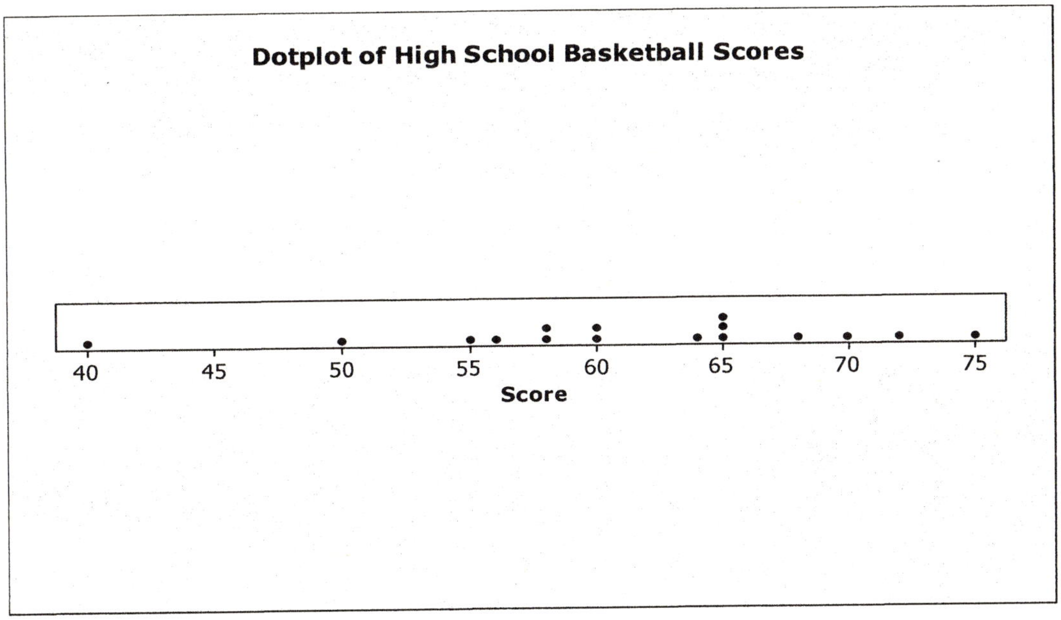

35. Use the dotplot in question 34 to uncover the lowest and highest scores.

ANSWER:
The lowest score was 40 and the highest was 75.

36. Use the dotplot in question 34 to determine the most common score? How many teams share that score?

ANSWER:
65; three teams share that score

QUESTIONS 37 THROUGH 40 ARE BASED ON THE FOLLOWING INFORMATION:
The data shown below are the heights (in inches) of the basketball players who were the first round picks by the professional NBA teams in a recent year.

83	83	75	80	76	80	81	84	79	80
84	86	72	82	82	79	81	79	80	73
90	82	81	75	77	80	79	76	85	

37. Construct a dotplot of the heights of these players.

ANSWER:

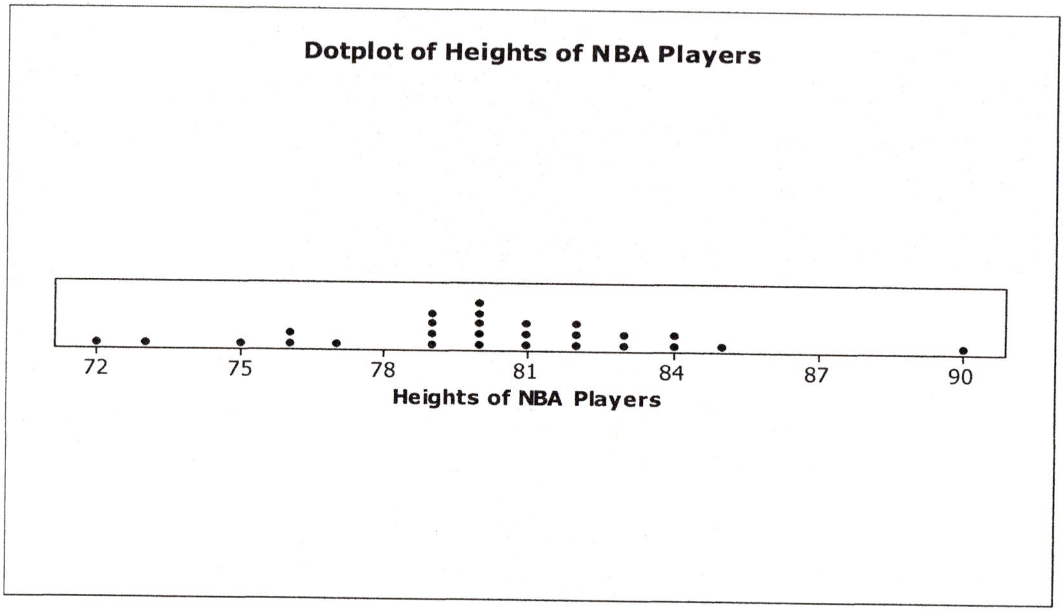

38. Use the dotplot in question 37 to uncover the shortest and the tallest players.

ANSWER:
The shortest player is 72 inches and the tallest player is 90 inches.

39. Use the dotplot in question 37 to determine the most common height and how many players share that height?

ANSWER:
The most common height is 80 inches, shared by 5 players.

40. What feature of the dotplot in question 37 illustrates the most common height?

ANSWER:
The height of column of dots illustrates the most common height.

Sections 2.3 through 2.6

True-False Questions

41. A histogram is used to summarize attributive data.
ANSWER: F

42. One major reason for constructing a graph of quantitative data is to display its distribution.
ANSWER: T

43. In a *J*-shaped histogram, there is one tail on the side of the class with the highest frequency.
ANSWER: F

44. A line graph of a cumulative frequency or cumulative relative frequency distribution is referred to as an ogive.
ANSWER: T

45. The frequency of a class is the number of pieces of data whose values fall within the boundaries of that class.
ANSWER: T

46. Frequency distributions are used in statistics to present large quantities of repeating values in a concise form.
ANSWER: T

47. If grouping data are used to form a frequency distribution, the class width is the difference between the upper and lower class boundaries.
ANSWER: T

48. If grouping data are used to form a frequency distribution, the class midpoint (sometimes called the class mark) is the numerical value that is exactly in the middle of each class. It is found by adding the class boundaries and dividing by 2.
ANSWER: T

49. A histogram is a bar graph that represents a frequency distribution of categorical data.
ANSWER: F

50. A bimodal distribution has two high-frequency classes separated by classes with lower frequencies. It is not necessary for the two high frequencies to be the same.
ANSWER: T

51. Relative frequency can be expressed as a common fraction, in decimal form, but not as a percentage.
ANSWER: F

52. The histogram of a sample should have a distribution shape very similar to that of the population from which the sample was drawn.
ANSWER: T

53. An ogive is a line graph of a cumulative frequency or cumulative relative frequency distribution.
ANSWER: T

54. Every ogive starts on the left with a cumulative relative frequency of zero at the lower class boundary of the first class and ends on the right with a cumulative relative frequency of 100% at the upper class boundary of the last class.
ANSWER: T

55. Measures of central tendency measure the spread of a set of data about its center.
ANSWER: F

56. For every set of data, the value of the median will always be one of the original items of data.
ANSWER: F

57. In a sample of size *n*, the median of the sample is $(n+1)/2$.
ANSWER: F

58. The midrange for a set of data is found by subtracting the lowest valued data *L* from the highest valued data *H*.
ANSWER: F

59. The mean, median and mode are the most common measures of dispersion (spread).
ANSWER: F

60. Measures of central tendency are numerical values that locate, in some sense, the center of a set of data.
ANSWER: T

61. The mean, median and mode for the set of data {3, 5, 3, 8, 6} are all the same value.
ANSWER: F

62. The mean of a sample always divides the data into two equal halves (half larger and half smaller in value than itself).
ANSWER: F

63. A measure of central tendency is a quantitative value that describes how widely the data are dispersed about a central value.
ANSWER: F

64. For any distribution, the sum of the deviations from the mean equals zero.
ANSWER: T

65. Measures of central tendency are attribute data that locate, in some sense, the center of a set of data.
ANSWER: F

66. The term average is often associated with all measures of central tendency.
ANSWER: T

67. The population mean, μ (lowercase mu in the Greek alphabet), is the mean of all x values in the entire population.
ANSWER: T

68. The median is the value of the data that occupies the middle position when the data are ranked in order according to size.
ANSWER: T

69. The sample median is represented by $\bar{x}$.
ANSWER: F

70. The midrange is the number exactly midway between a lowest value data *L* and a highest value data *H*. It is found by averaging the low and high values.
ANSWER: T

71. The sample mean is represented by $\tilde{x}$ (read "x-tilde").
ANSWER: F

72. The population median is represented by *M* (the uppercase mu in the Greek alphabet).
ANSWER: T

73. When *n* is odd, the depth of the median, $d(\bar{x})$ will always be an integer.
ANSWER: T

74. When *n* is even, the depth of the median, $d(\bar{x})$ will always be an integer or a half-number.
ANSWER: F

75. According to your book, if two or more values in a sample are tied for the highest frequency (number of occurrences), we say there is no mode.
ANSWER: T

76. The midrange is the range of the middle two values.
ANSWER: F

77. There are several kinds of measures ordinarily known as averages and each gives a different picture of the figures it is called on to represent.
ANSWER: T

78. The mean absolute deviation for any set of data is zero since $\sum(x-\bar{x})$ is always zero.
ANSWER: F

79. The standard deviation is the positive square root of the variance.
ANSWER: T

80. The sum of the squares of the deviations from the mean $\sum(x-\bar{x})^2$, will sometimes be negative.
ANSWER: F

81. The standard deviation for the set of values 5, 5, 5, 5, and 5 is 5.
ANSWER: F

82. The sample variance, s^2, is the mean of the squared deviations of *x* values from the sample mean $\bar{x}$, calculated using $n-1$ as the divisor.
ANSWER: T

83. The measures of dispersion include the *range, variance*, and *standard deviation.*
ANSWER: T

84. The unit of measure for the variance is the same as the unit of measure for the data.
ANSWER: F

85. There is no limit to how widely spread out the data can be; therefore, measures of dispersion can be very large.
ANSWER: T

86. Although the mean deviation is always zero, it is a useful statistic in some occasions.
ANSWER: F

87. The range is the difference in value between the highest-valued (*H*) and the lowest-valued (*L*) data.
ANSWER: T

88. Although mean absolute deviation is not used very frequently, it is a measure of dispersion.
ANSWER: T

89. The sample variance, s^2 is the mean of the deviations of *x* values from the sample mean $\overline{x}$.
ANSWER: F

90. The standard deviation of a sample is the square of the sample variance.
ANSWER: F

91. If a rounded value of $\overline{x}$ is used, then $\sum(x-\overline{x})$ will not always be exactly zero. It will, however, be reasonably close to zero.
ANSWER: T

92. In a box-and-whisker display, the length of the "box" is the same as the interquartile range.
ANSWER: T

93. Each set of data has four quartiles; they divide the ranked data into four equal quarters.
ANSWER: F

94. The numerical value midway between the first quartile and the third quartile is referred to as the midquartile.
ANSWER: T

95. Each set of data has 100 percentiles; they divide the ranked data into 100 equal subsets.
ANSWER: F

96. The median, the midrange, and the midquartile are always the same value, since each is a middle value.
ANSWER: F

97. The interquartile range is the difference between the first and third quartiles; it is the range of the middle 50% of the data.
ANSWER: T

98. The standard score (or z-score) identifies the position a particular value of x has relative to the mean, measured in standard deviations; that is, $z = (x - \overline{x})/s$.
ANSWER: T

99. On a test Jim scored at the 50th percentile and Jean scored at the 25th percentile; therefore, Jim's test score was twice Jean's test score.
ANSWER: F

100. The unit of measure for the standard score is always in standard deviations.
ANSWER: T

101. Data must be ranked before calculating many of the measures of position.
ANSWER: T

102. Each set of data has four quartiles.
ANSWER: F

103. Measures of position are used to describe the position a specific data value possesses in relation to the mean of the data.
ANSWER: F

104. Measures of position are used to describe the position a specific data value possesses in relation to the rest of the data.
ANSWER: T

105. Quartiles and percentiles are two of the most popular measures of dispersion.
ANSWER: F

106. The median, the second quartile, and the 50th percentile are all the same.
ANSWER: T

107. The first quartile, Q_1, is a number such that at most 25 of the data values are smaller in value than Q_1 and at most 75 of the data values are larger.
ANSWER: F

108. The median, the midrange, and the midquartile are not necessarily the same value. Each is the middle value, but by different definitions of "middle."
ANSWER: T

109. Percentiles are values of the variable that divide a set of ranked data into 100 equal subsets.
ANSWER: T

110. Each set of data has 100 percentiles.
ANSWER: F

111. The 30th percentile, P_{30}, is a value such that at most 30% of the data are smaller in value than P_{30} and at most 70% of the data are larger.
ANSWER: T

112. The first quartile and the 25th percentile are the same.
ANSWER: T

113. The mean, median, the second quartile, and the 50th percentile are all the same.
ANSWER: F

114. The midquartile is a measure of central tendency.
ANSWER: T

115. The 5-number summary divides a set of data into four subsets, with one-quartile of the data in each subset.
ANSWER: T

116. The median, the midrange, and the midquartile are the same, since each is the middle value.
ANSWER: F

117. The midquartile is the numerical value midway between the first quartile and the third quartile.
ANSWER: T

118. The interquartile range is the average of the first and third quartiles.
ANSWER: F

119. The interquartile range is the range of the middle 50% of the data.
ANSWER: T

120. The interquartile range is very unique in the sense that it is a measure of central tendency as well as a measure of dispersion.
ANSWER: F

121. Since the *z*-score is a measure of relative position with respect to the mean, it can be used to help us compare two raw scores that come from separate populations.
ANSWER: T

122. The midquartile, defined as the average of the first and third quartiles, is a measure of position, simply because quartiles are one of the most popular measures of position.
ANSWER: F

Multiple-Choice Questions

123. At a large company, the majority of the employees earn from $20,000 to $30,000 per year. Middle management employees earn between $30,000 and $50,000 per year while top management earn between $50,000 and $100,000 per year. A histogram of all salaries would have which of the following shapes?

A) Symmetrical
B) Uniform
C) Skewed to right
D) Skewed to left
ANSWER: C

124. Which of the following statements is false regarding an ogive?

A) The horizontal scale identifies the upper class boundaries.
B) The vertical scale identifies either the cumulative frequencies or the cumulative relative frequencies.
C) Every ogive starts on the left with a cumulative relative frequency of one at the upper class boundary of the first class.
D) None of the above
ANSWER: C

125. Which of the following statements is false regarding an ogive?

A) It is a line graph of a cumulative frequency or cumulative relative frequency distribution.
B) Its horizontal scale is always based on the lower class boundaries.
C) Its vertical scale identifies either the cumulative frequencies or the cumulative relative frequencies.
D) None of the above
ANSWER: B

126. Which of the following are graphic representations of sets of data?

A) A descriptive and meaningful title
B) A proper identification of the vertical scale
C) A proper identification of the horizontal scale
D) All of the above
ANSWER: D

127. Which of the following statements is false?

A) Relative frequencies are often useful in a presentation because nearly everybody understands fractional parts when expressed as percents.
B) Relative frequencies are particularly useful when comparing the frequency distributions of two different size sets of data.
C) The histogram of a sample should have a distribution shape that is bimodal.
D) A stem-and-leaf display contains all the information needed to create a histogram.
ANSWER: C

128. The following set of data represents letter grades on term papers in a rhetoric class:

A, A, A, B, B, B, B, C, C, C, C, C, C, C, C, C, C, C, D, D, D, F.

Select the most appropriate measure of central tendency for the data described.

A) Mean
B) Median
C) Mode
D) Midrange
ANSWER: C

Chapter 2

129. The following set of data represents the ages of students in a small seminar: 20, 21, 22, 25, 26, 27, and 68. Select the most appropriate measure of central tendency for the data described.

A) Mean
B) Median
C) Mode
D) Midrange
ANSWER: **B**

130. The following set of data represents the temperature high for seven consecutive days in February in Chicago: 22, 14, 26, 27, 35, 38, and 41. Select the most appropriate measure of central tendency for the data described.

A) Mean
B) Median
C) Mode
D) Midrange
ANSWER: **A**

131. Which of the following is not affected by extreme values?

A) Median
B) Tenth percentile
C) Third quartile
D) All of the above
ANSWER: **D**

132. The measure most affected by extreme values is the:

A) mean
B) second quartile
C) first quartile
D) midquartile
ANSWER: **A**

133. Which of the following is *not* a measure of central tendency?

A) Mean
B) Median
C) Midrange
D) None of the above
ANSWER: **D**

134. Which of the following statements is not true?

A) When *n* is odd, the depth of the median, $d(\tilde{x})$, will always be an integer.

B) When *n* is even, the depth of the median, $d(\tilde{x})$, will always be a half-number.

C) The midrange is a measure of position.
D) None of the above.
ANSWER: **C**

Descriptive Analysis and Presentation of Single-Variable Data

135. The following data set represents shirt sizes for girls' field hockey team:

S, S, S, M, M, M, M, M, M, M, M, M, M, L, L, L, L, L, XL, XL

Select the most appropriate measure of central tendency for the data described.

A) Mean
B) Median
C) Mode
D) Midrange
ANSWER: **C**

136. Adding 5 to each value in a data set would *not* change which of the following measures?

A) Mode
B) Mean
C) Mid-range
D) Standard deviation
ANSWER: **D**

137. Which of the following is a correct statement?

A) The interquartile range is found by taking the difference between the first and third quartiles and dividing that value by 2.
B) The standard deviation is expressed in terms of the original units of measurement but the variance is not.
C) The values of the standard deviation may be either positive or negative, while the value of the variance will always be positive.
D) The mean absolute deviation is a measure of spread that is used too frequently.
ANSWER: **B**

138. Which of the following is a correct statement?

A) The mean is a measure of the deviation in a data set.
B) The standard deviation is a measure of dispersion.
C) The range is a measure of central tendency.
D) The median is a measure of dispersion.
ANSWER: **B**

139. The difference between the largest and smallest values in an ordered array is called the:

A) standard deviation
B) variance
C) interquartile range
D) range
ANSWER: **D**

140. Which of the following is the weakest measure of dispersion?

A) Range
B) Mean absolute deviation
C) Standard deviation
D) Variance
ANSWER: **A**

Chapter 2

141. Which of the following statements is false?

A) The measures of dispersion include the *range, variance*, and *standard deviation*.
B) The numerical values of measures of dispersion describe the amount of spread, or variability that is found among the data values.
C) Closely grouped data have relatively small measures of dispersion values, and more widely spread-out data have larger values.
D) None of the above

ANSWER: D

142. Which of the following statements is false?

A) $\sum(x-\bar{x})$ is always zero even if a rounded value of $\bar{x}$ is used.
B) The standard deviation of a sample, *s*, is the positive square root of the variance.
C) The unit of measure for the standard deviation is the same as the unit of measure for the data.
D) The unit of measure for the variance is *units squared* of the unit of measure for the data.

ANSWER: A

143. Which of the following types of graphs would *not* be good for qualitative data?

A) Box-and-whiskers display
B) Circle graph
C) Bar graph
D) Pareto diagram

ANSWER: A

144. For a normal distribution, a value that is two standard deviations below the mean would be closer to which of the following?

A) Third percentile
B) First quartile
C) Fortieth percentile
D) Median

ANSWER: A

145. Which of the following statements is correct?

A) Measures of position are used to describe the position a specific data value possesses in relation to the rest of the data.
B) Quartiles and percentiles are two of the most popular measures of position.
C) Quartiles are values of the variable that divide the ranked data into 4 equal subsets called quarters.
D) All of the above.

ANSWER: D

146. Which is the depth of Q_1 for a ranked set of 40 exam scores?

A) 9.5
B) 10.0
C) 10.5
D) 11.0

ANSWER: C

147. Which of the following statements is false?

A) The first quartile, Q_1, is a number such that at most 25% of the data are smaller in value than Q_1 and at most 75% are larger.
B) The second quartile is the mean.
C) The third quartile, Q_3, is a number such that at most 75% of the data are smaller in value than Q_3, and at most 25% are larger.
D) None of the above

ANSWER: B

148. Which is the depth of the 65th percentile for a ranked set of 50 student ages?

A) 32.5
B) 33.0
C) 33.5
D) 34.0

ANSWER: B

149. If the 70th percentile for a set of exam scores is 82, what does this mean?

A) At most 70% of the exam scores are smaller in value than 82
B) At most 82% of the exam scores are smaller in value than 70
C) At least 70% of the exam scores are larger in value than 82
D) At least 82% of the exam scores are larger in value than 70

ANSWER: A

150. The 5-number summary divides a set of data into how many subsets?

A) 6
B) 5
C) 4
D) 3

ANSWER: C

151. Which of the following statements is false?

A) The 5-number summary is more informative when it is displayed on a diagram drawn to scale. A computer-generated graphic display that accomplishes this is known as the box-and-whiskers display.
B) The position of a specific value in a set of data can be measured in terms of the mean and variance using the standard score, commonly called the *z*-score.
C) The *z*-scores are typically range in value from approximately -3.00 to +3.00.
D) None of the above

ANSWER: B

152. Which is the depth of the 5th percentile for a ranked set of 35 student weights?

A) 1.50
B) 2.00
C) 2.50
D) 3.00

ANSWER: B

Chapter 2

Short-Answer Questions

153. Explain the difference between a *J*-shaped histogram and a skewed histogram.

ANSWER:
J-shaped histogram has only one tail with the highest frequency as an end class. A skewed histogram has tails on both sides of the class with the highest frequency, with one tail being considerably longer.

154. If a histogram is constructed for the following frequency distribution, what shape would it have?

Class Boundaries	*Frequency*
$20 \le x < 30$	5
$30 \le x < 40$	15
$40 \le x < 50$	20
$50 \le x < 60$	18
$60 \le x < 70$	13
$70 \le x < 80$	10
$80 \le x < 90$	5
$90 \le x \le 100$	1

ANSWER:
Skewed to right or positively skewed

155. What is the largest possible value needed on the vertical axis of a relative frequency histogram?

ANSWER:
One

156. A relative frequency distribution was constructed for a sample of size n = 120. The relative frequency for the third class was 0.15. How many items of data fell into the third class?

ANSWER:
18

157. A relative frequency distribution was constructed for a sample of size n = 150. The relative frequency for the second class was 0.067. How many items of data fell into this class?

ANSWER:
10

158. In an ogive, what does the vertical scale identify?

ANSWER:
The vertical scale identifies either the cumulative frequencies or the cumulative relative frequencies.

159. In an ogive, what does the horizontal scale identify?

ANSWER:
The horizontal scale identifies the upper class boundaries. Until the upper boundary of a class has been reached, you cannot be sure you have accumulated all the data in that class. Therefore, the horizontal scale for an ogive is always based on the upper class boundaries.

160. Explain what is wrong with the statement, "The mean is always the best measure of central tendency."

ANSWER:
It depends on the type of data, and what would be an appropriate measure of central tendency.

161. A company found that the mean number of sales for the 20 salesmen during the past month was 8.5. What was the total number of sales for the salesmen?

ANSWER:
170

162. For a particular sample $\bar{x} = 14.7$ and $s = 3.5$. A new sample is formed by subtracting 2 from each value in the original sample. Find $\bar{x}$ for this new sample.

ANSWER:
$\bar{x} = 12.7$

163. Explain why it is possible to find the mean for the data of a quantitative variable, but not for a qualitative variable.

ANSWER:
Quantitative variable results in numbers for which arithmetic is meaningful; qualitative variable does not

164. Find the median height of cheerleaders of a college basketball team: 66, 69, 65, 63 and 67 inches.

ANSWER:
$\tilde{x}$ = 66 inches

165. Explain why the standard deviation is not always less than the variance and give an example.

ANSWER:
If $s < 1$, then s will be larger than s^2; example: $s = 0.25$, then $s^2 = 0.0625$.

166. Which of the three measures of variability, range, standard deviation, and variance, does *not* preserve the same unit of measurement as the observations themselves?

ANSWER:
Variance

167. If a sample has a standard deviation of 4.5, what is its variance?

ANSWER:
20.25

168. For a particular sample $\bar{x} = 14.7$ and s = 3.5. A new sample is formed by subtracting 2 from each value in the original sample. Find s for this new sample.

ANSWER:
s = 3.5

169. For a particular sample of size n = 10, the sample variance is 4.8 and $\bar{x} = 0.5$. For this sample, find $\sum x^2$.

ANSWER:
45.7

170. Why the sum of the deviations, $\sum(x-\bar{x})$, is always zero?

ANSWER:
$\sum(x-\bar{x})$, is always zero because the deviations of x values smaller than the mean (which are negative values) cancel out x values larger than the mean (which are positive).

171. What does mean absolute deviation tell us?

ANSWER:
It tells us the mean "distance" the data values are from the mean.

172. Explain the meaning of the following statement "The data value x = 30 has a deviation value of 6."

ANSWER:
The value x = 30 is 6 larger than the mean.

173. Explain the meaning of the following statement "The data value x = 80 has a deviation value of -15."

ANSWER:
The value x = 80 is 15 smaller than the mean.

174. A particular standardized test has a mean score of 455 with a standard deviation of 112. A student scored 575 on this test. Determine the student's z-score.

ANSWER:
1.07

175. On a standardized test, a student's z-score was near zero. What does this tell us about the student's actual score on the test?

ANSWER:
The actual score was near the mean.

176. For a particular sample, the mean is 4.74, and the standard deviation is 3.10. What score in the sample has a *z*-score equal to −0.40?

ANSWER:
3.5

177. What statistical measure gives the range of the middle 50% of the data?

ANSWER:
Interquartile range

178. An aptitude test is known to have a mean score of 37.75 with a standard deviation equal to 3.5. A company requires a standard score of at least 1.5 for employment as one of its requirements. What must your test score be in order to be considered for employment?

ANSWER:
43 or larger

179. A normal distribution has a mean equal to 55.0 and a standard deviation equal to 7.5. Find the value of the midquartile.

ANSWER:
55.0

180. For a particular sample $\bar{x} = 4.2$, one item in the sample is $x = 4.8$. This item has a *z*-score at 2.50. Find the sample standard deviation.

ANSWER:
$s = 0.24$

181. For a particular sample $\bar{x} = 4.4$, an item in the sample is $x = 3.4$, and the *z*-score of this item is equal to −1.25. Find the sample variance.

ANSWER:
$s^2 = 0.64$

182. Determine your raw score on a test that has a sample mean of 65 and a sample variance of 121 if your instructor told you that your standard score is 1.50.

ANSWER:

$$z = \frac{x - \bar{x}}{s} \Rightarrow 1.50 = \frac{x - 65}{11} \Rightarrow x = 81.5$$

183. In general, the median, the midrange, and the midquartile are not necessarily the same value. Each is the middle value, but by different definitions of "middle". What property does the distribution need for these three measures to all be the same value?

ANSWER:
The distribution of the data needs to be symmetric for these three measures to all be the same value.

184. What does it mean to say that x = 163 has a standard score of +1.60?

ANSWER:
It means that x = 163 is 1.60 standard deviations above the mean.

185. Determine your raw score on a test that has a sample mean of 74 and a sample standard deviation of 12 if your instructor told you that your standard score is -0.50.

ANSWER:

$$z = \frac{x - \overline{x}}{s} \Rightarrow -0.50 = \frac{x - 74}{12} \Rightarrow x = 68$$

186. What does it mean to say that a particular value of x has a z score of -1.94?

ANSWER:
It means that value of x is 1.94 standard deviations below the mean.

187. In general, the standard score is a measure of what?

ANSWER:
The standard score is a measure of the number of standard deviations from the mean.

Applied and Computational Questions

QUESTIONS 188 THROUGH 193 ARE BASED ON THE FOLLOWING INFORMATION:
The frequency distribution below gives the weight loss in pounds for 90 patients.

Class Number	*Class Boundaries*	*f*
1	$0.0 \le x < 5.0$	5
2	$5.0 \le x < 10.0$	12
3	$10.0 \le x < 15.0$	16
4	$15.0 \le x < 20.0$	27
5	$20.0 \le x < 25.0$	19
6	$25.0 \le x < 30.0$	9
7	$30.0 \le x \le 35.0$	2

188. What is the upper class boundary of the fifth class?

ANSWER:
25.0

189. What is the class width?

ANSWER:
5.0

190. What is the class mark of the third class?

ANSWER:
12.5

191. What is the value of $\sum f$?

ANSWER:
90

192. What is the sample size?

ANSWER:
90

193. Convert the above table to a relative frequency distribution.

ANSWER:

Class Number	*Class Boundaries*	*Relative frequency*
1	$0.0 \le x < 5.0$	0.056
2	$5.0 \le x < 10.0$	0.133
3	$10.0 \le x < 15.0$	0.178
4	$15.0 \le x < 20.0$	0.300
5	$20.0 \le x < 25.0$	0.211
6	$25.0 \le x < 30.0$	0.100
7	$30.0 \le x \le 35.0$	0.022

QUESTIONS 194 THROUGH 198 ARE BASED ON THE FOLLOWING INFORMATION:
A sample of families living in a large, suburban subdivision resulted in the following frequency distribution, where: x = number of children in the family.

x	*f*
0	8
1	11
2	23
3	21
4	13
5	7
6	2

194. What does the "3" represent?

ANSWER:
3 children per family for 21 families

195. What does the "7" represent?

ANSWER:
7 families with 5 children each

196. How many families were used to form this sample?

ANSWER:
85

197. How many children are included in this sample?

ANSWER:
219

198. Determine the mean number of children per family in the sample.

ANSWER:
$\bar{x}=219/85=2.58$

199. A sample of twenty-five snow blowers of a given brand were filled with gasoline (one gallon) and allowed to run until the tank was empty. The times (in minutes) that the snow blowers operated were recorded as follows:

65	70	60	65	67	68	63	62	63	70	72	66	63
66	66	62	70	58	60	60	60	62	67	71	65	

Form a frequency distribution consisting of 5 classes.

ANSWER:

Class Boundaries	*Frequency*
$58 \le x < 61$	5
$61 \le x < 64$	6
$64 \le x < 67$	6
$67 \le x < 70$	3
$70 \le x \le 73$	5

QUESTIONS 200 THROUGH 204 ARE BASED ON THE FOLLOWING INFORMATION:
The frequency distribution below gives the daily high temperature for 40 consecutive winter days in northern Wisconsin.

Class Boundaries	*f*
$0 \le x < 3$	2
$3 \le x < 6$	4
$6 \le x < 9$	7
$9 \le x < 12$	10
$12 \le x < 15$	8
$15 \le x < 18$	6
$18 \le x \le 21$	3

200. In how many days was the daily high temperature between 9 and 12 degrees?

ANSWER:
10 days

201. Convert the above frequency distribution to a relative frequency distribution.

ANSWER:

Class Boundaries	*Relative frequency*
$0 \le x < 3$	0.050
$3 \le x < 6$	0.100
$6 \le x < 9$	0.175
$9 \le x < 12$	0.250
$12 \le x < 15$	0.200
$15 \le x < 18$	0.150
$18 \le x \le 21$	0.075

202. What is the proportion of days in which the daily high temperature was between 15 and 18?

ANSWER:
0.15

203. Construct the cumulative frequency distribution.

ANSWER:

Class Boundaries	*Cumulative frequency*
$0 \le x < 3$	2
$3 \le x < 6$	6
$6 \le x < 9$	13
$9 \le x < 12$	23
$12 \le x < 15$	31
$15 \le x < 18$	37
$18 \le x \le 21$	40

204. Construct the cumulative relative frequency.

ANSWER:

Class Boundaries	*Cumulative Relative frequency*
$0 \le x < 3$	0.050
$3 \le x < 6$	0.150
$6 \le x < 9$	0.325
$9 \le x < 12$	0.575
$12 \le x < 15$	0.775
$15 \le x < 18$	0.925
$18 \le x \le 21$	1.000

205. The following frequency distribution gives the pay ranges (in thousands of dollars) for all middle management personnel in large company.

Class Boundaries	*f*
$20 < x < 30$	4
$30 \le x < 40$	27
$40 \le x < 50$	29
$50 \le x < 60$	25
$60 \le x < 70$	17

Describe what shape a histogram of this data would have.

ANSWER:
Skewed to left (or negatively skewed)

QUESTIONS 206 THROUGH 209 ARE BASED ON THE FOLLOWING INFORMATION:
The ages of 50 students who are attending a community college in Iowa are shown below:

20	20	19	21	21	22	19	19	21	19
18	21	19	18	22	21	24	20	24	17
21	19	22	19	18	20	23	19	19	20
19	20	21	22	21	20	22	20	21	20
21	19	21	21	19	19	20	19	19	19

206. Prepare an ungrouped frequency distribution of these ages.

ANSWER:

Age	17	18	19	20	21	23	23	24
Frequency	1	3	16	10	12	5	1	2

207. Prepare an ungrouped relative frequency distribution of the same data.

ANSWER:

Age	17	18	19	20	21	23	23	24
Rel. Freq.	0.02	0.06	0.32	0.20	0.24	0.10	0.02	0.04

208. Prepare a frequency histogram of these data.

ANSWER:

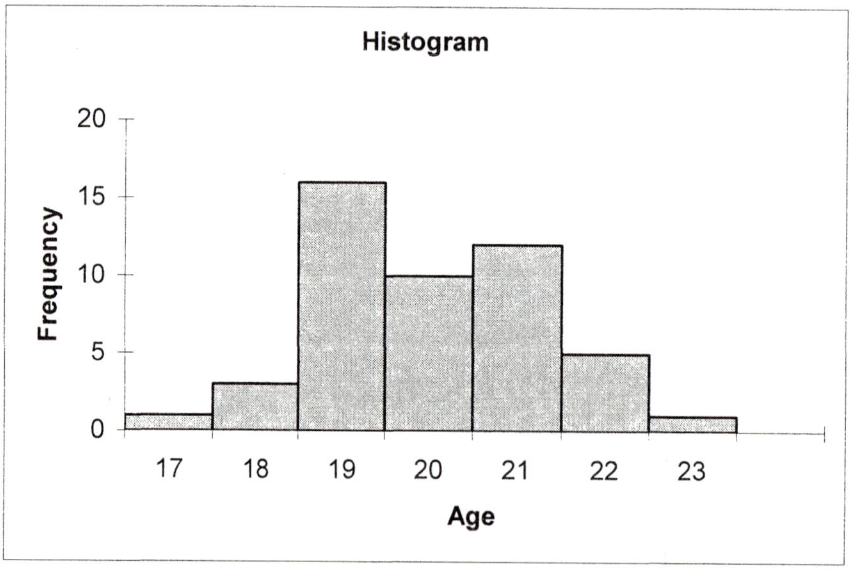

209. Prepare a cumulative relative frequency distribution of the same data.

ANSWER:

Age	17	18	19	20	21	23	23	24
Cum. rel. freq.	0.02	0.08	0.40	0.60	0.84	0.94	0.96	1.00

210. Briefly discuss the basic guidelines to follow in constructing a grouped frequency distribution.

ANSWER:
(a) Each class should be the same width.
(b) Classes (sometimes called bins) should be set up so that they do not overlap and so that each data belongs to exactly one class.
(c) Five to twelve classes are most desirable. (The square root of *n* is a reasonable guideline for the number of classes with samples of fewer than 125 data.)
(d) Use a system that takes advantage of a number pattern to guarantee accuracy.
(e) When it is convenient, an even class width is often advantageous.

211. The terms "symmetrical, uniform, skewed, J-shaped, bimodal, and normal" are usually used to describe histograms. Discuss each term briefly.

ANSWER:
Symmetrical: Both sides of this distribution are identical (halves are mirror images).
Uniform (rectangular): Every value appears with equal frequency.
Skewed: One tail is stretched out longer than the other. The direction of skewness is on the side of the longer tail.
J-Shaped: There is no tail on the side of the class with the highest frequency.
Bimodal: The two most populous classes are separated by one or more classes. This situation often implies that two populations are being sampled.
Normal: A symmetrical distribution is mounded up about the mean and becomes sparse at the extremes.

QUESTIONS 212 AND 213 ARE BASED ON THE FOLLOWING INFORMATION:
The following frequency distribution provides the number of managers and their annual salaries (in $1000):

Annual Salary ($1000)	15-25	25-35	35-45	45-55	55-65
Number of Managers	24	74	52	38	12

212. Prepare a cumulative frequency distribution for this frequency distribution.

ANSWER:

Class Boundaries	Cumulative Frequency
$15 \le x \le 25$	24
$15 \le x \le 25$	98
$15 \le x \le 25$	150
$15 \le x \le 25$	188
$15 \le x \le 25$	200

213. Prepare a cumulative relative frequency distribution for this frequency distribution.

ANSWER:

Class Boundaries	Cumulative Frequency
$15 \leq x \leq 25$	0.12
$15 \leq x \leq 25$	0.49
$15 \leq x \leq 25$	0.75
$15 \leq x \leq 25$	0.94
$15 \leq x \leq 25$	1.00

QUESTIONS 214 THROUGH 217 ARE BASED ON THE FOLLOWING INFORMATION:
The players on a professional soccer team scored 40 goals during last season.

Player	1	2	3	4	5	6	7	8	9	10	11	12	13
Goals	2	7	3	2	2	5	2	1	6	2	3	2	3

214. If you want to show the number of goals scored by each player, would it be more appropriate to display this information on a bar graph or a histogram? Explain.

ANSWER:
In order to show the number of goals scored by each player, it would be more appropriate to display this information on a bar graph

215. Construct the appropriate graph for question 214.

ANSWER:

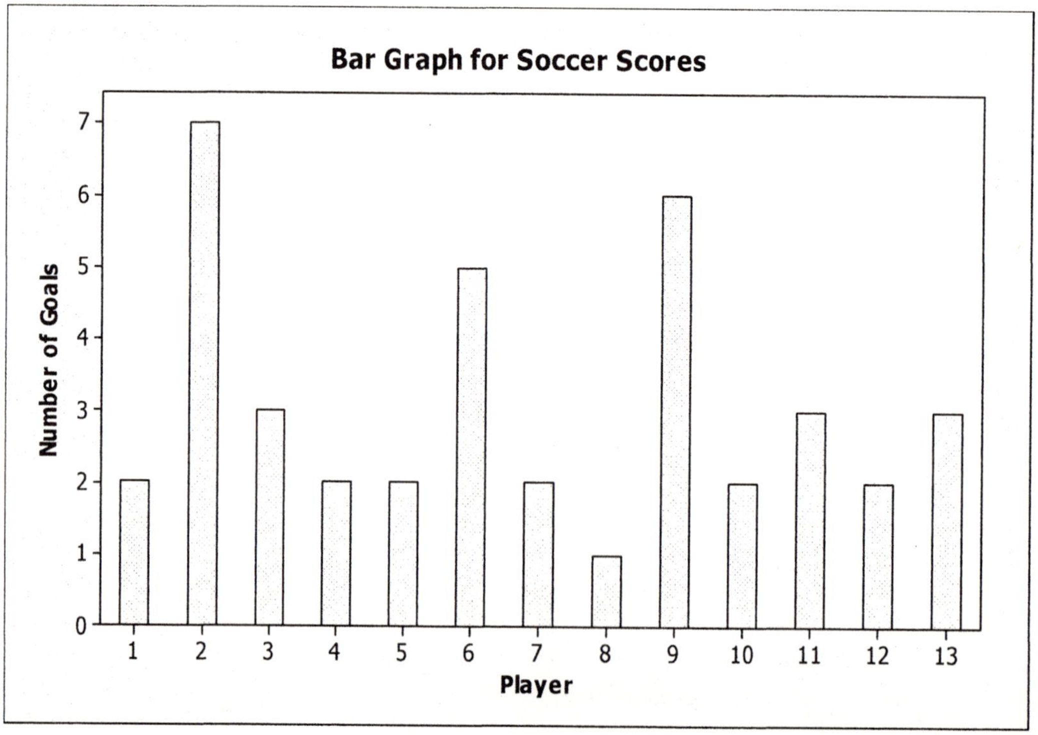

216. If you wanted to show (emphasize) the distribution of scoring by the team, would it be more appropriate to display this information on a bar graph or a histogram? Explain.

ANSWER:
If we want to emphasize the distribution of scoring by the team, it would be more appropriate to display this information on a histogram.

217. Construct the appropriate graph for question 216.

ANSWER:

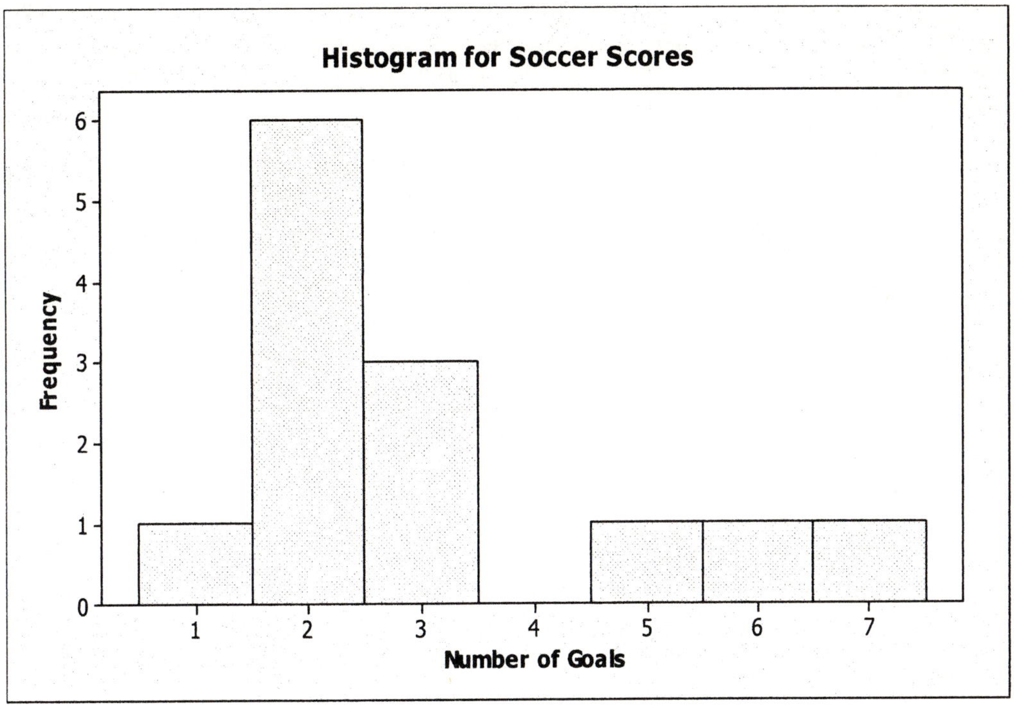

QUESTIONS 218 AND 219 ARE BASED ON THE FOLLOWING INFORMATION:
A sample of twenty-five snow blowers of a given brand were filled with gasoline (one gallon) and allowed to run until the tank was empty. The times (in minutes) that the snow blowers operated were recorded as follows:

65	70	60	65	67	68	63	62	63	70	72	66	63
66	66	62	70	58	60	60	60	62	67	71	65	

218. Construct a stem-and-leaf display.

ANSWER:

Stems	*Leaves*
5	8
6	0 0 0 0 2 2 2 3 3 3 5 5 5 6 6 6 7 7 8
7	0 0 0 1 2

219. Find the mean, median, mode, and midrange.

ANSWER:
Mean = 64.84, Median = 65.0, Mode = 60.0, Midrange = 65.0

220. Nine households had the following number of children per household: 2, 0, 2, 2, 1, 2, 4, 3, 2. Find the mean, median, mode, and midrange for these data.

ANSWER:
Mean = 2, Median = 2, Mode = 2, Midrange = 2

QUESTIONS 221 THROUGH 223 ARE BASED ON THE FOLLOWING INFORMATION:
The commuting distance was determined for each of 10 employees at Acme manufacturing. One of the employees lives in another town and has a large commuting distance. The 10 distances were as follows: 5, 10, 7, 15, 10, 12, 8, 120, 20, 18.

221. Find the mean distance.

ANSWER:
22.5

222. Find the median distance.

ANSWER:
11

223. Which measurement, A or B, is most representative for the data? Why?

ANSWER:
Median; since median is not affected by outliers.

224. Consider the following sample: 26, 49, 9, 42, 60, 11, 43, 26, 30, and 14. Find the mean.

ANSWER:
31.0

225. Consider the following sample: 26, 49, 9, 42, 60, 11, 43, 26, 30, and 14. Find the median.

ANSWER:
28.0

226. Consider the following sample: 26, 49, 9, 42, 60, 11, 43, 26, 30, and 14. Find the midrange.

ANSWER:
34.5

227. For a particular sample of 50 scores on a statistics exam, the following results were obtained:

Mean = 78	Midrange = 72	Third quartile = 94	Mode = 84
Median = 80	Standard deviation = 11	Range = 52	First quartile = 68

What score was earned by more students than any other score? Why?

ANSWER:
84; since it is the mode

228. If a sample with a mean of 10.5 and a standard deviation of 2.30 has every item multiplied by 10, find the mean of the new sample.

ANSWER:
105

229. For a particular sample, the mean is 3.7 and the standard deviation is 1.2. A new sample is formed by adding 6.3 to every item of data in the original sample. Find the mean of the new sample.

ANSWER:
10.0

230. Find the mean, median, mode, and midrange for the following data:

x	f
10	2
11	4
12	7
15	3
20	1

ANSWER:
Mean = 12.5, Median = 12, Mode = 12, Midrange = 15

231. A student computed the mean of a particular sample to be 40.0. After computing the mean, he discovered that he forgot to include the number 36 in the sample. When this number was included, the sample mean changed to 39.5. What is the sample size when the number 36 is correctly included in the sample?

ANSWER:
$n = 8$

<u>**QUESTIONS 232 THROUGH 238 ARE BASED ON THE FOLLOWING INFORMATION:**</u>
Starting with a sample of two values 70 and 100, add three data values to your sample to obtain a new sample with certain statistics.

232. What are the three data values such that the new sample has a mean of 100? Justify your answer.

ANSWER:
Many different answers are possible. The sum of the five numbers needs to be 500; therefore we need any three numbers that total 330, such as 100, 110,120.Thus, the new sample mean $\bar{x} = 500 / 5 = 100$.

233. What are the three data values such that the new sample has a median of 70? Justify your answer.

ANSWER:
Many different answers are possible. Need two numbers smaller than 70 and one number larger than 70. For example, we may choose 50, 60, and 80.Thus the five numbers are 50, 60, 70, 80, 100, and the median is 70.

234. What are the three data values such that the new sample has a mode of 87? Justify your answer.

ANSWER:
Many different answers are possible. Need multiple 87's. For example, we may choose 87, 87 and 95. Thus, the five numbers are 70, 87, 87, 95, 100, and the mode = 87.

235. What are the three data values such that the new sample has a midrange of 70? Justify your answer.

ANSWER:
Many different answers are possible. Need any two numbers that total 140 for the extreme values *L* and *H*, where one is 100 or larger. For example, we may choose the numbers 40, 50, and 60. Thus the five numbers are 40, 50, 60, 70, 100, and midrange = $(L+H)/2 = (40+100)/2 = 70$.

236. What are the three data values such that the new sample has a mean of 100 and a median of 70? Justify your answer.

ANSWER:
Many different answers are possible. Need two numbers smaller than 70 and one number larger than 70 so that their total is 330. For example, we may choose the numbers 65, 65, and 200. Thus the five numbers are 65, 65, 70, 100, 200. Hence, $\bar{x} = 500/5 = 100$, and the median is 70.

237. What are the three data values such that the new sample has a mean of 100 and a mode of 87? Justify your answer.

ANSWER:
Many different answers are possible. Need two numbers of 87 and a number large enough so that the total of all five numbers is 500. Therefore the three numbers are 87, 87,156. The five numbers are 70, 87, 87, 100, 156. Thus the mode = 87, and $\bar{x} = 500 / 5 = 100$.

238. What are the three data values such that the new sample has a mean of 100, a median of 70, and a mode of 87? Justify your answer.

ANSWER:
Many different answers are possible. There must be two 87's in order to have a mode of 87, and there can only be two data values larger than 70 in order for 70 to be the median, which is impossible since 100 is one of the numbers, and that makes three of the five numbers larger than 70.

239. The Next Door Store kept track of the number of paying customers it had during the noon hour each day for 100 days. The following are the resulting statistics rounded to the nearest integer: Mean = 95, Median = 97, Mode = 98, First quartile = 85, Third quartile = 107, Midrange = 93, Range = 56, and Standard deviation = 12. The Next Door Store served what number of paying customers during the noon hour more often than any other number? Explain how you determined your answer.

ANSWER:
98 customers; this is the mode.

240. A statistics test was given with the following results:

80, 69, 92, 75, 88, 37, 98, 92, 90, 81, 32, 50, 59, 66, 67, 66

Find the range, standard deviation, and variance for the scores.

ANSWER:
Range = 66, s = 19.64, s^2 = 385.85

QUESTIONS 241 THROUGH 248 ARE BASED ON THE FOLLOWING INFORMATION:
Starting with a sample of two values 75 and 105, add three data values to your sample to obtain a new sample with certain statistics.

241. What are the three data values such that the new sample has a mean of 110 (Hint: Many different answers are possible). Justify your answer.

ANSWER:
$\sum x$ needs to be 550; therefore, need any three numbers that total 370, such as 110, 120, and 140. Hence, the mean $\bar{x} = \sum x / n$ = 550 / 5 = 110

242. What are the three data values such that the new sample has a median of 75 (Hint: Many different answers are possible). Justify your answer.

ANSWER:
Need two numbers smaller than 75 and one number larger. For example, choose the numbers 60, 70, and 80. Hence, the five data values are 60, 70, 75, 80, 105, and d($\tilde{x}$) = (n+1)/2 = (5+1)/2 = 3rd value; therefore the median $\tilde{x}$ = 75.

243. What are the three data values such that the new sample has a mode of 85 (Hint: Many different answers are possible). Justify your answer.

ANSWER:
Choose three numbers, each is 85. Hence the five data values are 75, 85, 85, 85, 105, and the mode = 85.

244. What are the three data values such that the new sample has a midrange of 80 (Hint: Many different answers are possible). Justify your answer.

ANSWER:
Need any two numbers that total 160 for the extreme values where one is 105 or larger. For example, choose the values 40, 50, and 120. Hence the five data values are 40, 50, 75, 105, and 120. Therefore, midrange = (L+H)/2 = (40+120)/2 = 80.

245. What are the three data values such that the new sample has a mean of 110 and a median of 75 (Hint: Many different answers are possible). Justify your answer.

ANSWER:
Need two numbers smaller than 75 and one larger than 75 so that their total is 365. For example, choose the values 65, 70, and 230. Hence the five data values are 65, 70, 75, 105, and 230. Hence the mean $\bar{x} = \sum x / n$ = 550 / 5 = 110, and d($\tilde{x}$) = (n+1)/2 = (5+1)/2 = 3rd value; therefore the median $\tilde{x}$ = 75.

246. What are the three data values such that the new sample has a mean of 110 and a mode of 80 (Hint: Many different answers are possible). Justify your answer.

ANSWER:
Need two numbers of 80 and a third number large enough so that the total of all five values is 550. Then the third number must be 210. Hence, the five values are 75, 80, 80, 105, and 210. Hence, the mean $\bar{x} = \sum x / n$ = 550 / 5 = 110, and the mode = 80.

247. What are the three data values such that the new sample has a mean of 110 and a midrange of 80 (Hint: Many different answers are possible). Justify your answer.

ANSWER:
We started with the data values 75 and 105. A mean of 110 requires the five data values to total 550 and a midrange of 80 requires the total of the lowest value *L* and the highest value *H* to be 160. The sum of 75 and 105 is 180; hence, the total of the other three remaining numbers is 370. Since *L* + *H* must be 160, then the fifth number must be 210, which would then become *H* and change the value of the midrange. So, this situation is impossible!

248. What are the three data values such that the new sample has a mean of 110, a median of 75, and a mode of 85 (Hint: Many different answers are possible). Justify your answer.

ANSWER:
There must be two 85's in order to have a mode of 85, and there can only be two data values larger than 75 in order for 75 to be the median, but since 105 is one of the starting numbers, then we have three data values larger than 75; namely 85, 85, and 105. As a result, 75 can't be the median. So, this situation is impossible!.

249. A sample of twenty-five snow blowers of a given brand were filled with gasoline (one gallon) and allowed to run until the tank was empty. The times (in minutes) that the snow blowers operated were recorded as follows:

65	70	60	65	67	68	63	62	63	70	72	66	63
66	66	62	70	58	60	60	60	62	67	71	65	

Find the standard deviation and the range.

ANSWER:
Standard deviation *s* = 3.91, Range=14

250. A group of children had the following heights in inches: 45, 46, 42, 56, 37, 50, 51, 50, 47, 47. Find the range, standard deviation, and variance for the scores.

ANSWER:
Range = 19, s = 5.22, and s^2 = 27.211

<u>QUESTIONS 251 AND 252 ARE BASED ON THE FOLLOWING INFORMATION:</u>
Consider the following sample: 26, 49, 9, 42, 60, 11, 43, 26, 30, and 14.

251. Find the variance.

ANSWER:
294.89

252. Find the standard deviation

ANSWER:
17.17

QUESTIONS 253 AND 254 ARE BASED ON THE FOLLOWING INFORMATION:
For a particular sample of 50 scores on a statistics exam, the following results were obtained:

Mean = 78	Midrange = 72	Third quartile = 94	Mode = 84
Median = 80	Standard deviation = 11	Range = 52	First quartile = 68

253. What was the highest score earned on the exam?

ANSWER:
98 [Recall that: Midrange = $(L+H)/2$, and Range = H-L]

254. What was the lowest score earned on the exam?

ANSWER:
46

255. If a sample with a mean of 10.5 and a standard deviation of 2.30 has every item multiplied by 10, find the variance of the new sample.

ANSWER:
529

QUESTIONS 256 AND 257 ARE BASED ON THE FOLLOWING INFORMATION:
For a particular sample, the mean is 3.7 and the standard deviation is 1.2. A new sample is formed by adding 6.3 to every item of data in the original sample.

256. Find the standard deviation of the new sample.

ANSWER:
1.20

257. Find the variance of the new sample.

ANSWER:
1.44

258. For the following three samples, for which sample is the data most closely grouped about the sample mean? Give a written explanation that supports your conclusion.

Sample 1: 15, 16, 19, 21, 28;
Sample 2: 44, 49, 50, 51, 57; and
Sample 3: 122.8, 123.7, 124.6, 130.5, 135.8.

ANSWER:
Since the sample standard deviation s measures dispersion about the mean, we compute s of each sample. Sample 1, s = 5.17; Sample 2, s = 4.66; Sample 3, s = 5.54. Since sample 2 has the smallest standard deviation, data most closely grouped about its mean.

259. The mean for 50 pressure readings equals 5.5, and the sum of the squares of the readings equals 1622.75. Find the standard deviation of these pressure readings.

ANSWER:

$$s = \sqrt{[\sum x^2 - n\overline{x}^2]/(n-1)} = \sqrt{[1622.75 - 50(5.5)^2]/49} = \sqrt{2.25} = 1.5$$

260. A set of 25 measurements has a mean of 24.5 and a standard deviation equal to 4.0. Find $\sum x^2$.

ANSWER:

$$s = \sqrt{[\sum x^2 - n\overline{x}^2]/(n-1)} \Rightarrow 4.0 = \sqrt{[\sum x^2 - 25(24.5)^2]/24} \Rightarrow \sum x^2 = 15{,}390.25$$

QUESTIONS 261 AND 262 ARE BASED ON THE FOLLOWING INFORMATION:
Consider the following data: 21, 41, 41, 36, 39, 23, 30, 30, 34, 31, 26, 25, 29, 28, 36.

261. Find the mean.

ANSWER:
$\overline{x} = 31.3$

262. Find the standard deviation.

ANSWER:
$s = 6.3$

263. Consider the following two sets of data:

Set 1:	45	55	50	48	52
Set 2:	35	50	65	47	53

Both sets have the same mean $\overline{x} = 50$. Compare the following measures for both sets: $\sum(x-\overline{x})$, $\sum|x-\overline{x}|$, $SS(x)$, and range. Comment on the meaning of these comparisons.

ANSWER:
Set 1:

x	$x-\overline{x}$	$\lvert x-\overline{x}\rvert$	$(x-\overline{x})^2$
45	-5	5	25
55	+5	5	25
50	0	0	0
48	-2	2	4
52	+2	2	4
250	0	14	58

Set 2:

x	$x-\bar{x}$	$\lvert x-\bar{x}\rvert$	$(x-\bar{x})^2$
35	-15	15	225
50	0	0	0
65	+15	15	225
47	-3	3	9
53	+3	3	9
250	0	36	468

Comparisons:

	$\sum x$	$\sum(x-\bar{x})$	$\sum\lvert x-\bar{x}\rvert$	$\sum(x-\bar{x})^2$	Range
Set1	250	0	14	58	10
Set2	250	0	36	468	30

The values of $\sum\lvert x-\bar{x}\rvert$, $SS(x)$ [recall $SS(x)=\sum(x-\bar{x})^2$], and range all reflect the fact that there is more variability in the data forming set 2 than in the data of set 1. $\sum(x-\bar{x})=0$ for both sets of data (in fact this is always true for any data).

QUESTIONS 264 AND 265 ARE BASED ON THE FOLLOWING INFORMATION:
A sample of twenty-five snow blowers of a given brand were filled with gasoline (one gallon) and allowed to run until the tank was empty. The times (in minutes) that the snow blowers operated were recorded as follows:

65	70	60	65	67	68	63	62	63	70	72	66	63
66	66	62	70	58	60	60	60	62	67	71	65	

264. Find the first quartile

ANSWER:
$Q_1 = 62$

265. Find the ninetieth percentile.

ANSWER:
$P_{90} = 70$

266. For a particular sample of 50 scores on a statistics exam, the following results were obtained: Mean = 78, Midrange = 72, third quartile = 94, Mode = 84, Median = 80, Standard deviation = 11, Range = 52, and first quartile = 68. How many students scored between 68 and 94 on the exam?

ANSWER:
25

QUESTIONS 267 THROUGH 269 ARE BASED ON THE FOLLOWING INFORMATION:
Consider the following sample of size n = 65, ordered from smallest to largest:

124	127	128	129	133	134	137	139	141	143
147	148	156	159	163	166	169	170	173	179
199	201	207	210	213	217	219	222	225	228
234	238	244	259	261	262	263	264	266	268
279	280	286	298	299	305	306	307	311	313
320	328	333	345	350	351	361	362	363	364
378	388	390	400	417					

267. Prepare a five-number summary for this set of data.

ANSWER:
L = 124, $Q_1 = 169$, $\tilde{x} = 244$, $Q_3 = 311$, $H = 417$

268. Find the 80th percentile.

ANSWER:
P_{80} =330.5

269. Find the 29th percentile.

ANSWER:
P_{29} = 173115.

QUESTIONS 270 AND 271 ARE BASED ON THE FOLLOWING INFORMATION:
Consider the following sample of size, n = 60 ordered from smallest to largest:

24	27	28	29	33	34	37	39	41	43	47	48
56	59	63	66	69	70	73	79	99	21	27	10
13	17	19	22	25	28	34	38	44	59	61	62
63	64	66	68	79	80	86	98	99	35	36	37
11	13	20	28	33	45	50	51	61	62	63	64

270. Prepare a five-number summary for this set of data.

ANSWER:
L =10, $Q_1 = 28$, $\tilde{x} = 44.5$, $Q_3 = 63.25$, H = 99

271. Find the 20th percentile.

ANSWER:
P_{20} =27

272. Consider the sample 9, 11, 17, 23, 26, 38, 47. Find the z-score for the data point of "11."

ANSWER:
$\bar{x} = 24.4286$ and $s = 13.9745$. Then, $z = (x - \bar{x}) / s$ = (11.0 – 24.4286) / 13.9745 = -0.96

273. In which of these situations (A, B, or C) is the x-value lowest in relation to the sample from which it comes? These samples come from three different populations.

Situation A: $x = 6$, $\bar{x} = 20.0$, $s = 9.0$
Situation B: $x = 350$, $\bar{x} = 400.0$, $s = 20.0$
Situation C: $x = 1.6$, $\bar{x} = 2.00$, $s = 0.30$

ANSWER:
In situations A, B, and C; z = -1.56, -2.50, and –1.33, respectively. In situation B we see the lowest z-score of –2.50. Therefore, the x-value in B is lowest in relation to the sample from which it comes.

274. Find the first quartile and the third quartile for the following data:

2.1, 2.1, 2.2, 2.4, 2.5, 2.5, 2.5, 2.5, 2.6, 2.6, 2.6, 2.7, 2.7,
2.7, 2.8, 2.9, 3.0, 3.0, 3.2, 3.2, 3.3, 3.3, 3.5, 3.6, 4.0

ANSWER:
$Q_1 = 2.5$, $Q_3 = 3.2$

275. Consider the following measurements of ozone concentration (in ppm):

11.1, 11.5, 11.9, 12.0, 11.6, 12.2, 11.9, 12.5, 12.8, 19.0,
10.9, 11.6, 12.7, 5.0, 11.5, 12.6, 19.5, 12.7, 4.0, 19.1

The mean equals 12.31 and the variance equals 14.2884. Find the standard score for the smallest and largest data values.

ANSWER:
The smallest data value is 4.0, and its z-score is -2.198, while the largest data value is 19.5 and its z-score is 1.902.

276. Use the following stem-and-leaf display to find the tenth percentile for the distribution of lengths:

Stems	*Leaves*
2.1	0 2 1
2.3	3 6 5 2 1 1
2.4	1 1
2.5	0 1 2
2.7	7 7 8
3.1	2 2 4
3.5	1 1 2 1 1

ANSWER:
$P_{10} = 2.12$

277. The interquartile range (IQR) of a set of measurements is defined to be the difference between the upper and lower quartiles. Find the IQR for the following HLT scores that measure the degree of hostility: 80, 70, 63, 92, 81, 76, 78, 88, 70, 83, 74, 77, and 85.

ANSWER:
$\text{IQR} = Q_3 - Q_1 = 83 - 74 = 9$

278. The following subscripted x's represent a sample of size n = 67 which has been ranked from smallest (x_1) to largest (x_{67}): x_1, x_2, $x_3, \ldots x_{65}$, x_{66}, x_{67}.. Prepare a 5-number summary for this sample in terms of the subscripted x's.

ANSWER:

$L = x_1, \quad Q_1 = x_{17}, \quad \tilde{x} = x_{34}, \quad Q_3 = x_{51}, \quad H = x_{67}$

279. What does it mean to say that x = 152 have a standard score of +1.5?

ANSWER:
152 is one and one-half standard deviations above the mean.

280. What does it mean to say that a particular value of x has a z-score of –2.1?

ANSWER:
The score is 2.1 standard deviations below the mean.

281. In general, the standard score is a measure of what?

ANSWER:
The standard score is a measure of the number of standard deviations from the mean.

<u>QUESTIONS 282 THROUGH 293 ARE BASED ON THE FOLLOWING INFORMATION:</u>
Below are the ACT scores attained by the 25 members of a local high school graduating class.

23	26	25	19	33	21	21	22	21	27
19	25	18	23	22	30	27	27	23	16
21	19	20	30	22					

282. Draw a dotplot of the ACT scores.

ANSWER:

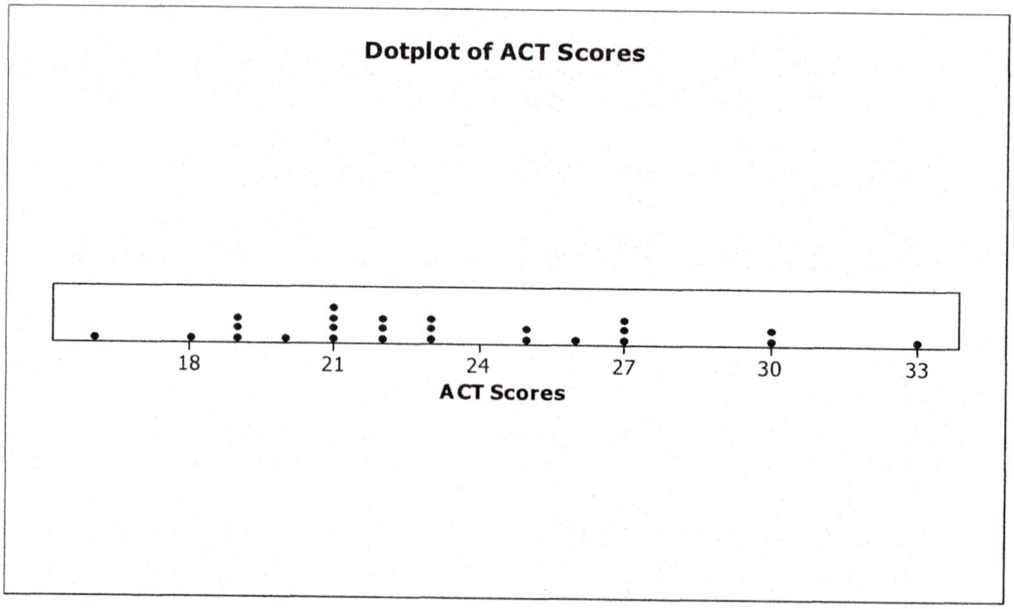

283. Using the concept of depth, describe the position of 26 in the set of 25 ACT scores in two different ways.

ANSWER:
The data values in ascending are:

16	18	19	19	19	20	21	21	21	21
22	22	22	23	23	23	25	25	26	27
27	27	30	30	33					

Therefore, the value 26 is in the 19th position from L = 16, and in the 7th position from H = 33.

284. Find P_5 for the ACT scores.

ANSWER:
nk / 100 =(25)(5) / 100 = 1.25. Hence, $d(P_5)$ = 2, and P_5 = 18

285. Find P_{10} for the ACT scores.

ANSWER:
nk / 100 =(25)(10) / 100 = 2.5. Hence $d(P_{10})$ = 3, and P_{10} = 19

286. Find P_{20} for the ACT scores.

ANSWER:
nk / 100 =(25)(20) / 100 = 5. Hence $d(P_{20})$ = 5.5, and P_{20} =(19+20)/2 = 19.5

287. Find P_{99} for the ACT scores.

ANSWER:
Since k = 99 > 50, subtract 99 from 100 and use 100 – k in place of k to determine the depth, which is then counted from the largest-valued data H. Therefore, n(100 – k) / 100 = 25 (1) / 100 = 0.25; then $d(P_{99})$ = 1, and P_{99} = 33

288. Find P_{90} for the ACT scores.

ANSWER:
Since k = 90 > 50, subtract 90 from 100 and use 100 – k in place of k to determine the depth, which is then counted from the largest-valued data H. Therefore, P_{90} : n(100 – k) / 100 = 25 (10) / 100 = 2.5; then $d(P_{90})$ = 3, and P_{90} = 30

289. Find P_{80} for the ACT scores.

ANSWER:
Since k = 80 > 50, subtract 80 from 100 and use 100 – k in place of k to determine the depth, which is then counted from the largest-valued data H. Therefore, n(100 – k) / 100 = 25 (20) / 100 = 5; then $d(P_{80})$ = 5.5, and P_{80} = (27+27) / 2 = 27

290. Find the first quartile, Q_1, for the ACT scores.

ANSWER:
$nk / 100 = (25)(25) / 100 = 6.25$. Hence $d(Q_1) = 7$, and $Q_1 = 21$

291. Find the second quartile, Q_2 for the ACT scores.

ANSWER:
$nk / 100 = (25)(50) / 100 = 12.5$. Hence $d(Q_2) = 13$, and $Q_2 = 22$

292. Find the third quartile, Q_3, for the ACT scores.

ANSWER:
Since $k = 75 > 50$, subtract 75 from 100 and use $100 - k$ in place of k to determine the depth, which is then counted from the largest-valued data H. Therefore, $n(100 - k) / 100 = 25 (25) / 100 = 6.25$; then $d(Q_3) = 7$, and $Q_3 = 26$.

293. Use Minitab to find the 5-number summary and draw a box-and-whiskers display.

ANSWER:
The five number summary reported by Minitab are: $L = 16$, $Q_1 = 20.5$, $Q_2 = 22$, $Q_3 = 26.5$, and $H = 33$.
Note that the values of Q_1 and Q_3 reported by Minitab are slightly different compared to our earlier calculations that showed $Q_1 = 21$ and $Q_3 = 26$.

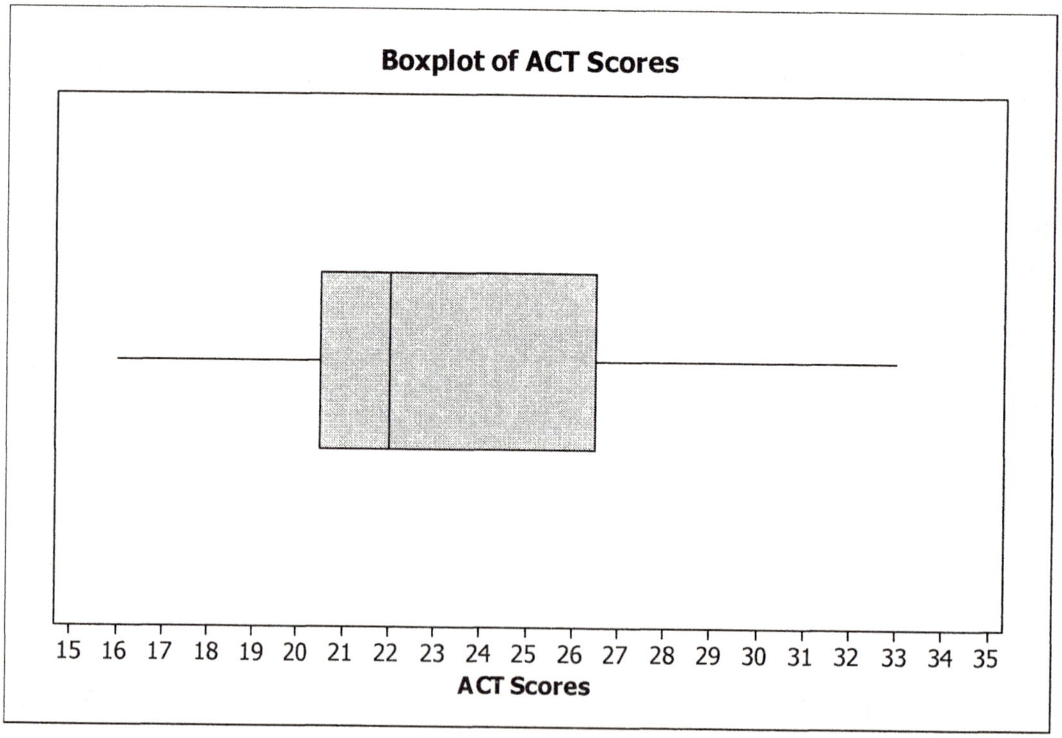

294. The Next Door Store kept track of the number of paying customers it had during the noon hour each day for 100 days. The following are the resulting statistics rounded to the nearest integer:

Mean = 95	Median = 97	Mode = 98
First quartile = 85	Third quartile = 107	Midrange = 93
Range = 56	Standard deviation = 12	

On how many days were there between 85 and 107 paying customers during the noon hour? Explain how you determined your answer.

ANSWER:
50 days; since 50% of the 100 days fall between the first and third quartiles.

QUESTIONS 295 THROUGH 300 ARE BASED ON THE FOLLOWING INFORMATION:
The annual salaries (in $100) of high school teachers employed at one of the high schools in Kent County, Michigan are listed below:

600	440	461	419	397	477	464	275	507	497
332	373	440	373	501	382	377	301	323	383

295. Draw a dotplot of the salaries.

ANSWER:

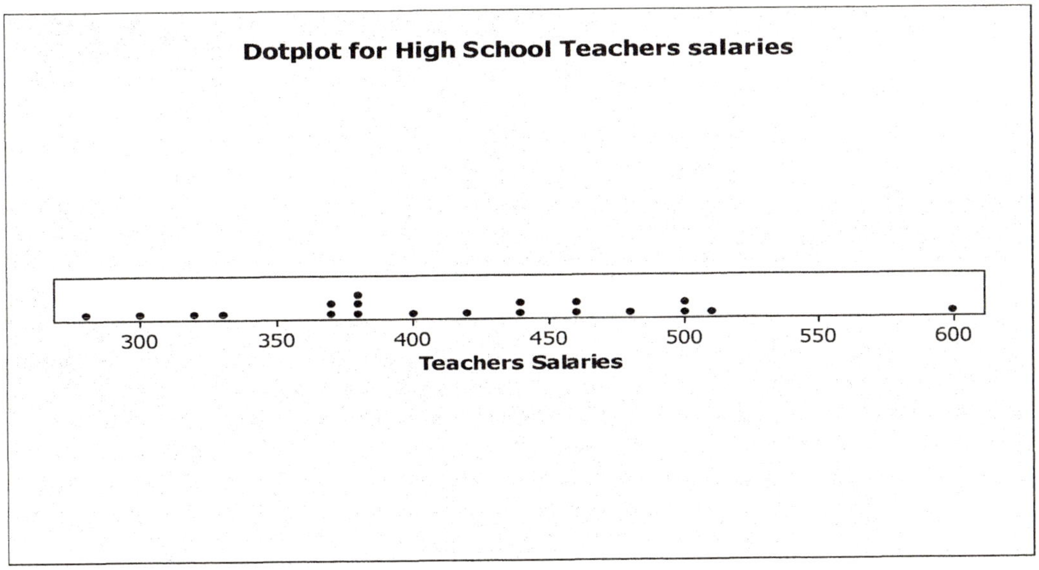

296. Using the concept of depth, describe the position of 332 in the set of 20 salaries in two different ways.

ANSWER:
The data values in ascending are:

275	301	323	332	373	373	377	382	383	397
419	440	440	461	464	477	497	501	507	600

Therefore, the value 332 is in the 4th position from L = 270, and in the 17th position from H = 33.

297. Find the first quartile for these salaries, and interpret the result.

ANSWER:
nk / 100 =(20)(25) / 100 =5.0. Hence $d(Q_1)$ =5.5, and Q_1 = (373+373)/2 = 373 or $37,300. This means that at most 25% of high school teachers' salaries are lower than $37,300 and at most 75% are higher.

298. Find the third quartile for these salaries, and interpret the result.

ANSWER:
Since *k* = 75 > 50, subtract 75 from 100 and use 100 – *k* in place of *k* to determine the depth, which is then counted from the largest-valued data *H*. Therefore,
n(100 – *k*) / 100 = 20 (25) / 100 = 5.0; then $d(Q_3)$ = 5.5, and Q_3 = (464+477)/2 = 470.5 or $47,050.
This means that at most 75% of high school teachers' salaries are lower than $47,50 and at most 25% are higher.

299. Find the midquartile for these salaries, and interpret the result.

ANSWER:
Midquartile = $(Q_1 + Q_3) / 2$ = (373 + 470.5) / 2 = 421.75 or $42,175.
This means that the salary midway between the first and third quartile is $42,175.

300. Find the interquartile range for these salaries, and interpret the result.

ANSWER:
Interquartile range = $Q_3 - Q_1$ = 470.5 – 373 = 97.5 or $9,750.
This means that the range of the middle 50% of the salaries is $9,750.

Sections 2.7 and 2.8

True-False Questions

301. *Chebyshev's* Theorem says that within two standard deviations of the mean, you will always find at least 89% of the data.
ANSWER: F

302. The Empirical Rule can be used to determine whether or not a set of data is approximately normally distributed.
ANSWER: T

303. For a bell-shaped distribution, the range will be approximately equal to six standard deviations.
ANSWER: T

304. The standard deviation is a kind of yardstick by which we can compare the variability of one set of data with another.
ANSWER: T

305. The standard deviation, as a measure of variation (dispersion), can be understood by examining two statements that tell us how the standard deviation relates to the data: the Empirical Rule and Chebyshev's Theorem.
ANSWER: T

306. The Empirical Rule applies specifically to a normal (bell-shaped) distribution, but it is frequently applied as an interpretive guide to any mounded distribution.
ANSWER: T

307. The Empirical Rule applies to any distribution, regardless of its shape, as an interpretive guide to the distribution.
ANSWER: F

308. The Empirical Rule can be used to determine whether or not a set of data is approximately normally distributed.
ANSWER: T

309. The normal probability plot is an ogive drawn on probability paper.
ANSWER: T

310. The normal probability plot is a Dotplot drawn on probability paper.
ANSWER: F

311. In the event that the data do not display an approximately normal distribution, Chebyshev's Theorem gives us information about how much of the data will fall within intervals centered at the mean for all distributions.
ANSWER: T

312. Graphs in which the frequency scale starts at zero tend to emphasize the size of the numbers involved.
ANSWER: T

313. Graphs that are chopped off may tend to emphasize the variation in the numbers without regard to the actual size of the numbers.
ANSWER: T

314. Truncating scales on graphs often leads to misleading visual impressions.
ANSWER: T

Multiple-Choice Questions

315. Which of the following is *not* a correct statement?

A) The mean absolute deviation is a measure of dispersion.
B) Chebyshev's Theorem applies only to non-normal distributions.
C) The sum of ($x-\overline{x}$) will always be zero.
D) The calculation of the range does not consider all values.
ANSWER: B

316. According to the Empirical Rule, if the variable is normally distributed, then within one standard deviation of the mean, there well be approximately:

A) 75% of the data.
B) 85% of the data.
C) 95% of the data.
D) None of the data.
ANSWER: D

317. The proportion of any distribution that lies within four standard deviations of the mean is:

A) 93.75% or more.
B) 93.75% or less.
C) 6.25% or more.
D) 6.25% or less.
ANSWER: A

Short-Answer Questions

318. According to Chebyshev's Theorem, what percent of a set of data will be more than three standard deviations from the mean?

ANSWER:
About 11%

319. According to the Empirical Rule, at least what percent of a set of data will lie within two standard deviations from the mean?

ANSWER:
Approximately 95%

320. A sample has a mean of 100.0 and a standard deviation of 15.0. According to Chebyshev's Theorem, at least 8/9 of all of the data will lie between what two values?

ANSWER:
55.0 and 145.0

321. A sample of size 50 has a mean of 60.0 and a standard deviation of 10.0. According to Chebyshev's Theorem, at least what percent of the data is between 10 and 110?

ANSWER:
96%

322. A sample of size 100 from a normal population has a mean of 110 and a standard deviation of 10.0. Using the Empirical Rule, about how many items of the sample will be above 130?

ANSWER:
Approximately 2 to 3 items

323. Complete the following statement: According to the Empirical Rule, ________ of the data for any distribution will occur within one standard deviations of the mean of the distribution.

ANSWER:
68%

324. The lifetimes of electronic components have a mean equal to 2.5 years and a standard deviation equal to 0.2 years. Within what time interval will at least 75% of the lifetimes fall?

ANSWER:
2.1 years to 2.9 years

325. A set of measurements has a mean equal to 35.5 and a standard deviation equal to 3.0. At least what percent of the data falls between 31.0 and 40.0?

ANSWER:
55.6%

326. For a normal distribution, a value that is one standard deviation above the mean would be approximately the same as what percentile?

ANSWER:
Eighty-fourth percentile

327. According to Chebyshev's Theorem, how many standard deviations on both sides of the mean do you need to go so that at least 96% of the distribution is covered?

ANSWER:
Five

328. According to Chebyshev's Theorem at least 75% of all the data in a particular sample lies between 74.5 and 82.9. Find the sample mean for this sample.

ANSWER:
$\bar{x} = 78.7$

329. According to Chebyshev's Theorem at least 75% of all the data in a particular sample lies between 74.5 and 82.9. Find the sample standard deviation for this sample.

ANSWER:
$s = 2.1$

Applied and Computational Questions

330. For a particular sample of 50 scores on a statistics exam, the following results were obtained:

Mean = 78	Midrange = 72	Third quartile = 94	Mode = 84
Median = 80	Standard deviation = 11	Range = 52	First quartile = 68

How many students scored within three standard deviations of the mean $(\bar{x} \pm 3s)$? Explain how you determined your answer.

ANSWER:
45. According to Chebyshev's Theorem, the proportion of any distribution that lies within 3 standard deviations of the mean is at least 89%. Therefore we should expect at least 45 of the 50 students to have scores within 3 standard deviations of the mean.

331. The bar graph below compares the mean time in seconds for seven-year-old girls to complete a certain task to the mean time in seconds for seven-year-old boys to complete the same task. There is statistical deception here. Explain what is deceptive about the bar graph.

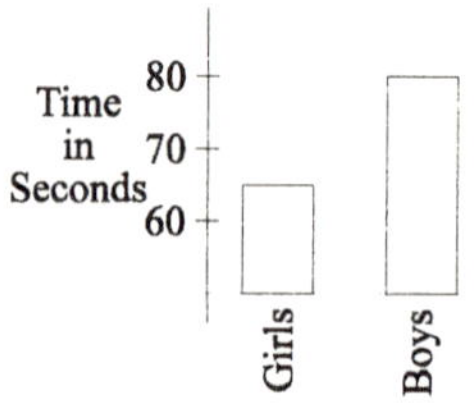

ANSWER:
From the graph it appears that the time for the boys is twice the time for the girls. However, the time for the boys is 80 seconds while the time for the girls is 65 seconds. The deception is caused by the vertical scale not starting at zero.

332. A large sample is selected from a normal distribution. The middle 99.7% of the sample data falls between 24.2 and 69.2. Estimate the sample mean and the sample standard deviation.

ANSWER:
$\bar{x} - 3s = 24.2$, and $\bar{x} + 3s = 69.2 \Rightarrow \bar{x} = 46.7$, and $s = 7.5$

QUESTIONS 333 AND 334 ARE BASED ON THE FOLLOWING INFORMATION:
The average clean-up time for a crew of a medium-size firm is 80.0 hours and the standard deviation is 6.5 hours. Assuming that the Empirical Rule is appropriate.

333. What proportion of the time will it take the clean-up crew 93.0 or more hours to clean the plant?

ANSWER:
$z = (93 - 80) / 6.5 = 2$. Therefore, 93.0 is 2 standard deviations above the mean. Hence, 2.5% of the time more than 93.0 hours will be required.

334. The total clean-up time will fall within what interval 95% of the time?

ANSWER:
95% of the time, the total clean-up time will fall within 2 standard deviations of the mean; that is 80.0 ± 2 (6.5) or from 67 to 93 hours.

QUESTIONS 335 AND 336 ARE BASED ON THE FOLLOWING INFORMATION:
Chebyshev's Theorem can be stated in an equivalent form to that given in your book. For example, to say "at least 75% of the data fall within two standard deviations of the mean" is equivalent to stating that "at most, 25% will be more than two standard deviations away from the mean".

335. At most, what percentage of a distribution will be three or more standard deviations from the mean?

ANSWER:
At most 11%

336. At most, what percentage of a distribution will be four or more standard deviations from the mean?

ANSWER:
At most 6.25%

337. The Next Door Store kept track of the number of paying customers it had during the noon hour each day for 100 days. The following are the resulting statistics rounded to the nearest integer:

Mean = 95	Median = 97	Mode = 98
First quartile = 85	Third quartile = 107	Midrange = 93
Range = 56	Standard deviation = 12	

For how many of the 100 days was the number of paying customers within three standard deviations of the mean $(\bar{x} \pm 3s)$? Explain how you determined your answer.

ANSWER:
According to Chebyshev's Theorem, the proportion of any distribution that lies within 3 standard deviations of the mean is at least 89%. Therefore, we should expect in at least 89 of the 100 days that the number of paying customers was within three standard deviations of the mean.

<u>QUESTIONS 338 AND 339 ARE BASED ON THE FOLLOWING INFORMATION:</u>
The mean lifetime of a certain tire is 50,000 miles and the standard deviation is 2,500 miles.

338. If we assume the mileages are normally distributed, approximately what percentage of all such tires will last between 42,500 and 57,500 miles?

ANSWER:
According to the Empirical Rule, approximately 99.7% of all such tires will last between 42,500 and 57,500 miles (i.e., within three standard deviations of the mean).

339. If we assume nothing about the shape of distribution, approximately what percentage of all such tires will last between 42,500 and 57,500 miles?

ANSWER:
According to Chebyshev's Theorem, at least 89% of all such tires will last between 42,500 and 57,500 miles (i.e., within three standard deviations of the mean).

Section 2.9

Short-Answer Questions

340. Suppose we have a large set of data and we wish to construct a frequency distribution. Explain under what conditions we would use an ungrouped frequency distribution and under what conditions we would use a grouped frequency distribution.

ANSWER:
Ungrouped frequency distribution is used if data consists of a few distinct values that repeat many times. A grouped frequency distribution is for data with many different values.

Applied and Computational Questions

341. The following frequency distribution gives the pay ranges (in thousands of dollars) for all middle management personnel in large company.

Class Boundaries	*f*
$20 < x < 30$	4
$30 \le x < 40$	27
$40 \le x < 50$	29
$50 \le x < 60$	25
$60 \le x < 70$	17

Find the mean pay.

ANSWER:
$44,705.88

QUESTIONS 342 THROUGH 344 ARE BASED ON THE FOLLOWING INFORMATION:
The frequency distribution below gives the class size for 70 sections of mathematics at a particular college.

Class Boundaries	*f*
$21 \le x < 24$	4
$24 \le x < 27$	7
$27 \le x < 30$	9
$30 \le x < 33$	15
$33 \le x < 36$	17
$36 \le x < 39$	9
$39 \le x < 42$	7
$42 \le x \le 45$	2

342. Prepare the extensions table.

ANSWER:

x	f	xf	$x^2 f$
22.5	4	90	2025
25.5	7	178.5	4551.75
28.5	9	256.5	7310.25
31.5	15	472.5	14883.75
34.5	17	586.5	20234.25
37.5	9	337.5	12656.25
40.5	7	283.5	11481.75
43.5	2	87	3784.5

$\sum f = 70$ $\sum xf = 2292$ $\sum x^2 f = 76927.5$

343. Find the sample mean.

ANSWER:
$\bar{x} = \sum xf / \sum f = 2292 / 70 = 32.74$

344. Find the sample standard deviation.

ANSWER:

$$SS(x) = \sum x^2 f - [(\sum xf)^2 / \sum f] = 76927.5 - (2292^2 / 70) = 1880.87$$

$s^2 = SS(x)/(\sum f - 1) = 1880.87 / 69 = 27.259$, and $s = \sqrt{s^2} = \sqrt{27.259} = 5.22$

<u>QUESTIONS 345 AND 346 ARE BASED ON THE FOLLOWING INFORMATION:</u>
Consider the following frequency distribution:

Class Boundaries	*f*
$20 \le x < 25$	2
$25 \le x < 30$	4
$30 \le x < 35$	4
$35 \le x < 40$	3
$40 \le x \le 45$	2

345. Find the mean.

ANSWER:
$\bar{x} = 32.2$

346. Find the standard deviation.

ANSWER:
$s = 6.4$

347. A survey of attorneys asked the number of children each had fathered. The results are summarized by this ungrouped frequency distribution:

Number of Children	0	1	2	3	4	6
Number of Attorneys	16	12	26	14	4	3

Calculate the sample mean, variance, and standard deviation for the number of children the doctors had fathered.

ANSWER:

x	f	xf	$x^2 f$
0	16	0	0
1	12	12	12
2	26	52	104
3	14	42	126
4	4	16	64
6	3	18	108
$\sum$	75	140	414

$\bar{x} = \sum xf / \sum f$ = 414 / 75 = 5.52

$SS(x) = \sum x^2 f - [(\sum xf)^2 / \sum f] = 414 - (140^2 / 75) = 152.667$

$s^2 = SS(x)/(\sum f - 1)$ = 152.667 / 74 = 2.063, and $s = \sqrt{s^2} = \sqrt{2.063} = 1.436$

348. The mean of the following distribution equals 4.0477. Find the missing frequency.

Class Boundaries	*f*
2.1 ≤ x < 3.0	2
3.0 ≤ x < 3.9	2
3.9 ≤ x < 4.8	
4.8 ≤ x < 5.7	1
5.7 ≤ x ≤ 6.6	1

ANSWER:
3

349. The following distribution of commuting distances, in miles, was obtained for a sample of Michigan State University employees.

Distance	*Frequency*
1.0 – 3.0	2
3.0 – 5.0	6
5.0 – 7.0	12
7.0 – 9.0	50
9.0 – 11.0	35
11.0 – 13.0	15
13.0 – 15.0	5

Find the mean and standard deviation for the commuting distances.

ANSWER:

x	f	xf	$x^2 f$
2	2	4	8
4	6	24	96
6	12	72	432
8	50	400	3200
10	35	350	3500
12	15	180	2160
14	5	70	980
$\sum$	125	1100	10,376

$$\bar{x} = \sum xf / \sum f = 1100/125 = 8.8$$

$$SS(x) = \sum x^2 f - [(\sum xf)^2 / \sum f] = 10{,}376 - (1100^2/125) = 696.0$$

$$s^2 = SS(x)/(\sum f - 1) = 696.0/124 = 5.6129$$

$$s = \sqrt{s^2} = \sqrt{5.6129} = 2.3692$$

QUESTIONS 350 THROUGH 353 ARE BASED ON THE FOLLOWING INFORMATION:
The monthly weight losses (in pounds) for a sample of 100 overweight people who exercise regularly at least two hours per week are as follows:

Weight Loss	7.5	7.7	8.0	8.1	8.2	8.8
Frequency	3	8	29	39	16	5

350. Prepare the extensions table.

ANSWER:

x	f	xf	$x^2 f$
7.5	3	22.5	168.75
7.7	8	61.6	474.32
8.0	29	232.0	1856.0
8.1	39	315.9	2558.79
8.2	16	131.2	1075.84
8.8	5	44.0	387.2
	$\sum f = 100$	$\sum xf = 807.2$	$\sum x^2 f = 6520.9$

351. Find the monthly mean weight loss.

ANSWER:

$$\bar{x} = \frac{\sum xf}{\sum f} = \frac{807.2}{100} = 8.072$$

352. Find the variance for the weight loss.

ANSWER:

$$s^2 = \frac{\sum x^2 f - \frac{(\sum xf)^2}{\sum f}}{\sum f - 1} = \frac{6520.9 - \frac{(807.2)^2}{100}}{100-1} = \frac{5.1816}{99} = 0.0523$$

353. Find the standard deviation for the weight loss

ANSWER:

$$s = \sqrt{s^2} = \sqrt{0.0523} = 0.2287$$

QUESTIONS 354 THROUGH 357 ARE BASED ON THE FOLLOWING INFORMATION:

The following distribution of commuting distances (in miles) was obtained for a sample of Ford Company employees:

Distance	Frequency
1.0 – 3.0	2
3.0 – 5.0	7
5.0 – 7.0	14
7.0 – 9.0	60
9.0 – 11.0	40
11.0 – 13.0	18
13.0 – 15.0	6
15.0 – 17.0	3

354. Prepare the extensions table.

ANSWER:

x	f	xf	$x^2 f$
2.0	2	4	8
4.0	7	28	112
6.0	14	84	504
8.0	60	480	3840
10.0	40	400	4000
12.0	18	216	2592
14.0	6	84	1176
16.0	3	48	768
	$\sum f = 150$	$\sum xf = 1344$	$\sum xf^2 = 13{,}000$

355. Find the mean for the commuting distance.

ANSWER:

$$\bar{x} = \frac{\sum xf}{\sum f} = \frac{1344}{150} = 8.96$$

356. Find the variance for the commuting distance.

ANSWER:

$$s^2 = \frac{\sum x^2 f - \frac{(\sum xf)^2}{\sum f}}{\sum f - 1} = \frac{13,000 - \frac{(1344)^2}{150}}{150-1} = \frac{957.76}{149} = 6.4279$$

357. Find the standard deviation for the commuting distance.

ANSWER:

$s = \sqrt{s^2} = \sqrt{6.4279} = 2.5353$

Chapter 3

Descriptive Analysis and Presentation of Bivariate Data

Sections 3.1 and 3.2

True-False Questions

1. The scatter diagram is an appropriate display of bivariate data when both variables are quantitative.
ANSWER: T

2. In problems that deal with two quantitative variables, we will present the sample data pictorially on a *scatter diagram*.
ANSWER: T

3. Bivariate data refers to the values of two different variables that are obtained from the same population element.
ANSWER: T

4. When bivariate data result from two quantitative variables, the data are often arranged on a cross-tabulation or contingency table.
ANSWER: F

5. The total of the marginal totals in a contingency table is the *grand total* and is equal to n, the *sample size*.
ANSWER: T

6. Bivariate data refers to the values of two different variables that are obtained from two different populations.
ANSWER: F

7. When bivariate data result from two qualitative variables, the data are often arranged on a cross-tabulation or contingency table.
ANSWER: T

8. The frequencies in a contingency table can easily be converted to percentages of the grand total by dividing each frequency by the grand total and multiplying the result by 100.
ANSWER: T

9. The frequencies in a contingency table can easily be converted to percentages of the grand total by dividing each frequency by 100 and multiplying the result by the grand total.
ANSWER: F

10. The frequencies in a contingency table can be expressed as percentages of the row totals by dividing each row entry by the row's total and multiplying the results by 100.
ANSWER: T

11. The frequencies in a contingency table can easily be converted to percentages of the grand total by dividing each frequency by the row or column total and multiplying the result by 100.
ANSWER: F

12. The frequencies in a contingency table can be expressed as percentages of the column totals by dividing each column entry by that column's total and multiplying the result by 100.
ANSWER: T

13. The frequencies in a contingency table can be expressed as percentages of the column totals by dividing each column entry by the grand total and multiplying the result by 100.
ANSWER: F

14. When bivariate data result from one qualitative and one quantitative variable, the quantitative values are viewed as separate samples, each set identified by levels of the qualitative variable.
ANSWER: T

15. The scatter diagram is a plot of all the ordered pairs of bivariate data on a coordinate axis system. The input variable x is plotted on the horizontal axis, and the output variable y is plotted on the vertical axis.
ANSWER: T

16. When the bivariate data are the result of two attribute variables, it is customary to express the data mathematically as ordered pairs (x, y).
ANSWER: F

17. In problems that deal with two quantitative variables, we present the sample data pictorially on a scatter diagram.
ANSWER: T

Multiple-Choice Questions

18. In bivariate data, where both response variables are quantitative ordered pairs (*x*, *y*), what name do we give to the variable *x*?

    A) Attribute variable
    B) Dependent variable
    C) Output variable
    D) Independent variable
    ANSWER: D

19. Which of the following would *not* be appropriate when considering two qualitative variables?

    A) Contingency table
    B) Two histograms
    C) Two bar graphs
    D) Two circle graphs
    ANSWER: B

20. For which of the following situations is it appropriate to use a scatter diagram?

    A) Presenting two qualitative variables
    B) Presenting one qualitative and one quantitative variable
    C) Presenting two quantitative variables
    D) All of the above.
    ANSWER: C

21. Which of the following statements is false?

    A) When you construct a scatter diagram, it is convenient to construct scales so that the range of y values along the vertical axis is equal to or slightly shorter than the range of the x values along the horizontal axis. This creates a "window of data" that is approximately circled shape.
    B) The scatter diagram is a plot of all the ordered pairs of bivariate data on a coordinate axis system. The input variable *x* is plotted on the horizontal axis, and the output variable *y* is plotted on the vertical axis.
    C) When you construct a scatter diagram, it is convenient to construct scales so that the range of y values along the vertical axis is shorter than the range of the *x* values along the horizontal axis. This creates of a "window of data" that is approximately square.
    D) None of the above.
    ANSWER: A

22. Which of the following statements is false?

    A) The total of the marginal totals in a contingency table is the *grand total* and is equal to *n*, the *sample size*.
    B) The frequencies in a contingency table can be expressed as percentages of the row totals by dividing each row entry by the grand total and multiplying the results by 100.
    C) In problems that deal with two quantitative variables, we present the sample data pictorially on a scatter diagram.
    D) None of the above.
    ANSWER: B

23. Which of the following statements is false?

A) Two attribute variables can form bivariate data.
B) Two numerical variables can form bivariate data.
C) One attribute variable and another numerical variable can form bivariate data.
D) None of the above.
ANSWER: D

24. Which of the following statements is false?

A) When you construct a scatter diagram, it is convenient to construct scales so that the range of *y* values along the vertical axis is equal to or slightly shorter than the range of the *x* values along the horizontal axis. This creates of a "window of data" that is approximately square.
B) The scatter diagram is a plot of all the ordered pairs of bivariate data on a coordinate axis system. The input variable *x* is plotted on the horizontal axis, and the output variable *y* is plotted on the vertical axis.
C) When you construct a scatter diagram, it is convenient to construct scales so that the range of *y* values along the vertical axis is longer than the range of the *x* values along the horizontal axis. This creates of a "window of data" that is approximately square.
D) None of the above.
ANSWER: C

Short-Answer Questions

25. Given a data set that contains both qualitative data and quantitative data, describe an appropriate way to analyze this type of data.

ANSWER:
Quantitative values are separate samples and each set identified by levels of the qualitative variable.

26. In an experiment, a fixed amount of fertilizer was applied to each of 10 plots, and the corresponding yield in pounds of corn was measured. Identify the independent and dependent variables in this experiment.

ANSWER:
Independent variable = amount of fertilizer, Dependent variable = yield of corn.

27. Briefly discuss the three combinations of variable types that can form bivariate data.

ANSWER:
(a) Both variables are qualitative (attribute).
(b) One variable is qualitative (attribute) and the other is quantitative (numerical).
(c) Both variables are quantitative (both numerical).

28. When the bivariate data are the result of two quantitative variables, it is customary to express the data mathematically as ordered pairs (x, y). What do the variables x and y represent?

ANSWER:
x is the input variable (sometimes called the independent variable) and y is the output variable (sometimes called the dependent variable).

29. When the bivariate data are the result of two quantitative variables, it is customary to express the data mathematically as ordered pairs (x, y). Why are the data said to be ordered?

ANSWER:
The data are said to be ordered because one value, x, is always written first.

30. When the bivariate data are the result of two quantitative variables, it is customary to express the data mathematically as ordered pairs (x, y). Why the data are called paired?

ANSWER:
The data are called paired because for each x value, there is a corresponding y value from the same source.

31. Consider the two variables, a person's height and weight. Which variable, height or weight, would you use as the input variable when studying their relationship? Explain why.

ANSWER:
The variable height would be used as the input variable, because a person's weight depends on his/her height.

Applied and Computational Questions

<u>**QUESTIONS 32 THROUGH 34 ARE BASED ON THE FOLLOWING INFORMATION:**</u>
Shown below is a scatter diagram for high school GPAs (x) versus college GPAs (y). The sample was selected from freshmen who had completed two semesters at a small college.

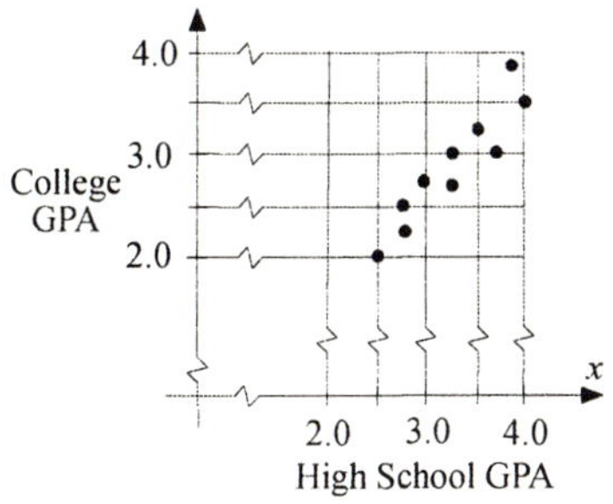

32. What is the sample size?

ANSWER:
10

33. What is the smallest value reported for the output variable?

ANSWER:
2.0

34. What is the largest value reported for the input variable?

ANSWER:
4.0

35. A survey of 15 doctors and 15 nurses was conducted, and one question related to their smoking habit. The following coding was used: Doctor (D), Nurse (N), Smoker (S), Nonsmoker (NS). The following results were obtained:

Respondent	D	D	D	N	N	D	D	D	N	D
Smoking Habit	S	NS	NS	S	S	NS	NS	NS	S	NS

Respondent	D	N	N	N	D	N	N	D	D	D
Smoking Habit	NS	NS	S	NS	S	NS	NS	NS	NS	NS

Respondent	D	N	N	N	D	D	N	N	N	N
Smoking Habit	NS	S	NS	NS	S	S	NS	S	NS	NS

Summarize the data into a 2 × 2 cross-tabulation table.

ANSWER:

	Respondent		
Smoking?	Doctor	Nurse	**Row total**
Yes	4	6	10
No	11	9	20
Column total	15	15	30

QUESTIONS 36 THROUGH 39 ARE BASED ON THE FOLLOWING INFORMATION:
A large survey of doctors and nurses was conducted, and one of the topics investigated was their smoking habit, i.e., whether they were smokers or not. The following results were obtained:

	Respondent		
Smoking?	Doctor	Nurse	**Row total**
Yes	100	320	420
No	850	2205	3055
Column total	950	2525	3475

36. Convert this table to a table of percentages based on the grand total (entire sample).

ANSWER:

	Respondent		
Smoking?	Doctor	Nurse	**Row %**
Yes	2.88	9.21	12.09
No	24.46	63.45	87.91
Column %	27.34	72.66	100.0

37. Convert this table to a table of percentages based on the column totals.

ANSWER:

	Respondent		
Smoking?	Doctor	Nurse	**Row %**
Yes	10.53	12.67	12.09
No	89.47	87.33	87.91
Column %	100.0	100.0	100.0

38. Convert this table to a table of percentages based on the row totals.

ANSWER:

	Respondent		
Smoking?	Doctor	Nurse	**Row %**
Yes	23.81	76.19	100.0
No	27.82	72.18	100.0
Column %	27.34	72.66	100.0

39. What is the percentage of smokers in this sample?

ANSWER:
12.09%

40. A study was done on undergraduate students. Of the males sampled, 80 were in the college of liberal arts and sciences, 40 were in the college of commerce, and 10 were in the college of engineering. For the females sampled, 70 were in the college of liberal arts and sciences, 16 were in the college of commerce, and 34 were in the college of engineering. For this sample construct a complete contingency table showing percents based on the entire sample. Have rows represent gender and columns represent colleges.

ANSWER:

		College		
Gender	A&S	Commerce	Engineering	**Row total**
Male	32%	16%	4%	52%
Female	28%	6.4%	13.6%	48%
Column total	60%	22.4%	17.6%	100.0%

41. Shown below is a scatter diagram for high school GPAs (x) versus college GPAs (y). The sample was selected from freshmen that had completed two semesters at a small college.

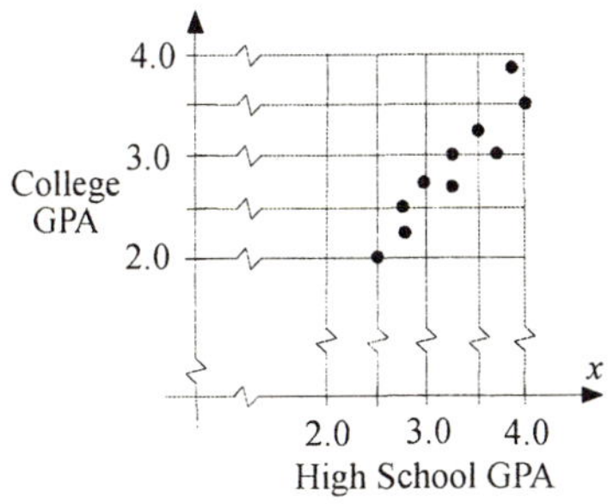

Match the items described in Column I with the terms in Column II.

Column I	**Column II**
1. Population	a. College GPA
2. Sample	b. High school GPA
3. Input variable	c. All freshmen at the college having completed two semesters.
4. Output variable	d. The students whose college GPA's are shown in the scatter diagram.

ANSWER:
(1, c), (2, d), (3, b), (4, a)

Chapter 3

QUESTIONS 42 THROUGH 44 ARE BASED ON THE FOLLOWING INFORMATION:
In a national survey of 400 business and 400 leisure-travelers, each were asked where they would most like "more space."

	On Airplane	Hotel Room	All Other
Business	280	80	40
Leisure	200	134	66

42. Express the table as percentages of the grand total.

ANSWER:

	On Airplane	Hotel Room	All Other	**Row %**
Business	35%	10%	5%	**50%**
Leisure	25%	16.75%	8.25%	**50%**
Column %	**60%**	**26.75%**	**13.25%**	**100%**

43. Express the table as percentages of the row totals. Why might one prefer the table to be expressed that way?

ANSWER:

	On Airplane	Hotel Room	All Other	**Row %**
Business	70%	20%	10%	**100%**
Leisure	50%	33.5%	16.5%	**100%**

One might prefer the table to be expressed that way because business and leisure travelers are treated as separate distributions.

44. Express the table as percentages of the column totals. Why might one prefer the table to be expressed that way?

ANSWER:

	On Airplane	Hotel Room	All Other
Business	58.33%	37.38%	37.74%
Leisure	41.67%	62.62%	62.26%
Column %	**100%**	**100%**	**100%**

One might prefer the table to be expressed that way because each category (Airplane, Room, Other) is treated as a separate distribution.

QUESTIONS 45 THROUGH 58 ARE BASED ON THE FOLLOWING INFORMATION:
A statewide survey was conducted to investigate the relationship between viewers' preferences for ABC, CBS, NBC, or PBS for new information and their political party affiliation. The results are shown in tabular form:

	Viewers' Preferences				
Political Affiliation	ABC	CBS	NBC	CNN	FOX
Democrat	242	185	305	418	208
Republican	503	235	510	260	270
Other	190	70	125	372	107

45. How many viewers were surveyed?

ANSWER:
4000

46. Why is this bivariate data? Name the two variables. What type of variable is each one?

ANSWER:
This is bivariate data since the values of the two variables - television network viewers' preferences and political affiliation - are obtained from the same population element. Both variables are attitude variables.

47. Express the table as percentages of the grand total.

ANSWER:

Viewers' Preferences

Political Affiliation	ABC	CBS	NBC	CNN	FOX	**Row %**
Democrat	6.05%	4.625%	7.625%	10.45%	5.2%	**33.95%**
Republican	12.575%	5.875%	12.75%	6.5%	6.75%	**44.45%**
Other	4.75%	1.75%	3.125%	9.3%	2.675%	**21.6%**
Column %	**23.375%**	**12.25%**	**23.5%**	**26.25%**	**14.625%**	**100%**

48. Express the table as percentages of row totals.

ANSWER:

Viewers' Preferences

Political Affiliation	ABC	CBS	NBC	CNN	FOX	**Row %**
Democrat	17.820%	13.623%	22.459%	30.781%	15.317%	**100%**
Republican	28.290%	13.217%	28.684%	14.623%	15.186%	**100%**
Other	21.991%	8.102%	14.467%	43.056%	12.384%	**100%**

49. Express the table as percentages of column totals.

ANSWER:

Viewers' Preferences

Political Affiliation	ABC	CBS	NBC	CNN	FOX
Democrat	25.882%	37.755%	32.447%	39.809%	35.556%
Republican	53.797%	47.959%	54.255%	24.762%	46.154%
Other	20.321%	14.286%	13.298%	35.429%	18.290%
Column %	**100%**	**100%**	**100%**	**100%**	**100%**

50. How many preferred to watch CBS?

ANSWER:
490

51. What percentage of the viewers were Republicans?

ANSWER:
44.45%

52. What percentage of the Democrats preferred ABC?

ANSWER:
17.82%

53. What percentage of the viewers were Republicans and preferred CNN?

ANSWER:
6.5%

54. What percentage of the viewers who preferred ABC were Democrats?

ANSWER:
25.882%

55. What percentage of the Republicans preferred Fox?

ANSWER:
15.186%

56. What percentage of the viewers who preferred Fox were Republicans?

ANSWER:
46.154%

57. What percentage of the viewers were neither Democrats nor Republicans and preferred Fox?

ANSWER:
18.29%

58. What percentage of the viewers preferred NBC?

ANSWER:
23.5%

QUESTIONS 59 THROUGH 62 ARE BASED ON THE FOLLOWING INFORMATION:
Can a man's height be predicted from his father's height? The heights of some father-son pairs are listed; *x* is the father's height and *y* is the son's height.

x	70	70	74	72	68	70	68	71	69	70	71
y	70	72	72	72	71	71	70	69	70	71	71

x	70	71	71	70	74	68	72	71	72	73	67
y	71	72	72	69	73	69	70	73	73	72	70

59. Draw two dotplots using the same scale and showing the two sets of data side by side.

ANSWER:

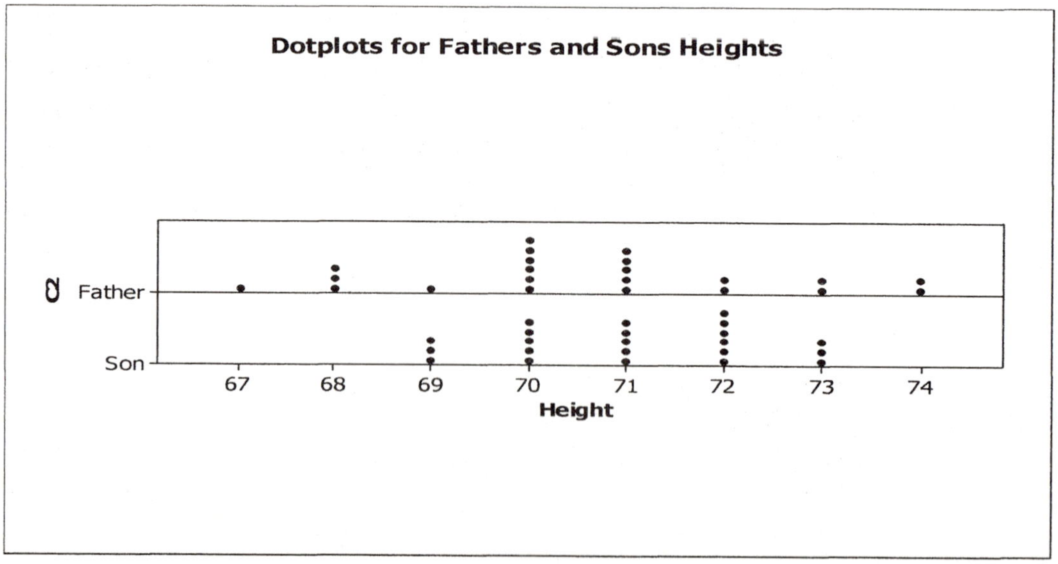

60. What can you can conclude from seeing the two sets of heights as separate sets in question 59? Explain.

ANSWER:
The father heights are more spread out than the son heights. No sons were as short as the shortest fathers and no sons were as tall as the tallest fathers.

61. Draw a scatter diagram of these data as ordered pairs.

ANSWER:

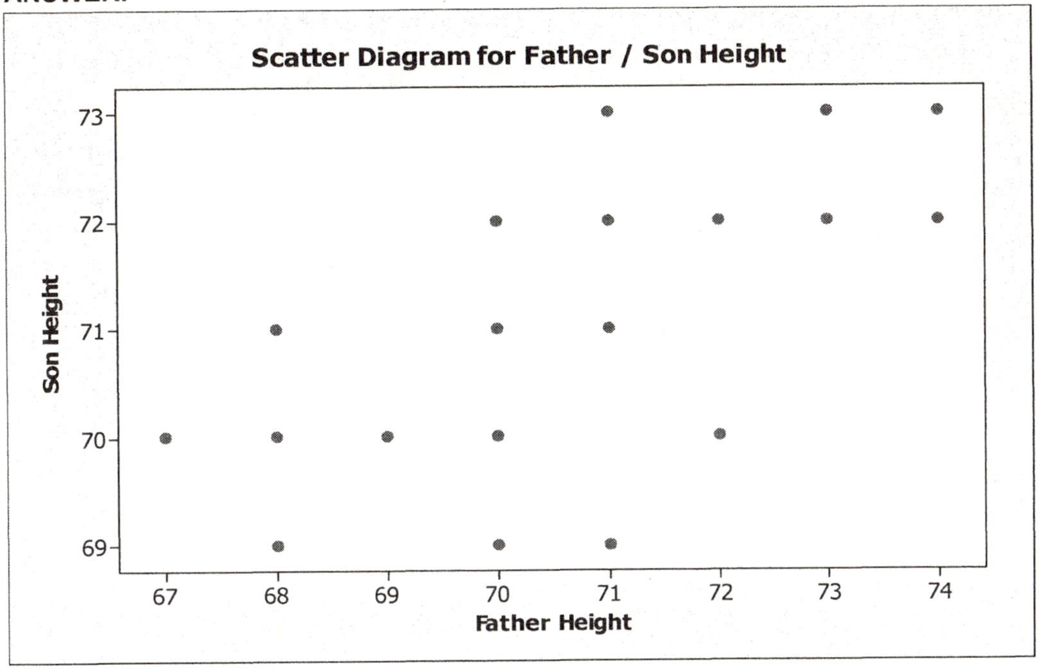

62. What can you conclude from seeing the data presented as ordered pairs in question 61? Explain.

ANSWER:
As fathers' heights increased, the sons' heights also tended to increase.

QUESTIONS 63 THROUGH 67 ARE BASED ON THE FOLLOWING INFORMATION:
Fear of being in the dentist's chair is an emotion felt by many people of all ages. A survey of 100 individuals in five age groups was conducted about this fear. These results are shown in the table below:

	Elementary	Jr. High	Sr. High	College	Adult
Fear	42	33	30	32	26
Do Not Fear	58	67	70	68	74

63. Find the marginal totals.

ANSWER:

	Elementary	Jr. High	Sr. High	College	Adult	**Row Total**
Fear	42	33	30	31	24	**160**
Do Not Fear	58	67	70	69	76	**340**
Column Total	**100**	**100**	**100**	**100**	**100**	**500**

64. Express the frequencies as percentages of the grand total.

ANSWER:

	Elementary	Jr. High	Sr. High	College	Adult	**Row %**
Fear	8.4%	6.6%	6%	6.2%	4.8%	**32%**
Do Not Fear	11.6%	13.4%	14%	13.8%	15.2%	**68%**
Column %	**20%**	**20%**	**20%**	**20%**	**20%**	**100%**

65. Express the frequencies as percentages of each age group; marginal totals

ANSWER:

	Elementary	Jr. High	Sr. High	College	Adult
Fear	42%	33%	30%	31%	24%
Do Not Fear	58%	67%	70%	69%	76%
Column %	**100%**	**100%**	**100%**	**100%**	**100**

66. Express the frequencies as percentages of those who fear and those who do not fear.

ANSWER:

	Elementary	Jr. High	Sr. High	College	Adult	**Row %**
Fear	26.25%	20.625%	18.75%	19.375%	15%	**100%**
Do Not Fear	17.059%	19.706%	20.588%	20.294%	22.353%	**100%**

67. Draw a bar graph based on age groups.

ANSWER:

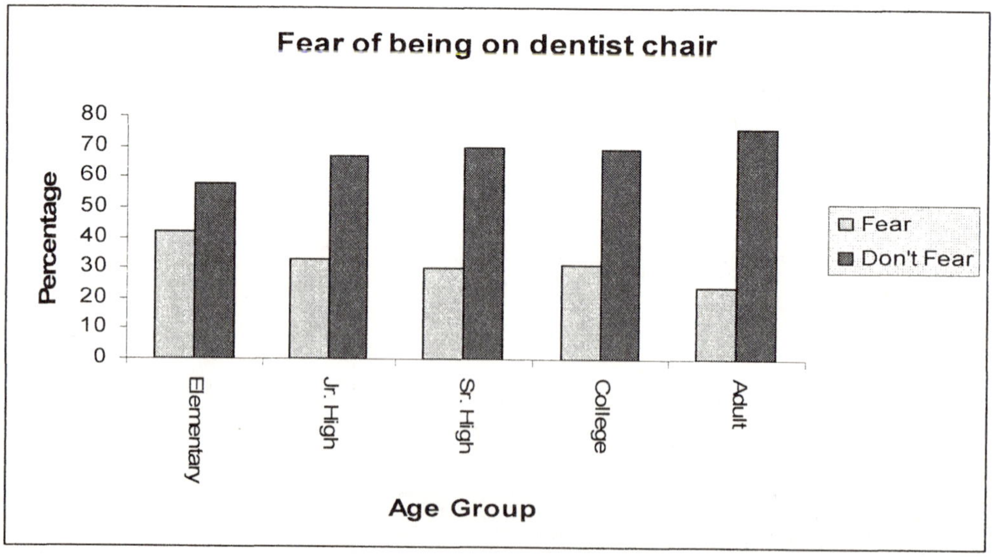

Sections 3.3 and 3.4

True-False Questions

68. If the value of the coefficient of linear correlation, r, is near –1 for two variables, then the variables are not related.
ANSWER: F

69. If there is high positive linear correlation between two variables, then there is a strong relationship between the two variables.
ANSWER: T

70. If two variables are not linearly correlated, then they are not related.
ANSWER: F

71. When both variables from a bivariate set of data are quantitative, the appropriate measure of linear relationship is the coefficient of linear correlation.
ANSWER: T

72. The equation for the line of best fit relating the height (x) and weight (y) for freshman women attending a particular college was found to be $\hat{y} = -187.4 + 4.82x$. This equation could be used to predict weights of senior women attending this college.
ANSWER: F

73. The signs of r and b_1 are always the same; that is, r and b_1 are both either positive or negative.
ANSWER: T

74. Prediction equations are always linear equations.
ANSWER: F

75. A set of data for which $r = -1$ or $r = +1$ will be such that $\Sigma(y-\hat{y})^2$ equals zero.
ANSWER: T

76. The closer the absolute value of r is to one, the better will be the predictions made using the equation of the line of best fit, provided the prediction is made for x values between the smallest value of x and the largest value of x in the observed data.
ANSWER: T

77. If the data points form a straight horizontal or vertical line, there is strong correlation.
ANSWER: F

78. Although the correlation coefficient measures the strength of a linear relationship, it does not tell us about the mathematical relationship between the two variables.
ANSWER: T

79. Perfect positive linear relationship occurs when all the points fall exactly above the straight line.
ANSWER: F

Chapter 3

80. The primary purpose of *linear correlation analysis* is to measure the strength of a linear relationship between two variables.
ANSWER: **T**

81. Correlation analysis is a method of obtaining the equation that represents the relationship between two variables.
ANSWER: **F**

82. The linear correlation coefficient is used to determine the equation that represents the relationship between two variables.
ANSWER: **F**

83. A correlation coefficient of zero means that the two variables are perfectly correlated.
ANSWER: **F**

84. Whenever the slope of the regression line is zero, the correlation coefficient will also be zero.
ANSWER: **T**

85. When *r* is positive, b_1 will be either positive or negative.
ANSWER: **F**

86. The slope of the regression line represents the amount of change expected to take place in *y* when *x* increases by one unit.
ANSWER: **T**

87. When the calculated value of *r* is positive, the calculated value of b_1 may be negative.
ANSWER: **F**

88. Correlation coefficients range between 0 and +1.
ANSWER: **F**

89. The value being predicted is called the output or predicted value.
ANSWER: **T**

90. The line of best fit is used to predict the average value of *y* that can be expected to occur at a given value of *x*.
ANSWER: **T**

91. The primary purpose of linear correlation analysis is to measure the strength of a linear relationship between two variables.
ANSWER: **T**

92. If as *x* increases there is no definite shift in the values of *y*, we say there is no correlation or no relationship between *x* and *y*.
ANSWER: **T**

93. If as *x* increases there is a definite shift in the values of *y*, we say there is a positive correlation between the two variables.
ANSWER: **F**

94. The linear correlation coefficient, r, always has a value between -1 and +1. A value of +1 signifies a perfect positive correlation, and a value of -1 shows a perfect negative correlation.
ANSWER: T

95. If as x decreases there is a definite shift in the values of y, we say there is a negative correlation between the two variables.
ANSWER: F

96. If as x increases there is a definite shift in the values of y, we say there is a correlation between the two variables.
ANSWER: T

97. Remember that a strong correlation between two variables does imply causation.
ANSWER: F

98. When the calculated value of the linear correlation coefficient r is close to zero, we conclude that there is little or no linear correlation.
ANSWER: T

99. As the calculated value of the linear correlation coefficient r changes from 0.0 toward -1.0, it indicates an increasingly weaker linear correlation between the two variables.
ANSWER: F

100. The equation of the line of best fit is determined by its slope (b_1) and its y-intercept (b_0).
ANSWER: T

101. The slope, b_1, represents the predicted change in y per unit increase in x.
ANSWER: T

102. The line of best fit will not always pass through the centroid, the point ($\bar{x}$, $\bar{y}$).
ANSWER: F

103. The y-intercept is the value of y where the line of best fit intersects the y-axis.
ANSWER: T

104. Correlation analysis is a method of obtaining the equation that represents the relationship between two variables.
ANSWER: F

105. Whenever the slope of the regression line is zero, the correlation coefficient will also be zero.
ANSWER: T

106. When the linear correlation coefficient r is positive, the slope of the regression line b_1 will also be positive.
ANSWER: T

107. The linear correlation coefficient is used to determine the equation that represents the relationship between two variables.
ANSWER: F

108. A correlation coefficient of zero means that the two variables are perfectly correlated.
ANSWER: F

109. The slope of the regression line represents the average amount of change expected to take place in y when x increases by one unit.
ANSWER: T

110. When the calculated value of linear correlation coefficient r is negative, the calculated value of the slope of the regression line b_1 may be positive.
ANSWER: F

Multiple-Choice Questions

111. Select the most likely answer for the coefficient of linear correlation for the following two variables: x = the number of hours spent studying for a test, and y = the number of points earned on the test

A) $r = 1.20$
B) $r = 0.70$
C) $r = -0.85$
D) $r = 0.05$
ANSWER: B

112. Select the most likely answer for the coefficient of linear correlation for the following two variables: x = the weight, in pounds, of a college student, and y = the grade point average for the student

A) $r = 0.98$
B) $r = 0.65$
C) $r = 0.07$
D) $r = -0.65$
ANSWER: C

113. Select the most likely value for the coefficient of linear correlation for the following two variables: x = the number of police patrol cars cruising in a given neighborhood, and y = the number of burglaries committed in the neighborhood

A) $r = 1.14$
B) $r = 0.78$
C) $r = -0.13$
D) $r = -0.75$
ANSWER: D

114. Select the most likely value for the coefficient of linear correlation for the following two variables: x = height in inches of college students, and y = IQ's of these college students

A) $r = -0.87$
B) $r = 0.65$
C) $r = -0.02$
D) $r = 0.47$
ANSWER: C

115. Suppose we used the equation, $y = -2x + 3$, to generate eight ordered pairs, (x, y). Then using these ordered pairs to compute the coefficient of linear correlation, what value should we expect to obtain for r?

A) +3
B) –1
C) +1
D) –2
ANSWER: **B**

116. A strong linear relationship ($r = 0.97$) exists between the two variables x and y in the table. The equation of the least squares line is $\hat{y} = 15.75 - 0.55x$. For what values of x should we use this equation to make predictions?

x	5	7	8	10	11	12
y	5.5	8	8	9	10	11

A) Any positive value of x
B) Values of x less than or equal to 12
C) Values of x less than or equal to 5
D) Values of x between 5 and 12 inclusive.
ANSWER: **D**

117. Shown below is a scatter diagram for high-school GPAs (x) versus college GPAs (y). The sample was selected from freshmen who had completed two semesters at a small college.

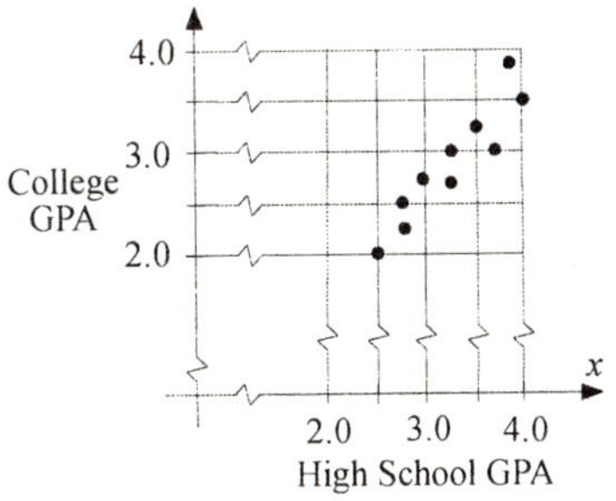

What can we say about the slope of the line of best fit?

A) The slope is positive.
B) The slope is near zero.
C) The slope is negative.
D) The slope is exactly zero.
ANSWER: **A**

Chapter 3

118. The sum of the squares for x is given by which of the following equations?

A) $SS(x) = \Sigma x^2$
B) $SS(x) = (\Sigma x)^2$
C) $SS(x) = (\Sigma x)^2 - \frac{\Sigma x^2}{n}$
D) $SS(x) = \Sigma(x - \bar{x})^2$

ANSWER: D

119. Suppose we find the equation of the line of best fit for a set of bivariate data. If we use $x = \bar{x}$ in the equation, what value should we expect for $\hat{y}$?

A) $\hat{y} = b_0$
B) $\hat{y} = b_1$
C) $\hat{y} = \bar{y}$
D) Cannot predict the value of $\hat{y}$.

ANSWER: C

120. Which of the following statements is false?

A) The correlation between x and y is positive when y tends to increase as x increases, and negative when y tends to decrease as x decreases.
B) If the ordered pairs (x, y) tend to follow a straight-line path, there is a linear correlation. The preciseness of the shift in y as x increases determines the strength of the linear correlation.
C) Perfect linear correlation occurs when all the points fall exactly along a straight line.
D) None of the above.

ANSWER: A

121. Which of the following statements is false?

A) The linear correlation coefficient, r, always has a value between -1 and +1.
B) The coefficient of linear correlation, r, is the numerical measure of the strength of the linear relationship between two variables.
C) If the data form a straight horizontal or vertical line, there is a weak correlation, since one variable has no significant effect on the other.
D) None of the above.

ANSWER: C

122. Which of the following statements is false?

A) The lurking variable is a variable that has an important effect on the relationship between the variables of a study but is not included in the study.
B) If there is a strong linear correlation between two variables, then one can definitely conclude that there is a direct cause-and-effect relationship between the two variables.
C) As the calculated value of the linear correlation coefficient r changes from 0.0 toward +1 or -1.0, it indicates an increasingly stronger linear correlation between the two variables.
D) None of the above.

ANSWER: B

123. Which of the following statements is true?

A) The value of the linear correlation coefficient ranges between 0 and +1.
B) The value being predicted in regression analysis is called the input variable.
C) The line of best fit is used to predict the average value of *y* that can be expected to occur at a given value of *x*.
D) All of the above

ANSWER: **C**

Short-Answer Questions

124. What is the primary purpose of linear correlation analysis?

ANSWER:
Measuring the strength of a linear relationship between two variables.

125. Explain why the following statement is false: "If the value of the coefficient of linear correlation, *r*, is near zero for two variables, then the variables are never related".

ANSWER:
They may be related, but not linearly.

126. What is the largest value that the coefficient of linear correlation can ever equal?

ANSWER:
1.0

127. What difficulty is encountered in computing *r* for the following data?

x	1	2	3	4	5
y	6	6	6	6	6

ANSWER:
Division by zero, since the standard deviation of *y* equals zero.

128. What difficulty is encountered in computing *r* for the following data?

x	1	2	3	4	5
y	5	3	2	3	5

ANSWER:
Plotted data would be curvilinear rather than linear and *r* would show little or no relationship.

129. Suppose that a rectangle is drawn around a set of plotted points and that the length of the rectangle is 30 millimeters. If the data indicate a positive correlation, what value must the width of the rectangle have in order that the estimate of *r* equals 0.7?

ANSWER:
9mm

Chapter 3

130. For a particular set of bivariate data, the equation of the line of best fit is $\hat{y} = -73.1 + 5.12x$, and $\overline{y}$ = 11.80. Find the value of $\overline{x}$.

ANSWER:
$\overline{x} = 16.58$

131. What is the main purpose of regression analysis?

ANSWER:
The main purpose of regression analysis is to make predictions.

132. A student correctly computed the coefficient of linear correlation for two variables and found the value to be $r = -0.02$. The student's conclusion was that since the value of *r* is near zero, the two variables are not related. Comment on this conclusion.

ANSWER:
Variables are not *linearly* related, but some other type of relationship may exist.

133. A study investigating the relationship between speed (miles per hour) and gas rate (miles per gallon) covered speeds ranging from 20 mph to 70 mph. Speed was the independent variable, and gas rate was the dependent variable. The equation of the line of best fit was $\hat{y}$ = 355 - 0.1x. Estimate the average miles per gallon for cars of type tested traveling at 50 mph.

ANSWER:
30.5 mpg

134. What does this graph tell you?

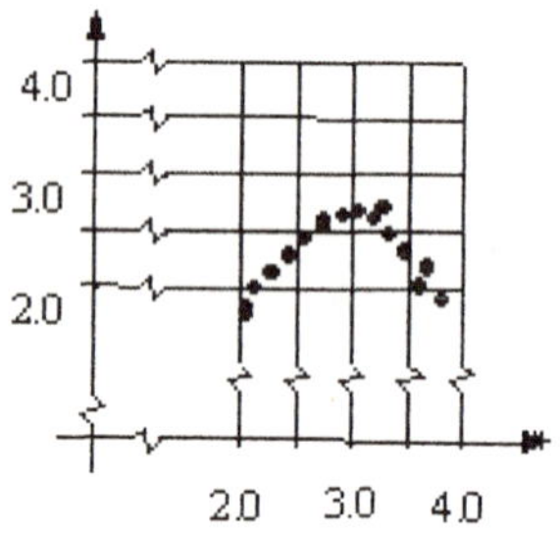

ANSWER:
While there is a relationship, it is not linear (rather quadratic).

135. When making predictions on the line of best fit, explain what is wrong with utilizing data from 20 years ago to make predictions for today.

ANSWER:
Data may not be relevant to today's world.

136. A strong linear relationship exists between the two variables in the table ($r = -0.95$). The equation of the least squares line is $\hat{y} = 15.75 - 0.55x$. For what values of x should we use this equation to make predictions?

x	6	7	10	12	14	15
y	13	11	10	10	8	7

ANSWER:
$6 \leq x \leq 15$

137. For a particular set of bivariate data, the equation of the line of best fit is $\hat{y} = -82.4 + 6.28x$, and $\bar{y} = 11.8$. Find $\bar{x}$ for this data.

ANSWER:
$\bar{x} = 15$

138. Diane collected a set of bivariate data and calculated r, the linear correlation coefficient. The resulting value was – 1.54. Diane proclaimed that this indicated that there was no correlation between the two variables since the value of r was not between –1.0 and +1.0. Amy argues that –1.54 was impossible and that only values of r near zero implied no correlation. Who is correct? Justify your answer.

ANSWER:
Amy is correct, since the linear correlation coefficient r must take a value between –1.0 and +1.0, and that only values of r near zero implied no correlation.

139. How would you interpret the findings of a correlation study that reported a linear correlation coefficient of -1.25? Why?

ANSWER:
There must be a calculation or typographical error, since the linear correlation coefficient, r, always has a value between -1 and +1.

140. How would you interpret the findings of a correlation study that reported a linear correlation coefficient of +0.095?

ANSWER:
A linear correlation coefficient of +0.095 indicates that there is very little or no linear correlation.

141. Explain why it makes sense for a set of data to have a correlation coefficient of zero when the scatter diagram of the data shows a very definite pattern.

ANSWER:
The scatter diagram may suggest a non-linear relationship between the two variables. The correlation coefficient measures the strength of a linear relationship; therefore a value near zero indicates no linear relationship.

142. Briefly discuss the difference between the purpose of regression analysis and the purpose of correlation.

ANSWER:
In regression analysis, we seek a relationship between the variables. The equation that represents this relationship may be the answer that is desired, or it may be the means to the prediction that is desired. In correlation analysis, we measure the strength of the linear relationship between the two variables.

143. Determine whether the following question requires correlation analysis or regression analysis to obtain an answer: "Is there a correlation between the grades a student obtained in high school and the grades he or she attained in college?"

ANSWER:
Correlation

144. Determine whether the following question requires correlation analysis or regression analysis to obtain an answer: "What is the relationship between the weight of a package and the cost of mailing it first class?"

ANSWER:
Regression

145. Determine whether the following question requires correlation analysis or regression analysis to obtain an answer: "Is there a linear relationship between a person's height and shoe size? "

ANSWER:
Correlation

146. Determine whether the following question requires correlation analysis or regression analysis to obtain an answer: "What is the relationship between the number of worker-hours and the number of units of production completed?"

ANSWER:
Regression

147. Determine whether the following question requires correlation analysis or regression analysis to obtain an answer: "Is the score obtained on a certain aptitude test linearly related to a person's ability to perform a certain job?"

ANSWER:
Correlation

Applied and Computational Questions

148. Find $\sum x$, $\sum y$, $\sum x^2$, $\sum y^2$, and $\sum xy$ for the following bivariate data:

x	0	1	2	4
y	2	6	7	1

ANSWER:
$\sum x = 7, \quad \sum y = 25, \quad \sum x^2 = 21, \quad \sum y^2 = 189, \quad \sum xy = 60$

149. The diastolic blood pressure, *x*, and the systolic blood pressure, *y*, were recorded for 13 females. Find the correlation coefficient for these data:

x	76	70	82	90	68	60	62	60	62	72	68	80	74
y	122	102	118	126	108	130	104	118	130	116	102	122	120

ANSWER:
$r = 0.18$

150. For a group of army inductees, the weight, *x*, and exercise capacity, *y*, were recorded for 10 individuals. For the following results, give the values for SS(xy), SS(*x*), SS(*y*), and *r*.

x	180	150	200	155	225	175	130	250	160	190
y	30	25	20	30	15	28	30	20	26	20

ANSWER:
SS(*xy*) = −1351,SS(*x*) = 11852.5, SS(*y*) = 256.4, and $r = -0.77$

151. The coefficient of linear correlation equals 0.8 for a set of bivariate data. The standard deviation for the *x* variable is 20.5, and the standard deviation for the *y* variable equals 30.2. Find the value for $\frac{\sum(x-\bar{x})(y-\bar{y})}{n-1}$.

ANSWER:
495.28

152. Compute the value of the coefficient of linear correlation for the following data and interpret the value obtain.

x	1.2	1.8	2.4	3.6	3.8
y	0.1	0.4	0.7	1.3	1.4

ANSWER:
$r = 1$ is perfect positive correlation and data points are collinear.

153. Compute the value of the coefficient of linear correlation for the following data and interpret the value obtained.

x	71.2	67.1	61.1	51.3	47.8
y	17.9	21.8	27.5	36.8	40.2

ANSWER:
$r = -1$; a perfect negative correlation and data points are collinear.

154. Compute the value of the coefficient of linear correlation for the following data and then interchange the values of *x* and *y* and compute the value of the coefficient of linear correlation for the changed data. How do the two values compare?

x	2	5	9	14
y	2.4	4.1	5.9	8.6

ANSWER:
The values are equal. They are both 0.99924.

155. Based on the following bivariate data, find the value of k so that the value of the coefficient of linear correlation r will be exactly zero.

x	2	4	7
y	3	5	k

ANSWER:
$k = 3.25$

156. Based on the following bivariate data, find the value of k so that the value of the coefficient of linear correlation r will be exactly +1.0.

x	5	k	7
y	8	9.5	11

ANSWER:
$k = 6$

QUESTIONS 157 THROUGH 164 ARE BASED ON THE FOLLOWING INFORMATION:
Consider the following bivariate data, extensions, and totals:

x	y	x^2	xy	y^2
2	14	4	28	196
3	13	9	39	169
4	11	16	44	121
5	8	25	40	64
5	9	25	45	81
7	4	49	28	16
7	3	49	21	9
33	62	177	245	656

157. Find SS(x).

ANSWER:
21.43

158. Find SS(y).

ANSWER:
106.86

159. Find SS(xy).

ANSWER:
–47.29

160. Find the linear correlation coefficient, r.

ANSWER:
–0.99

161. Find the slope b_1.

ANSWER:
–2.21

162. Find the *y*-intercept b_0.

ANSWER:
19.26

163. Find the equation of the line of best fit.

ANSWER:
$\hat{y}$ = 19.26 – 2.21*x*

164. Interpret the slope in question (161)

ANSWER:
As *x* increases by one unit, *y* decreases by about 2.2 units, on average.

165. For a group of army inductees, the weight, *x*, and exercise capacity, *y*, were recorded for 10 individuals. Based on the results in the table, find the equation of the line of best fit.

x	180	150	200	155	225	175	130	250	160	190
y	30	25	20	30	15	28	30	20	26	20

ANSWER:
$\hat{y}$ = 45.1 – 0.114*x*

166. Find the value of $\Sigma(y-\hat{y})^2$, if $\hat{y}$ = –0.6 + 3.2*x* is given as the equation of the line of best fit for the following data.

x	1	2	3	4	5
y	5	1	13	9	17

ANSWER:
57.6

167. Students were given a reading competency test (scores range from 0 to 48) and also a math competency test (scores range from 50 to 100). Find SS(*xy*), SS(*x*), and the equation of the line of best fit for the data.

Reading (x)	40	36	42	29	44	35	38	42	45	40
Math (y)	78	80	90	60	95	70	77	83	90	80

ANSWER:
SS(*x*) = 206.9, SS(*xy*) = 414.7, and $\hat{y}$ = 1.93 + 2.0x.

168. The moisture content of a chemical compound is determined for different relative humidity values. Treat the humidity as the independent variable and the moisture content as the dependent variable and find the equation of the line of best fit.

Humidity	30	45	60	50	80	65	75	20
Moisture Content	8	10	12	7	15	10	12	8

ANSWER:
$\hat{y} = 4.9 + 0.1x$

169. Using the following bivariate data, find the equation of the line of best fit and use it to predict the value of *y* when *x* = 7.

x	2	5	9	13
y	65	108	215	251

ANSWER:
$\hat{y} = 2.95 + 18.0x$
The predicted value of *y* when *x* is 7 is *y* = 155.5.

170. For a particular set of bivariate data, the equation of the line best fit is $\hat{y} = 2.9 + 6.8x$ and SS(*xy*) = 78.2. For this data, find SS(*x*).

ANSWER:
SS(*x*) = 11.5

171. For a particular set of bivariate data, the equation of the line best fit is $\hat{y} = 3.5 + 7.2x$ and SS(*x*) = 10.1. For this data, find SS(*xy*).

ANSWER:
SS(*xy*) = 72.72

QUESTIONS 172 THROUGH 175 ARE BASED ON THE FOLLOWNG INFORMATION:
An experimental psychologist asserts that the older a child is, the fewer irrelevant answers he or she will give during a controlled experiment. To investigate this claim, the following data were collected.

Age (x)	2	3	4	5	6	7	8	9	10
# Irrelevant Answers (y)	12	14	9	7	11	8	6	9	5

172. Construct a scatter diagram for these data.

ANSWER:

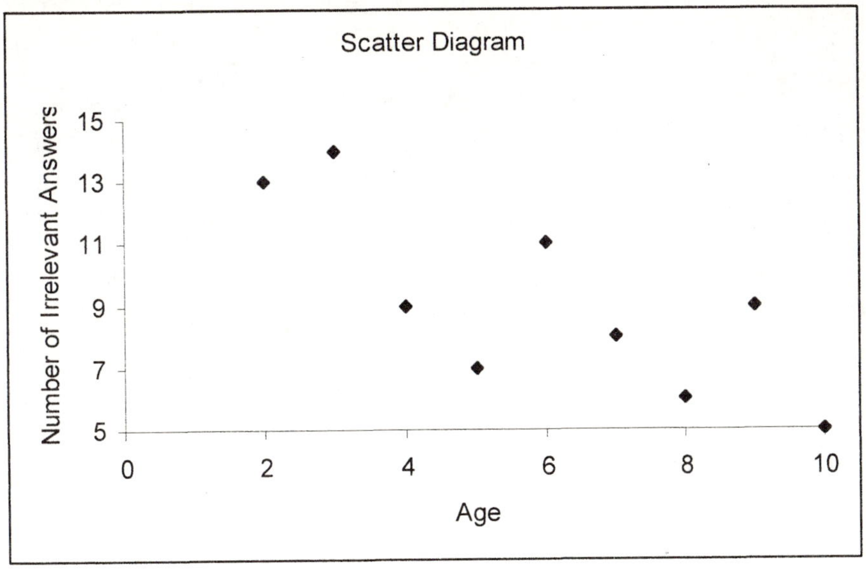

173. Use the scatter diagram in question 172 to estimate the coefficient of linear correlation *r*.

ANSWER:
r is about –0.75

174. Calculate *r* for these data.

ANSWER:

$\sum x = 54, \quad \sum y = 82, \quad \sum xy = 440, \quad \sum x^2 = 384, \quad \sum y^2 = 822$

$SS(xy) = \sum xy - (\sum x \sum y)/n = 440 - (54)(82)/9 = -52$

$SS(x) = \sum x^2 - (\sum x)^2/n = 384 - (54)^2/9 = 60$

$SS(y) = \sum y^2 - (\sum y)^2/n = 822 - (82)^2/9 = 74.8889$

$r = SS(xy)/\sqrt{SS(x) \cdot SS(y)} = -52/\sqrt{(60)(74.8889)} = -0.7757$

175. In general, what does *r* measure?

ANSWER:
The coefficient of linear correlation *r* is the numerical measure of the strength of the linear relationship between two variables.

QUESTIONS 176 THROUGH 178 ARE BASED ON THE FOLLOWING INFORMATION:
In a study involving children's fear related to being examined by a physician, the age and the score each child made on the Child Medical Fear Scale (CMFS) were:

Age (x)	8	9	9	9	9	9	10	10	11	11
CMFS score (y)	30	25	25	29	35	42	28	27	32	35

176. Construct a scatter diagram of these data.

ANSWER:

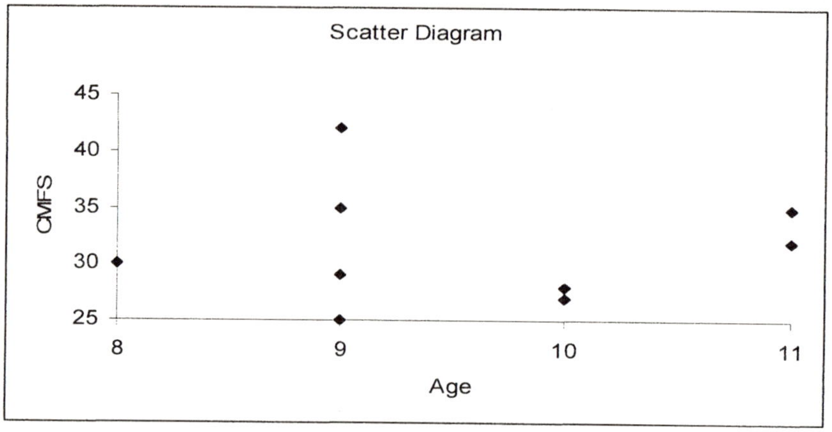

177. Calculate SS(*x*), SS(*y*), and SS(*xy*).

ANSWER:

$\sum x = 95,\ \ \sum y = 308,\ \ \sum xy = 2931,\ \ \sum x^2 = 911,\ \ \sum y^2 = 9742$

$SS(x) = \sum x^2 - (\sum x)^2 / n = 911 - (95)^2 / 10 = 8.5$

$SS(y) = \sum y^2 - (\sum y)^2 / n = 9742 - (308)^2 / 10 = 255.6$

$SS(xy) = \sum xy - (\sum x \sum y) / n = 2931 - (95)(308) / 10 = 5.0$

178. Calculate the coefficient of linear correlation *r* and interpret its meaning.

ANSWER:

$r = SS(xy) / \sqrt{SS(x) \cdot SS(y)} = 5.0 / \sqrt{(8.5)(255.6)} = 0.107$

There is a very weak positive linear relationship between the age of a child and the score each child made on the CMFS.

<u>QUESTIONS 179 AND 180 ARE BASED ON THE FOLLOWING INFORMATION:</u>

Ali used linear regression to help him understand his monthly telephone bill. The line of best fit was $\hat{y} = 25.75 + 1.32x$; *x* is the number of long-distance calls made during a month and *y* is the total telephone cost for a month. In terms of number of long distance calls and cost:

179. Explain the meaning of the *y*-intercept.

ANSWER:

The *y*-intercept of $25.75 is the amount of the total monthly telephone cost when *x*, the number of long distance calls, is equal to zero. That is, when no long distance calls are made, there is still the monthly phone charge of $25.75.

180. Explain the meaning of the slope.

ANSWER:

The slope of $1.32 is the rate at which the total phone bill will increase for each additional long distance call; it is related to average cost of the long distance calls.

QUESTIONS 181 AND 182 ARE BASED ON THE FOLLOWING INFORMATION:
Consider the following data, which give the weight (in thousands of pounds) x and gasoline mileage (miles per gallon) y for ten different automobiles.

x	2.0	2.4	2.6	2.9	3.2	3.5	3.8	4.2	4.6	5.2
y	45	40	42	39	44	36	34	28	18	13

The following data summary values are given:
$\sum x = 34.4, \quad \sum y = 339, \quad \sum xy = 1072.3, \quad \sum x^2 = 127.7, \quad \sum y^2 = 12575$

181. Calculate $SS(x)$, $SS(y)$, and $SS(xy)$.

ANSWER:
$SS(x) = \sum x^2 - (\sum x)^2 / n = 127.7 - (34.4)^2 / 10 = 9.364$
$SS(y) = \sum y^2 - (\sum y)^2 / n = 12575 - (339)^2 / 10 = 1082.9$
$SS(xy) = \sum xy - (\sum x \cdot \sum y) / n = 1072.3 - (34.4)(339) / 10 = -93.86$

182. Find Pearson's product moment r, and interpret its meaning.

ANSWER:
$r = SS(xy) / \sqrt{SS(x) \cdot SS(y)} = -93.86 / \sqrt{(9.364)(1082.9)} = -0.932$
There is a strong negative linear relationship between the weight (in thousands of pounds) and gasoline mileage (miles per gallon) for different automobiles.

QUESTIONS 183 AND 184 ARE BASED ON THE FOLLOWING INFORMATION:
The following data were generated using the equation $y = 2x + 3$.

x	0	1	2	3	4
y	3	5	7	9	11

183. Draw a scatter diagram of these data. What did you notice?

ANSWER:

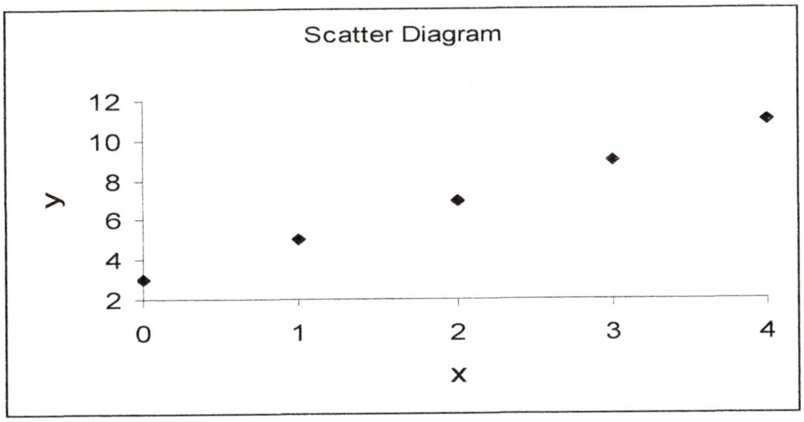

The scatter diagram of these data results in five points that fall perfectly on a straight line.

184. Find the correlation coefficient and the equation of the line of best fit.

ANSWER:

$n = 5$, $\sum x = 10, \sum y = 35, \sum x^2 = 30, \sum y^2 = 285, \sum xy = 90$

$SS(x) = \sum x^2 - (\sum x)^2 / n = 30 - (10)^2 / 5 = 10.0$

$SS(y) = \sum y^2 - (\sum y)^2 / n = 285 - (35)^2 / 5 = 40.0$

$SS(xy) = \sum xy - (\sum x \cdot \sum y) / n = 90 - (10)(35) / 5 = 20$

$r = SS(xy) / \sqrt{SS(x) \cdot SS(y)} = 20.0 / \sqrt{(10)(40)} = 1.00$

$b_1 = SS(xy) / SS(x) = 20.0 / 10.0 = 2.0$

$b_0 = [\sum y - b_1 \cdot \sum x] / n = [35 - (2)(10)] / 5 = 3$

The equation of best fit is: $\hat{y} = 3.0 + 2.0x$.

QUESTIONS 185 THROUGH 189 ARE BASED ON THE FOLLOWING INFORMATION:
A medical researcher studied the relationship between two variables: a person's current age (x) and the expected number of years remaining (y). The following data for a sample of ten people were recorded:

x	64	66	68	70	72	74	76	78	80	82
y	16.6	15.2	13.8	12.6	11.5	10.2	9.3	8.5	7.1	6.2

The following data summary values are given:
$n = 10$, $\sum x = 730, \sum y = 111, \sum x^2 = 53620, \sum y^2 = 1339.68, \sum xy = 7915$

185. Draw a scatter diagram for these data.

ANSWER:

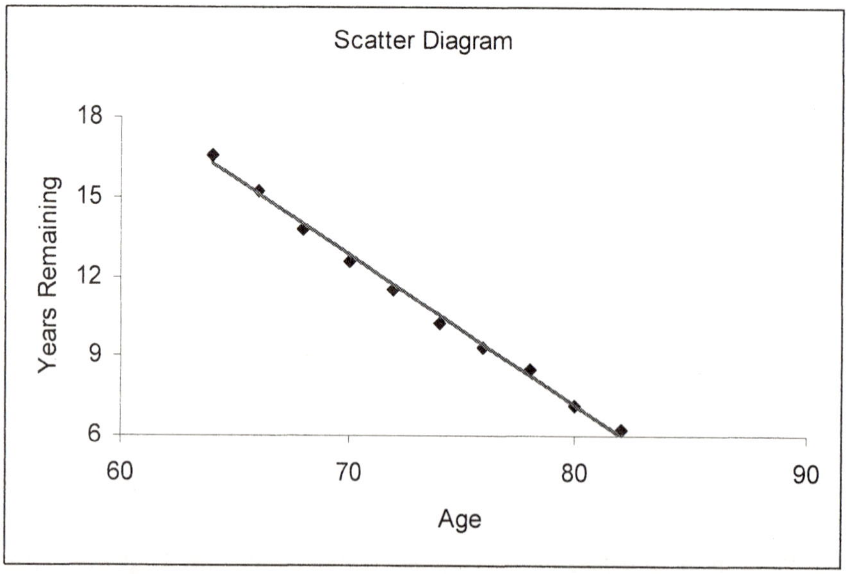

186. Calculate the equation of best fit.

ANSWER:

$SS(x) = \sum x^2 - (\sum x)^2 / n = 53620 - (730)^2 / 10 = 330.0$

$SS(xy) = \sum xy - (\sum x \cdot \sum y) / n = 7915 - (730)(111) / 10 = -188.0$

$b_1 = SS(xy) / SS(x) = -188.0 / 330.0 = -0.5697$

$b_0 = [\sum y - b_1 \cdot \sum x] / n = [111 - (-0.5697)(730)] / 10 = 52.6881$

The equation of best fit is: $\hat{y} = 52.6881 - 0.5697x$.

187. Draw the line of best fit on the scatter diagram.

ANSWER:
The line of best fit is shown on the scatter diagram in question 185.

188. What are the expected years remaining for a person who is 75 years old? Find the answer in two different ways: Use the equation from question 186 and use the line on the scatter diagram in question 185.

ANSWER:

Using the equation of best fit: $\hat{y} = 52.6881 - 0.5697(75) = 9.96$. Using the graph of the line of best fit shown in question 185: $\hat{y} \approx 9.7$

189. Are you surprised that the data all lie so close to the line of best fit? Explain why the ordered pairs follow the line of best fit so closely.

ANSWER:
The apparent linear relationship should not be a surprise. One's *age* and *years remaining* should total a fixed value, life expectancy.

QUESTIONS 190 THROUGH 193 ARE BASED ON THE FOLLOWING INFORMATION:
A dietician conducted a study to compare calories (x) and fat (y) in 18 of the most popular fast-food items. The results of the study are shown below:

x	120	200	220	230	270	290	310	340	360
y	7	13	11	12	10	8	26	28	8

x	370	420	440	450	460	540	550	640	740
y	36	20	20	22	22	55	25	40	20

The following data summary values are given:

$n = 18, \sum x = 6950, \sum y = 383, \sum x^2 = 3{,}126{,}300, \sum y^2 = 10{,}885, \sum xy = 168{,}490$

190. Draw a scatter diagram of these data.

ANSWER:

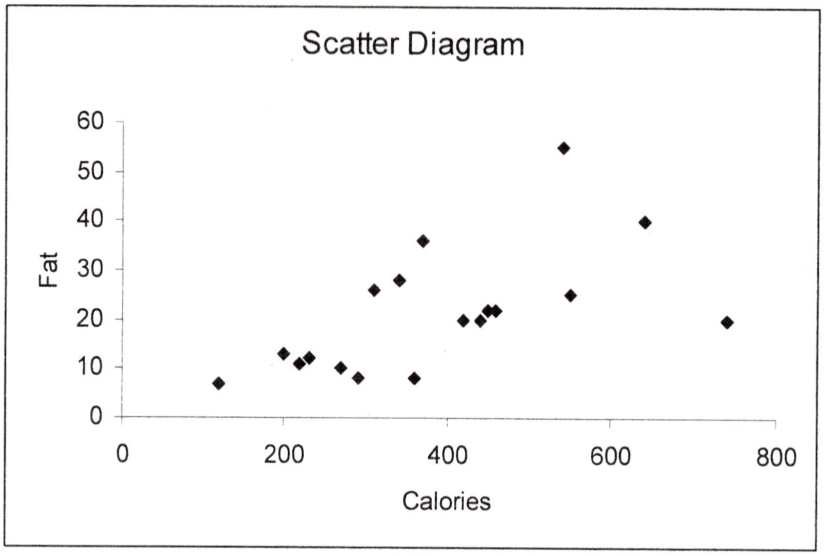

191. Calculate the linear coefficient, *r*.

ANSWER:

$SS(x) = \sum x^2 - (\sum x)^2 / n = 3{,}126{,}300 - (6950)^2 / 18 = 517{,}177.8$

$SS(y) = \sum y^2 - (\sum y)^2 / n = 10{,}885 - (383)^2 / 18 = 2{,}735.611$

$SS(xy) = \sum xy - (\sum x \cdot \sum y) / n = 168{,}490 - (6950)(383) / 18 = 20{,}609.444$

$r = SS(xy) / \sqrt{SS(x) \cdot SS(y)} = 20{,}609.444 / \sqrt{(517{,}177.8)(2{,}735.611)} = 0.5479$

192. Find the equation of the line of best fit.

ANSWER:

$b_1 = SS(xy) / SS(x) = 20{,}609.444 / 517{,}177.8 = 0.0398$

$b_0 = [\sum y - b_1 \cdot \sum x] / n = [383 - (0.0398)(6950)] / 18 = 5.911$

The equation of best fit is: $\hat{y} = 5.911 + 0.0398x$.

193. Explain the meaning of the answers to questions 190, 191, and 192.

ANSWER:
There is slight correlation between fast-food calories and the corresponding amount of fat. Generally, if the calories increase, so does the fat content.

194. Briefly discuss all possible situations that may be true about the relationship between two variables *x* and *y* if there is a strong linear correlation between them.

ANSWER:
One of the following situations may be true situations may be true about the relationship between the two variables:
(a) There is a direct cause-and-effect relationship between the two variables.
(b) There is a reverse cause-and-effect relationship between the two variables.
(c) Their relationship may be caused by a third variable.
(d) Their relationship may be caused by the interactions of several other variables.
(e) The apparent relationship may be strictly a coincidence.

QUESTIONS 195 THROUGH 198 ARE BASED ON THE FOLLOWING INFORMATION:
The number of hours studied, *x*, is compared to the exam score, *y* as shown below:

x	2	5	6	3	4	6	5	2	3
y	58	95	92	85	80	85	88	75	65

195. Use computer to calculate $\sum x,\ \sum y,\ \sum xy,\ \sum x^2,$ and $\sum y^2$.

ANSWER:
$\sum x = 36$, $\sum y = 723$, $\sum xy = 3013$, $\sum x^2 = 164$, and $\sum y^2 = 59{,}297$

196. Calculate the sums SS(*x*), SS(*y*), and SS(*xy*).

ANSWER:

$$SS(x) = \sum x^2 - \frac{(\sum x)^2}{n} = 164 - \frac{(36)^2}{9} = 20$$

$$SS(y) = \sum y^2 - \frac{(\sum y)^2}{n} = 59{,}297 - \frac{(723)^2}{9} = 1216$$

$$SS(xy) = \sum xy - \frac{(\sum x)(\sum y)}{n} = 3013 - \frac{(36)(723)}{9} = 121$$

197. Calculate the linear correlation coefficient, *r*.

ANSWER:

$$r = \frac{SS(xy)}{\sqrt{SS(x)\cdot SS(y)}} = \frac{121}{\sqrt{(20)(1216)}} = 0.0.776$$

198. What does the value of *r* in Question 197 tell you?

ANSWER:
There is a strong linear correlation between the number of hours studied for a test and the test scores. In other words, studying for an exam pays off.

QUESTIONS 199 THROUGH 201 ARE BASED ON THE FOLLOWING INFORMATION:
A study was conducted to investigate the relationship between the resale price, y (in hundreds of dollars), and the age, x (in years) of midsize American automobiles. The equation of the line of best fit was determined to be $\hat{y} = 195.6 - 21.5x$.

199. Find the resale value of such a car when it is four years old.

ANSWER:
$\hat{y} = 195.6 - 21.5(4) = 109.6$ or $10,960.

200. Find the resale value of such a car when it is seven years old.

ANSWER:
$\hat{y} = 195.6 - 21.5(7) = 45.1$ or $4,510.

201. What is the average annual decrease in the resale price of these cars?

ANSWER:
(21.5)($100) = $2,150

202. Start with the point (4, 4) and add at least four ordered pairs, (x, y), to make a set of ordered pairs such that the correlation of x and y is 0.0.

ANSWER:
One possible answer is (4, 4), (1, 4), (2, 4), (0, 3), (0, 5).

203. Start with the point (4, 4) and add at least four ordered pairs, (x, y), to make a set of ordered pairs such that the correlation of x and y is +1.0.

ANSWER:
One possible answer is (4, 4), (0, 0), (1, 1), (2, 2), (3, 3), (5, 5).

204. Start with the point (4, 4) and add at least four ordered pairs, (x, y), to make a set of ordered pairs such that the correlation of x and y is -1.0.

ANSWER:
One possible answer is (4, 4), (7, 1), (1, 7), (6, 2), (2, 6), (5, 3), (3, 5).
Start with the point (4, 4) and add at least four ordered pairs, (x, y), to make a set of ordered pairs such that the correlation of x and y is between -0.25 and 0.0.

205. Start with the point (4, 4) and add at least four ordered pairs, (x, y), to make a set of ordered pairs such that the correlation of x and y is between +0.5 and +0.8.

ANSWER:
One possible answer is (4, 4), (2, 4), (1, 3), (2, 2), (0, 1).

Chapter 4

Probability

Sections 4.1 and 4.2

True-False Questions

1. If A is any event of a sample space S, then $P(A)$ represents the relative frequency with which event A can be expected to occur.
ANSWER: T

2. If A is any event of a sample space S and if $P(A)$ is computed using $P(A) = n(A)/n(S)$, then $n(A)$ may never equal zero.
ANSWER: F

3. If A is any event of a sample space S and if the probability of event A is denoted by $P(A)$, then the probability of A is a theoretical probability.
ANSWER: F

4. Under certain conditions, it is possible that the sum of the probabilities of all the sample points in a sample space is less than one.
ANSWER: F

5. If A is any event of a sample space S, then $P(A)$ is a numerical value between -1 and 1, inclusive.
ANSWER: F

6. The probability of an event is a whole number.
ANSWER: F

7. The concepts of probability and relative frequency as related to an event are very similar.
ANSWER: T

8. The sample space is the theoretical population for probability problems.
ANSWER: T

9. The sample points of a sample space are equally likely events.
ANSWER: F

10. The value found for experimental probability will always be exactly equal to the theoretical probability assigned to the same event.
ANSWER: F

11. The empirical probability that event *A* will occur is the relative frequency with which event A can be expected to occur, and this probability is denoted by P' (A).
ANSWER: T

12. The probability of an event may be obtained in three different ways: (1) empirically, (2) theoretically, and (3) objectively.
ANSWER: F

13. The experimental, or empirical probability P' (A) of an event *A* is the ratio $n(A)$ of number of times *A* occurred to the number *n* of trials.
ANSWER: T

14. The theoretical method for obtaining the probability of an event uses a sample space in which each possible outcome has a certain probability of occurring, but the probabilities of all outcomes do not necessarily have the same value.
ANSWER: F

15. A sample space is a listing of all possible outcomes from the experiment being considered.
ANSWER: T

16. A probability is always a numerical value larger than zero but smaller than one.
ANSWER: F

17. The sum of the probabilities for all outcomes of an experiment is equal to exactly one.
ANSWER: T

18. The number of times an event can be expected to occur in *n* trials is always less than or equal to the total number of trials, *n*.
ANSWER: T

19. The Law of Large Numbers tells us that the larger the number of experimental trials *n*, the larger the empirical probability P' (*A*) is expected to be compared to the true of theoretical probability $P(A)$.
ANSWER: F

20. The Law of Large Numbers states that "as the number of times an experiment is repeated increases, the ratio of the number of successful occurrences to the number of trials will tend to approach the theoretical probability of the outcome for an individual trial."
ANSWER: T

21. Odds are a way of expressing probabilities by expressing the number of ways an event can happen compared to the number of ways it can't happen.
ANSWER: T

Multiple-Choice Questions

22. Which of the following statements is false?

A) With the theoretical method for obtaining the probability of an event, the sample space must contain equally likely sample points.
B) The theoretical probability *P*(*A*) of an event *A* is the ratio of the number *n*(*A*) of points that satisfy the definition of event *A* to the number of trials *n*.
C) Prime symbol of the probability of an event *A*; namely $P'(A)$, is not used with theoretical probabilities – it is used only for empirical probabilities.
D) None of the above.
ANSWER: B

23. Which of the following statements is false?

A) When a probability experiment can be thought of as a sequence of events, a Dotplot often is a very helpful way to picture the sample space.
B) When a probability question provides information about the events in the form of the number of items per set, the percentage of each set, or the probability of the various events, a Venn diagram often is a very helpful way to picture the sample space.
C) A subjective probability generally results from personal judgment, and the accuracy of such probability depends on the individual's ability to correctly assess the situation.
D) None of the above.
ANSWER: A

24. Which of the following probabilities is suitable in establishing proper life insurance rates?

A) Empirical probability
B) Theoretical probability
C) Subjective probability
D) All of the above.
ANSWER: A

25. Which of the following statements is false If the odds in favor of an event *A* are *a* to *b*?

A) The odds against event *A* are *b* to *a*.
B) The probability that event will occur is $P(A) = a / (a + b)$.
C) The probability that event *A* will not occur is $P(\text{not } A) = b / (a + b)$.
D) None of the above
ANSWER: D

26. Which of the following statements is false?

A) An empirical probability and an observed proportion are the same thing.
B) An observed proportion and a relative frequency are the same thing.
C) A relative frequency and an empirical probability are the same thing.
D) None of the above.
ANSWER: D

27. If the odds favoring rain tomorrow are 3 to 1, then the probability of rain tomorrow is

A) 1.00
B) 0.75
C) 0.50
D) 0.25

ANSWER: B

Short-Answer Questions

28. State whether the probability in the following situation is being determined empirically, theoretically, or subjectively: "A box contains 30 red beads and 70 blue beads. Jessica is going to randomly select one bead from the box and is interested in determining the relative frequency that the bead will be blue. She determines a relative frequency of 0.700".

ANSWER:
Theoretically

29. State whether the probability in the following situation is being determined empirically, theoretically, or subjectively: "Abby takes a test, and based on feeling, assigns a relative frequency of 0.8 that her grade will be an A."

ANSWER:
Subjectively

30. State whether the probability in the following situation is being determined empirically, theoretically, or subjectively: "In order to determine the relative frequency of obtaining a sum of 17 when three dice are tossed, Heidi tosses three dice 200 times and observe that the sum of 17 occurs 5 times. She obtains a relative frequency of 0.025."

ANSWER:
Empirically

31. State whether the probability in the following situation is being determined empirically, theoretically, or subjectively: "Lily is interested in determining the relative frequency of being dealt blackjack, which is an ace and a ten or an ace and a face card. She correctly reasons that there are 64 possible blackjacks and 1326 possible two-card hands. She then computes the relative frequency of being dealt blackjack as approximately 0.048."

ANSWER:
Theoretically

32. A computer program produces a random integer between 0 and 9 (inclusive). Find the probability that the integer is a number greater than 5.

ANSWER:
0.40

33. A computer program produces a random integer between 0 and 9 (inclusive). Find the probability that the integer is a number less than 7.

ANSWER:
0.70

34. After examining 5000 records of children of age 5, a dentist finds that 2235 had at least one cavity on their first dental check-up. What empirical probability would the dentist assign to the event that a 5-year-old would have at least one cavity on his/her first dental check-up?

ANSWER:
0.447

35. Three identical slips of paper with the numbers 1, 2, and 3 (one number on each slip) are placed in a box. One slip is randomly selected, and then, without replacement, a second slip is selected. Find the probability that the sum of the two numbers is even.

ANSWER:
1/3

36. Explain why the following statement is false: "If a sample space *S* has 5 sample points and if event *A* contains exactly 1 of these sample points, then it must follow that $P(A) = 0.20$".

ANSWER:
If sample points are not treated equally likely, then $P(A)$ is not necessarily 0.20.

37. Explain why the following statement is true: if *A* is an event of a sample space *S*, then it is possible that $P(A) = 1$.

ANSWER:
If $A = S$, then $P(A) = 1$.

38. Heidi is interested in determining the probability that a randomly selected student in her statistics class earned a passing grade (A, B, C, or D) on the first test. She reasons that each student earned either a passing grade (*P*) or a failing grade (*F*) and constructs the sample space $S = \{P,F\}$. Are the sample points equally likely or not equally likely?

ANSWER:
Not equally likely

39. Amy is interested in determining the probability that a randomly selected card from a standard deck of 52 will be a club. She reasons that the deck contains clubs (*C*), spades (*S*), diamonds (*D*), and hearts (*H*). She constructs the sample space $S = \{C, S, D, H\}$. Determine if the sample points are equally likely or not equally likely.

ANSWER:
Equally likely

40. A sample space is composed of three outcomes, called *A*, *B*, and *C*. Outcome *B* is twice as probable as *A*, and *C* is twice as probable as *B*. Find the probabilities of the events of *A*, *B*, and *C*.

ANSWER:
$P(A) = 1/7$, $P(B) = 2/7$, $P(C) = 4/7$

41. A meteorologist predicts that there will be a measurable amount of precipitation or no precipitation on a given day. The sample space is *S* = {precipitation, no precipitation}. Event *A* is defined to be *A* = {precipitation}. A student uses *P*(*A*) = *n*(*A*)/*n*(*S*) to obtain *P*(*A*) = 0.50 . Explain why this is *not* correct.

ANSWER:
The formula *P*(*A*) = *n*(*A*)/*n*(*S*) cannot be used since the sample points are not equally likely to occur.

42. If the odds in favor of an event *B* are *x* to *y*, what is the probability that event *B* will occur?

ANSWER:
$P(B) = x / (x + y)$

43. If the odds in favor of an event *A* are 2 to 3, what is the probability that event *A* will not occur?

ANSWER:
P(not *A*) = 0.60

44. One single-digit number is to be selected randomly. List the sample space.

ANSWER:
S = {0, 1, 2, 3, 4, 5, 6, 7, 8, 9}

45. Explain why an empirical probability, an observed proportion, and a relative frequency are actually three different names for the same thing.

ANSWER:
All three are calculated by dividing the experimental count by the sample size.

46. A single die is rolled once. What is the probability that the number on top is an odd number?

ANSWER:
3 / 6

Applied and Computational Questions

47. A two-stage experiment is performed, in which the first stage a coin is tossed and heads (*H*) or tails (*T*) is observed. In the second stage, a single card is randomly selected from a standard deck of 52 cards, and the suit of clubs (*C*), spades (*S*), diamonds (*D*), or hearts (*H*) is observed. List the sample space for this experiment.

ANSWER:
S = {(*H*, *C*), (*H*, *S*), (*H*, *D*), (*H*, *H*), (*T*, *C*), (*T*, *S*), (*T*, *D*), (*T*, *H*)}

48. You are tossing two coins and want to determine the probability that exactly one head is obtained. Construct the sample space and determine the probability for getting exactly one head.

ANSWER:
S = {*HH*, *HT*, *TH*, *TT*}; *P*(one head) = *P*(*HT*) + *P*(*TH*) = 0.25 + 0.25 = 0.50

QUESTIONS 49 THROUGH 53 ARE BASED ON THE FOLLOWING INFORMATION:
A sample of 240 undergraduates is randomly selected from a state university in Michigan. For the male students, 80 were in the College of A&S, 40 were in the College of Business (COB), and 10 were in the College of Engineering (COE). For the female students, 60 were in the College of A&S, 16 were in the COB, and 34 were in the COE.

49. Construct a contingency table for the given information.

ANSWER:

	A & S	*COB*	*COE*	Total
Male	80	40	10	130
Female	60	16	34	110
Total	140	56	44	240

50. If one student is randomly selected, find the probability that the student is a male in the College of Business.

ANSWER:
0.167

51. If one student is randomly selected, find the probability that the student is a female in the College of Engineering.

ANSWER:
0.142

52. If one student is randomly selected, find the probability that the student is a male.

ANSWER:
0.542

53. If one student is randomly selected, find the probability that the student is a student in the College of A&S.

ANSWER:
0.583

QUESTIONS 54 THROUGH 58 ARE BASED ON THE FOLLOWING INFORMATION:
In a sample of 300 undergraduates, 90 males and 65 females were in the College of A&S, 45 males and 36 females were in the College of Business (COB), and 30 males and 34 females were in the College of Education (COE).

54. Construct a contingency table for the given information.

ANSWER:

	A & S	*COB*	*COE*	Total
Male	90	45	30	165
Female	65	36	34	135
Total	155	81	64	300

55. If one student is randomly selected, find the probability that the student is a female in the College of Business.

ANSWER:
0.12

56. If one student is randomly selected, find the probability that the student is a male in the College of Education.

ANSWER:
0.10

57. If one student is randomly selected, find the probability that the student is a female.

ANSWER:
0.45

58. If one student is randomly selected, find the probability that the student is a student in the College of A&S.

ANSWER:
0.52

59. Suppose that a box of marbles contains an equal number of red and white marbles but twice as many blue marbles as red marbles. Draw one marble from the box and observe its color. Assign probabilities to the elements in the sample space.

ANSWER:
Let $P(R)$ = a, then $P(W)$ = a, and $P(B)$ = 2a. Hence, a + a + 2a = 1, which implies a = 0.25. Therefore, $P(R) = 0.25$, $P(W) = 0.25$, $P(B) = 0.50$.

60. Events A, B, and C are defined on sample space S. Their corresponding sets of sample points do not intersect and their union is S. Further, event B is twice as likely to occur as event A, and event C is twice as likely to occur as event B. Determine the probability of each of these three events.

ANSWER:
Let $P(A)$ = a, then $P(B)$ = 2a, and $P(C)$ = 4a. Hence, a + 2a + 4a = 1, which implies a = 1/7. Therefore, $P(A) = 1/7$, $P(B) = 2/7$, $P(C) = 4/7$.

QUESTIONS 61 THROUGH 63 ARE BASED ON THE FOLLOWING INFORMATION:
A medical clinic in Chicago classifies the patients' files by gender and by type of diabetes (A or B). The number of patients in each classification is shown below.

	Type of Diabetes	
Gender	A	B
Male (M)	50	40
Female (F)	70	40

One file is selected at random.

61. Find the probability that the selected individual is female.

ANSWER:
$P(F)$ = (70 +40) / 200 = 0.55

62. Find the probability that the selected individual is Type B.

ANSWER:
$P(B)$ = (40 +40) / 200 = 0.40

63. Find the probability that the selected individual is Type *A* male.

ANSWER:
$P(A \text{ and } M) = 50 / 200 = 0.25$

QUESTIONS 64 THROUGH 68 ARE BASED ON THE FOLLOWNG INFORMATION:
A single die is rolled once. Assume that the die is fair.

64. List the sample space.

ANSWER:
$S = \{1, 2, 3, 4, 5, 6\}$

65. Find the probability the number on top is a 4.

ANSWER:
1/ 6

66. Find the probability the number on top is an even number.

ANSWER:
3 / 6

67. Find the probability the number on top is less than 3.

ANSWER:
2 / 6

68. Find the probability the number on top is no greater than 5.

ANSWER:
5 / 6

QUESTIONS 69 THROUGH 71 ARE BASED ON THE FOLLOWING INFORMATION:
An experiment consists of drawing one marble from a box that contains a mixture of red, white, and blue marbles.

69. List the sample space.

ANSWER:
$S = \{R, W, B\}$

70. Can we be sure that each outcome in the sample space is equally likely? Explain.

ANSWER:
No; since no information is given in regard to the proportion of marbles for each color.

71. If two marbles are drawn from the box, list the sample space.

ANSWER:
$S = \{RR, RW, RB, WR, WW, WB, BR, BW, BB\}$

QUESTIONS 72 THROUGH 76 ARE BASED ON THE FOLLOWING INFORMATION:
A group of files in a medical clinic classifies the patients by gender and by the type of diabetes (I or II). The cross-tabulation (contingency table) below gives the number in each classification.

Gender	*Type of Diabetes* I	II
Male	42	21
Female	49	28

Assume that one file is selected at random.

72. Express the frequencies in the table as proportions of the grand total.

ANSWER:

Gender	*Type of Diabetes* I	II	**Row**
Male	0.30	0.15	**0.45**
Female	0.35	0.20	**0.55**
Column	**0.65**	**0.35**	**1.00**

73. Find the probability that the selected individual is female.

ANSWER:
P(Female) = 0.55

74. Find the probability that the selected individual is Type II.

ANSWER:
P(Type II) = 0.35

75. Find the probability that the selected individual is a male and Type I.

ANSWER:
P(Male and Type I) = 0.30

76. Find the probability that the selected individual is a female and Type II.

ANSWER:
P(Female and Type II) = 0.20

QUESTIONS 77 THROUGH 82 ARE BASED ON THE FOLLOWING INFORMATION:
Researchers have for a long time been interested in the relationship between cigarette smoking and lung cancer. The following table shows the percentages of adult females observed in a recent study.

Cigarette smoking

Lung cancer	Smokes (C)	Does not smoke (D)
Gets cancer (A)	0.08	0.02
Does not get cancer (B)	0.15	0.75

Suppose an adult female is randomly selected from this particular population.

77. What is the probability that she smokes and gets cancer?

ANSWER:
$P(C \text{ and } A) = 0.08$

78. What is the probability that she smokes?

ANSWER:
$P(C) = 0.08 + 0.15 = 0.23$

79. What is the probability that she does not get cancer?

ANSWER:
$P(B) = 0.15 + 0.75 = 0.90$

80. What is the probability that she does not smoke and does not get cancer?

ANSWER:
$P(D \text{ and } B) = 0.75$

81. What is the probability that she gets cancer knowing she smokes?

ANSWER:
$P(A \text{ given } C) = 0.15 / 0.23 = 0.652$

82. What is the probability that she does not get cancer, knowing she does not smoke?

ANSWER:
$P(B \text{ given } D) = 0.75 / 0.77 = 0.974$

83. Events A, B, and C are defined on sample space S. Their corresponding sets of sample points do not intersect and their union is S. Furthermore, event B is twice as likely to occur as event A, and event C is twice as likely to occur as event B. Determine the probability of each of the three events.

ANSWER:
Given information: $P(A) + P(B) + P(C) = 1$. Let $P(A) = p$, then $P(B) = 2p$ and $P(C) = 4p$. Now, $p + 2p + 4p = 1$, then $p = 1/7$. Therefore, $P(A) = 1/7$, $P(B) = 2/7$, and $P(C) = 4/7$.

QUESTIONS 84 THROUGH 86 ARE BASED ON THE FOLLOWING INFORMATION:
The odds for a student to pass a statistics class with an "A" grade are 3 to 7.

84. What is the probability the student will pass the class with an "A" grade?

ANSWER:
$P(A) = 3 / 10$ or 0.30

85. What are the odds against passing the class with an "A" grade?

ANSWER:
Odds against passing the class with an "A" grade are 7 to 3 (or 7:3).

86. What is the probability the student will not pass the class with an "A" grade?

ANSWER:
$P(\text{not } A) = 7 / 10$ or 0.70

Sections 4.3 and 4.4

True-False Questions

87. If A is an event of a sample space with $P(A) = P(\overline{A})$, then $P(A) = 0.50$.
ANSWER: T

88. If A is an event of a sample space S and if $P(\overline{A}) = 0$, then $A = S$.
ANSWER: T

89. Suppose A, B, and C are three nonempty events of a sample space S, all of which have no sample points in common, then it is possible that $A = \overline{B}$.
ANSWER: F

90. If A and B are any two events of a sample space S, then the addition rule is: $P(A \text{ or } B) = P(A) + P(B) - P(A \text{ and } B)$.
ANSWER: T

91. If A and B are any two events of a sample space S, then $P(A) = P(A \text{ and } B) - P(B)$.
ANSWER: F

92. The probabilities of complementary events always sum to 1.0.
ANSWER: T

93. A compound event formed by use of the word *and* requires the use of the addition rule.
ANSWER: F

94. A conditional probability is the relative frequency with which an event A can be expected to occur under the condition that additional pre-existing information is known about some other event, B.
ANSWER: T

95. If the results of a probability experiment can be any integer from 0 to 20, then the probability of each integer is 0.05.
ANSWER: F

96. The complement of an event A, denoted by $\overline{A}$, is the set of all sample points in the sample space that do not belong to event A.
ANSWER: T

Multiple-Choice Questions

97. If A is any event of a sample space S with $P(A) = q$, then $P(\overline{A})$ is equal to

A) $q - 1$.
B) $1 / q$.
C) $q + 1$.
D) $1 - q$.
ANSWER: D

98. A sample space is composed of three outcomes, called *A*, *B*, and *C*. Outcome *A* is twice as probable as *B*, and *B* is twice as probable as *C*. The probabilities of *A*, *B*, and *C* would be:

A) *P*(*A*) = 0.5; *P*(*B*) = 0.33; *P*(*C*) = 0.167.
B) *P*(*A*) = 0.4; *P*(*B*) = 0.4; *P*(*C*) = 0.2.
C) *P*(*A*) = 0.57; *P*(*B*) = 0.286; *P*(*C*) = 0.143.
D) Insufficient information given to determine answer.
ANSWER: C

99. Suppose *A* and *B* are two nonempty events of a sample space *S*, then *P*(*B*) always equals to:

A) *P*(*B* | *A*).
B) *P*(*B* and *A*) + *P*(*B* and $\bar{A}$).
C) *P*($\bar{B}$) – 1.
D) *P*(*B* or *A*)$\cdot P(B \text{ or } \bar{A})$.
ANSWER: B

100. If *P*(*A*) = 0.80, *P*(*B*) =0.70 and *P*(*A* or *B*) =0.90, then *P*(*A* and *B*) is:

A) 0.10.
B) 0.14.
C) 0.60.
D) 0.72.
ANSWER: C

101. If *P*(*A*) = 0.45, *P*(*B*) = 0.35 and *P*(*A* and *B*) =0.25, then *P*(*A* | *B*) is:

A) 1.4.
B) 1.8.
C) 0.714.
D) 0.556.
ANSWER: C

102. If *P*(*A*) = 0.60, *P*(*B*) = 0.63, and *P*(*A* and *B*) = 0.73, then *P*(*A* or *B*) is:

A) 1.23.
B) 0.50.
C) 0.13.
D) 0.10.
ANSWER: B

103. Which of the following statements is always correct?

A) *P*(*A* and *B*) = *P*(*A*)· *P*(*B*)
B) *P*(*A* or *B*) = *P*(*A*) + *P*(*B*)
C) *P*(*A* or *B*) = *P*(*A*) + *P*(*B*) - *P*(*A* and *B*)
D) *P*(*A*) = *P*(*B*|*A*)
ANSWER: C

104. If events A and B are defined on a sample space, with $P(A) = 0.25$ and $P(B \mid A) = 0.18$, then the probability that A and B can both occur at the same time is

A) 0.250
B) 0.180
C) 0.070
D) 0.045
ANSWER: **D**

105. If events A and B are defined on a sample space, with $P(A) = 0.5$ and $P(A \text{ and } B) = 0.8$, then the probability that event B will occur given that event A has already occurred is

A) 0.80
B) 0.50
C) 0.30
D) impossible to find $P(B \mid A)$
ANSWER: **D**

Short-Answer Questions

106. Five cards are randomly selected from a standard deck. Let A be the event that all five selected cards are the same suit. Using probability rules, $P(A)$ can be computed to be 0.002. Find the probability that all the cards are not the same suit.

ANSWER:
0.998

107. Events A and B are defined on a common sample space. If $P(A) = 0.20$, $P(B) = 0.40$, and $P(A \text{ or } B) = 0.56$, find $P(A \text{ and } B)$

ANSWER:
0.04

108. If the probability that event A occurs during an experiment is 0.62, what is the probability that event A doesn't occur during that experiment?

ANSWER:
$P(\bar{A}) = 1 - P(A) = 1 - 0.62 = 0.38$

109. If the results of a probability experiment can be any integer from 15 to 30 and the probability that the integer is less than 20 is 0.58, what is the probability the integer will be 20 or more?

ANSWER:
Let A = The integer is less than 20, then $\bar{A}$ = The integer is 20 or more.
$P(\bar{A}) = 1 - P(A) = 1 - 0.58 = 0.42$

110. If $P(A) = 0.35$, $P(B) = 0.55$, and $P(A \text{ and } B) = 0.1$, find $P(A \text{ or } B)$.

ANSWER:
$P(A \text{ or } B) = P(A) + P(B) - P(A \text{ and } B) = 0.35 + 0.55 - 0.1 = 0.80$

111. If $P(A) = 0.54$, $P(B) = 0.29$, and $P(A \text{ and } B) = 0.17$, find $P(A \text{ or } B)$.

ANSWER:
$P(A \text{ or } B) = P(A) + P(B) - P(A \text{ and } B) = 0.54 + 0.29 - 0.17 = 0.66$

112. If $P(A) = 0.35$, $P(B) = 0.45$, and $P(A \text{ or } B) = 0.65$, find $P(A \text{ and } B)$.

ANSWER:
$P(A \text{ or } B) = P(A) + P(B) - P(A \text{ and } B) \Rightarrow 0.65 = 0.35 + 0.45 - P(A \text{ and } B) \Rightarrow P(A \text{ and } B) = 0.15$

113. If $P(A) = 0.35$, $P(A \text{ or } B) = 0.85$, and $P(A \text{ and } B) = 0.2$, find $P(B)$.

ANSWER:
$P(A \text{ or } B) = P(A) + P(B) - P(A \text{ and B}) \Rightarrow 0.85 = 0.35 + P(B) - 0.20 \Rightarrow P(B) = 0.70$

Applied and Computational Questions

QUESTIONS 114 THROUGH 117 ARE BASED ON THE FOLLOWING INFORMATION:
Twenty percent of the trees in a particular forest have a disease, 30% of the trees are too small to be used for lumber, and 40% are too small to be used for lumber or have a disease. What percent of the trees are too small to be used for lumber and have a disease?

114. What percent of the trees are too small to be used for lumber and have a disease?

ANSWER:
10%

115. What percent of the trees are not too small to be used for lumber and do not have a disease?

ANSWER:
60%

116. If a tree is too small to be used for lumber, what is the probability it has a disease?

ANSWER:
0.333

117. If a tree has a disease, what is the probability it is not too small to be used for lumber?

ANSWER:
0.50

QUESTIONS 118 THROUGH 121 ARE BASED ON THE FOLLOWING INFORMATION:
Five hundred individuals were classified into three age groups: Group 1, 20-29; Group 2, 30-39; Group 3, 40 and over. In addition, they were placed into three groups depending on their diastolic blood pressure (DBP), as shown below.

		Age Group 1	2	3
	Below 70	20	30	20
DBP	70-90	60	140	60
	Above 90	20	80	70

118. Find the probability that a randomly selected individual in this study was in age group 3 or had a DBP above 90.

ANSWER:
0.50

119. Find the probability that a randomly selected individual in this study was in age group 1 or had a DBP below 70.

ANSWER:
0.30

120. If a randomly selected individual in this study was in group 2, what is the probability that she/he has a DBP between 70 and 90?

ANSWER:
0.56

121. If a randomly selected individual in that study had a DBP between 70 and 90, what is the probability that she/he was in group 1?

ANSWER:
0.23

QUESTIONS 122 THROUGH 124 ARE BASED ON THE FOLLOWING INFORMATION:
The probability that a first-time tourist to the city of Chicago will visit the Art Institute is 0.4, will visit the Museum of Science and Industry is 0.3, and will visit both is 0.1. Assume a first-time tourist to Chicago is randomly selected.

122. Find the probability that the tourist will visit the Art Institute or the Museum of Science and Industry.

ANSWER:
0.6

123. Find the probability that the tourist will visit neither of these attractions.

ANSWER:
0.4

124. Find the probability that the tourist will visit one, but not both, of these attractions.

ANSWER:
0.5

QUESTIONS 125 THROUGH 127 ARE BASED ON THE FOLLOWING INFORMATION:
The probability that a first-time tourist to the city of Toledo will visit the Art Museum is 0.5, will visit the Toledo Zoo is 0.4, and will visit both is 0.25. Assume a first-time tourist to Toledo is randomly selected.

125. Find the probability that the tourist will visit the Art Museum or the Toledo Zoo.

ANSWER:
0.65

126. Find the probability that the tourist will visit neither of these attractions.

ANSWER:
0.35

127. Find the probability that the tourist will visit one, but not both, of these attractions.

ANSWER:
0.40

QUESTIONS 128 THROUGH 131 ARE BASED ON THE FOLLOWING INFORMATION:
Sixty percent of the applicants at a "high tech" firm have a college degree, 45% have at least three years experience in the high tech industry, and 35% have both a college degree and three years experience in the high tech industry. An applicant is randomly chosen.

128. Find the probability that the applicant has a college degree or has had at least three years of experience in the high tech industry.

ANSWER:
0.70

129. Find the probability that the applicant has no college degree.

ANSWER:
0.70

130. Find the probability that the applicant has less than three years experience in the high tech industry.

ANSWER:
0.55

131. Find the probability that the applicant has at least three years experience in the high tech industry and no college degree.

ANSWER:
0.10

QUESTIONS 132 AND 133 ARE BASED ON THE FOLLOWING INFORMATION:
Five hundred people are classified based on their smoking habits and whether or not they have prominent wrinkles. The results are shown below:

	Prominent Wrinkles	*No Prominent Wrinkles*
Heavy smoker	120	60
Light or nonsmoker	75	245

One individual is randomly selected from that group of 500 people.

132. Given that the individual is a heavy smoker, what is the probability that he/she does not have prominent wrinkles?

ANSWER:
0.333

133. What is the probability that the selected individual is a heavy smoker or has prominent wrinkles?

ANSWER:
0.510

134. A large system is composed of many different components. The probability that a type 1 component fails is 0.95. Given that a type 1 component fails, the probability that the system fails is 0.90. What is the probability that a type 1 component fails and that the system also fails?

ANSWER:
0.855

135. The probability that an individual will contract a particular disease is 0.005. Past experience reveals that the probability that an individual who contracts the disease will make a complete recovery is 0.68. Find the probability that a randomly selected individual contracts the disease and does not make a complete recovery.

ANSWER:
0.0016

QUESTIONS 136 AND 137 ARE BASED ON THE FOLLOWING INFORMATION:
Records at a particular bank show that if a customer at the bank is randomly selected, the probability that the customer has a savings account at the bank is 0.42, the probability that the customer has a checking account at the bank is 0.74, and the probability that the customer has both is 0.28. A customer is randomly selected.

136. Find the probability that he/she has a checking account given that the customer has a savings account.

ANSWER:
0.667

137. Find the probability that he/she has a savings account given that the customer has a checking account.

ANSWER:
0.378

QUESTIONS 138 THROUGH 142 ARE BASED ON THE FOLLOWING INFORMATION:
Suppose A and B are events of a sample space S with $P(A) = 0.36$, $P(B) = 0.24$, and $P(A \text{ and } B) = 0.06$.

138. Find $P(A \text{ or } B)$.

ANSWER:
0.54

139. Find $P(A|B)$.

ANSWER:
0.25

140. Find $P(A \mid \overline{B})$.

ANSWER:
0.395

141. Find $P(B|A)$.

ANSWER:
0.395

142. Find $P(B \mid \overline{A})$.

ANSWER:
0.281

QUESTIONS 143 THROUGH 145 ARE BASED ON THE FOLLOWING INFORMATION:
A published article in a medical journal stated that one out of every ten American women will get breast cancer. It also states that of those who does, one out of four will die of it.

143. Find the probability that a randomly selected American woman will never get breast cancer.

ANSWER:
Let C represent a women gets breast cancer, and C represent she dies of it. $P(C) = 0.1$; then $P(\overline{C}) = 1.0 - 0.1 = 0.9$

144. Find the probability that a randomly selected American woman will get breast cancer and not die of it.

ANSWER:
Let C represent a women gets breast cancer, and C represent she dies of it. $P(D|C) = 0.25$; $P(\overline{D}|C) = 1 - P(D|C) = 0.75$, Then, $P(C \text{ and } \overline{D}) = P(C)\ P(\overline{D}|C) = (0.10)(0.75) = 0.075$.

145. Find the probability that a randomly selected American woman will get breast cancer and die from it.

ANSWER:
Let C represent a women gets breast cancer, and C represent she dies of it. $P(C \text{ and } D) = P(C)\ P(D|C) = (0.1)(0.25) = 0.025$

QUESTIONS 146 THROUGH 151 ARE BASED ON THE FOLLOWING INFORMATION:
A shipment of grapefruit arrived containing the following proportions of types: 10% pink seedless, 20% white seedless, 30% pink with seeds, 40% white with seeds. A grapefruit is selected random from the shipment.

146. Find the probability that it is seedless.

ANSWER:
P(seedless) = 0.10 + 0.20 = 0.30

147. Find the probability that it is pink.

ANSWER:
P(pink) = 0.10 + 0.30 = 0.40

148. Find the probability that it is pink and seedless.

ANSWER:
P(pink and seedless) = 0.10

149. Find the probability that it is pink or seedless.

ANSWER:
P(pink or seedless) = P(pink) + P(seedless) – P(pink and seedless)
= 0.40 + 0.30 – 0.10 = 0.60

150. Find the probability that it is pink, given that it is seedless.

ANSWER:
P (pink | seedless) = 0.10 / 0.30 = 0.333

151. Find the probability that it is seedless, given that it is pink.

ANSWER:
P(seedless | pink) = 0.10 / 0.40 = 0.25

152. Bianca wants to become a police officer. She must pass a physical exam and then a written exam. Records show the probability of passing the physical exam is 0.75 and that once the physical is passed the probability of passing the written exam is 0.60. What is the probability that Bianca passes both exams?

ANSWER:
Let A represent passing physical exam and B represent passing written exam.
$P(A) = 0.75$, and $P(B|A) = 0.60$. Then, $P(A \text{ and } B) = P(B|A) \cdot P(A) = (0.60)(0.75) = 0.45$

QUESTIONS 153 THROUGH 164 ARE BASED ON THE FOLLOWING INFORMATION:
Five hundred viewers were asked if they were satisfied with TV coverage of the London Terror on July 7th, 2005. The cross-tabulation (contingency table) below gives the number in each classification.

	Gender	
Opinion	Male	Female
Satisfied	92	133
Dissatisfied	75	200

One viewer is to be randomly selected from those surveyed.

153. Convert this 2 X 2 contingency table to a probability table.

ANSWER:

Opinion	*Gender* Male (M)	Female (F)	**Row Probability**
Satisfied (S)	0.184	0.266	**0.45**
Dissatisfied (D)	0.150	0.400	**0.55**
Column Probability	**0.334**	**0.666**	**1.00**

154. Find *P*(Satisfied).

ANSWER:
$P(S) = 0.45$

155. Find *P*(Satisfied | Female).

ANSWER:
$P(S \mid F) = 0.266 / 0.666 = 0.399$

156. Find *P*(Satisfied | Male).

ANSWER:
$P(S \mid M) = 0.184 / 0.334 = 0.551$

157. Find *P*(Male).

ANSWER:
$P(M) = 0.334$

158. Find *P*(Male | Dissatisfied).

ANSWER:
$P(M \mid D) = 0.15 / 0.55 = 0.273$

159. Find *P*(Female | Satisfied).

ANSWER:
$P(F \mid S) = 0.266 / 0.45 = 0.591$

160. Find *P*(Female and Satisfied).

ANSWER:
$P(F \text{ and } S) = 0.266$

161. Find *P*(Male and Dissatisfied).

ANSWER:
$P(M \text{ and } D) = 0.15$

162. Find *P*(Female | Dissatisfied).

ANSWER:
$P(F \mid D) = 0.40 / 0.55 = 0.727$

163. Find *P*(Male | Satisfied).

ANSWER:
$P(M \mid S) = 0.184 / 0.45 = 0.409$

164. Show that P(Male | Satisfied) + P(Female | Satisfied) = 1.0.

ANSWER:
P(Male | Satisfied) + P(Female | Satisfied) = 0.409 + 0.591 = 1.0

165. Events A and B are defined on a sample space, with $P(A) = 0.8$ and $P(B \mid A) = 0.3$. What is the probability that A and B can both occur at the same time?

ANSWER:
$P(A \text{ and } B) = P(A) \cdot P(B \mid A) = (0.8)(0.3) = 0.24$

166. Events A and B are defined on a sample space, with $P(A \mid B) = 0.5$ and $P(B) = 06$. What is the probability that A and B can both occur at the same time?

ANSWER:
$P(A \text{ and } B) = P(B) \cdot P(A \mid B) = (0.6)(0.5) = 0.30$

167. Events A and B are defined on a sample space, with $P(A) = 0.8$ and $P(A \text{ and } B) = 0.4$. Find the probability that event B will occur given that event A has already occurred.

ANSWER:
$P(B \mid A) = P(A \text{ and } B) / P(A) = 0.4 / 0.8 = 0.5$

168. Events A and B are defined on a sample space, with $P(B) = 0.36$ and $P(A \text{ and } B) = 0.5$. Find the probability that event A will occur given that event B has already occurred.

ANSWER:
It is impossible to find $P(A \mid B)$ in this situation since $P(A \text{ and } B)$ cannot exceed $P(B)$.

169. Suppose that A and B are two events defined on a common sample space and that the following probabilities are known: $P(A) = 0.4$, $P(B) = 0.3$, and $P(B \mid A) = 0.2$. Find $P(A \text{ or } B)$.

ANSWER:
Since $P(A \text{ and } B) = P(A) \cdot P(B \mid A) = (0.4)(0.2) = 0.08$, then $P(A \text{ or } B) = P(A) + P(B) - P(A \text{ and } B) = 0.4 + 0.3 - 0.08 = 0.62$.

170. Suppose that A and B are events defined on a common sample space and that the following probabilities are known: $P(A \text{ or } B) = 0.75$, $P(B) = 0.5$, and $P(A \mid B) = 0.25$. Find $P(A)$.

ANSWER:
Since $P(A \text{ and } B) = P(B) \cdot P(A \mid B) = (0.5)(0.25) = 0.125$, then
$P(A \text{ or } B) = P(A) + P(B) - P(A \text{ and } B) \Rightarrow 0.75 = P(A) + 0.5 - 0.125 \Rightarrow P(A) = 0.375$

171. Suppose that A and B are events defined on a common sample space and that the following probabilities are known: $P(A) = 0.5$, $P(A \text{ and } B) = 0.16$, and $P(A \mid B) = 0.4$. Find $P(A \text{ or } B)$.

ANSWER:
$P(A \text{ and } B) = P(B) \cdot P(A \mid B) \Rightarrow 0.16 = P(B) \cdot (0.4) \Rightarrow P(B) = 0.4$. Then,
$P(A \text{ or } B) = P(A) + P(B) - P(A \text{ and } B) = 0.5 + 0.4 - 0.16 = 0.74$

Sections 4.5 through 4.7

True-False Questions

172. If A and B are any two mutually exclusive events of a sample space S, then the occurrence of B means that A will occur.
ANSWER: F

173. Suppose A, B, and C are three nonempty events of a sample space S, all of which have no outcomes in common, then it is possible that $P(A) = 0.4$, $P(B) = 0.5$, and $P(C) = 0.6$.
ANSWER: F

174. If A and B are any two mutually exclusive events of a sample space S, then if A has occurred, B may also occur.
ANSWER: F

175. If A and B are both nonempty events of a sample space S, and A and B are mutually exclusive, then A and B are dependent.
ANSWER: T

176. If A and B are two nonempty events of a sample space S, that have no outcomes in common, then $P(A|B) = 1$.
ANSWER: F

177. If A and B are any two independent events of a sample space S, then A and B may be mutually exclusive.
ANSWER: F

178. If A and B are any two independent events of a sample space S, then $P(A \text{ and } B) = P(A) \cdot P(B|A)$.
ANSWER: T

179. If two events are mutually exclusive, they are also independent.
ANSWER: F

180. If events A and B are mutually exclusive, the sum of their probabilities must be exactly one.
ANSWER: F

181. If the sets of sample points belonging to two different events do not intersect, the events are mutually exclusive or dependent.
ANSWER: T

182. If $P(A) = 0.3$, $P(B) = 0.6$, and $P(A \text{ and } B) = 0.18$, then A and B are independent events.
ANSWER: T

183. If $P(A) = 0.2$, $P(B) = 0.5$, and $P(A \text{ and } B) = 0.05$, then A and B are mutually exclusive events.
ANSWER: F

184. If $P(A) = 0.4$, $P(B) = 0.3$, and $P(A \text{ and } B) = 0.15$, then $P(B | A) = 0.45$.
ANSWER: F

185. If events *A* and *B* are independent, they must be mutually exclusive.
ANSWER: F

186. Mutually exclusive events are non-empty events defined on the same sample space with each event excluding the occurrence of the other. In other words, they are events that share no common elements.
ANSWER: T

187. Mutually exclusive is an important probability concept.
ANSWER: F

188. Mutually exclusive events cannot be independent.
ANSWER: T

189. If events *A* and *B* are not independent, they must be mutually exclusive.
ANSWER: F

190. If events *A* and *B* are independent, they must be mutually exclusive.
ANSWER: F

191. Two events are independent if the occurrence (or nonoccurrence) of one gives us no information about the likeliness of occurrence for the other.
ANSWER: T

192. If the occurrence of one event does have an effect on the probability for occurrence of the other event, we say that the two events are mutually exclusive.
ANSWER: F

193. $P(A \text{ and } B) = P(A) \cdot P(B)$ can be used as the definition of independence of events *A* and *B*.
ANSWER: F

194. If the occurrence of one event does have an effect on the probability for occurrence of the other event, we say that the two events are dependent.
ANSWER: T

195. If events *A* and *B* are not mutually exclusive, they must be independent.
ANSWER: F

196. If events *A* and *B* are not mutually exclusive, they may be either independent or dependent.
ANSWER: T

197. Mutually exclusive events may be dependent or independent.
ANSWER: F

Multiple-Choice Questions

198. Which of the following defines a sample space that has sample points in common?

A) $P(A) = 0.60$ and $P(B) = 0.70$
B) $P(A) = 0.35$ and $P(B) = 0.65$
C) $P(A) = 0.60$ and $P(B) = 0.40$
D) $P(A) = 0.30$, $P(B) = 0.40$, and $P(B) = 0.30$
ANSWER: **A**

199. If A and B are events of a sample space S with A and B mutually exclusive, then $P(A) + P(B)$:

A) must equal 1.
B) could equal l.
C) would equal $P(A) \cdot P(B)$.
D) greater than 1.
ANSWER: **B**

200. Suppose A and B are two independent events of a sample space S with $P(A) = 0.30$ and $P(B) = 0.50$, then $P(A \text{ and } B)$ is

A) 0.80.
B) 0.60.
C) 0.20.
D) 0.15.
ANSWER: **D**

201. Suppose A and B are events of a sample space S with $P(A) = 0.22$, $P(B) = 0.40$, and $P(A \text{ and } B) = 0.04$, then $P(A \mid \overline{B})$ is

A) 0.462.
B) 0.300.
C) 0.182.
D) 0.100.
ANSWER: **B**

202. If $P(A) = 0.20$, $P(B) = 0.40$ and $P(A \text{ and } B) = 0.08$, then A and B are:

A) dependent events.
B) independent events.
C) mutually exclusive events.
D) complementary events.
ANSWER: **B**

203. If A and B are mutually exclusive events with $P(A) = 0.40$, then $P(B)$:

A) can be any value between 0 and 1.
B) cannot be larger than 0.40.
C) cannot be larger than 0.60.
D) cannot be determined with the information given.
ANSWER: **C**

204. If A and B are independent events with $P(A) = 0.35$ and $P(A \mid B) = 0.35$, then $P(B)$:

A) equals 0.35.
B) equals 0.70.
C) equals 0.65.
D) cannot be determined with the information given.
ANSWER: D

205. Two events A and B are said to be independent if:

A) $P(A \text{ and } B) = P(A) \cdot P(B)$.
B) $P(A \text{ and } B) = P(A) + P(B)$.
C) $P(A \mid B) = P(B)$.
D) $P(B \mid A) = P(A)$.
ANSWER: A

206. Two events A and B are said to mutually exclusive if:

A) $P(A \mid B) = 1$.
B) $P(B \mid A) =1$.
C) $P(A \text{ and } B) = 1$.
D) $P(A \text{ and } B) = 0$.
ANSWER: D

207. Which of the following statements is true?

A) $P(A \text{ and } B) = P(A) \cdot P(B)$ can be used as the definition of independence of events A and B.
B) $P(A \text{ and } B) = P(A) \cdot P(B)$ cannot be used as a test for independence of events A and B
C) $P(A \text{ and } B) = P(A) \cdot P(B)$ can be used as the definition of mutually exclusive events
D) None of the above
ANSWER: D

208. Which of the following statements is true?

A) Mutually exclusive is a probability concept by definition.
B) $P(A \text{ and } B) = 0$ can be used as a definition of mutually exclusive events.
C) $P(A \text{ and } B)$ can be used as a test for mutually exclusive events.
D) All of the above.
ANSWER: C

209. Suppose that A and B are mutually exclusive events, and that $P(A) = 0.4$ and $P(B) = 0.3$. Then, $P(A \text{ and } B)$ will be

A) 0.0
B) 0.4
C) 0.3
D) 0.7
ANSWER: A

210. Which of the following is true if A and B are mutually exclusive events?

A) $P(A \mid B) = 0$
B) $P(B \mid A) = 0$
C) $P(A \text{ and } B) = 0$
D) All of the above.
ANSWER: **D**

211. Which of the following statement is false?

A) If two events are mutually exclusive, this means that the two events cannot occur together; that is, they have no intersection.
B) If two events are independent, this means that the occurrence of either event does not affect the probability of the other event.
C) Either (*A*) or (*B*), but not both, is true.
D) Both (*A*) and (*B*) are true.
ANSWER: **C**

212. Which of the following statements is false?

A) If two events are mutually exclusive, then they are not independent.
B) If two events are independent, then they are not mutually exclusive.
C) Both (*A*) and (*B*) are true.
D) Both (*A*) and (*B*) are false.
ANSWER: **D**

213. Which of the following statements is true?

A) If two events are not mutually exclusive, then they may be either dependent or independent.
B) If two events are not independent, then they may be either mutually exclusive or not mutually exclusive.
C) Both (*A*) and (*B*) are true.
D) Both (*A*) and (*B*) are false.
ANSWER: **C**

Short-Answer Questions

214. Explain why events A and B cannot be mutually exclusive if they are defined on a common sample space with $P(A) = 0.56$ and $P(B) = 0.61$.

ANSWER:
If A and B were mutually exclusive, $P(A \text{ or } B) = 0.56 + 0.61 = 1.17$ which is impossible.

215. Explain why $P(B$ occurring when A has already occurred) = 0 when events A and B are mutually exclusive.

ANSWER:
Since A and B are mutually exclusive events, the occurrence of either event excludes the occurrence of the other. Now, since A has already occurred, then B cannot possibly occur.

216. Events A and B are mutually exclusive events defined on a common sample space. If $P(A) = 0.4$ and $P(A \text{ or } B) = 0.9$, find $P(B)$.

ANSWER:
0.5

217. Events A and B are defined on a common sample space. If $P(A) = 0.7$, $P(B) = 0.6$, and A and B are independent events, find $P(A \text{ or } B)$.

ANSWER:
0.88

218. A box contains five red, three blue, and two white poker chips. Two are selected without replacement. Find the probability that both are the same color.

ANSWER:
0.311

219. Explain why nonempty, mutually exclusive events A and B must be dependent.

ANSWER:
If A and B are independent, $P(A \text{ and } B) = P(A) \cdot P(B)$ Since A and B are mutually exclusive, $P(A \text{ and } B) = 0$. Thus, $0 = P(A) \cdot P(B)$. This is impossible since $P(A) \neq 0$ and $P(B) \neq 0$. Therefore, A and B are dependent.

220. If A and B are independent events, and $P(A) = 0.7$ and $P(B) = 0.6$, find $P(A \text{ and } B)$

ANSWER:
$P(A \text{ and } B) = P(A) \cdot P(B) = (0.7)(0.6) = 0.42$

221. If events A and B are independent, and $P(A) = 0.6$ and $P(B) = 0.5$, find $P(A \text{ and } B)$.

ANSWER:
$P(A \text{ and } B) = P(A) \cdot P(B) = (0.6)(0.5) = 0.30$

222. If A and B are independent events, and $P(A) = 0.8$ and $P(B) = 0.1$, find $P(B | A)$.

ANSWER:
$P(B | A) = P(B) = 0.1$

223. If A and B are independent events, and $P(A) = 0.8$ and $P(A \text{ and } B) = 0.4$, find $P(B)$.

ANSWER:
$P(A \text{ and } B) = P(A) \cdot P(B) \Rightarrow 0.4 = (0.8) \cdot P(B) \Rightarrow P(B) = 0.5$

224. If A and B are independent events, and $P(B) = 0.3$ and $P(A \text{ and } B) = 0.4$, find $P(A)$.

ANSWER:
It is impossible to find $P(A)$ since $P(A \text{ and } B)$ cannot exceed $P(B)$.

225. If A and B are independent events, and $P(A) = 0.5$ and $P(B) = 0.3$, find $P(A | B)$.

ANSWER:
$P(A | B) = P(A) = 0.5$

Applied and Computational Questions

226. Events A and B are events of a sample space S with $P(A)$ = 0.32, $P(B)$ = 0.11, and $P(A$ and $B)$ = 0.08. Are A and B independent events? You must give a written explanation. A simple answer of "yes" or "no" will receive no credit.

ANSWER:

$$P(A \mid B) = \frac{P(A \cap B)}{P(B)} = \frac{0.08}{0.11} = 0.727 \text{ , but } P(A) = 0.32$$

Since $P(A \mid B) \neq P(A)$, A and B are not independent events.

QUESTIONS 227 AND 228 ARE BASED ON THE FOLLOWING INFORMATION:
Five cards are randomly selected from a standard deck of 52 cards.

227. Find the probability that all five cards are red if they are selected without replacement.

ANSWER:
(26/52)(25/51)(24/50)(23/49)(22/48) = 0.0253

228. Find the probability that all five cards are red if they are selected with replacement.

ANSWER:
$(26/52)^5 = 0.0313$

QUESTIONS 229 THROUGH 234 ARE BASED ON THE FOLLOWING INFORMATION:
Events A, B, and C are events of a sample space S with A and C mutually exclusive, B and C mutually exclusive, $P(A)$ = 0.32, $P(B)$ = 0.11, $P(A$ and $B)$ = 0.08, and $P(C)$ = 0.42.

229. Find $P(A$ or $B)$.

ANSWER:
0.35

230. Find $P(A$ or $C)$.

ANSWER:
0.74

231. Find $P(B$ or $C)$.

ANSWER:
0.53

232. Find $P(\overline{A})$.

ANSWER:
0.68

233. Find $P(\overline{C})$.

ANSWER:
0.58

234. Find $P(A \text{ and } C)$.
ANSWER:
0.0

QUESTIONS 235 AND 236 ARE BASED ON THE FOLLOWING INFORMATION:
A box contains 12 red marbles and 8 blue marbles. Three marbles are randomly selected, one at a time.

235. Find the probability that all three are blue if they are selected with replacement.

ANSWER:
$(8/20)^3 = 0.064$

236. Find the probability that all three are blue if they are selected without replacement.

ANSWER:
(8/20)(7/19)(6/18)=0.049

QUESTIONS 237 THROUGH 240 ARE BASED ON THE FOLLOWING INFORMATION:
Let the sample space be the set of all students currently enrolled at your college. Suppose a student is randomly selected. Define the events *A*, *B*, *C*, *D*, and *E* as follows:

A: the student is over six feet tall,
B: the student owns an automobile,
C: the student wears size 13 shoes,
D: the student has natural blonde hair, and
E: the student has a GPA over 3.00.

237. Determine if events *A* and *B* are dependent or independent.

ANSWER:
Independent

238. Determine if events *A* and *C* are dependent or independent.

ANSWER:
Dependent

239. Determine if events *D* and *E* are dependent or independent.

ANSWER:
Independent

240. Determine if events *A* and *D* are dependent or independent.

ANSWER:
Independent

241. For what values of *k* will *A* and *B* be dependent events?

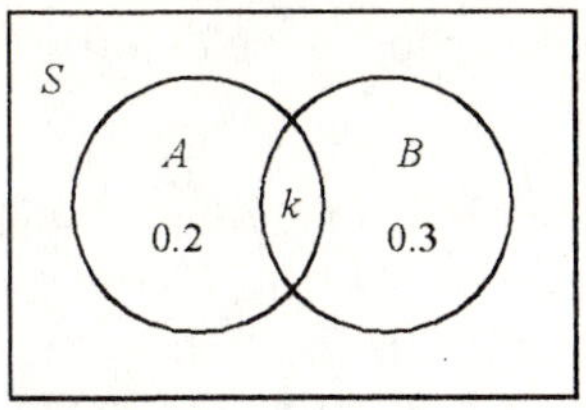

ANSWER:
$0.0 \leq k < 0.06$ or $0.06 < k \leq 0.62$

242. For what values of *k* will *A* and *B* be independent events?

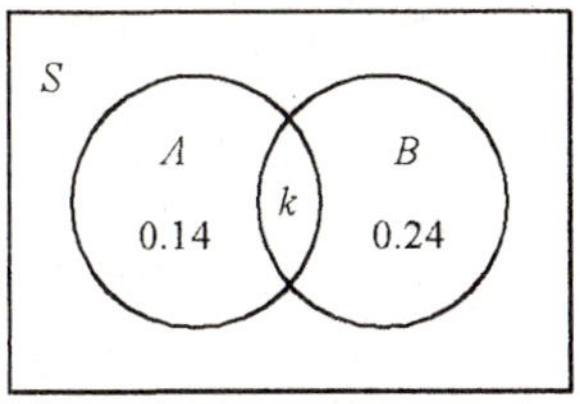

ANSWER:
$k = 0.14$ or 0.24

<u>QUESTIONS 243 THROUGH 248 ARE BASED ON THE FOLLOWING INFORMATION:</u>
A and *B* are two independent events of a sample space *S* with $P(A) = 0.25$ and $P(B) = 0.48$.

243. Find $P(A \text{ and } B)$.

ANSWER:
0.12

244. Find $P(A \text{ or } B)$.

ANSWER:
0.61

245. Find $P(A|B)$.

ANSWER:
0.25

246. Find $P(A \mid \overline{B})$.

ANSWER:
0.25

247. Find $P(B|A)$.

ANSWER:
0.48

248. Find $P(B \mid \overline{A})$.

ANSWER:
0.48

249. One letter is randomly selected from the word TOOT, and one letter is randomly selected from the word BOOT. Find the probability that the same letter is selected from both words.

ANSWER:
(1/2)(1/2) + (1/2)(1/4) = 0.375

250. Professor Brown gives her students a maximum of three attempts to pass a final examination in her statistics course. She has found that the probability of passing on the first attempt is 0.40, the probability of passing on the second attempt is 0.65, and the probability of not passing on the third attempt is 0.15. Find the probability that a randomly selected student of hers will pass the final examination.

ANSWER:
1 – (0.60)(0.35)(.15) = 0.9685

QUESTIONS 251 AND 252 ARE BASED ON THE FOLLOWING INFORMATION:
An experiment consists of selecting a marble from box one and placing it in box two, and then a marble is selected from box two and its color is noted. Box one contains two red, three blue, and five white marbles, and box two contains six red, two blue, and two white marbles.

251. Find the probability that the first and second selected marbles were both red.

ANSWER:
0.028

252. Find the probability that the marble selected from box two was red.

ANSWER:
0.564

253. A box contains 3 defective units and 17 non-defective units. Two units are selected from the box without replacement. What is the probability that both units are defective given that the first one selected was defective?

ANSWER:
(3/20)(2/19)=0.0158

QUESTIONS 254 THROUGH 258 ARE BASED ON THE FOLLOWING INFORMATION:
A hospital classifies some of the patients' files by gender and by type of care received (Intensive Care Unit (ICU) and Surgical Unit). The number of patients in each classification is presented below:

	Type of Care	
Gender	ICU	Surgical Unit
Male (M)	25	39
Female (F)	21	15

One of these patients is randomly selected.

254. Are the events "being a female" and "being in the ICU" mutually exclusive?

ANSWER:
No, they can occur at the same time; i.e., a patient can be both female and in ICU.

255. Are the events "being in the ICU" and "being in the surgical unit" mutually exclusive?

ANSWER:
Yes, they cannot occur at the same time; i.e., a patient cannot be in ICU and the Surgical Unit at the same time.

256. Find *P*(ICU or Female).

ANSWER:
P(ICU or Female) = 46/100 + 36/100 – 21/100 = 61/100 = 0.61

257. Find *P*(Surgical unit or Male).

ANSWER:
P(Surgical unit or Male) = 54/100 + 64/100 – 39/100 = 79/100 = 0.79

258. Find *P*(ICU and Male).

ANSWER:
P(ICU and Male) = 25/100 = 0.25

QUESTIONS 259 THROUGH 262 ARE BASED ON THE FOLLOWING INFORMATION:
Assume $P(A) = 0.4$ and $P(B) = 0.5$ and A and B are independent events.

259. Find $P(A \text{ and } B)$.

ANSWER:
$P(A \text{ and } B) = P(A) \cdot P(B) = (0.4)(0.5) = 0.20$

260. Find $P(B \mid A)$.

ANSWER:
$P(B \mid A) = P(B) = 0.5$

261. Find $P(A \mid B)$.

ANSWER:
$P(A \mid B) = P(A) = 0.4$

262. Find $P(A \text{ or } B)$.

ANSWER:
$P(A \text{ or } B) = P(A) + P(B) - P(A \text{ and } B) = 0.4 + 0.5 - 0.20 = 0.70$

QUESTIONS 263 THROUGH 265 ARE BASED ON THE FOLLOWING INFORMATION:
Suppose that $P(A) = 0.3$, $P(B) = 0.6$, and $P(A \text{ and } B) = 0.18$.

263. What is $P(A \mid B)$?

ANSWER:
$P(A \text{ and } B) = P(B) \cdot P(A \mid B) \Rightarrow 0.18 = (0.6) \cdot P(A \mid B)$; therefore, $P(A \mid B) = 0.3$.

264. What is $P(B \mid A)$?

ANSWER:
$P(A \text{ and } B) = P(A) \cdot P(B \mid A) \Rightarrow 0.18 = (0.3) \cdot P(B \mid A)$; therefore, $P(B \mid A) = 0.6$.

265. Are A and B independent? Justify your answer in three different ways.

ANSWER:
Yes, A and B are independent since
$P(A \mid B) = 0.3 = P(A)$, or
$P(B \mid A) = 0.6 = P(B)$, or
$P(A \text{ and } B) = 0.18 = P(A) \cdot P(B)$

QUESTIONS 266 THROUGH 268 ARE BASED ON THE FOLLOWING INFORMATION:
A husband and wife make their decisions independently of each other, and then they compare their decisions. If they agree, the decision is made; if they do not agree, then further consideration is necessary before a decision is reached. Assume each has a history of making the right decision 70% of the time.

266. What is the probability that they together make the right decision on the first try?

ANSWER:
Let A = right decision, B = wrong decision
$P(\text{right decision}) = P(\text{A1 and A2}) = (0.7) \cdot (0.7) = 0.49$

267. What is the probability that they together make the wrong decision on the first try?

ANSWER:
Let A = right decision, B = wrong decision
$P(\text{wrong decision}) = P(\text{B1 and B2}) = (0.3) \cdot (0.3) = 0.09$

268. What is the probability that they together delay the decision for further study?

ANSWER:
Let A = right decision, B = wrong decision
$P(\text{delay the decision}) = P[(\text{A1 and B2}) \text{ or } (\text{B1 and A2})]$
$= P(\text{A1 and B2}) + P(\text{B1 and A2})$
$= (0.7)(0.3) + (0.3)(0.7) = 0.21 + 0.21 = 0.42$

QUESTIONS 269 THROUGH 271 ARE BASED ON THE FOLLOWING INFORMATION:
A box contains 50 parts, of which 6 are defective and 44 are nondefective. Assume that two parts are selected *without replacement*.

269. Find $P(\text{both are defective})$.

ANSWER:
Let D = defective, N = nondefective.
$P(\text{D1}) = 6/50 = 0.12$, $P(\text{D2}) = 5/49 = 0.1020$,
$P(\text{N1}) = 44/50 = 0.88$, $P(\text{N2}) = 43/49 = 0.8776$. Then,
$P(\text{both defective}) = P(\text{D1and D2}) = P(\text{D1}) \cdot P(\text{D2}) = (0.12)\ (0.1020) = 0.0122$

270. Find *P*(exactly one is defective).

ANSWER:
Let D = defective, N = nondefective.
P(D1) = 6/50 = 0.12, *P*(D2) = 5/49 = 0.1020,
P(N1) = 44/50 = 0.88, *P*(N2) = 43/49 = 0.8776. Then,
P(exactly one defective) = *P*[(D1 and N2) or (N1 and D2)]
= *P*(D1) · *P*(N2) + *P*(N1) *P*(D2)
= (0.12)(0.8776) +(0.88)(0.1020) = 0.1951

271. Find *P*(neither is defective).

ANSWER:
Let D = defective, N = nondefective.
P(D1) = 6/50 = 0.12, *P*(D2) = 5/49 = 0.1020,
P(N1) = 44/50 = 0.88, *P*(N2) = 43/49 = 0.8776. Then,
P(neither defective) = *P*(N1 and N2) = *P*(N1) · *P*(N2) = (0.88)(0.8776) = 0.7723

QUESTIONS 272 THROUGH 277 ARE BASED ON THE FOLLOWING INFORMATION:
Let *P*(*A*) = 0.3, *P*(*B*) = 0.4, and events *A* and *B* are mutually exclusive.

272. Find *P*(*A* and *B*).

ANSWER:
P(*A* and *B*) = 0.0 (they are mutually exclusive)

273. Find *P*(*A* or *B*).

ANSWER:
P(*A* or *B*) = *P*(*A*) + *P*(*B*) = 0.3 + 0.4 = 0.7

274. Find *P*(*A* or $\bar{B}$).

ANSWER:
P(*A* or $\bar{B}$) = *P*($\bar{B}$) = 1 – *P*(*B*) = 1 – 0.4 = 0.6 (*A* is a subset of $\bar{B}$ since *A* and *B* are mutually exclusive.)

275. Find *P*(*A* | *B*).

ANSWER:
P(*A* | *B*) = 0.0 (they are mutually exclusive)

276. Find *P*(*A*| $\bar{B}$).

ANSWER:
P(*A*| $\bar{B}$) = *P*(*A* and $\bar{B}$) / *P*($\bar{B}$) = *P*(*A*) / *P*($\bar{B}$) = 0.3 / 0.6 = 0.5
(recall that *A* is a subset of $\bar{B}$ since *A* and *B* are mutually exclusive.)

277. Are events *A* and *B* independent? Explain.

ANSWER:
No; mutually exclusive events are disjoint, therefore they must be dependent.

278. A company that manufactures windows has three factories. Factory 1 produces 30% of the company's windows, Factory 2 produces 60%, and Factory 3 produces 10%. One percent of the windows produced by Factory 1 are mislabeled, 0.5% of those produced by Factory 2 are mislabeled, and 2% of those produced by Factory 3 are mislabeled. If you purchase one window manufactured by this company, what is the probability that the window is mislabeled?

ANSWER:
Let *F* represent factory where window was produced, with *i* = 1, 2, 3, and *M* represent mislabeled. Then, *M* = (*M* and F_1) or (*M* and F_2) or (*M* and F_3), and hence

$$P(M) = P(F_1)\cdot P(M \mid F_1) + P(F_2)\cdot P(M \mid F_2) + P(F_3)\cdot P(M \mid F_3)$$
$$= (0.30)(0.01) + (0.60)(0.005) + (0.10)(0.02) = 0.008$$

QUESTIONS 279 THROUGH 281 ARE BASED ON THE FOLLOWING INFORMATION:
Two hundred employees were polled about worker satisfaction.

	Male		Female		
	Skilled	*Unskilled*	*Skilled*	*Unskilled*	**Total**
Satisfied	70	30	5	20	125
Unsatisfied	30	20	15	10	75
Total	100	50	20	30	200

One employee is selected at random.

279. Find the probability that an unskilled worker is satisfied with work.

ANSWER:
P(satisfied | unskilled) = (30+20) / (50+30) = 0.625

280. Find the probability that a skilled woman employee is satisfied with work.

ANSWER:
P(satisfied | skilled woman) = 5 / 20 = 0.25

281. Is satisfaction for women employees independent of their being skilled or unskilled?

ANSWER:
Compare *P*(satisfied | skilled woman) to *P*(satisfied | unskilled woman)
P(satisfied | skilled woman) = 5 / 20 = 0.25
P(satisfied | unskilled woman) = 20 / 30 = 0.667
Since these two probabilities are not equal, therefore the two events are not independent. That is, satisfaction for women employees depends on their being skilled or unskilled.

QUESTIONS 282 THROUGH 285 ARE BASED ON THE FOLLOWING INFORMATION:
Suppose a certain ophthalmic trait is associated with eye color. One hundred and fifty randomly selected individuals are studied with results as follows:

	Eye Color			
Trait	Blue	Brown	Other	**Total**
Yes	35	15	10	60
No	10	55	25	90
Total	45	70	35	150

282. What is the probability that a person selected at random has blue eyes?

ANSWER:
P(blue eyes) = 45 / 150 = 0.30

283. What is the probability that a person selected at random has the trait?

ANSWER:
P(yes) = 60 / 150 = 0.40

284. Are events A (has blue eyes) and B (has the trait) independent? Justify your answer.

ANSWER:
If independent; then $P(A \text{ and } B) = P(A) \cdot P(B)$
$P(A \text{ and } B) = 35 / 150 = 0.233$, and $P(A) \cdot P(B) = (45/150) \cdot (60/150) = 0.12$; therefore A and B are not independent events.

285. How are the two events A (has blue eyes) and C (has brown eyes) related (independent, mutually exclusive, complementary, or all-inclusive)? Explain why or why not each term applies.

ANSWER:
Blue eyes and brown eyes are mutually exclusive events. They are not complementary since not everyone was classified as having brown or blue eyes. Since they are mutually exclusive, they cannot be independent events.

286. In United States, professional basketball championship is often decided by two teams playing each other in a seven-game series. Suppose that team A is the better team, and the probability it will beat team B in any one game is 0.7. What is the probability that team A will win the series?

ANSWER:
P(team A wins best of 7 game series)
= P(A wins in 4 games) + P(A wins in 5 games) + P(A wins in 6 games) + P(A wins in 7 games)
$= 1 \cdot (0.7)^4 + 4 \cdot (0.7)^4 (0.3)^1 + 10 \cdot (0.7)^4 (0.3)^2 + 20 \cdot (0.7)^4 (0.3)^3$
$= 0.2401 + 0.2881 + 0.2161 + 0.1297 = 0.874$

QUESTIONS 287 THROUGH 289 ARE BASED ON THE FOLLOWING INFORMATION:
Events A and B are defined on a sample space. Assume that $P(A) = 0.2$ and $P(B) = 0.4$.

287. If A and B are mutually exclusive, what is $P(A \text{ or } B)$?

ANSWER:
$P(A \text{ or } B) = P(A) + P(B) = 0.2 + 0.4 = 0.6$

288. If A and B are independent, what is $P(A \text{ or } B)$?

ANSWER:
$P(A \text{ and } B) = P(A) \cdot P(B) = (0.2)(0.4) = 0.08$, then
$P(A \text{ or } B) = P(A) + P(B) - P(A \text{ and } B) = 0.2 + 0.4 - 0.08 = 0.52$

289. If A and B are mutually exclusive, what is $P(\text{A and } B)$?

ANSWER:
$P(A \text{ and } B) = 0$

QUESTIONS 290 THROUGH 293 ARE BASED ON THE FOLLOWING INFORMATION:
Suppose that *A* and *B* are mutually exclusive events, and that $P(A) = 0.4$ and $P(B) = 0.3$.

290. Find $P(\bar{A})$.

ANSWER:
$P(\bar{A}) = 1 - P(A) = 1 - 0.4 = 0.6$

291. Find $P(\bar{B})$.

ANSWER:
$P(\bar{B}) = 1 - P(B) = 1 - 0.3 = 0.7$

292. Find *P*(*A* or *B*).

ANSWER:
Since *A* and *B* are mutually exclusive, then *P*(*A* or *B*) = *P*(*A*) + *P*(*B*) = 0.4 + 0.3 = 0.7.

293. Find *P*(*A* and *B*).

ANSWER:
Since *A* and *B* are mutually exclusive, then *P*(*A* and *B*) = 0.

294. Give an example to demonstrate the fact that "If events *A* and *B* are mutually exclusive, they cannot be independent".

ANSWER:
Let *P*(*A*) = 0.3 and *P*(*B*) = 0.4. If *A* and *B* are mutually exclusive events, then *P*(*A* and *B*) = 0, and then $P(A \mid B) = 0.0$. Since we are given *P*(*A*) = 0.3, we see that the occurrence of *B* has an effect on the probability of *A*, therefore *A* and *B* cannot be independent events.

295. Give an example to demonstrate the fact that" If events *A* and *B* are not mutually exclusive, they may be either independent or dependent".

ANSWER:
Let *P*(*A*) = 0.3, and *P*(*B*) = 0.5. If events *A* and *B* are not mutually exclusive, it must be true that *P*(*A* and *B*) is greater than zero. Now if *P*(*A* and *B*) happens to be exactly 0.15, then events *A* and *B* are independent since *P*(*A* and *B*) = 0.15 = $P(A) \cdot P(B)$. But, if *P*(*A* and *B*) is any other positive value, say 0.12, than events *A* and *B* are not independent, since *P*(*A* and *B*) = 0.12 $\neq P(A) \cdot P(B) = 0.15$. Therefore, our conclusion is that If the events *A* and *B* are not mutually exclusive, they may be either independent or dependent, and additional information is needed in order to determine which.

QUESTIONS 296 THROUGH 298 ARE BASED ON THE FOLLOWING INFORMATION:
Suppose that *P*(*A*) = 0.5, *P*(*B*) = 0.7, and *P*(*A* and *B*) = 0.35.

296. What is $P(A \mid B)$?

ANSWER:
$P(A \mid B) = P(A \text{ and } B) / P(B) = (0.35) / (0.7) = 0.5$

297. What is $P(B \mid A)$?

ANSWER:
$P(B \mid A) = P(A \text{ and } B) / P(A) = (0.35) / (0.5) = 0.7$

298. Are events A and B independent?

ANSWER:
Yes, A and B are independent events since the following three equalities are satisfied: $P(A \mid B) = P(A)$, $P(B \mid A) = P(B)$, and $P(A \text{ and } B) = P(A) \cdot P(B)$ (need to satisfy only one equality, since if one is true, the other two must be true).

QUESTIONS 299 THROUGH 310 ARE BASED ON THE FOLLOWING INFORMATION:
An aquarium at a pet store contains 50 orange swordfish (27 females and 23 males) and 30 green swordtails (14 females and 16 males). You randomly net one of the fish.

299. Summarize the given frequency data using a 2 x 2 contingency table.

ANSWER:

Color of fish

Gender	Orange (O)	Green (G)	**Row Total**
Male (M)	23	16	**39**
Female (F)	27	14	**41**
Column Total	**50**	**30**	**80**

300. Express the frequencies in question 299 as relative frequencies.

ANSWER:

Color of fish

Gender	Orange (O)	Green (G)	**Row**
Male (M)	0.2875	0.200	**0.4875**
Female (F)	0.3375	0.175	**0.5125**
Column	**0.625**	**0.375**	**1.0**

301. What is the probability that it is an orange swordfish?

ANSWER:
$P(O) = 0.625$

302. What is the probability that it is a male fish?

ANSWER:
$P(M) = 0.4875$

303. What is the probability that it is an orange female swordfish?

ANSWER:
$P(O \text{ and } F) = 0.3375$

304. What is the probability that it is a female or a green swordtail?

ANSWER:
$P(F \text{ or } G) = P(F) + P(G) - P(F \text{ and } G) = 0.5125 + 0.375 - 0.175 = 0.7125$

305. What is the probability that it is a male or an orange swordtail?

ANSWER:
$P(M \text{ or } O) = P(M) + P(O) - P(M \text{ and } O) = 0.4875 + 0.625 - 0.2875 = 0.825$

306. What is the probability that it is a male, knowing that it is a green swordtail?

ANSWER:
$P(M \mid G) = 0.200 / 0.375 = 0.533$

307. What is the probability that it is a female, knowing that it is an orange swordfish?

ANSWER:
$P(F \mid O) = 0.3375 / 0.625 = 0.54$

308. Are the events "male" and "female" mutually exclusive? Explain.

ANSWER:
Yes; since a fish cannot be male and female at the same time

309. Are the events "male" and "swordfish" mutually exclusive? Explain.

ANSWER:
No; a fish can be both male and swordfish, that is, a male swordfish.

310. Are the events "gender" and "color of fish" independent? Explain.

ANSWER:
Since, for example, $P(F \mid O) = 0.54$ and $P(F) = 0.5125$, then $P(F \mid O) \neq P(F)$. Therefore, events the two events are not independent.

311. Give an example to demonstrate the fact that "If events *A* and *B* are independent and both have nonzero probabilities, they cannot be mutually exclusive."

ANSWER:
Let $P(A) = 0.3$ and $P(B) = 0.5$. If *A* and *B* are independent, then $P(A \text{ and } B) = P(A) \cdot P(B) = 0.15$, which is greater than zero. This means there is an intersection between events *A* and *B*, and the events cannot be mutually exclusive.

QUESTIONS 312 THROUGH 314 ARE BASED ON THE FOLLOWING INFORMATION:
Suppose that $P(A) = 0.20$, $P(B) = 0.40$, and $P(A \text{ and } B) = 0.15$.

312. What is $P(A \mid B)$?

ANSWER:
$P(A \mid B) = 0.15 / 0.40 = 0.375$

313. What is $P(B \mid A)$?

ANSWER:
$P(B \mid A) = 0.15 / 0.20 = 0.75$

314. Are *A* and *B* independent?

ANSWER:
A and *B* are not independent events since, for example, $P(A \mid B) = 0.375 \neq P(A) = 0.20$

QUESTIONS 315 THROUGH 321 ARE BASED ON THE FOLLOWING INFORMATION:
One student is selected at random from a group of 150 known to consist of 105 full-time (60 female and 45 male) students and 45 part-time (30 female and 15 male) students. Event *A* is "the student selected is full-time," event *B* is "the student selected is part-time", event *M* is "the selected student is male", and event *F* is "the selected student is female".

315. Summarize the given frequency data using a 2 x 2 contingency table.

ANSWER:

Student Status

Gender	Full-time (A)	Part-time (B)	**Row Total**
Male (M)	45	15	**60**
Female (F)	60	30	**90**
Column Total	**105**	**45**	**150**

316. Express the frequencies in question 315 as relative frequencies.

ANSWER:

Student Status

Gender	Full-time (A)	Part-time (B)	Row
Male (M)	0.3	0.1	0.4
Female (F)	0.4	0.2	0.6
Column	0.7	0.3	1.0

317. Are events *A* and *F* independent? Justify your answer.

ANSWER:
Since $P(A \text{ and } F) = 0.4 \neq P(A) \cdot P(F) = (0.7)(0.6) = 0.42$, then events *A* and *F* are not independent.

318 Are events *B* and *M* independent? Justify your answer.

ANSWER:
Since $P(B \text{ and } M) = 0.1 \neq P(B) \cdot P(M) = (0.3)(0.4) = 0.12$, then events *B* and *M* are not independent.

319. Based on your answers to questions 317 and 318, what is your conclusion?

ANSWER:
We conclude that student status (as full-time or part-time) does not depend on the gender of the student.

320. Find $P(A \mid F)$.

ANSWER:
$P(A \mid F) = 0.4 / 0.6 = 0.667$

321. Find $P(M \mid B)$.

ANSWER:
$P(M \mid B) = 0.1 / 0.3 = 0.333$

322. Give an example to demonstrate the fact that" If events A and B are not independent, they can be either mutually exclusive or not mutually exclusive".

ANSWER:
Let $P(A) = 0.3$ and $P(B) = 0.5$. If A and B are not independent events, it must be that $P(A$ and $B)$ is different than 0.15; the value it would be if they were independent $[P(A) \cdot P(B) = (0.3)(0.5) = 0.15]$. Now if $P(A$ and $B)$ happens to be exactly 0.00, then events A and B are mutually exclusive, but if $P(A$ and $B)$ is any other positive value, say 0.13, then events A and B are not mutually exclusive. Therefore, our conclusion is that if the events A and B are not independent, they could be either mutually exclusive or not, and some other information is needed to make that determination.

QUESTIONS 323 AND 324 ARE BASED ON THE FOLLOWING INFORMATION:
A box contains 40 parts, of which 5 are defective and 35 are nondefective. Assume that 2 parts are selected without replacement. Let event D_1 = first part is defective, event D_2 = second part is defective, event N_1 = first part is not defective, and event N_2 = second part is not defective.

323. Find the probability that both parts are defective.

ANSWER:
P(both parts are defective) = $P(D_1$ and $D_2) = (5/40)(4/39) = 0.0128$

324. Find the probability that exactly one part is defective.

ANSWER:
P(exactly one part is defective) = $P(D_1$ and $N_2) + P(N_1$ and $D_2)$
$= (5/40)(35/39) + (35/40)(5/39) = 0.2244$

QUESTIONS 325 THROUGH 327 ARE BASED ON THE FOLLOWING INFORMATION:
Are college graduation rates low? A recent survey shows that the percentage of students who graduate within five years is 42% for public colleges and 55% for private colleges. One of the reasons for this might be that only 56% of the students attend full time.

325. What additional information do you need to determine the probability that a student selected at random is part time and will graduate within five years?

ANSWER:
We need to know whether or not the events part-time and graduate within five years are independent.

326. Is it likely that the two events cited in question 325 have the needed property? Explain.

ANSWER:
Clearly the events part-time and graduate within five years are not independent. Whether a student is part-time or full-time will make a difference in how soon he/she will graduate.

327. If appropriate, find the probability that a student selected at random is part time and will graduate within five years.

ANSWER:
Let event A = Student is part-time, event B = Student will graduate within five years, event C = Student attending public college, and D = Student attending private college. Then,
$P(A \text{ and } B) = P(A) \cdot P(B \mid C) + P(A) \cdot P(B \mid D) = (0.44)(0.42) + (0.44)(0.55) = 0.4268$

QUESTIONS 328 THROUGH 332 ARE BASED ON THE FOLLOWING INFORMATION:
Suppose that when a candidate comes to a campus interview for an administrative position at an academic institution, the probability that he or she will want the job (event A) after the interview is 0.70. Also, the probability that the institution wants the candidate (event B) is 0.35. In addition, assume that $P(A \mid B)$ is 0.90

328. Find $P(A \text{ and } B)$.

ANSWER:
$P(A \text{ and } B) = P(B) \cdot P(A \mid B) = (0.35)(0.90) = 0.315$

329. Find $P(B \mid A)$.

ANSWER:
$P(B \mid A) = P(A \text{ and } B) / P(A) = 0.315 / 0.70 = 0.45$

330. Are events A and B independent? Explain.

ANSWER:
Since $P(B \mid A) = 0.45 \neq P(B) = 0.35$, events A and B are not independent.

331. Are events A and B mutually exclusive? Explain.

ANSWER:
Since $P(A \text{ and } B) = 0.315 > 0$, events A and B are not mutually exclusive.

332. What would it mean to say A and B are mutually exclusive events in this particular situation?

ANSWER:
"Candidate wants the administrative position" and "institution wants candidate" could not both happen.

QUESTIONS 333 THROUGH 337 ARE BASED ON THE FOLLOWING INFORMATION:
The odds against throwing a pair of dice and getting a total of 5 are 8 to 1. The odds against throwing a pair of dice and getting a total of 11 are 11 to 1.

333. What are the odds in favor of throwing a pair of dice and getting a total of 5?

ANSWER:
1 to 8

334. What are the odds in favor of throwing a pair of dice and getting a total of 11?

ANSWER:
1 to 11

335. What is the probability of throwing a pair of dice and getting a total of 5?

ANSWER:
1 / (1 + 8) = 1 / 9

336. What is the probability of throwing a pair of dice and getting a total of 11?

ANSWER:
1 / (1 + 11) = 1 / 12

337. What is the probability of throwing the dice twice and getting a total of 5 on the first throw and 10 on the second throw?

ANSWER:
Clearly the events of getting a total of 5 on the first throw and 10 on the second throw are independent, so the special multiplication rule applies.
Therefore,
P(5 on first throw and 10 on second throw) = P(5 on first throw).P(10 on second throw)
= (1 / 9) (1 / 12) = 1 / 108 ≈ 0.0093

Chapter 5

Probability Distributions (Discrete Variables)

Sections 5.1 and 5.2

True-False Questions

1. A random variable may assume many values for each outcome of a probability experiment.
ANSWER: F

2. A quantitative random variable that can assume an uncountable number of values is referred to as continuous random variable.
ANSWER: T

3. The number of hours you studied for your final exams last semester is an example of a continuous random variable.
ANSWER: F

4. The number of speeding tickets you received last year is an example of a discrete random variable.
ANSWER: T

5. The number of hours you waited in line to register this semester is an example of a discrete random variable.
ANSWER: F

6. The number of automobile accidents you were involved in as a driver last year is an example of a discrete random variable.
ANSWER: T

7. The various values of a random variable form a list of mutually exclusive events.
ANSWER: T

8. A random variable is a variable that assumes a unique numerical value for each of the outcomes in the sample space of a probability experiment.
ANSWER: T

9. Continuous random variable is a quantitative random variable that can assume a countable number of values.
ANSWER: F

10. Numerical random variables can be subdivided into two classifications: discrete random variables and continuous random variables.
ANSWER: T

11. Discrete random variable is a qualitative random variable that can assume an uncountable number of values.
ANSWER: F

Multiple-Choice Questions

12. Which of the following probability experiments would result in a discrete random variable?

A) Observing the number of minutes required to walk a mile
B) Observing the number of light bulbs burned out on a display sign
C) Observing the number of inches tall of second grade students
D) Observing the number of pounds in each of 15 bags of apples
ANSWER: B

13. Which of the following would *not* be a continuous random variable?

A) Age of student upon graduation from college.
B) Number of attempts to make a field goal in football.
C) Number of miles driven on a trip.
D) Body temperature of small children.
ANSWER: B

14. Which of the following statements is false?

A) Discrete random variable is a quantitative random variable that can assume each countable number of values.
B) Continuous random variable is a quantitative random variable that can assume an uncountable number of values.
C) A random variable is called "random" because the value it assumes is the result of a chance, or random event.
D) The length of the cord on an electrical appliance is an example of a discrete random variable.
ANSWER: D

Short-Answer Questions

15. Classify the following as discrete or continuous random variables: The weight of bags of apples, with 10 apples in each bag.

ANSWER:
Continuous

16. Classify the following as discrete or continuous random variables: The number of times required for a modem to dial an internet provider before connecting.

ANSWER:
Discrete

17. Classify the following as discrete or continuous random variables: Out of 10 times connecting to an internet provider, the average number of attempts necessary before connecting.

ANSWER:
Continuous

18. Classify the following as discrete or continuous random variables: A pair of dice is rolled, and the sum to appear on the dice is recorded.

ANSWER:
Discrete

19. A bag contains nickels, dimes, and quarters (more than two of each). Two coins are randomly selected and their total value is noted. Describe what the random variable x represents.

ANSWER:
The random variable x represents the total value of the two coins.

20. A bag contains nickels, dimes, and quarters (more than two of each). Two coins are randomly selected and their total value in cents is noted. Find the possible values of the random variable x.

ANSWER:
The possible values of x are 10, 15, 20, 30, 35, and 50.

21. In order to monitor the quality of a production process, samples of size five are selected daily. The random variable of interest is the number of defectives in the five items selected. What values are possible for this random variable?

ANSWER:
0, 1, 2, 3, 4, or 5

22. A bridge hand of 13 cards is dealt from a standard deck. Let x represents the number of clubs in the hand. What values are possible for x?

ANSWER:
The possible values for x are whole numbers from 0 through 13, inclusive.

23. From census data, a census worker obtains information regarding the number of cars per family for a certain community in Indiana. Identify the random variable of interest, determine whether it is discrete or continuous, and list its possible values.

    ANSWER:
    The random variable is: number of cars per family. It is discrete with possible values: 0,1,2,3 ... *n*.

24. Is the distance you travel from home to school discrete or continuous random variable? Explain.

    ANSWER:
    The distance your travel from home to school is a continuous random variable, since it can assume an uncountable number of numerical values. In other words, distance is a measurement and can assume any value along a line interval including all possible fractions.

25. Is the number of textbooks you bought this semester discrete or continuous random variable? Explain.

    ANSWER:
    The number of textbooks you bought this semester is a discrete random variable, since it can only assume a countable number of numerical values. A value of 2.75, for example, would not make sense.

Applied and Computational Questions

QUESTIONS 26 AND 27 ARE BASED ON THE FOLLOWING INFORMATION:
Pairs of random numbers (x, y) are integers between 0 and 5 inclusive.

26. How many different pairs are possible?

    ANSWER:
    36

27. Suppose a random variable W is defined to equal the absolute value of the difference between x and y. How many distinct values are possible for W?

    ANSWER:
    6

QUESTIONS 28 AND 29 ARE BASED ON THE FOLLOWING INFORMATION:
"Are you getting a summer job?" A recent study reported that 68% of college students answered, "I have one"; 22% said "Maybe" and 10% said "No".

28. What is the variable involved, and what are the possible values?

    ANSWER:
    The variable is: summer job status; with 3 possible values: have one, maybe, no.

29. Is the variable in question 28 a random variable? Explain.

    ANSWER:
    This is not a random variable since it is an attribute (qualitative or categorical) variable.

QUESTIONS 30 AND 31 ARE BASED ON THE FOLLOWING INFORMATION:
Survey your friends about the number of siblings they have and the length of the last phone call they had with their boyfriend / girlfriend.

30. Identify the two random variables of interest and list their possible values.

ANSWER:
First variable is: number of siblings that a friend has, with possible values: 0, 1, 2, ..., *n*. Second variable: length of last phone call to boyfriend / girlfriend, with possible values: 0 to any number (e.g., 36, 52, 81, ...) and/or any number including fractions (e.g., 41.67, 59.04, 75.92, ...)

31. The two variables in question 30 are either discrete or continuous. Which are they and why?

ANSWER:
Number of siblings that a friend has is a discrete random variable, since it can only assume a countable number of numerical values. A value of 1.68, for example, would not make sense.

Length of last phone call to boyfriend / girlfriend is a continuous random variable, since it can assume an uncountable number of numerical values. In other words, length is a measurement and can assume any value along a line interval including all possible fractions.

Sections 5.3 and 5.4

True-False Questions

32. The histogram of a probability distribution uses the physical area of each bar to represent its assigned probability.
ANSWER: T

33. The mean, μ , of a discrete random variable *x* is found by multiplying each possible value of *x* by its own probability and then adding all the products together; that is, $\mu = \sum[xP(x)]$.
ANSWER: T

34. For every discrete random variable *x*, the variance is given by the formula: $\sigma^2 = npq$.
ANSWER: F

35. The formula $\mu = np$ may be used to find the mean of any discrete random variable *x*.
ANSWER: F

36. The sum of all probabilities in any discrete probability distribution is not always exactly one, since some of the probabilities may be slightly larger than one.
ANSWER: F

37. The sum of all the probabilities in any probability distribution is always exactly one.
ANSWER: T

38. A parameter is a statistical measure of some aspect of a population.
ANSWER: T

39. Sample statistics are represented by letters from the Greek alphabet.
ANSWER: F

40. The probability of event *A* or *B* is equal to the sum of the probability of event *A* and the probability of event *B* when *A* and *B* are mutually exclusive events.
ANSWER: T

41. Probability distribution of a discrete random variable is a distribution of the probabilities associated with each of the values the random variable can assume.
ANSWER: T

42. Regardless of the specific graphic representation of the probability distribution of a discrete random variable *x*, the values of *x* are plotted on the horizontal scale, and the probability *P*(*x*) associated with each value of *x* is plotted on the vertical scale.
ANSWER: T

43. A probability function is a rule that assigns probabilities to the values of the random variable of interest.
ANSWER: T

44. The formula $\mu = np$ may be used to compute the mean of many discrete populations.
ANSWER: F

45. $P(x) = x / 20$ for $x = 1, 2, 3, 4$, and 5 is a probability function of a discrete random variable x.
ANSWER: F

46. A probability function provides a probability of zero for all values of the random variable x other than the values specified as part of the domain.
ANSWER: T

47. A probability distribution of a discrete random variable x can be presented as a mathematical function but, unfortunately, it cannot be presented graphically.
ANSWER: F

48. The mean of the probability distribution of a discrete random variable, or the mean of a discrete random variable, is found in a manner somewhat similar to that used to find the mean of a frequency distribution.
ANSWER: T

49. The mean, μ, of a discrete random variable x is found by adding all possible values of x and dividing the total by the number of values that x assumes.
ANSWER: F

50. The mean of a discrete random variable is often referred to as its expected value.
ANSWER: T

51. The sum of all the probabilities in any probability distribution is always exactly 1.25.
ANSWER: F

Multiple-Choice Questions

52. Given that the numbers 1 through 6 are equally likely to occur, what is $P(x \leq 2)$?

A) Cannot be determined since we do not know the probability for each number.
B) 1/2
C) 1/3
D) 1/6
ANSWER: C

53. Consider the probability function $P(x) = \frac{6 - |x - 7|}{36}$ for $x = 2, 3, 4, 5, \ldots, 12$. Find the probability that x takes values between 6 and 8 (not inclusive).

A) 5/36
B) 6/36
C) 10/36
D) 16/36
ANSWER: B

Chapter 5

54. Consider the data in the table. Which answer is *not* true?

x	$P(x)$
1	0.60
2	0.20
3	0.15
4	0.05

A) This is a probability distribution.
B) The histogram of this distribution is skewed to the right.
C) The random variable is discrete.
D) $P(x \leq 3) = 0.15$
ANSWER: D

55. A ball is drawn from a box containing three balls, one red, one blue, and one green. The ball is returned and a second ball is drawn. A tree diagram is drawn to give the outcomes of the experiment with respect to the colors of the two balls. If x represent the number of red balls in the two selected, how many branches are assigned the value of $x = 1$?

A) 1
B) 2
C) 3
D) 4
ANSWER: D

56. A tree diagram is constructed for the experiment of tossing a coin three times. If x represents the number of tails in the three tosses, how many branches are assigned the value $x = 3$?

A) 0
B) 1
C) 2
D) 3
ANSWER: B

57. Which of the following statements is false?

A) The mean, μ, of a discrete random variable x is found by multiplying each possible value of x by its own probability and then adding all the products together.
B) The variance, σ^2, of a discrete random variable x is found by multiplying the square of each possible value of x by its own probability and then adding all the products together.
C) The variance, σ^2, of a discrete random variable x is found by multiplying each possible value of the squared deviation from the mean, $(x-\mu)^2$, by its own probability and then adding all the products together.
D) None of the above.
ANSWER: B

58. Which of the following statements is true?

A) A probability distribution of a discrete random variable x cannot be presented graphically.
B) A probability distribution of a discrete random variable x can be presented graphically as a line graph or probability histogram.
C) $P(x) = x / 12$ for $x = 1, 2, 3$, and 4 is a probability function of a discrete random variable x.
D) None of the above

ANSWER: **B**

Short-Answer Questions

59. Determine the value of the constant c in the following probability function: $P(x) = c$ for $x = 1, 2, 3, 4, 5$.

ANSWER:
$c = 0.20$

60. The values of a random variable x have a uniform probability distribution. If the random variable x has the values of 0, 1, 2, 3, and 4, what is the probability that the value of x is less than 2?

ANSWER:
0.40

61. Is $F(x) = \frac{4 - x}{5}$ for $x = 1, 2, 3, 4$, and 5 a probability function? Give a short explanation by writing a sentence or two.

ANSWER:
Since $F(5) = -1/5$, this is not a probability function. Probabilities can never be negative.

62. Explain why the following statement is false: "The mean of a probability distribution of a discrete random variable always has a value equal to one of the values of the random variable".

ANSWER:
Although variables are discrete, very likely the mean could be a non-discrete value and therefore, not equal to one of the variables.

63. In an experiment in which a single die is rolled once and the number of dots on the top surface is observed, let the random variable x represent the number observed. Find the probability distribution for x.

ANSWER:
$P(x) = 1/6$ for $x = 1, 2, 3, 4, 5, 6$

64. Determine the value of the constant c in the following probability function: $P(x) = c$ for $x =$ 0, 1, 2, 3.

ANSWER:
$c = 0.25$

65. A probability distribution has a mean equal to 10 and a standard deviation equal to 2. Find $\sum x^2 P(x)$.

ANSWER:
104

66. A probability distribution has a mean equal to 8 and a standard deviation equal to 5. Find $\sum x^2 P(x)$.

ANSWER:
89

67. Hope and Mike were discussing one entry in a probability distribution: $P(x) = 0.5$ when $x = -3$. Hope felt that this entry was okay since the $P(x)$ was a value between 0.0 and 1.0. Mike argued that this entry was impossible for a probability distribution since $x = -3$, and negative values are not possible in probability distributions. Who is correct, Hope or Mike? Justify your choice.

ANSWER:
Hope is correct, since negative values of x are possible but $P(x)$ must be a value between 0.0 and 1.0, since probabilities cannot be negative for any probability distribution.

68. Express the tossing of two coins as a probability distribution of x, the number of heads occurring.

ANSWER:

x	P(x)
0	0.25
1	0.50
2	0.25

69. Explain how the various values of x in a probability distribution form a set of mutually exclusive events?

ANSWER:
Each unique outcome is assigned a specific numerical value. In other words, the values of x in a probability distribution can never overlap.

70. Explain how the various values of x in a probability distribution form a set of "all inclusive" events.

ANSWER:
All possible outcomes are accounted for.

Applied and Computational Questions

71. Let x represent the number of times a one appears when a pair of dice is rolled once. Give the probability distribution for x.

ANSWER:
$P(0) = 0.694$, $P(1) = 0.278$, $P(2) = 0.028$

72. A card is selected from a standard deck of 52. Random variable x is defined to be 0, if an ace occurs; 1, if a two through ten occurs; and 2, if a face card (Jack, Queen, or King) occurs. Give the probability distribution for x.

ANSWER:
$P(0) = 0.077$, $P(1) = 0.692$, $P(2) = 0.231$

73. A small bag of M&M candies has the following assortment: red (10), blue (2), orange (5), brown (21), green (0), and yellow (18). Give the probability distribution for x.

ANSWER:
$P(\text{red}) = 0.185$; $P(\text{blue}) = 0.137$; $P(\text{orange}) = 0.093$; $P(\text{brown}) = 0.389$; $P(\text{green}) = 0.0$; $P(\text{yellow}) = 0.333$

74. Consider the function $T(x) = \frac{x+k}{12}$ for $x = 1, 2, 3, 4$. Find all values of k which make the function T a probability function.

ANSWER:
$k = 0.5$

75. Find all values of k so that the following is a probability distribution:

x	$P(x)$
1	0.15
2	$2k$
3	0.52
4	k

ANSWER:
$3k = 1 - (0.15 + 0.52) = 0.33$, which implies that $k = 0.11$

76. The function $P(x) = \frac{1+(x-3)^2}{10}$ for $x = 1$, 2, 3, and 4 is a probability function. Find the mean and standard deviation of this distribution.

ANSWER:
$\mu = 2.0$ and $\sigma = 1.2$

Chapter 5

77. Find the mean and standard deviation of the following probability distribution:

x	$P(x)$
1	0.3
2	0.5
3	0.2

ANSWER:
$\mu = 1.9$ and $\sigma = 0.7$

78. Compare the standard deviations of the following two probability distributions, both of which have a mean equal to 5.

Distribution A:

x	4	5	6
$P(x)$	0.1	0.8	0.1

Distribution B:

x	1	2	3	4	5	6	7	8	9
$P(x)$	0.05	0.05	0.1	0.2	0.2	0.2	0.1	0.05	0.05

ANSWER:
Standard deviation for distribution A = 0.45, Standard deviation for distribution B = 1.92

79. A probability distribution has a standard deviation equal to 2.5 and $\sum x^2 P(x) = 10.25$. Find the mean for this distribution.

ANSWER:
Mean = 2 or -2

80. Find the amount of the probability distribution within two standard deviations of the mean for rolling a pair of dice and observing the sum. Compare this with the bound given by Chebyshev's Theorem.

ANSWER:
94.4% of distribution are within 2 standard deviations σ of the mean μ. Chebyshev's Theorem: at least 75% of distribution within 2σ of μ

81. An arsenal contains several identical boxes of ammunition. If the number of defective bullets per box has the following distribution, find the mean and standard deviation for x.

x	0	1	2
$P(x)$	0.90	0.07	0.03

ANSWER:
Mean = 0.13, Standard deviation = 0.42

QUESTIONS 82 AND 83 ARE BASED ON THE FOLLOWING INFORMATION:
The following is a probability distribution.

x	$P(x)$
1	0.25
2	0.25
3	0.25
4	0.25

82. Find the mean and standard deviation of the probability distribution.

ANSWER:
$\mu = 2.5$ and $\sigma = 1.12$

83. Explain why it is a uniform distribution.

ANSWER:
This is a uniform distribution since the probability is the same for all possible values of x.

84. Census data for families with a combined income of $60,000 or more in Michigan show that 25% have no children, 30% have one child, 35% have two children, and 10% have three children. From this information, construct the probability distribution for x, where x represents the number of children per family for this income group.

ANSWER:

x	0	1	2	3
$P(x)$	0.25	0.30	0.35	0.10

85. Test the following function to determine whether it is a probability function.

$$P(x) = \frac{x^2 + 5}{80}; \quad \text{for } x = 1, 2, 3, 4, \text{ or } 5.$$

ANSWER:

x	$P(x)$
1	0.0750
2	0.1125
3	0.1750
4	0.2625
5	0.3750

Notice that each $P(x)$ is a value between 0.0 and 1.0, and the sum of all $P(x)$ values is exactly 1.0. Therefore, $P(x)$ is a probability function.

86. Given the probability function $P(x) = (6-x)/15$, for x = 1, 2, 3, 4, or 5. Find the mean and standard deviation.

ANSWER:

x	$P(x)$	$xP(x)$	$x^2P(x)$
1	5/15	5/15	5/15
2	4/15	8/15	16/15
3	3/15	9/15	27/15
4	2/15	8/15	32/15
5	1/15	5/15	25/15
$\sum$	1.0	35/15	105/15

$\mu = \sum[xP(x)] = 35/15 = 2.333$

$\sigma^2 = \sum[x^2P(x)] - (\sum[xP(x)])^2 = 105/15 - (35/15)^2 = 1.556$

$\sigma = \sqrt{\sigma^2} = \sqrt{1.556} = 1.247$

QUESTIONS 87 AND 88 ARE BASED ON THE FOLLOWING INFORMATION:
The random variable $\overline{x}$ has the following probability distribution.

$\overline{x}$	1	2	3	4	5
$P(\overline{x})$	0.5	0.2	0.1	0.1	0.1

87. Find the mean and standard deviation of $\overline{x}$.

ANSWER:

$\overline{x}$	$P(\overline{x})$	$\overline{x}P(\overline{x})$	$\overline{x}^2P(\overline{x})$
1	0.5	0.5	0.5
2	0.2	0.4	0.8
3	0.1	0.3	0.9
4	0.1	0.4	1.6
5	0.1	0.5	2.5
Sum	1.0	2.1	6.3

$\mu = \sum[\overline{x}P(\overline{x})] = 2.1$

$\sigma^2 = \sum[\overline{x}^2P(\overline{x})] - (\sum[\overline{x}P(\overline{x})])^2 = 6.3 - (2.1)^2 = 1.89$

$\sigma = \sqrt{\sigma^2} = \sqrt{1.89} = 1.3748$

88. What is the probability that $\bar{x}$ is between $\mu-\sigma$ and $\mu+\sigma$?

ANSWER:
$\mu-\sigma$ = 2.1 – 1.3748 = 0.7252, and $\mu+\sigma$ = 2.1 + 1.3748 = 3.4748. The interval from 0.7252 to 3.4748 encompasses the number 1, 2, and 3. The total probability associated with these values of $\bar{x}$ is 0.8.

QUESTIONS 89 THROUGH 94 ARE BASED ON THE FOLLOWING INFORMATION:
Consider the following function: *H*(*x*) = 0.25 for *x* = 1, 2, 3, and 4.

89. Express *H*(*x*) in distribution form.

ANSWER:

x	H(x)
1	0.25
2	0.25
3	0.25
4	0.25

90. Determine whether *H*(*x*) is a probability function.

ANSWER:
Since $0 \le \text{each } H(x) \le 1$ and $\sum_{\text{all outcomes}} H(x) = 1$, the function *H*(*x*) is a probability function.

91. Sketch a histogram of the probability distribution in question 89.

ANSWER:

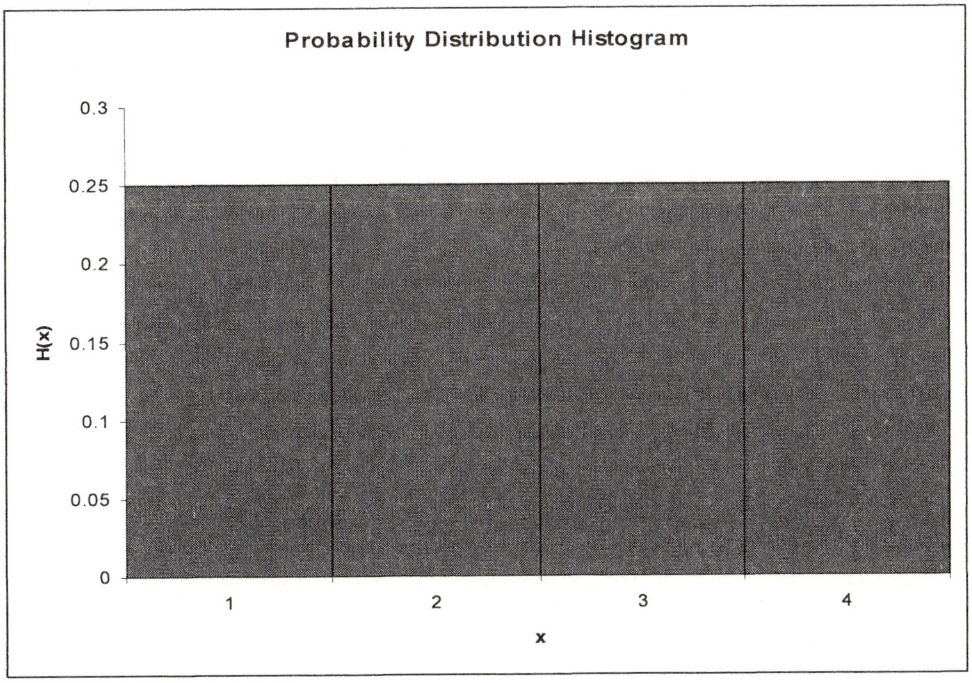

92. Describe the shape of the histogram in question 91.

ANSWER:
It is uniform or rectangular.

93. Determine the mean of the probability distribution in question 89.

ANSWER:
$\mu = \sum x \cdot H(x) =$ (1)(0.25) + (2)(0.25) + (3)(0.25) + (4)(0.25) = 2.5

94. Determine the variance and standard deviation of the probability distribution in question 89.

ANSWER:
$\sigma^2 = \sum x^2 \cdot H(x) - \mu^2 = (1)(0.25) + (4)(0.25) + (9)(0.25) + (16)(0.25) - (2.5)^2 = 1.25$
$\sigma = \sqrt{\sigma^2} = \sqrt{1.25} = 1.118$

QUESTIONS 95 THROUGH 99 ARE BASED ON THE FOLLOWING INFORMATION:
Consider the following function $P(x) = (x^2 + 5)/50$, for x = 1, 2, 3, or 4.

95. Express *P*(*x*) in distribution form.

ANSWER:

x	P(x)
1	0.12
2	0.18
3	0.28
4	0.42

96. Determine whether *P*(*x*) is a probability function.

ANSWER:
Since $0 \le \text{each } P(x) \le 1$ and $\sum_{\text{all outcomes}} P(x) = 1$, the function *P*(*x*) is a probability function.

97. Determine the mean of the probability function in question 95.

ANSWER:
$\mu = \sum x \cdot P(x) =$ (1)(0.12) + (2)(0.18) + (3)(0.28) + (4)(0.42) = 3.0

98. Determine the standard deviation of the probability function question 95.

ANSWER:
$\sigma^2 = \sum x^2 \cdot P(x) - \mu^2 = (1)(0.12) + (4)(0.18) + (9)(0.28) + (16)(0.42) - (3)^2 = 1.08$
$\sigma = \sqrt{\sigma^2} = \sqrt{1.08} = 1.039$

99. Sketch a histogram of the probability function in question 95.

ANSWER:

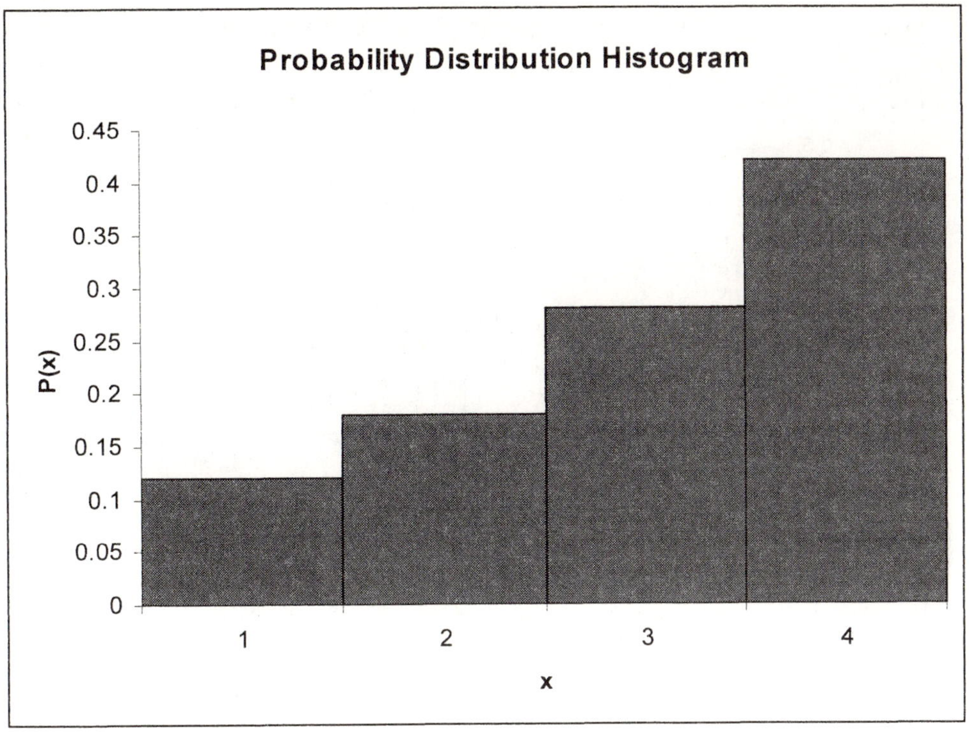

QUESTIONS 100 THROUGH 104 ARE BASED ON THE FOLLOWING INFORMATION:
Consider the following discrete probability distribution.

x	1	2	3	4	5
$P(x)$	0.30	0.20	0.25	0.15	0.10

100. Use a computer (or random numbers table) to generate a random sample of 25 observations drawn from the discrete probability distribution.

ANSWER:
Everyone's generated values will be different. Listed here is one such sample.

3	1	3	1	3	2	4	2	5	1
1	2	3	2	2	4	1	5	4	5
4	1	1	1	3					

101. Form a relative frequency distribution of the observed data (generated random data).

ANSWER:

x	1	2	3	4	5
Relative Frequency.	0.32	0.20	0.20	0.16	0.12

102. Construct a probability histogram of the given distribution.

ANSWER:

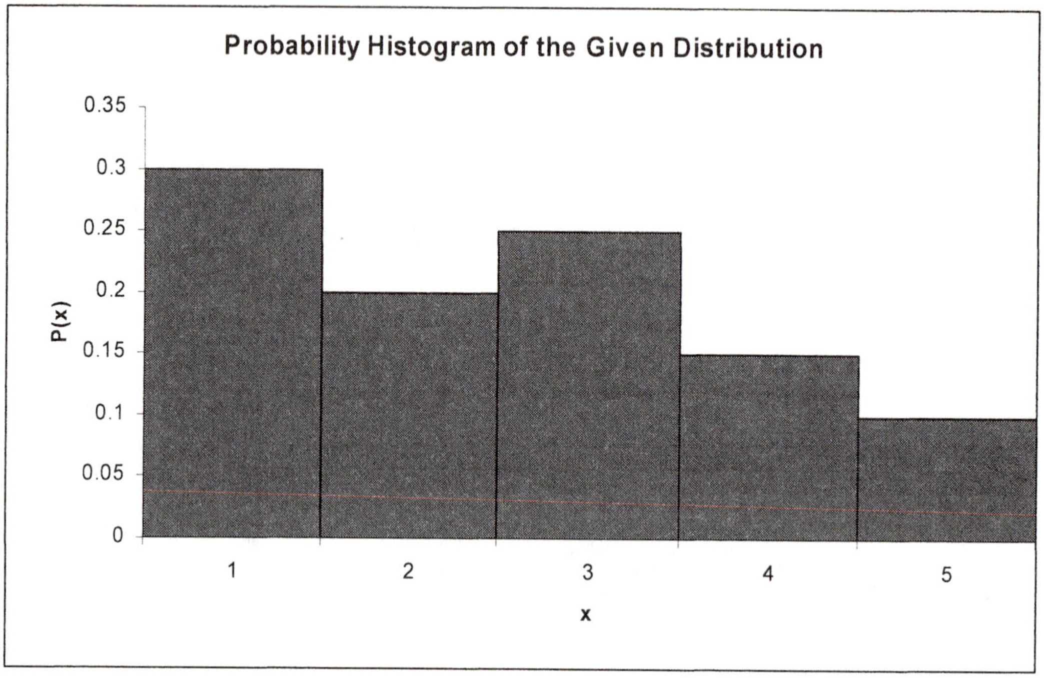

103. Construct a relative frequency histogram of the observed data using class marks of 1, 2, 3, 4, and 5.

ANSWER:

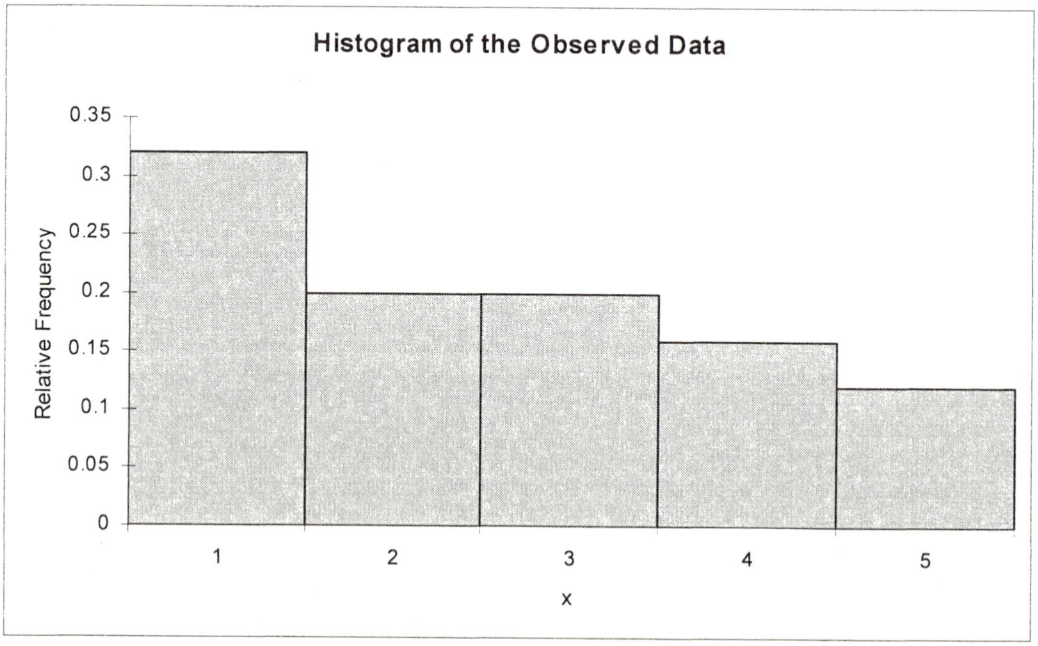

104. Compare the observed data with the theoretical distribution. Describe your conclusions.

ANSWER:
The distribution of the sample is somewhat similar to that of the given distribution. The two highest probabilities in the random data occurred at x = 1 and 3, matching the two highest probabilities for the given distribution. Also, the two lowest probabilities in the random data occurred at x = 4 and 5, matching the two lowest probabilities for the given distribution. Finally, the probability in the random data occurred at x = 2 is identical to that for the given distribution.

QUESTIONS 105 THROUGH 109 ARE BASED ON THE FOLLOWING INFORMATION:
Consider the following probability function $P(x) = x / 15$ for x =1, 2, 3, 4, or 5.

105. Form the probability distribution table.

ANSWER:

X	1	2	3	4	5
P(x)	1/15	2/15	3/15	4/15	5/15

106. Find $\sum[xP(x)]$ and $\sum[x^2P(x)]$.

ANSWER:
$\sum[xP(x)]$ = 3.667 and $\sum[x^2P(x)]$ = 15

107. Find the mean of the probability distribution.

ANSWER:
$\mu = \sum[xP(x)] = 3.667$

108. Find the variance of the probability distribution.

ANSWER:
$\sigma^2 = \sum[x^2P(x)] - \mu^2 = 15 - (3.667)^2 = 1.553$

109. What is the standard deviation of the probability distribution?

ANSWER:
$\sigma = \sqrt{\sigma^2} = \sqrt{1.553} = 1.246$

QUESTIONS 110 THROUGH 114 ARE BASED ON THE FOLLOWING INFORMATION:
Consider the following probability function $P(x) = x / 10$, for x = 1, 2, 3, 4.

110. Form the probability distribution table for $P(x)$.

ANSWER:

X	1	2	3	4
P(x)	0.1	0.2	0.3	0.4

111. Find $\sum[xP(x)]$ and $\sum[x^2P(x)]$.

ANSWER:
$\sum[xP(x)]$ = 3.0 and $\sum[x^2P(x)]$ = 10.0

112. Find the mean of the probability function.

ANSWER:
$\mu = \sum[xP(x)] = 3.0$

113. Find the variance of the probability function.

ANSWER:
$\sigma^2 = \sum[x^2P(x)] - \mu^2 = 10.0 - 9.0 = 1.0$

114. Find the standard deviation of the probability function.

ANSWER:
$\sigma = \sqrt{\sigma^2} = 1.0$

QUESTIONS 115 THROUGH 120 ARE BASED ON THE FOLLOWING INFORMATION:
The number of credits that full-time college students take on any given semester is a random variable represented by *x*. The probability distribution for *x* is

x	12	13	14	15	16
P(x)	0.4	0.2	0.2	0.1	0.1

115. Find the mean of the number of credits that full-time college students take in a given semester.

ANSWER:
$\mu = \sum[xP(x)] = 13.3$

116. Find the standard deviation of the number of credits that full-time college students take on a given semester.

ANSWER:
The variance $\sigma^2 = \sum[x^2P(x)] - \mu^2 = 178.7 - (13.3)^2 = 1.81$, then the standard deviation

$\sigma = \sqrt{\sigma^2} = 1.345$

117. How much of the probability distribution is within two standard deviations of the mean?

ANSWER:
$\mu \pm 2\sigma = 13.3 \pm 2(1.345) = (10.61, 15.99)$
The interval from 10.61 to 15.99 encompasses the values 12, 13, 14, and 15.

118. How much of the probability distribution is within one standard deviations of the mean?

ANSWER:
$\mu \pm \sigma = 13.3 \pm 1.345 = (11.955, 14.645)$
The interval from 11.955 to 14.645 encompasses the values 12, 13, and 14.

119. Find $P(\mu - 2\sigma \leq x \leq \mu + 2\sigma)$.

ANSWER:
$P(\mu - 2\sigma \leq x \leq \mu + s\sigma) = P(x = 12, 13, 14, 15) = 0.90$

120. Find $P(\mu - \sigma \leq x \leq \mu + \sigma)$.

ANSWER:
$P(\mu - \leq x \leq \mu + \sigma) = P(x = 12, 13, 14) = 0.80$

Sections 5.5 and 5.6

True-False Questions

121. The histogram for a binomial distribution that has a success probability close to one will be skewed to the right, and the histogram for a binomial distribution that has a success probability close to zero will be skewed to the left.
ANSWER: F

122. It is possible to obtain eight successes in a binomial probability experiment with six trials, provided the probability of a success on a single trial is greater than 0.5.
ANSWER: F

123. A binomial experiment always has at least three possible outcomes to each trial.
ANSWER: F

124. The binomial random variable *x* is the count of the number of successful trials that occur in *n* repeated (identical) independent trials; *x* may take on any integer value from 0 to *n*.
ANSWER: T

125. In any binomial probability experiment, independent trials mean that the result of one trial does not affect the probability of success of any other trial in the experiment.
ANSWER: T

126. The binomial coefficient $\binom{n}{x} = \frac{n!}{x!(n-x)!}$ is equivalent to the number of combinations, ${}_nC_x$, the symbol most likely on your calculator.
ANSWER: T

127. A binomial experiment always has three or more possible outcomes to each trial.
ANSWER: F

128. The formula $\mu = np$ may be used to compute the mean of a binomial distribution.
ANSWER: T

129. The binomial parameter *p* is the probability of one success occurring in *n* trials when a binomial experiment is performed, while 2*p* is the probability of two successes.
ANSWER: F

130. A convenient notation to identify the binomial probability distribution for a binomial experiment with *n* = 20 and *p* = 0.25 is *B*(20, 0.25).
ANSWER: T

131. The binomial random variable *x* is the count of the number of successful trials that occur in *n* trials. The random variable *x* may take on any real value from zero to *n*.
ANSWER: F

132. Each trial in a binomial probability experiment has two possible outcomes (success, failure) and that *P*(success) + *P*(failure) = 1.
ANSWER: T

133. A binomial experiment always has two or more possible outcomes to each trial.
ANSWER: F

134. The binomial parameter p is the possibility of one success occurring in n trials when a binomial experiment is performed.
ANSWER: F

135. The number of ways that exactly x can occur in a set of n trials is represented by the symbol $\binom{n}{x}$, which must always be a positive integer. This term is called the binomial coefficient and is found by using the formula $\binom{n}{x} = \frac{n!}{x!(n-x)!}$.

ANSWER: T

136. The number of hours you waited in line to register this semester is an example of a binomial random variable.
ANSWER: F

Multiple-Choice Questions

137. If a student inadvertently interchanged the values of p and q in a binomial probability experiment, which of the following would give the probability of x successes?

A) $\binom{n}{x} p^x q^{n-x}$

B) $\binom{n}{n-x} p^x q^{n-x}$

C) $\binom{n}{x} p^{n-x} q^x$

D) $\binom{x}{n} pq$

ANSWER: C

138. In a binomial probability experiment with P(success) = p, P(failure) = q, and eight trials, what is the probability of three successes?

A) $5p^3q^5$

B) $5p^5q^3$

C) $56p^3q^5$

D) $56p^5q^3$

ANSWER: C

Chapter 5

139. The binomial coefficient $\binom{10}{3}$ equals which of the following?

A) 10! / 3!
B) 120
C) 720
D) 30
ANSWER: B

140. Which of the following is a characteristic of a binomial probability experiment?

A) Each trial has at least two possible outcomes.
B) *P*(success) = *P*(failure)
C) The binomial random variable *x* is the count of the number of trials that occur.
D) The result of one trial does not affect the probability of success on any other trial.
ANSWER: D

141. Which of the following is *not* true regarding a binomial distribution for $n = 50$ and $p = 0.4$?

A) The mean equals 25.
B) The variance equals 0.24.
C) The highest probability occurs for $x = 50$.
D) The distribution is not symmetrical.
ANSWER: D

142. If a tree diagram is drawn for a binomial experiment having *n* trials, how many branches will it have?

A) 2^n
B) $2n$
C) n^2
D) Need to know the value of *n* before number of branches can be determined.
ANSWER: A

143. For a binomial distribution with five trials and equal probability of success per trial, what is the highest probability?

A) 0.2
B) 0.2%
C) 5%
D) 1
ANSWER: A

144. Suppose that the value of *n* in a binomial distribution is fixed, but we let the value of *p* vary. As the value of *p* increases from values near 0 to values close to 1, what conclusion can be made about the mean of the distribution?

A) The mean will decrease in value and become closer in value to 0.
B) The mean will increase in value and become closer in value to *n*.
C) The mean will not change in value.
D) No conclusion can be made about the value of the mean.
ANSWER: B

145. If a tree diagram is drawn for a binomial experiment having 4 trails, how many branches will it have?

A) 2
B) 4
C) 8
D) 16
ANSWER: D

Short-Answer Questions

146. Given a binomial probability experiment with six trials, in how many ways can we obtain two successes?

ANSWER:
We can obtain two successes in 15 ways

147. For a particular binomial distribution with $n = 4$, $P(0) = P(1)$. Find p.

ANSWER:
$p = 0.20$

148. For a particular binomial distribution, $n = 4$. If $P(2) = 0.346$ and $P(3) = 0.154$, find p.

ANSWER:
$p = 0.40$

149. How many times must a fair coin be flipped in order that the mean number of heads equals 25?

ANSWER:
50

150. For a particular binomial distribution, $n = 28$ and $p = 0.35$. For this distribution, find $\sum x \cdot P(x)$.

ANSWER:
Since $\Sigma xP(x) = \mu$ and $\mu = np = (28)(0.35) = 9.8$, then $\Sigma xP(x) = 9.8$.

151. For a particular binomial experiment, $n = 18$ and $p = 0.7$. For this experiment, find the value of $\sum[x \cdot P(x)]$.

ANSWER:
12.6

152. A particular binomial distribution is given by $P(x) = \binom{100}{x}(0.2)^x(0.8)^{100-x}$, for $x = 0$, 1, 2, 3, ……, 100. Find the mean and standard deviation of this distribution.

ANSWER:
$\mu = 20$ and $\sigma = 4.0$

Chapter 5

153. Briefly define a binomial probability experiment and discuss its properties.

ANSWER:
A binomial probability experiment is an experiment that is made up of repeated trials that possess the following properties:
(a) There are *n* repeated identical independent trials.
(b) Each trial has two possible outcomes (success, failure).
(c) P(success) = p, P(failure) = q and $p + q = 1$.
(d) The binomial random variable x is the count of the number of successful trials that occur; x may take on any integer value from zero to n.

154. State a very practical reason why the defective item in an industrial situation would be defined to be the "success" in a binomial experiment.

ANSWER:
The number of defective items in an industrial situation should be fairly small and therefore easier to count.

155. A carton containing 75 towels is inspected. Each towel is rated "first quality" or "irregular". After all 100 towels are inspected, the number of irregulars is reported as a random variable. Explain why x is a binomial variable

ANSWER:
x is a binomial variable since it satisfies the following properties:
n = 75 repeated identical independent trials (towels), there are only two outcome (first quality, irregular), p =P(success) = P(irregular), x = number of irregular towels that may take on any integer value from 0 to 75.

156. The employees at a Ford assembly plant are polled as they leave work. Each is asked, "What brand of automobile are you riding home in?" The random variable to be reported is the number of each brand mentioned. Is x a binomial random variable? Justify your answer.

ANSWER:
x is not a binomial random variable because there are more than two categories of outcomes. As the exercise is stated, each different brand (or make) of automobile is an outcome; therefore there are many different possible outcomes on each trial.

157. Four cards are selected, one at a time, from a standard deck of 52 cards. Let x represent the number of jacks drawn in the set of four cards. If this experiment is completed without replacement, explain why x is not a binomial random variable.

ANSWER:
x is not a binomial random variable because the trials are not independent. The probability of success (get a jack) changes from trial to trial. On the first trial it is 4 / 52. The probability of a jack on the second trial depends on the outcome of the first trial; it is 4 / 51 if a jack is not selected, and it is 3 / 51 if a jack was selected. The probability of a jack on any given trial continues to change when the experiment is completed without replacement.

158. Four cards are selected, one at a time, from a standard deck of 52 cards. Let x represents the number of queens drawn from the set of four cards. If this experiment is completed without replacement, explain why x is not a binomial random variable.

ANSWER:
x is a binomial random variable because the trials are independent. n = 4, the number of independent trials; two outcomes, success = queen and failure = not queen; p = P(queen) = 4/ 52 and q = P(not queen) = 48 / 52; x = number of queens drawn in 4 trials, and could be any integer number 0, 1, 2, 3 or 4. Further, the probability of success (get a queen) remains 4 / 52 for each trial throughout the experiment, as long as the card drawn on each trial is replaced before the next trial occurs.

159. Find the mean and standard deviation of x = number of heads seen in 100 tosses of a quarter.

ANSWER:
x is binomial random variable with n = 100 and p = 0.5. Then, the mean $\mu = np = 50$ and standard deviation $\sigma = \sqrt{npq} = \sqrt{(100)(0.5)(0.5)} = 5.0$

Applied and Computational Questions

160. Let x represent the flip upon which a head first occurs when a coin is flipped repeatedly. Find the probability that x is equal to or greater than 4.

ANSWER:
0.875

161. Thirty percent of hospital admissions for diabetic patients are related to problems with the kidneys. In a sample of 10 diabetic hospital admissions, what is the probability that none will be for a kidney problem?

ANSWER:
0.028

162. A manufacturer of matches puts 100 matches in each box of matches produced. One-tenth of one percent of the matches produced has flaws. If a box is randomly selected, what is the probability that it will have one or fewer matches with a flaw?

ANSWER:
0.995

QUESTIONS 163 AND 164 ARE BASED ON THE FOLLOWING INFORMATION:
In testing a new drug, researchers found that 5% of all patients using it will have a mild side effect. A random sample of 11 patients using the drug is selected.

163. Find the probability that exactly two will have this mild side effect.

ANSWER:
0.087

164. Find the probability that at least one will have this mild side effect.

ANSWER:
0.431

165. A quality control inspector has determined that 0.25% of all parts manufactured by a particular machine are defective. If 50 parts are randomly selected, find the probability that there will be at most one defective part.

ANSWER:
0.9930

166. A multiple-choice test has 30 questions each with five responses, one of which is correct. The lowest passing grade is 18. Find the probability of obtaining this grade by random guessing. Write your answer to seven decimal places.

ANSWER:
0.0000016

167. A fair die is rolled 10 times. Compute the probability that a "one" appears exactly once.

ANSWER:
0.323

168. If two dice are tossed six times, find the probability of obtaining a sum of 7 two or three times.

ANSWER:
0.255

QUESTIONS 169 AND 170 ARE BASED ON THE FOLLOWING INFORMATION:
Consider the probability distribution for *x*, the number of heads to occur when a coin is tossed four times.

x	0	1	2	3	4
P(*x*)	0.0625	0.250	0.375	0.250	0.0625

169. Find $\sum x^2P(x)-4$ and $np(1-p)$, and explain why they are equal.

ANSWER:
$\sum x^2 P(x)-4$ = 1.0, and $np(1 - p)$ = 1.0
Both represent the value of variance of a binomial random variable found by using two different formulas.

170. Find $\sum xP(x)$ and *np*. Why are they exactly the same?

ANSWER:
$\sum xP(x)=2.00$, $np=2$
They are exactly the same because they are the mean of a binomial random variable found by using two different formulas.

171. A binomial distribution is based on $n = 15$ trials and success probability $p = 0.4$. What is the probability that the binomial random variable equals its mean value?

ANSWER:
0.207

172. A coin is tossed 100 times. Find numbers *a* and *b* that are such that the number of heads to appear will be between *a* and *b* at least 89% of the time.

ANSWER:
$a = 35$, $b = 65$

173. A binomial distribution has a mean equal to 20 and a standard deviation equal to 4. Find *n* and *p*.

ANSWER:
$n = 100$, $p = 0.2$

174. Find the mean and standard deviation of the binomial distribution when $n = 60$ and $p = 1/6$. Note that this would correspond to the number of times a "one" would appear in 60 tosses of a fair die.

ANSWER:
Mean = 10, Standard deviation = 2.89

175. A manufacturer of matches puts 100 matches in each box of matches produced. One-tenth of one percent of the matches produced has a flaw. If a box is randomly selected, what is the mean and standard deviation of *x* where *x* is defined as the number of matches having a flaw in the box?

ANSWER:
Mean = 0.10, Standard deviation = 0.32

176. For the binomial distribution with $n = 48$ and $p = 1/3$, which of the possible values of x ($x = 0, 1, 2, 3, \cdots\cdots, 48$) lie between $\mu - 2\sigma$ and $\mu + 2\sigma$.

ANSWER:
10, 11, 12, 13, 14, 15, 16, 17, 18, 19, 20, 21, and 22

177. A machine produces parts of which 0.2% are defective. If a random sample of ten parts produced by this machine contains two or more defectives, the machine is shut down for repairs. Find the probability that the machine will be shut down for repairs based on this sampling plan.

ANSWER:
P(shut down) = $P(x \geq 2)$, where *x* represents the number of defective parts in the sample. By using the binomial formula with $n = 10$, and $p = 0.002$, we get $P(x = 0) = 0.9802$, and $P(x = 1) = 0.0196$. Hence, $P(x \geq 2) = 1.0 - [P(x = 0) + P(x = 1)] = 1.0 - (0.9802 + 0.0196) = 0.0002$.

178. For a particular binomial distribution, $\mu = 4$ and $\sigma = \sqrt{3}$. Find the values of *n* and *p*.

ANSWER:
n = 16 and *p* = 0.25

179. A binomial distribution has a mean of 12 and a standard deviation of 2.683. Find *n* and *p*.

ANSWER:
n = 30, *p* = 0.4

QUESTIONS 180 AND 181 ARE BASED ON THE FOLLOWING INFORMATION:
Four cards are selected, one at a time, from a standard deck of 52 cards. Let *x* represent the number of aces drawn in the set of 4 cards.

180. If this experiment is completed without replacement, explain why *x* is not a binomial random variable.

ANSWER:
x is not a binomial random variable because the trials are not independent. The probability of success (get an ace) changes from trial to trial. On the first trial it is 4/52. The probability of an ace on the second trial depends on the outcome of the first trial; it is 4/51 if an ace is not selected, and it is 3/51 if an ace was selected. The probability of an ace on any given trial continues to change when the experiment is completed without replacement.

181. If this experiment is completed with replacement, explain why *x* is a binomial random variable.

ANSWER:
x is a binomial random variable because the trials are independent. n = 4, the number of trials; two outcomes, success = ace and failure = not ace; *p* = *P*(ace) = 4/52 and *q* = *P*(not ace) = 48/52; *x* = n (aces drawn in 4 trials) and could be any number 0, 1, 2, 3 or 4. Further, the probability of success (get an ace) remains 4/52 for each trial throughout the experiment, as long as the card drawn on each trial is replaced before the next trial occurs.

QUESTIONS 182 THROUGH 186 ARE BASED ON THE FOLLOWING INFORMATION:
It was reported in a medical journal that about 70% of the individuals needing a kidney transplant find a suitable donor when they turn to registries of unrelated donors. Assume that a group of ten individuals needing a kidney transplant. Let *x* represent the number of individuals needing a kidney transplant who will find a suitable donor among the registries of unrelated donors. Consider a group of ten individuals needing a kidney transplant.

182. What is the distribution of *x*?

ANSWER:
The random variable *x* is binomial with *n* = 10 and *p* = 0.7.

183. Find the probability that all ten will find a suitable donor among the registries of unrelated donors.

ANSWER:
P(x = 10) = 0.028

184. Find the probability that exactly eight will find a suitable donor among the registries of unrelated donors.

ANSWER:
$P(x = 8) = 0.233$

185. Find the probability that at least eight will find a suitable donor among the registries of unrelated donors.

ANSWER:
$P(x = 8, 9, 10) = 0.233 + 0.121 + 0.028 = 0.382$

186. Find the probability that no more than five will find a suitable donor among the registries of unrelated donors.

ANSWER:
$P(x = 0,1, 2, 3, 4, 5) = 0 + 0 + 0.001 + 0.009 + 0.037 + 0.103 = 0.15$

187. Test the following function to determine whether or not it is a binomial probability function. List the distribution of probabilities.

$P(x) = \binom{4}{x}(0.75)^x(0.25)^{4-x}$ for $x = 0, 1, 2, 3, 4$

ANSWER:
By inspecting the function we see the binomial properties: Number of trials $n = 4$, probability of success $p = 0.75$ and probability of failure $q = 0.25$; $(p + q = 1)$, the two exponents x and $4-x$ add up to $n = 4$, and x can take on any integer value from 0 to $n = 4$; therefore x is binomial. The given function produces the following table:

x	P(x)
0	0.0039
1	0.0469
2	0.2109
3	0.4219
4	0.3164

This is a binomial probability function since each $P(x)$ is between 0 and 1, and $\sum P(x) = 1.0$

188. A recent study showed that only 20% of the women who lived with their boyfriends eventually walked down the aisle with them. In a sample of 15 women who have lived with a boyfriend in the past, what is the probability that 5 or fewer of them married the boyfriend?

ANSWER:
Let x represents the number of women who lived with their boyfriends and eventually married the boyfriend. The random variable x is $B(n = 15, p = 0.2)$. Using the table of binomial probabilities, we have: $P(x \leq 5) = 0.035 + 0.132 + 0.231 + 0.250 + 0.188 + 0.103 = 0.939$.

Chapter 5

189. If the binomial (q + p) is squared, the result is $(q+p)^2 = q^2 + 2qp + p^2$. For the binomial experiment with n = 2, the probability of no successes in two trials is q^2 (the first term in the expansion), the probability of one success in two trials is 2qp (the second term in the expansion), and the probability of two successes in two trials is p^2 (the third term). Find $(q+p)^3$ and compare its terms to the binomial probability for n = 3 trials.

ANSWER:

$(q+p)^3 = q^3 + 3q^2p + 3qp^2 + p^3$

$P(x = 0) = q^3$; $P(x = 1) = 3q^2p$; $P(x = 2) = 3qp^2$; $P(x = 3) = p^3$

190. The probability of success on a single trial of a binomial experiment is known to be 0.40. The random variable x, number of successes, has a mean value of 80. Find the number of trials involved in this experiment and the standard deviation of x.

ANSWER:

Given that p = 0.40 and μ = 80.

$\mu = np = 80$ implies that $n \cdot (0.4) = 80$; therefore n = 200

$\sigma = \sqrt{\text{npq}} = \sqrt{(200)\cdot(0.4)\cdot(0.6)} = \sqrt{48} = 6.9282$

191. In Florida, 40% of the people have a certain blood type. What is the probability that exactly 5 out of a randomly selected group of 15 Floridians will have that blood type?

ANSWER:

Let x represent the number of Floridians having that blood type. Note that x is $B(n = 15, p = 0.4)$. Using the binomial probabilities table we have, $P(x = 5) = 0.186$.

192. A binomial random variable is based on n = 20 and p = 0.3. Find $\sum x^2 P(x)$.

ANSWER:

$\mu = np = (20)(0.3) = 6.0$ and $\sigma^2 = npq = (20)(0.3)(0.7) = 4.2$

$\sigma^2 = \sum x^2 p(x) - \mu^2 \Rightarrow 4.2 = \sum x^2 P(x) - 6^2 \Rightarrow \sum x^2 P(x) = 40.2$

QUESTIONS 193 THROUGH 195 ARE BASED ON THE FOLLOWING INFORMATION:
A large shipment of TV sets is accepted upon delivery if an inspection of ten randomly selected TV sets yields no more than one defective TV.

193. Find the probability that this shipment is accepted if 5% of the total shipment is defective.

ANSWER:

$P(\text{accepted}) = P[x = 0, 1 \mid B(n = 10, p = 0.05)] = P(0) + P(1) = 0.599 + 0.315 = 0.914$

194. Find the probability that this shipment is not accepted if 10% of this shipment is defective.

ANSWER:

$P(\text{not accepted}) = P[x = 2,3,\ldots,10 \mid B(n = 10, p = 0.10)]$

$= 1 - P[x = 0, 1 \mid B(n = 10, p = 0.10)] = 1 - (0.349 + 0.387) = 0.264$

195. The binomial probability distribution is often used in situations similar to this one, namely, large populations sampled without replacement. Explain why the binomial yields a good estimate.

ANSWER:
Even though the *P*(defective) changes from trial to trial, if the population is very large, the probabilities are very similar. For example, suppose the population has 10,000 items and 50 are defective. *P*(defective) on the first trial is 50/10,000 = 0.0050; if after 10 trials 45 defectives have been selected, *P*(defective) will be 45/9990 = 0.0045.

QUESTIONS 196 THROUGH 199 ARE BASED ON THE FOLLOWING INFORMATION:
Suppose that you buy 25 plants from a nursery and the nursery claims that 95% of its plants survive when planted. Let x represent the number of plants that survive.

196. What is the distribution of x?

ANSWER:
x is binomial random variable with $n = 25$ and $p = 0.95$.

197. Use computer (or statistical software) to determine the probability that all 25 will survive.

ANSWER:
$P(x = 25) = 0.2774$

198. Use computer (or statistical software) to determine the probability that at most 21 will survive.

ANSWER:
$P(x \leq 21) = 0.0269$

199. Use computer (or statistical software) to determine the probability that at least 23 will survive.

ANSWER:
$P(x \geq 23) = 1 - P(x \leq 22) = 0.8729$

200. Find the mean and standard deviation of x = number of right-handed students in a classroom of 30 students. Assume that 10% of the population is left-handed.

ANSWER:
x is binomial random variable with $n = 30$ and $p = 0.9$. Then, the mean $\mu = np = 27$ and standard deviation $\sigma = \sqrt{npq} = \sqrt{(30)(0.9)(0.1)} = 1.643$.

QUESTIONS 201 THROUGH 206 ARE BASED ON THE FOLLOWING INFORMATION:
Assume that x is a binomial random variable, with $p = P(\text{success})$, n = number of trials, and x = number of successes in n trials. Use the binomial probabilities table available in your text to answer the questions below.

201. Determine the probability of $x = 5$, given that $n = 15$, $p = 0.05$.

ANSWER:
0.001

Chapter 5

202. Determine the probability of $x = 8$, given that $n = 13$, $p = 0.90$.

ANSWER:
0.006

203. Determine the probability of $x = 4$, given that $n = 10$, $p = 0.50$.

ANSWER:
0.205

204. Determine the probability of $x = 7$, given that $n = 8$, $p = 0.95$.

ANSWER:
0.279

205. Determine the probability of $x = 1$, given that $n = 6$, $p = 0.40$.

ANSWER:
0.187

206. Determine the probability of $x = 0$, given that $n = 4$, $x = 0$, $p = 0.01$.

ANSWER:
0.961

207. Let x be a random variable with the following probability distribution:

x	0	1	2	3	4
$P(x)$	0.42	0.33	0.10	0.09	0.06

Does x have a binomial distribution? Justify your answer.

ANSWER:
If this distribution were binomial, then n would be 4 and $P(x = 0) = 0.42$, would be q^4; that means $q = \sqrt[4]{0.42} = 0.805$. Also, $P(x = 4) = 0.06$, would be p^4, that means $p = \sqrt[4]{0.06} = 0.495$. Since $p + q = 0.495 + 0.805 = 1.30$, which did not add up to 1.0, the only conclusion is that this distribution is not binomial.

208. A machine produces parts of which 1% are defective. If a random sample of twenty parts produced by this machine contains two or more defectives, the machine is shut down for repairs. Find the probability that the machine will be shut down for repairs based on this sampling plan.

ANSWER:
P(machine will be shut down) = $P(x \geq 2)$, where x represents the number of defectives in the sample of n = 20. Since

$P(x=0) = \binom{20}{0}(0.01)^0(0.99)^{20} = 0.8179$, and $P(x=0) = \binom{20}{1}(0.01)^1(0.99)^{19} = 0.1652$,

then the probability that the machine will be shut down for repairs based on this sampling plan is given by $P(x \geq 2) = 1 - [P(x = 0) + P(x = 1)] = 1 - (0.8179 + 0.1652) = 0.0169$.

209. Find the mean and standard deviation of x = number of melon seeds that germinate when a package of 75 seeds is planted. The package states that the probability of germination is 0.92.

ANSWER:
x is binomial random variable with n = 75 and p = 0.92. Then, the mean $\mu = np = 69$ and standard deviation $\sigma = \sqrt{npq} = \sqrt{(75)(0.92)(0.08)} = 2.349$.

QUESTIONS 210 THROUGH 215 ARE BASED ON THE FOLLOWING INFORMATION:

Consider the following function: $P(x) = \binom{4}{x}(0.5)^x(0.5)^{4-x}$ for x = 0, 1, 2, 3, 4.

210. Test to determine whether or not $P(x)$ is a binomial probability function.

ANSWER:
By inspecting the function $P(x)$ we see it satisfies the following binomial properties: n = 4, p = 0.5, q = 0.5 (p + q = 1), the two exponents x and 4-x add up to n = 4, and x can take on any integer value from zero to n = 4; therefore $P(x)$ is a binomial probability function.

211 List the probability distribution of x.

ANSWER:

x	0	1	2	3	4
$P(x)$	0.0625	0.250	0.375	0.250	0.0625

212. Sketch a histogram of the probability distribution of x in question 211, and briefly describe its shape.

ANSWER:

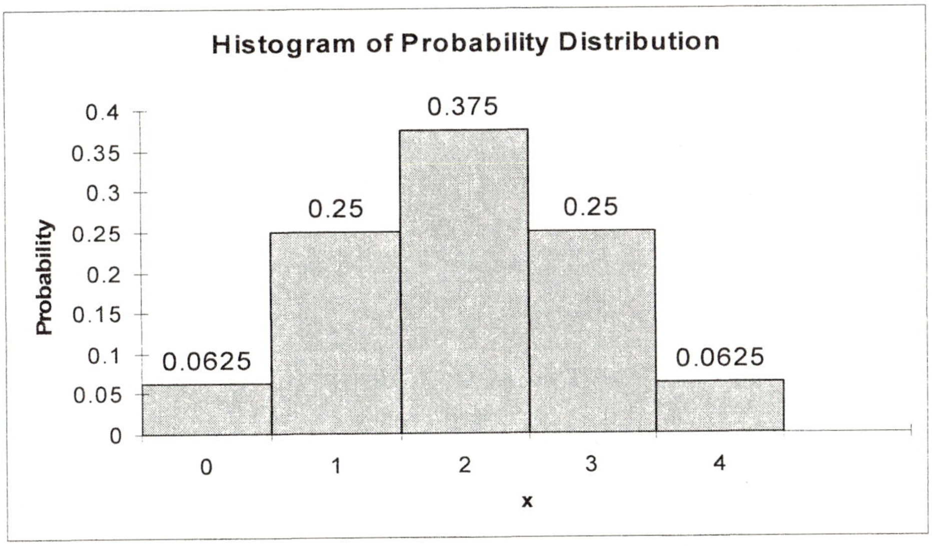

The histogram is symmetric, since both sides are identical (halves are mirror images around x = 3)

213. Calculate the mean and standard deviation of the probability distribution of x directly by using your answer to question 210.

ANSWER:

Mean: $\mu = \sum x \cdot P(x) = 2.0$

Variance: $\sigma^2 = \sum x^2 \cdot P(x) - \mu^2 = 5 - 4 = 1.0$, then standard deviation $\sigma = \sqrt{\sigma^2} = 1.0$.

214. Calculate the mean and standard deviation of the probability distribution of x by using your answer to question 210.

ANSWER:

Mean $\mu = np = (4)(0.5) = 2$, and standard deviation $\sigma = \sqrt{npq} = \sqrt{(4)(0.5)(0.5)} = 1$

215. Compare the results of questions 213 and 214.

ANSWER:

Same answers (mean $\mu = 2$, and standard deviation $\sigma = 1$).

216. If boys and girls are equally likely to be born, what is the probability that in a randomly selected family of four children, there will be no boys? (Find the answer using a formula).

ANSWER:

x = number of boys is a binomial random variable with $n = 4$ and $p = 0.5$.

$$P(x=0) = \binom{4}{0}(0.5)^0(0.50)^4 = 0.0625$$

217. Find the mean and standard deviation of x = number of cars found to have unsafe brakes among the 500 cars stopped at a roadblock for inspection. Assume that 5% of all cars have one or more unsafe brakes.

ANSWER:

x is binomial random variable with $n = 500$ and $p = 0.05$. Then, the mean $\mu = np = 25$ and standard deviation $\sigma = \sqrt{npq} = \sqrt{(500)(0.05)(0.95)} = 4.873$.

218. A binomial random variable x is based on 12 trials with the probability of success equal to 0.30. Find the probability that this variable will take on a value more than two standard deviations above the mean.

ANSWER:

x is binomial random variable with $n = 12$ and $p = 0.30$. Then, mean $\mu = np = 3.6$ and standard deviation $\sigma = \sqrt{npq} = \sqrt{(12)(0.30)(0.70)} = 1.587$.

Hence, $\mu + 2\sigma = 3.6 + 2(1.587) = 6.774$, and

$P(x > \mu + 2\sigma) = P(x > 6.7884) = P(x = 7, 8, 9, 10, 11, 12)$

$= 0.029 + 0.008 + 0.001 + 0 + 0 + 0 = 0.038$.

QUESTIONS 219 AND 220 ARE BASED ON THE FOLLOWING INFORMATION:
A doctor knows from experience that 20% of the patients to whom he gives a high blood pressure drug will have undesirable side effects. Assume the doctor gives that drug to ten of his patients.

219. Find the probability that among the ten patients to whom he gives the drug, at most two will have undesirable side effects.

ANSWER:
This is a binomial experiment with $n = 10$ and $p = 0.20$.
$P(x \leq 2) = P(x = 0) + P(x = 1) + P(x = 2) = 0.107 + 0.268 + 0.302 = 0.677$

220. Find the probability that among the ten patients to whom he gives the drug, at least two will have undesirable side effects.

ANSWER:
$P(x \geq 2) = 1 - [P(x = 0) + P(x = 1)] = 1 - (0.107 + 0.268) = 0.625$

QUESTIONS 221 THROUGH 223 ARE BASED ON THE FOLLOWING INFORMATION:
A random sample of 12 players from the active rosters of the 30 Major League Baseball teams is to be selected and tested for the use of illegal drugs.

221. If 10% of all the players are using illegal drugs at the time of the test, what is the probability that two or more test positive and fail the test?

ANSWER:
This is a binomial experiment with $n = 12$ and $p = 0.10$.
$P(x \geq 2) = 1 - [P(x = 0) + P(x = 1)] = 1 - (0.282 + 0.377) = 0.341$

222. If 20% of all the players are using illegal drugs at the time of the test, what is the probability that two or more test positive and fail the test?

ANSWER:
This is a binomial experiment with $n = 12$ and $p = 0.20$.
$P(x \geq 2) = 1 - [P(x = 0) + P(x = 1)] = 1 - (0.069 + 0.206) = 0.0.725$

223. If 30% of all the players are using illegal drugs at the time of the test, what is the probability that two or more test positive and fail the test?

ANSWER:
This is a binomial experiment with $n = 12$ and $p = 0.30$.
$P(x \geq 2) = 1 - [P(x = 0) + P(x = 1)] = 1 - (0.014 + 0.071) = 0.915$

QUESTIONS 224 THROUGH 226 ARE BASED ON THE FOLLOWING INFORMATION:
A large retailer has purchased 10,000 high quality videotapes. The retailer is assured by the supplier that the shipment contains no more than 1% defective tapes (according to agreed specifications). To check the supplier's claim, the retailer randomly selects 100 tapes and finds six of the 100 to be defective.

224. Assuming the supplier's claim is true, compute the mean and the standard deviation of the number of defective tapes in the sample.

ANSWER:
Mean = np =1, and standard deviation = $\sqrt{np(1-p)}$ = 0.995.

225. Based on your answer to question 224, is it likely that as many as six tapes would be found to be defective, if the claim is correct?

ANSWER:
No. If you were 3 standard deviations to the right of the mean, the value would be 3.985. It is unlikely you would observe 6 defects out of 100.

226. Suppose that six tapes are indeed found to be defective. Based on your answer to question 224, what might be a reasonable inference about the manufacturer's claim for this shipment of 10,000 tapes?

ANSWER:
You would have to infer that the manufacturer's claim is incorrect. Based on this observation, the supplier appears to have a higher defect rate than 1%.

QUESTIONS 227 THROUGH 231 ARE BASED ON THE FOLLOWING INFORMATION:
The service manager for a new appliances store reviewed sales records of the past 20 sales of new microwaves to determine the number of warranty repairs he will be called on to perform in the next 90 days. Corporate reports indicate that the probability any one of their new microwaves needs a warranty repair in the first 90 days is 0.05. The manager assumes that calls for warranty repair are independent of one another and is interested in predicting the number of warranty repairs he will be called on to perform in the next 90 days for this batch of new microwaves sold.

227. What type of probability distribution will most likely be used to analyze warranty repair needs on new microwaves in this situation?

ANSWER:
Binomial distribution

228. What is the probability that none of the 20 new microwaves sold will require a warranty repair in the first 90 days?

ANSWER:
$P(X = 0) = 0.3585$

229. What is the probability that exactly two of the 20 new microwaves sold will require a warranty repair in the first 90 days?

ANSWER:
$P(X = 2) = 0.1887$

230. What is the probability that at most two of the 20 new microwaves sold will require a warranty repair in the first 90 days?

ANSWER:
$P(X \leq 2) = 0.9245$

231. What is the probability that between two and four (inclusive) of the 20 new microwaves sold will require a warranty repair in the first 90 days?
ANSWER:
$P(2 \leq X \leq 4) = 0.2616$

Chapter 6

Normal Probability Distributions

Sections 6.1 Through 6.3

True-False Questions

1. Without the use of the standard normal tables, techniques of calculus must be used to find probabilities concerning a normal distribution.
   ANSWER: T

2. If the random variable z is the standard normal score, then the mean of the distribution of z is 0.
   ANSWER: T

3. If the random variable z is the standard normal score, then the standard deviation of the distribution of z is 1.
   ANSWER: T

4. The total area under the curve of the standard normal distribution is not necessarily 1.0.
   ANSWER: F

5. If the random variable z is the standard normal score, then z has a mean of one and a standard deviation of zero.
   ANSWER: F

6. The most common distribution of a continuous random variable is the binomial probability distribution.
   ANSWER: F

7. The area under the normal curve between $\mu - 2\sigma$ and $\mu + 2\sigma$ is about 0.95.
ANSWER: T

8. All normal probability distributions are symmetric about zero.
ANSWER: F

9. The total area under the curve of any normal distribution is 1.0.
ANSWER: T

10. The theoretical probability that a particular value of a continuous random variable will occur is exactly zero.
ANSWER: T

11. The unit of measure for the standard score is the same as the unit of measure of the data.
ANSWER: F

12. All normal distributions have the same general probability function and distribution.
ANSWER: T

13. Standard normal scores have a mean of zero and a standard deviation of one.
ANSWER: T

14. Probability distributions of all continuous random variables are normally distributed.
ANSWER: F

15. We are able to add and subtract the areas under the curve of a continuous distribution because these areas represent probabilities of independent events.
ANSWER: F

16. The most common distribution of a continuous random variable is the normal probability.
ANSWER: T

17. The area to the right of $z = 1.52$ is 0.4357
ANSWER: F

18. The normal probability distribution is considered the single most important probability distribution.
ANSWER: T

19. The most common distribution of a continuous random variable is the binomial probability.
ANSWER: F

20. Each different pair of values for the mean, μ and standard deviation, σ will result in a different normal probability distribution function. This means there are infinitely many probability distribution functions.
ANSWER: T

21. The Empirical Rule is a fairly crude measuring device; with it we are able to find probabilities associated only with any number multiple of the standard deviation from the mean.
ANSWER: F

22. The standard normal table can be used to find probabilities for all combinations of mean, μ and standard deviation, σ values.

ANSWER: **T**

23. All normal probability distributions are symmetric about zero.

ANSWER: **F**

24. All normal probability distributions have the same shape and distribution relative to the mean and standard deviation.

ANSWER: **T**

25. Probability distributions of all continuous random variables are normally distributed.

ANSWER: **F**

26. The unit of measure for the standard score is the same as the unit of measure of the data.

ANSWER: **F**

27. The total area under the curve of any normal distribution is 1.0.

ANSWER: **T**

28. The total area under the curve of any normal distribution is 100.

ANSWER: **F**

29. The Empirical Rule is a fairly crude measuring device; with it we are able to find probabilities associated only with whole-number multiples of the standard deviation (within one, two, or three standard deviations of the mean).

ANSWER: **T**

30. The total area under the curve of any continuous distribution is 1.0 as long as the distribution is symmetric around the mean value.

ANSWER: **F**

31. The theoretical probability that a particular value of a continuous random variable will occur is exactly zero.

ANSWER: **T**

Multiple-Choice Questions

32. Which of the following statements is false?

A) The total area under the curve of any normal distribution is 1.0.
B) Nearly all the area under the standard normal curve is between z = -3.00 and z = 3.00.
C) The symmetry of the normal distribution is a key factor in determining probabilities associated with values below (to the left of) the mean.
D) The z-score associated with the 50th percentile of the standard normal distribution is 1.0.

ANSWER: **D**

Chapter 6

33. The distribution that has a mean of zero and a standard deviation of one is called the

A) binomial probability distribution.
B) frequency distribution.
C) standard normal distribution.
D) uniform distribution.
ANSWER: C

34. Given a standard normal probability distribution, what can be said about the mean and standard deviation?

A) Mean = 1, standard deviation = any value
B) Mean = any value, standard deviation = 1
C) Mean = 0, standard deviation = any value
D) Mean = 0, standard deviation = 1
ANSWER: D

35. If the random variable z is the standard normal score, which of the following probabilities could easily be determined without referring to a table?

A) $P(z > 2.86)$
B) $P(z < 0)$
C) $P(z < -1.82)$
D) $P(z > -0.5)$
ANSWER: B

36. The area under the normal curve between $z = 0.0$ and $z = 2.0$ is

A) 0.9772.
B) 0.7408.
C) 0.1359.
D) 0.4772.
ANSWER: D

37. The area under the normal curve between $z = -1.0$ and $z = -2.0$ is

A) 0.3413.
B) 0.1359.
C) 0.4772.
D) 0.0228.
ANSWER: B

38. Which of the following statements is false?

A) There is unlimited number of normal probability distributions.
B) There is only one normal probability distribution.
C) Normal probability distributions have two parameters: μ (mean) and σ (standard deviation).
D) None of the above.
ANSWER: B

Short-Answer Questions

39. The random variable z is the standard normal score. Find the number k if $P(z > k) = 0.9750$.

ANSWER:
-1.96

40. The random variable z is the standard normal score. Find the number k if $P(-k < z < k) = 0.3900$.

ANSWER:
0.51

41. Find $P(z < -0.63)$.

ANSWER:
0.2643

42. Find $P(1.21 < z < 1.37)$.

ANSWER:
0.0278

43. Find $P(-0.31 < z < 1.31)$.

ANSWER:
0.5266

44. Find $P(z \geq -1.61)$.

ANSWER:
0.9463

45. Find z if the area to the right of z is 0.6736.

ANSWER:
-0.45

46. The random variable z is the standard normal score. Find z as shown in the diagram below given that the area of the shaded region is 0.4927.

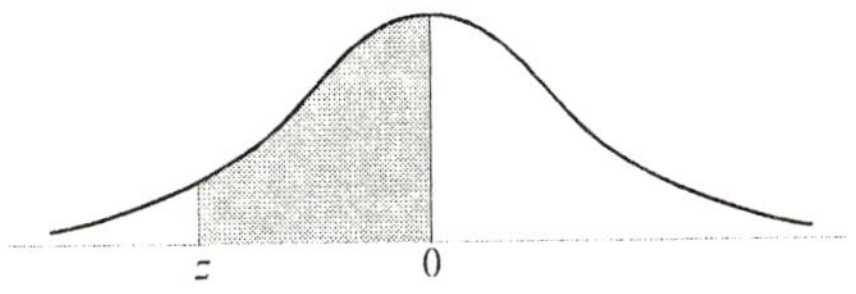

ANSWER:
-2.44

47. Find the probability of a randomly selected piece of data from a normal population will have a z-score between 0 and 1.25.

ANSWER:
0.3944

48. Find the probability that a randomly selected piece of data from a normal population will have a z-score greater than 1.25.

ANSWER:
0.1056

49. Find the probability that a randomly selected piece of data from a normal population will have a z-score less than 2.25.

ANSWER:
0.9878

50. Find the probability that a randomly selected piece of data from a normal population will have a z-score between 0 and -1.9.

ANSWER:
0.0287

51. Find the probability that a randomly selected piece of data from a normal population will have a z-score greater than -1.65.

ANSWER:
0.9505

52. Find the probability that a randomly selected piece of data from a normal population will have a z-score between -1.9 and 1.25.

ANSWER:
0.8657

53. Find the probability that a randomly selected piece of data from a normal population will have a z-score between 1.28 and 2.25.

ANSWER:
0.0881

54. Find z if the area to the left of z is 0.2981.

ANSWER:
-0.53

55. Find the number k if $P(z < k) = 0.1093$.

ANSWER:
-1.23

56. Find the number k if $P(z > k) = 0.0594$.

ANSWER:
1.56

57. Give the z-scores for the first, second, and third quartiles for the standard normal distribution.

ANSWER:
$Q_1 = -0.67$, $Q_2 = 0.00$, $Q_3 = 0.67$

58. A z-score of 1.28 corresponds to approximately what percentile of the standard normal distribution?

ANSWER:
90th percentile

59. What is the percentage of the total area under the normal curve within plus and minus three standard deviations of the mean?

ANSWER:
99.74%

60. "About one-third of the students entering a certain university drop out during or at the end of their first year." Does this statement illustrate percentage, proportion or probability?

ANSWER:
Proportion

61. "A recent survey reported that 56% of registered voters in Michigan are Democrats." Does this statement illustrate percentage, proportion or probability?

ANSWER:
Percentage

62. "The chance of receiving an "A" grade in this statistics class is 0.25". Does this statement illustrate percentage, proportion or probability?

ANSWER:
Probability

Applied and Computational Questions

63. Find the probability for $P(0.00 < z < 2.05)$.

ANSWER:
0.4798

64. Find the probability for $P(-2.10 < z < 2.54)$.

ANSWER:
0.4821 + 0.4945 = 0.9766

65. Find the probability for $P(z > 0.13)$.

ANSWER:
0.5000 – 0.0517 = 0.4483

66. Find the probability for $P(z < 1.84)$.

ANSWER:
0.5000 + 0.4671 = 0.9671

67. Find a value of z such that 40% of the distribution lies between it and the mean.

ANSWER:
There are two possible answers: z = -1.28 or + 1.28.

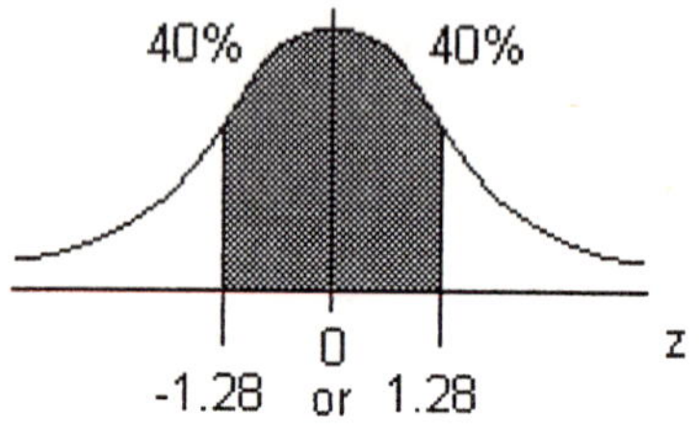

68. Find the standard z-score such that 80% of the distribution is to the left of this value.

ANSWER:
z = 0.84

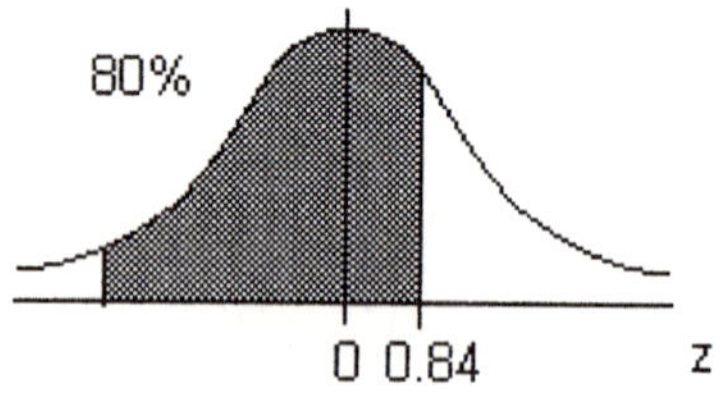

69. Find the standard z-score such that the area to the right of this value is 0.15.

ANSWER:
z = 1.04

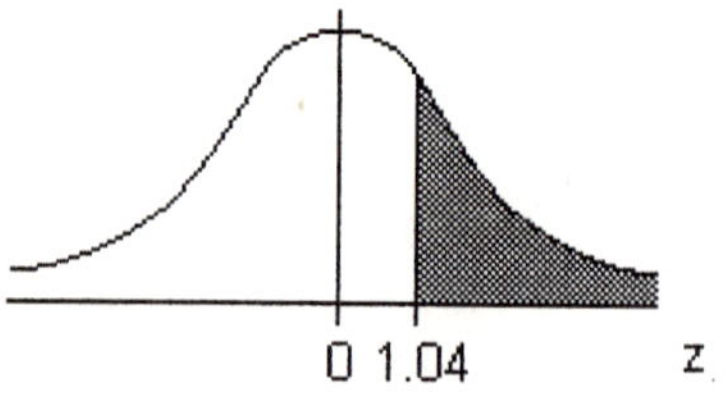

70. Find the two standard z-scores that bound the middle 50% of a normal distribution.

ANSWER:
$z = -0.67$ or $+ 0.67$

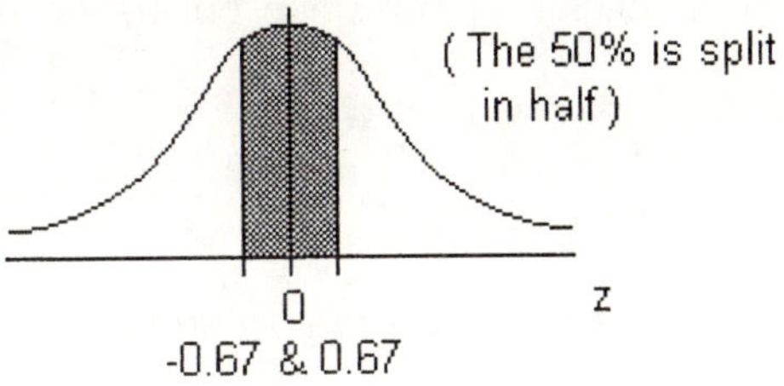

71. Find the two standard scores z such that the middle 90% of a normal distribution is bounded by them.

ANSWER:
$z = -1.65$ or $+ 1.65$

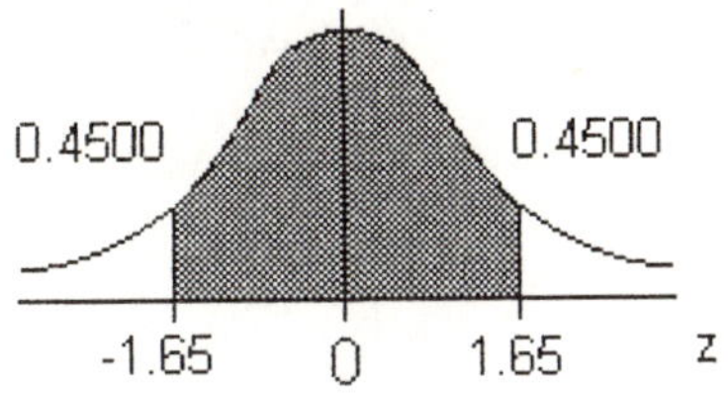

72. Find the two standard scores z such that the middle 98% of a normal distribution is bounded by them.

ANSWER:
$z = -2.33$ or $+ 2.33$

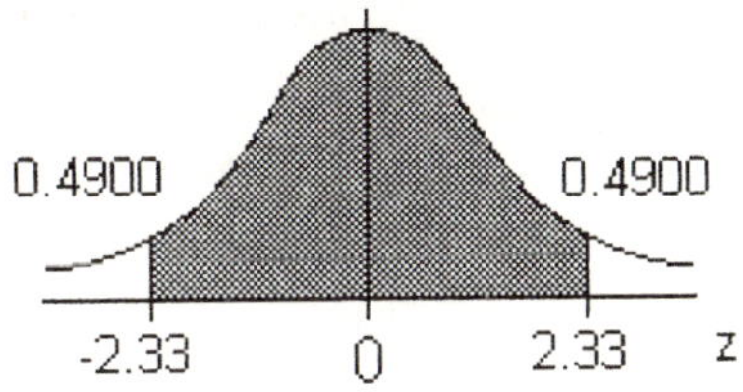

QUESTIONS 73 AND 74 ARE BASED ON THE FOLLOWING INFORMATION:
Given z is the standard normal variable:

73. Find the probability for $P(|z| > 1.75)$.

ANSWER:
$P(|z| > 1.75) = P(z < -1.75) + P(z > +1.75) = 2(0.5000 - 0.4599) = 0.0802$

74. Find the probability for $P(|z| < 2.05)$.

ANSWER:
$P(|z| < 2.05) = P(-2.05 < z < +2.05) = 2(0.4798) = 0.9596$

Chapter 6

75. Briefly discuss the properties of the standard normal distribution.

ANSWER:
(a) The total area under the normal curve is equal to one.
(b) The distribution is mounded and symmetric with respect to the vertical line drawn through $z = 0$; it extends indefinitely in directions, approaching but never touching the horizontal axis.
(c) The distribution has a mean of 0 and a standard deviation of 1.
(d) The mean divides the area in half - 0.50 on each side.
(e) Nearly all the area is between $z = -3.00$ and $z = 3.00$

76. Find the *z*-score associated with the 80th percentile of the standard normal distribution. What does this value tell you?

ANSWER:
$z = 0.84$. This says that the 80th percentile in a normal distribution is 0.84 standard deviations above the mean.

77. Find the *z*-scores that bound the middle 75% of the standard normal distribution.

ANSWER:
$z = -1.15$ and $+1.15$

78. Find the area under the standard normal curve to the right of $z = 2.12$.

ANSWER:
$P(z > 2.12) = 0.50 - 0.483 = 0.017$

79. Find the area under the standard normal curve to the left of $z = 1.93$.

ANSWER:
$P(z<1.93) = 0.50 + 0.4732 = 0.9732$

80. Find the area under the standard normal curve between -1.48 and the mean.

ANSWER:
$P(-1.78 < z< 0.0) = 0.4306$

81. Find the area under the standard normal curve to the left of $z = -1.33$.

ANSWER:
$P(z<-1.33) = 0.50 - 0.4082 = 0.0918$

82. Find the area under the standard normal curve between $z = -1.52$ and $z =1.25$.

ANSWER:
$P(-1.52 < z < 1.25) = 0.4357 + 0.3944 = 0.8301$

QUESTIONS 83 THROUGH 86 ARE BASED ON THE FOLLOWING INFORMATION:
A piece of data picked at random from a normal population.

83. Find the probability that it will have a standard score (z) that lies between 0 and 0.95.

ANSWER:
$P(0 < z < 0.95) = 0.3289$

84. Find the probability that it will have a standard score (z) that lies to the right of 0.95.

ANSWER:
$P(z > 0.95) = 0.50 - 0.3289 = 0.1711$

85. Find the probability that it will have a standard score (z) that lies to the left of 0.95.

ANSWER:
$P(z < 0.95) = 0.50 + 0.3289 = 0.8289$

86. Find the probability that it will have a standard score (z) that lies between -0.95 and 0.95.

ANSWER:
$P(-0.95 < z < 0.95) = 0.3289 + 0.3289 = 0.6578$

87. Find $P(0.0 < z < 2.65)$.

ANSWER:
0.4960

88. Find $P(-2.09 < z < 2.14)$.

ANSWER:
$0.4817 + 0.4838 = 0.9566$

89. Find $P(z > 0.39)$.

ANSWER:
$0.50 - 0.1517 = 0.3483$

90. Find $P(z < 1.51)$.

ANSWER:
$0.50 + 0.4345 = 0.9345$

91. Find the area under the standard normal curve between $z = 0.25$ and $z = 2.75$.

ANSWER:
$P(0.25 < z < 2.75) = 0.4970 - 0.0897 = 0.4073$

92. Find a value of z such that 43.7% of the distribution lies between it and the mean. (Hint: There are two possible answers.)

ANSWER:
Since $P(0 < z < 1.53) = 0.437 = P(-1.53 < z < 0)$, then $z = \pm 1.53$.

93. Find the two standard scores z such that the middle 75.4% of a normal distribution is bounded by them.

ANSWER:
Since $P(-1.16 < z < 1.16) = 0.377 + 0.377 = 0.754$, then $z = \pm 1.16$.

94. Find the two standard scores z such that the middle 82.3% of a normal distribution is bounded by them.

ANSWER:
Since $P(-1.35 < z < 1.35) = 0.4115 + 0.4115 = 0.823$, then $z = \pm 1.35$.

95. Find the two standard scores z such that the middle 90% of a normal distribution is bounded by them.

ANSWER:
Since $P(-1.645 < z < 1.645) = 0.45 + 0.45 = 0.90$, then $z = \pm 1.645$.

96. Find the two standard scores z such that the middle 95% of a normal distribution is bounded by them.

ANSWER:
Since $P(-1.96 < z < 1.96) = 0.475 + 0.475 = 0.95$, then $z = \pm 1.96$.

97. Find the two standard scores z such that the middle 99% of a normal distribution is bounded by them.

ANSWER:
Since $P(-2.575 < z < 2.575) = 0.495 + 0.495 = 0.99$, then $z = \pm 2.575$

98. Find the z-scores that bound the middle 55% of the standard normal distribution.

ANSWER:
Since $P(-.76 < z < 0.76) = 0.2764 + 0.2764 = 0.5528 \simeq 0.55$, then $z = \pm 0.76$

99. Find the z-score for the first quartile of the standard normal distribution.

ANSWER:
Since $P(z < -0.67) = 0.2486 \simeq 0.25$, then the z-score for the first quartile of the standard normal distribution is -0.67.

100. Find the z-score for the second quartile of the standard normal distribution.

ANSWER:
Since $P(z < 0.0) = 0.50$, then the z-score for the second quartile of the standard normal distribution is 0.

101. Find the z-score for the third quartile of the standard normal distribution.

ANSWER:
Since $P(z < 0.67) = 0.50 + 2486 = 0.7486 \simeq 0.75$, then the z-score for the third quartile of the standard normal distribution is 0.67

QUESTIONS 102 AND 103 ARE BASED ON THE FOLLOWING INFORMATION:
z is the standard normal variable, with mean of 0 and standard deviation of 1.

102. Find the value of k such that $P(|z| > 1.88) = k$.

ANSWER:
$P(|z| > 1.88) = P(z < -1.88) + P(z > +1.88) = 2P(z > +1.88) = 2(0.5000 - 0.4699) = 0.0602$, then $k = 0.0602$.

103. Find the value of k such that $P(|z| < 2.28) = k$.

ANSWER:
$P(|z| < 2.28) = P(-2.28 < z < +2.28) = 2\ P(0 < z < +2.28) = 2(0.4887) = 0.9774$, then $k = 0.9774$.

QUESTIONS 104 AND 105 ARE BASED ON THE FOLLOWING INFORMATION:
z is the standard normal variable, with mean of 0 and standard deviation of 1.

104. Find the value of c such that $P(|z| > c) = 0.0204$.

ANSWER:
$P(|z| > c) = P(z < -c) + P(z > +c) = 2P(z > +c) = 0.0204$, then $P(z > +c) = 0.0102$. Hence, $P(0 < z < c) = 0.5000 - 0.0102 = 0.4898$, and $c = 2.32$.

105. Find the value of c such that $P(|z| < c) = 0.8948$.

ANSWER:
$P(|z| < c) = P(-c < z < +c) = 2P(0 < z < +c) = 0.8948$, then $P(0 < z < +c) = 0.4474$, and $c = 1.62$.

Sections 6.4 and 6.5

True-False Questions

106. Assume that x is a normally distributed random variable with a mean of μ and standard deviation of σ. If x is converted to the standard score z, then given any three of the values of x, μ, σ, and z, we can always find the fourth value.
ANSWER: T

107. If the random variable z is the standard normal score, then $z(0.30) > z(0.20)$.
ANSWER: F

108. If the random variable z is the standard normal score, then $z(0.65) = -z(0.35)$.
ANSWER: T

109. If the random variable z is the standard normal score, then $z(0.35) = -z(0.35)$.
ANSWER: F

110. If the random variable z is the standard normal score, then $z(0.50) = -z(0.50)$.
ANSWER: T

111. In the statement $z(0.33) = 0.44$, the number 0.33 represents a value for z and the number 0.44 represents the area to the right of 0.33.
ANSWER: F

112. When using the notation $z(\alpha)$, the number α in parenthesis is the measure of the area to the right of the z-score.
ANSWER: T

113. $z(0.15)$ is the algebraic name for the z such that the area to the left and under the standard normal curve is exactly 0.15.
ANSWER: F

114. When using the notation $z(0.05)$, the number in parentheses is the measure of the area to the right of the z-score.
ANSWER: T

115. The value of $z(0.75) + z(.25)$ is 1.0.
ANSWER: F

116. The value of $z(0.60) + z(0.40)$ is 0.0.
ANSWER: T

117. Standard normal scores have a mean of one and a standard deviation of zero.
ANSWER: F

118. The standard normal distribution is the normal distribution of the standard variable z (called the "standard score" or "z-score").
ANSWER: T

119. In the notation $z(0.05)$, the number in parentheses is the measure of the area to the left of the z-score.
ANSWER: F

120. A standard notation used to abbreviate "normal distribution with mean μ and standard deviation σ is $N(\mu, \sigma)$.
ANSWER: T

121. $z(\alpha)$ is the value of z such that the area to the right of z and under the standard normal curve is exactly α.
ANSWER: T

122. The middle 0.90 of the standard normal distribution is bounded by -1.96 and 1.96.
ANSWER: F

Multiple-Choice Questions

123. The random variable x is normally distributed with a mean of 75 and a standard deviation of 15.0. For this distribution, the twenty-third percentile, P_{23}, is

A) 65.7.
B) 63.9.
C) 86.1.
D) 84.3.
ANSWER: B

124. If x is normally distributed random variable with a standard score of z, a mean of μ, and a standard deviation of σ, then x is equal to:

A) $(z-\mu)/\sigma$
B) $(z-\sigma)/\mu$
C) $\mu-\sigma z$
D) $\mu+\sigma z$
ANSWER: D

125. What is the value for $z(0.67)$?

A) +0.67
B) −0.44
C) 0.2486
D) −0.17
ANSWER: B

126. If the random variable z is the standard normal score, then $z(0.2611)$ is equal to

A) $+z(0.2389)$.
B) $+z(0.7611)$.
C) $-z(0.7389)$.
D) $-z(0.1026)$.
ANSWER: C

Chapter 6

127. If the random variable z is the standard normal score, then $z(0.2324)$ is equal to

A) + 0.6324.
B) + 0.7324.
C) $-z(0.7676)$.
D) $-z(0.2324)$.
ANSWER: C

128. Using the symbolic notation $z(\alpha)$, identify the value for α.

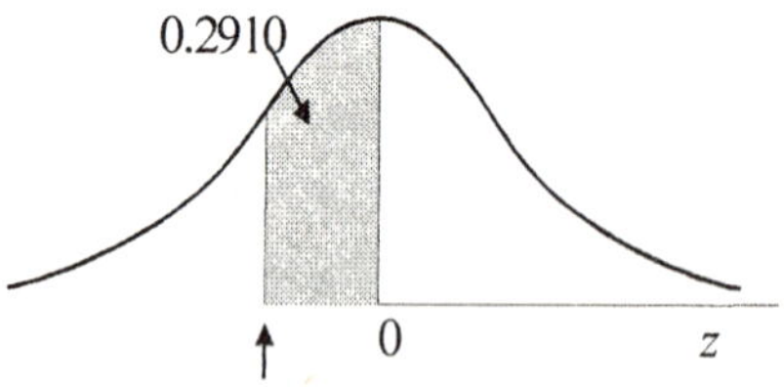

A) $z(0.2910)$
B) $z(0.2090)$
C) $z(0.8100)$
D) $z(0.7090)$
ANSWER: D

129. If the random variable z is the standard normal distribution, then $z(0.75)$ is equal to

A) P_{25} for the distribution.
B) P_{50} for the distribution.
C) P_{75} for the distribution.
D) 0.2734.
ANSWER: A

130. Using the symbolic notation $z(\alpha)$, identify the value for α.

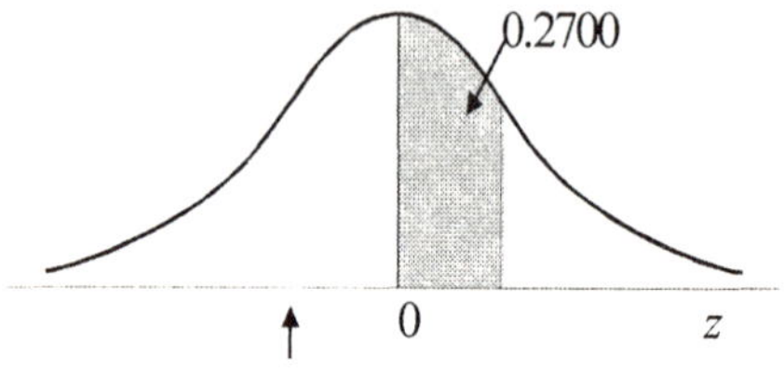

A) $z(0.1064)$
B) $z(0.2300)$
C) $z(0.5064)$
D) $z(0.7400)$
ANSWER: B

131. The value of $z(0.80) + z(0.20)$ is

A) 1.0.
B) 0.6.
C) 0.0.
D) None of the above.
ANSWER: C

132. The value of $z(0.70) - z(0.30)$ is

A) 1.04.
B) -1.04.
C) 0.52.
D) -0.52.
ANSWER: **B**

133. The value of $z(0.10) - z(0.90)$ is

A) 2.56.
B) -2.56.
C) 1.28.
D) -1.28.
ANSWER: **A**

Short-Answer Questions

134. The mean of a normal probability distribution is 500 and the standard deviation is 10. About 68% of the observations lie between what two values?

ANSWER:
490 and 510

135. Find the area between $z(0.79)$ and $z(0.43)$.

ANSWER:
0.36

136. Find the value of $z(0.73) + z(0.27)$.

ANSWER:
0.0

137. The mean of a normal probability distribution is 400 and the standard deviation is 10. About 95% of the observations lie between what two values?

ANSWER:
380 and 420

138. Use the standard normal table and the definition of $z(\alpha)$ notation to find $z(0.18)$.

ANSWER:
$z(0.18) = 0.92$

139. Find the area under the normal curve for z between $z(0.95)$ and $z(0.05)$.

ANSWER:
The area to the right of $z(0.95)$ is 0.95; the area to the right of $z(0.05)$ is 0.05; therefore the area between them is found by 0.95 - 0.05 and it is 0.90.

140. Use the standard normal table and the definition of $z(\alpha)$ notation to find $z(0.78)$.

ANSWER:
$z(0.78) = -0.77$

141. Find $z(0.063) - z(0.975)$.

ANSWER:
$z(0.063) - z(0.975) = 1.53 - (-1.96) = 3.49$

Applied and Computational Questions

QUESTIONS 142 AND 143 ARE BASED ON THE FOLLOWING INFORMATION:
X has a normal distribution with a mean of 47.5 and a standard deviation of 5.0.

142. Solve this equation for a: $P(X < a) = 0.95$.

ANSWER:
$a = 55.75$

143. Solve this equation for a: $P(X < a) = 0.05$.

ANSWER:
$a = 39.25$

144. A traffic study at one point on an interstate highway shows that vehicle speeds are normally distributed with a mean of 61.3 mph and a standard deviation of 3.3 mph. If a vehicle is randomly checked, find the probability that its speed is between 55.0 mph and 60.0 mph.

ANSWER:
0.3202

145. Two-year college students have mathematics competency scores that are normally distributed with a mean of 35 (the maximum possible score is 48). The 90th percentile is 40. Find the standard deviation of the math competency scores.

ANSWER:
3.9

146. Scores on a computer science aptitude test are normally distributed. The standard deviation of the distribution is 6.0, and the 95th percentile for the test is 92. Find the mean score for this test.

ANSWER:
Mean = 82.1

147. If heights of a certain group of adult males are normally distributed with a mean of 68.2 inches and a standard deviation of 4.1 inches, find the 25th percentile, P_{25}, for this distribution.

ANSWER:
65.5 inches

<u>QUESTIONS 148 AND 149 ARE BASED ON THE FOLLOWING INFORMATION:</u>
A machine cuts circular filters from large rolls of material. The diameters of the filters are normally distributed with a mean equal to 2.00 cm and a standard deviation equal to 0.02 cm.

148. Find the 95th percentile for the distribution of filter diameters.

ANSWER:
$P_{95} = 2.033$

149. If specifications call for the filters to have diameters between 1.95 cm and 2.03 cm, about what percent would be expected to not meet specifications?

ANSWER:
7.3%

<u>QUESTIONS 150 AND 151 ARE BASED ON THE FOLLOWING INFORMATION:</u>
Reading comprehension scores for junior high students in a school district are normally distributed with a mean of 80.0 and a standard deviation 5.0.

150. What percent have scores greater than 87.5?

ANSWER:
6.68%

151. What percent have scores between 75 and 85?

ANSWER:
68.26%

<u>QUESTIONS 152 AND 153 ARE BASED ON THE FOLLOWING INFORMATION:</u>
The times required to assemble a product part are normally distributed with a mean of 47.5 minutes and a standard deviation equal to 8.5 minutes.

152. What percent of the assembly workers require: more than one hour?

ANSWER:
7.08%

153. What percent of the assembly workers require: less than one-half hour?

ANSWER:
1.97%

<u>QUESTIONS 154 THROUGH 156 ARE BASED ON THE FOLLOWING INFORMATION:</u>
The random variable x has a normal distribution with mean of 75.0 and standard deviation of 2.5.

154. Find $P(x < 70.0)$.

ANSWER:
0.0228

155. Find $P(72.5 < x < 80.0)$.

ANSWER:
0.8185

Chapter 6

156. Find $P(x > 82.5)$.

ANSWER:
0.0013

157. Waiting times to see a doctor at a large clinic are normally distributed with a mean of 68.2 minutes and a standard deviation of 14.8 minutes. Find the probability that the waiting time to see a doctor is less than 45.0 minutes.

ANSWER:
0.0582

158. Scores on a particular test are normally distributed with a mean of 126 points. Find the standard deviation if for these scores, $P_{90} = 160.0$.

ANSWER:
26.6

159. For a particular normal distribution, $Q_3 = 27.8$ and $P_{40} = 24.2$. Find the mean and standard deviation of this distribution.

ANSWER:
$\mu = 25.2$ and $\sigma = 3.9$

160. A normal distribution has a mean of 65.0 and a standard deviation of 2.5. Use the z notation to represent the point that corresponds to 70.0 on the above nonstandard normal distribution.

ANSWER:
$z(0.0228)$

161. Use the standard normal table to find the values of z: $z(0.9940)$.

ANSWER:
–2.51

162. Use the standard normal table to find the values of z: $z(0.2054)$.

ANSWER:
0.54

163. Use the standard normal table to find the values of z: $z(0.3315)$.

ANSWER:
0.96

164. A machine cuts circular filters from large rolls of material. Specifications call for the filters to have diameters between 1.95 cm and 2.05 cm. If the diameters of the filters are normally distributed with a mean equal to 2.00 cm, then the machine needs to be fine-tuned to give what standard deviation so that only 1% of the filters do not meet specifications? (Give the answer to three decimal places.)

ANSWER:
0.019 cm

165. Use the standard normal table to find the values of z: $z(0.7881)$.

ANSWER:
-0.80

166. The value $z(0.25)$ associated with the standard normal distribution would correspond to what value associated with the nonstandard normal distribution having a mean equal to 90 and a standard deviation of 10?

ANSWER:
96.7

QUESTIONS 167 THROUGH 171 ARE BASED ON THE FOLLOWING INFORMATION:
A piece of data picked at random from a normally distributed population.

167. Find the probability that the piece of data will have a standard score less than 2.00.

ANSWER:
$0.5000 + 0.4772 = 0.9772$

168. Find the probability that the piece of data will have a standard score greater than -1.40.

ANSWER:
$0.5000 + 0.4192 = 0.9192$

169. Find the probability that the piece of data will have a standard score less than -1.75.

ANSWER:
$0.5000 - 0.4599 = 0.0401$

170. Find the probability that the piece of data will have a standard score less than 1.25.

ANSWER:
$0.5000 + 0.3944 = 0.8944$

171. Find the probability that the piece of data will have a standard score greater than -1.58.

ANSWER:
$0.5000 + 0.4429 = 0.9429$

QUESTIONS 172 THROUGH 177 ARE BASED ON THE FOLLOWING INFORMATION:
Assume that x is a normally distributed random variable with a mean of 70 and a standard deviation of 10.

172. Find $P(x > 70)$.

ANSWER:
$P(x > 70) = P(z > 0.0) = 0.5000$

173. Find $P(70 < x < 82)$.

ANSWER:
$P(70 < x < 82) = P(0.0 < z < 1.20) = 0.3849$

174. Find $P(67 < x < 93)$.

ANSWER:
$P[67 < x < 93) = P(-0.30 < z < 2.30) = 0.1179 + 0.4893 = 0.6072$

175. Find $P(75 < x < 92)$.

ANSWER:
$P(75 < x < 92) = P(0.50 < z < 2.20) = 0.4861 – 0.1915 = 0.2946$

176. Find $P(48 < x < 88)$.

ANSWER:
$P(48 < x < 88) = P(-2.20 < z < 1.80) = 0.4861 + 0.4641 = 0.9502$

177. Find $P(x < 48)$.

ANSWER:
$P(x < 48) = P(z < -2.20) = 0.5000 – 0.4861 = 0.0139$

QUESTIONS 178 AND 179 ARE BASED ON THE FOLLOWING INFORMATION:
For a particular age group of adult females, the distribution of cholesterol readings, in mg/dl, is normally distributed with a mean of 180 and a standard deviation of 12.

178. What percentage of this population would have readings exceeding 210?

ANSWER:
$P(x > 210) = P(z > 2.5) = 0.5000 – 0.4938 = 0.0062$, or 0.62%

179. What percentage would have readings less than 156?

ANSWER:
$P(x < 156) = P(z < -2) = 0.5000 – 0.4772 = 0.0228$ or 2.28%

180. The weights of ripe watermelons grown at Mr. Howard's farm are normally distributed with a standard deviation of 2.4 lbs. Find the mean weight of Mr. Howard's ripe watermelons if approximately 4% weigh less than 15 lb.

ANSWER:
The z-score is $z = -1.75$ (since the area to the left of z is 0.04). Now using the formula: $z = (x - \mu)/\sigma$, we have $-1.75 = (15 - \mu)/2.4$. Solving for μ, we get: $\mu = 15 - (-1.75)(2.4) = 19.2$ lbs.

QUESTIONS 181 THROUGH 183 ARE BASED ON THE FOLLOWING INFORMATION:
The waiting time x at a fast-food restaurant during lunch time is approximately normally distributed with a mean of 4.5 min and a standard deviation of 1.2 min.

181. Find the probability that a randomly selected customer has to wait less than 2.7 min.

ANSWER:
$P(x < 2.7) = P(z < -1.50) = 0.5000 – 0.4332 = 0.0668$

182. Find the probability that a randomly selected customer has to wait more than 6.8 min.

ANSWER:
$P(x > 6.8) = P(z > 1.92) = 0.5000 - 0.4726 = 0.0274$

183. Find the value of the 75th percentile for x.

ANSWER:
75th percentile is a value such that 75% of the data is less than this value; therefore the z-score of this value is to the right of 0 such that the area between 0 and z is 0.25. Hence, the corresponding z value is $z = +0.67$. Now, the formula $z = (x - 4.5) / 1.2$ implies $0.67 = (x - 4.5) / 1.2$. Solving for x, we get $x = (0.67)(1.2) + 4.5 = 5.304$ minutes.

184. A machine is programmed to fill 16-oz bottles. However, the variability inherent in any machine causes the actual amounts of fill to vary. The distribution is normal with a standard deviation Of 0.02 oz. What must the mean amount μ be in order that only 5% of the bottles receive less than 16 oz?

ANSWER:
The area to the left of z is 0.05, therefore $z = -1.65$. Then, the formula $z = (x - \mu)/\sigma$ reduces to $-1.65 = (16 - \mu) / 0.02$. Solving for μ, we get $\mu = 16 - (-1.65)(0.02) = 16.033$ oz.

QUESTIONS 185 THROUGH 188 ARE BASED ON THE FOLLOWING INFORMATION:
The z notation, $z(\alpha)$, combines two related concepts, the z-score and the area to the right of z, into a mathematical symbol.

185. If $z(A) = 0.10$, identify the letter as being a z-score or being an area, and then with the aid of a diagram explain what both the given number and the letter represent on the standard curve.

ANSWER:
A is an area. z is 0.10 and the area to the right of $z = 0.10$ is $0.5000 - 0.0398 = 0.4602$.

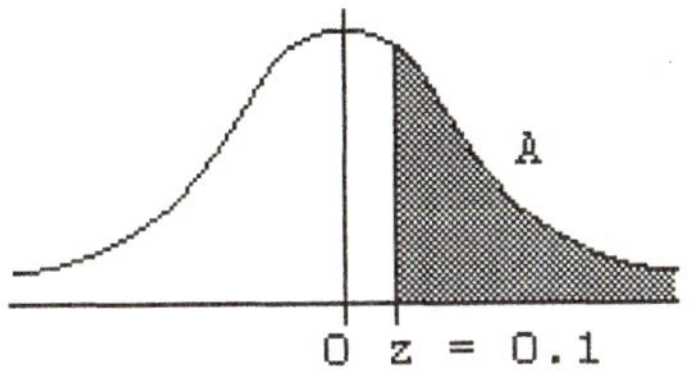

186. If $z(0.10) = B$, identify the letter as being a z-score or being an area, and then with the aid of a diagram explain what both the given number and the letter represent on the standard curve.

ANSWER:
B is a z-score. 0.10 is the area to the right of $z = B$. Use 0.4000 to look up the z-score on the table of standard normal distribution, $z = B = 1.28$

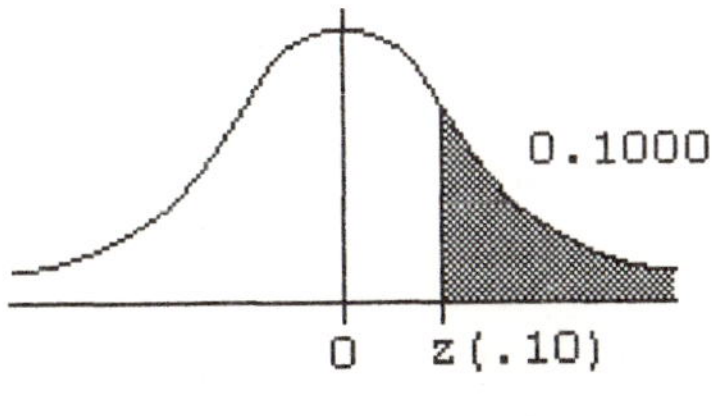

187. If $z(C) = -0.05$, identify the letter as being a z-score or being an area, and then with the aid of a diagram explain what both the given number and the letter represent on the standard curve.

ANSWER:
C is an area. *z* is –0.05 and the area to the right of $z = -0.05$ is $0.5000 + 0.0199 = 0.5199$.

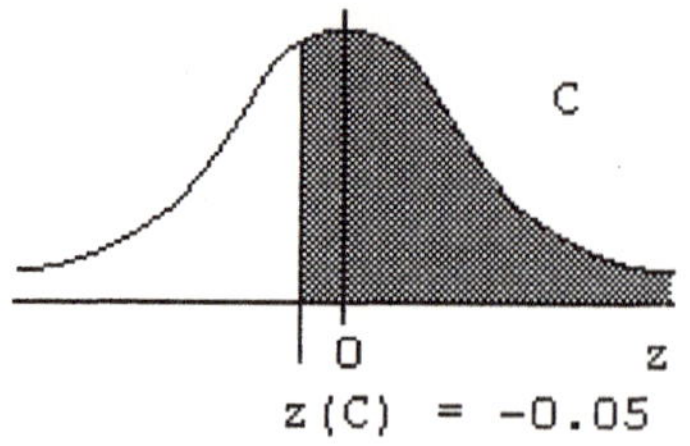

188. If $-z(0.05) = D$, identify the letter as being a z-score or being an area, and then with the aid of a diagram explain what both the given number and the letter represent on the standard curve.

ANSWER:
D is a z-score, and 0.05 is the area to the left of $z = D$. Then, *D* is to the left of zero [negative]. Use 0.4500 to look it up; $z = D = -1.65$.

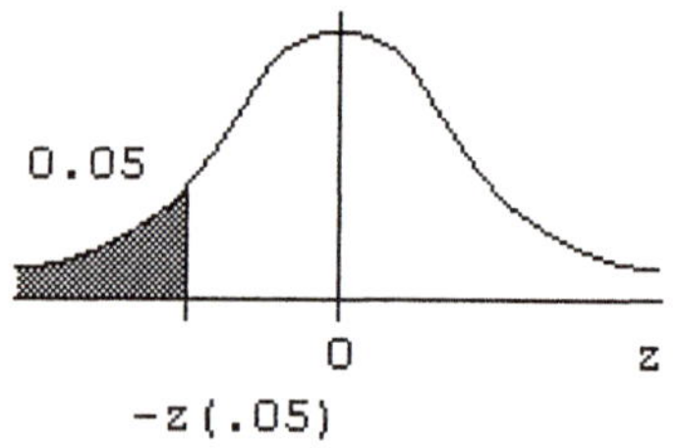

QUESTIONS 189 AND 190 ARE BASED ON THE FOLLOWING INFORMATION:
Assume that the average annual salary for a worker in the United States is $27,500, and that the annual salaries for Americans are normally distributed with a standard deviation equal to $6250.

189. What percentage of Americans earn below $18,000?

ANSWER:
$P(x < 18{,}000) = P(z < -1.52) = 0.5000 - 0.4357 = 0.0643$ or 6.43%

190. What percentage of Americans earn above $40,000?

ANSWER:
$P(x > 40{,}000) = P(z > 2.0) = 0.5000 - 0.4772 = 0.0228$ or 2.28%

QUESTIONS 191 THROUGH 194 ARE BASED ON THE FOLLOWING INFORMATION:
Understanding the z notation $z(\alpha)$ requires us to know whether we have a z-score or an area. Different expressions use the z notation in a variety of ways, some typical and some not so typical.

191. If $z(0.08)$, find the value asked for and then with the aid of a diagram explain what your answer represents.

ANSWER:
$z(0.08)$ is a z-score. $z(0.08) = 1.41$ [Use 0.4200 to look it up]

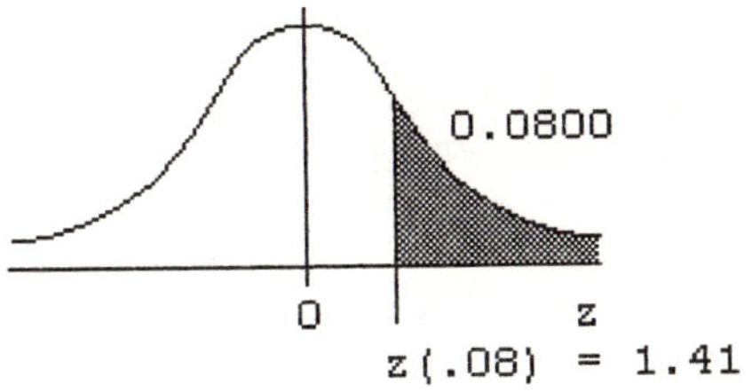

192. If the area between $z(0.98)$ and $z(0.02)$, find the value asked for and then with the aid of a diagram explain what your answer represents.

ANSWER:
Area between $z(0.98)$ and $z(0.02) = 0.98 - 0.02 = 0.96$

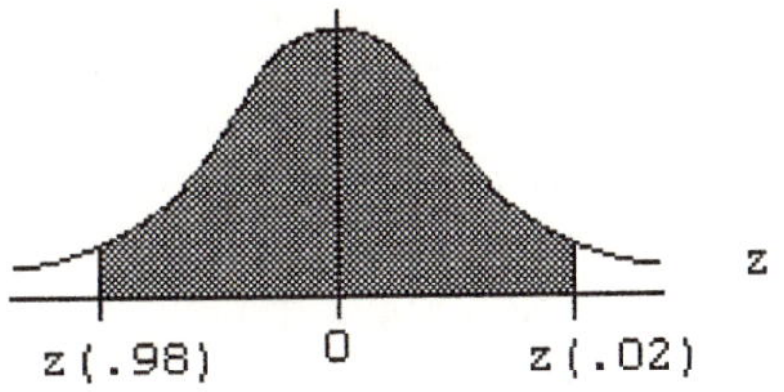

193. If $z(1.00 - 0.01)$, find the value asked for and then with the aid of a diagram explain what your answer represents.

ANSWER:
$z(1.00 - 0.01) = z\ (0.99) = -2.33$

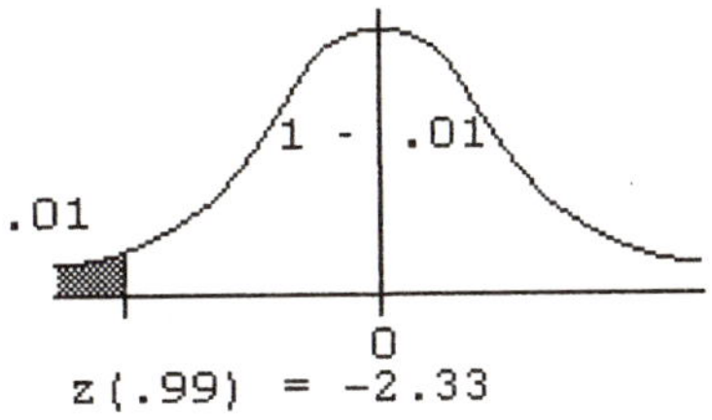

194. If $z(0.025) - z(0.975)$, find the value asked for and then with the aid of a diagram explain what your answer represents.

ANSWER:
$z(0.025) - z(0.975) = 1.96 - (-1.96) = 3.92$

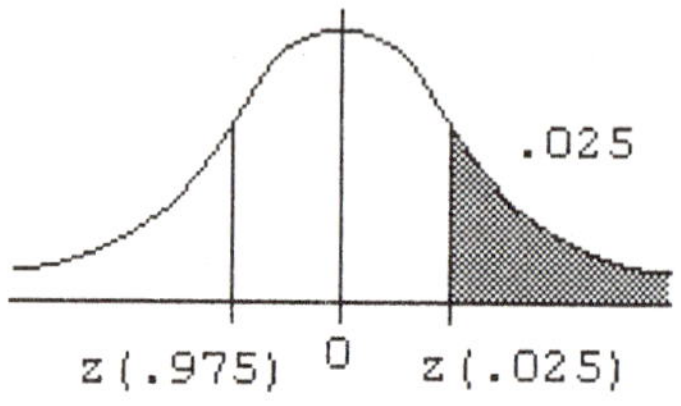

QUESTIONS 195 AND 196 ARE BASED ON THE FOLLOWING INFORMATION:
The length of life of a certain type of washer is approximately normally distributed with a mean of 6.2 years and a standard deviation of 1.4 years.

195. If this machine is guaranteed for two years, what is the probability that the machine you purchased will require replacement under guarantee?

ANSWER:
$P(x < 2) = P(z < -3.0) = 0.5000 - 0.4987 = 0.0013$

196. What period of time should the manufacturer give as a guarantee if it is willing to replace only 0.5% of the machines?

ANSWER:
The area to the left of z is 0.005, hence $z = -2.58$. Then, $-2.58 = (x - 6.2) / 1.4$. Solving for x, we get $x = (-2.58)(1.4) + 6.2 = 2.588$ years.

QUESTIONS 197 THROUGH 200 ARE BASED ON THE FOLLOWING INFORMATION:
The grades of an examination whose mean is 82 and whose standard deviation is 14 are normally distributed.

197. Anyone who scores below 55 will be retested. What percentage does this represent?

ANSWER:
$P(x < 55) = P(z < -1.93) = 0.5000 - 0.4732 = 0.0268$

198. The top 10% are to receive a special commendation. What score must be surpassed to receive this special commendation?

ANSWER:
The area to the right of z is 0.10, hence $z = 1.28$. Then, $1.28 = (x - 82)/14$. Solving for x, we get $x = (1.28)(14) + 82 = 99.92$.

199. Find the grade such that only 1% will score above it.

ANSWER:
The area to the right of z is 0.01. Hence $z = 2.33$. Then, $2.33 = (x - 82)/14$. Solving for x we get, $x = (2.33)(14) + 82 = 114.62$.

200. Find the interquartile range for the grades on this examination.0

ANSWER:
The interquartile range is the difference between Q_1 and Q_3; namely, Q_3 - Q_1. Q_1 has a z-score of –0.67 and Q_3 has a z-score of +0.67. Then, –0.67 = (Q_1 - 82) / 14. Solving for Q_1 we get, Q_1 = (-0.67)(14) + 82 = 72.62 +0.67 = (Q_3 - 82) / 14. Solving for Q_3 we get, Q_3 = (+0.67)(14) + 82 = 91.38. Therefore, the interquartile range = Q_3 - Q_1 = 91.38 – 72.62 = 18.76.

201. A vending machine can be regulated to dispense an average of μ oz of coffee per cup. If the ounces dispensed per cup are normally distributed with a standard deviation of 0.2 oz, find the setting for μ that will allow 8-oz cup to hold (without overflowing) the amount dispensed 99% of the time.

ANSWER:
The area to the left of z is 0.99. Hence z = 2.33. Then, 2.33 = (8 – μ) / 0.2. Solving for μ, we get μ = 8 - (2.33)(0.2) = 7.534.

202. The amount of time, x, spent commuting daily, one way, to college by students is believed to have a mean of 25 min with a standard deviation of 10 min. If the length of time spent commuting is approximately normally distributed, find the time, x, that separates the 25% who spend the most time commuting from the rest of the commuters.

ANSWER:
The area to the right of z is 0.25. Therefore z = 0.67. Then, 0.67 = (x –25)/10. Solving for x, we get x = (0.67)(10.0) + 25.0 = 31.7 min.

<u>QUESTIONS 203 THROUGH 207 ARE BASED ON THE FOLLOWING INFORMATION:</u>
The SAT scores attained by the students in New York City are approximately normally distributed with a mean of 500 and a standard deviation of 80.

203. Find the percentage of students who score between 550 and 650.

ANSWER:
$P(550 \leq x \leq 650) = P(0.63 \leq z \leq 1.88)$ = 0.4699 – 0.2357 = 0.2342 or 23.42%

204. Find the percentage of students who score less than 700.

ANSWER:
$P(x > 700) = P(z > 2.5)$ = 0.5000 – 0.4938 = 0.0062 or 0.62%

205. Find the 3rd quartile.

ANSWER:
Q_3 has a z-score of +0.67. Then, +0.67 = (Q_3 - 500)/80. Solving for Q_3 we get, Q_3 = (+0.67)(80) + 500 = 553.6.

206. Find the 15th percentile, P_{15}.

ANSWER:
The 15th percentile, P_{15}, has a *z*-score of –1.04. Then, -1.04 = (P_{15} - 500)/80. Solving for P_{15} we get, P_{15} = (-1.04)(80) + 500 = 416.8.

207. Find the 95th percentile P_{95}.

ANSWER:
The 95th percentile, P_{95}, has a *z*-score of +1.65. Then, +1.65 = (P_{95} - 500)/80. Solving for P_{95} we get, P_{95} = (+1.65)(80) + 500 = 632.

QUESTIONS 208 THROUGH 214 ARE BASED ON THE FOLLOWING INFORMATION:
It is known that college students sleep an average of 6 hours per night with a standard deviation equal to 1.8 hours. A student is selected at random.

208. Find the probability that he/she sleeps between 6 and 9 hours.

ANSWER:
$P(6 < x < 9) = P(0 < z < 1.67) = 0.4525$

209. Find the probability that he/she sleeps less than 5 hours.

ANSWER:
$P(x < 5) = P(z < -0.56) = 0.50 - 0.2123 = 0.2877$

210. Find the probability that he/she sleeps between 8 and 10 hours.

ANSWER:
$P(8 < x < 10) = P(1.11 < z < 2.22) = 0.4868 - 0.3665 = 0.1203$

211. Approximately 80% of college students sleep less than "*w*" hours per night. What is the value of *w*?

ANSWER:

$$P(x < w) = 0.80 \Rightarrow P\left(z < \frac{w-6}{1.8}\right) = 0.80 \Rightarrow \frac{w-6}{1.8} = 0.84 \Rightarrow w = 6 + (0.84)(1.8) \approx 7.5 \text{ hours.}$$

212. Approximately 30% of college students sleep less than "*w*" hours per night. What is the value of *w*?

ANSWER:

$$P(x < w) = 0.30 \Rightarrow P\left(z < \frac{w-6}{1.8}\right) = 0.30 \Rightarrow \frac{w-6}{1.8} = -0.52 \Rightarrow w = 6 + (-0.52)(1.8) \approx 5 \text{ hours.}$$

213. Approximately 10% of college students sleep at least "*w*" hours per night. What is the value of *w*?

ANSWER:

$$P(x \geq w) = 0.10 \Rightarrow P\left(z \geq \frac{w-6}{1.8}\right) = 0.10 \Rightarrow \frac{w-6}{1.8} = 1.28 \Rightarrow w = 6 + (1.28)(1.8) \approx 8.3 \text{ hours.}$$

214. Approximately 25% of college students sleep at most "*w*" hours per night. What is the value of *w*?

ANSWER:

$P(x \le w) = 0.25 \Rightarrow P\left(z \le \frac{w-6}{1.8}\right) = 0.25 \Rightarrow \frac{w-6}{1.8} = -0.67 \Rightarrow w = 6 + (-0.67)(1.8) \approx 4.8$ hours.

QUESTIONS 215 THROUGH 220 ARE BASED ON THE FOLLOWING INFORMATION:
Assume that *x* is normally distributed random variable with a mean of 30 and a standard deviation of 6.

215. Find $P(x < 30)$.

ANSWER:
$P(x < 30) = P(z < 0) = 0.50$

216. Find $P(30 < x < 40)$.

ANSWER:
$P(30 < x < 40) = P(0 < z < 1.67) = 0.4525$

217. Find $P(26 < x < 42)$.

ANSWER:
$P(26 < x < 42) = P(-0.67 < z < 2.0) = 0.2486 + 0.4772 = 0.7258$

218. Find $P(32 < x < 47)$.

ANSWER:
$P(32 < x < 47) = P(0.33 < z < 2.83) = 0.4977 – 0.1293 = 0.3684$

219. Find $P(21 < x < 37)$.

ANSWER:
$P(21 < x < 37) = P(-1.50 < z < 1.17) = 0.4332 + 0.3790 = 0.8122$

220. Find $P(x < 50)$.

ANSWER:
$P(x < 50) = P(z < 3.33) = 0.50 + 0.4996 = 0.9996$

QUESTIONS 221 THROUGH 231 ARE BASED ON THE FOLLOWING INFORMATION:
Suppose that daycare costs are normally distributed with a mean equal to $10,000 a year and a standard deviation equal to $2,000.

221. What percentage of daycare centers will cost between $8,000 and $12,000?

ANSWER:
$P(8{,}000 < x < 12{,}000) = P(-1.0 < z < 1.0) = 0.3413 + 0.3413 = 0.6826$

222. What percentage of daycare centers will cost between \$6,000 and \$14,000?

ANSWER:
$P(6{,}000 < x < 14{,}000) = P(-2.0 < z < 2.0) = 0.4772 + 0.4772 = 0.9544$

223. What percentage of daycare centers will cost between \$4,000 and \$16,000?

ANSWER:
$P(4{,}000 < x < 16{,}000) = P(-3.0 < z < 3.0) = 0.4987 + 0.4987 = 0.9974$

224. Compare the results of questions 221, 222, and 223 with the Empirical Rule. Explain the relationship.

ANSWER:
The Empirical Rule states that "If a variable is normally distributed, then: within one standard deviation of the mean there will be approximately 68% of the data; within two standard deviations of the mean there will approximately 95% of the data; and within three standard deviations of the mean there will be approximately 99.7% of the data." The answers to questions 221, 222, and 223 are 0.6826, 0.9544, and 0.9974, respectively. Since 0.6826 $\approx$ 68%, 0.9544 $\approx$95%, and 0.9974 $\approx$ 99.7%, our results are the same as stated in the Empirical Rule.

225. What percentage of daycare centers will cost between \$7,400 and \$11,000?

ANSWER:
$P(7{,}400 < x < 11{,}000) = P(-1.30 < z < 0.50) = 0.4032 + 0.1915 = 0.5947$

226. What percentage of daycare centers will cost between \$5,600 and \$12,800?

ANSWER:
$P(5{,}600 < x < 12{,}800) = P(-2.20 < z < 1.40) = 0.4861 + 0.4192 = 0.9053$

227. What percentage of daycare centers will cost between \$3,800 and \$14,600?

ANSWER:
$P(3{,}800 < x < 14{,}600) = P(-3.10 < z < 2.30) = 0.4990 + 0.4893 = 0.9883$

228. Approximately 80% of daycare costs are less than "*w*" dollars per year. What is the value of *w*?

ANSWER:

$$P(x < w) = 0.80 \Rightarrow P\left(z < \frac{w - 10{,}000}{2{,}000}\right) = 0.80 \Rightarrow \frac{w - 10{,}000}{2{,}000} = 0.84$$
$$\Rightarrow w = 10{,}000 + (0.84)(2{,}000) = \$11{,}680.$$

229. Approximately 30% of daycare costs are less than "*w*" dollars per year. What is the value of *w*?

ANSWER:

$$P(x < w) = 0.30 \Rightarrow P\left(z < \frac{w - 10{,}000}{2{,}000}\right) = 0.30 \Rightarrow \frac{w - 10{,}000}{2{,}000} = -0.52$$
$$\Rightarrow w = 10{,}000 + (-0.52)(2{,}000) = \$8{,}960.$$

230. Approximately 10% of daycare costs are at least "*w*" dollars per year. What is the value of *w*?

ANSWER:

$P(x \geq w) = 0.10 \Rightarrow P\left(z \geq \frac{w-10,000}{2,000}\right) = 0.10 \Rightarrow \frac{w-10,000}{2,000} = 1.28$

$\Rightarrow w = 10,000 + (1.28)(2,000) = \$12,560.$

231. Approximately 25% of daycare costs are at most "*w*" dollars per year. What is the value of *w*?

ANSWER:

$P(x \leq w) = 0.25 \Rightarrow P\left(z \leq \frac{w-10,000}{2,000}\right) = 0.25 \Rightarrow \frac{w-10,000}{2,000} = -0.67$

$\Rightarrow w = 10,000 + (-0.67)(2,000) = \$8,660$

QUESTIONS 232 THROUGH 237 ARE BASED ON THE FOLLOWING INFORMATION:
Final score averages are typically approximately normally distributed with a mean of 75 and a standard deviation of 13. Your professor says that the top 8% of the class will receive an A; The next 20%, a B; the next 42%, a C; the next 18% a D; and the bottom 12% an F.

232. What average must you exceed to obtain an A?

ANSWER:

$P(x \geq A) = 0.08 \Rightarrow P\left(z \geq \frac{A-75}{13}\right) = 0.08 \Rightarrow \frac{A-75}{13} = 1.41$

$\Rightarrow A = 75 + (1.41)(13) = 93.33 \approx 93.3$

233. What average must you exceed to obtain a B?

ANSWER:

$P(x \geq B) = 0.28 \Rightarrow P\left(z \geq \frac{B-75}{13}\right) = 0.28 \Rightarrow \frac{B-75}{13} = 0.58$

$\Rightarrow B = 75 + (0.58)(13) = 82.54 \approx 82.5$

234. What average must you exceed to receive a grade better than a C?

ANSWER:

$P(x \geq C) = 0.70 \Rightarrow P\left(z \geq \frac{C-75}{13}\right) = 0.70 \Rightarrow \frac{C-75}{13} = -0.52$

$\Rightarrow C = 75 + (-0.52)(13) = 68.24 \approx 68.2$

235. What average must you obtain to pass the course? (You'll need a "D" grade or better.)

ANSWER:

$P(x \geq D) = 0.88 \Rightarrow P\left(z \geq \frac{D-75}{13}\right) = 0.88 \Rightarrow \frac{D-75}{13} = -0.1.175$

$\Rightarrow D = 75 + (-1.175)(13) = 59.725 \approx 59.7$

236. Find the 90th percentile for the variable "final averages".

ANSWER:

$P(x < 90^{th}$ percentile) = 0.90 $\Rightarrow P\left(z < \dfrac{\text{90th percentile} - 75}{13}\right) = 0.90$

$\Rightarrow \dfrac{\text{90th percentile} - 75}{13} = 1.28$

$\Rightarrow 90^{th}$ percentile = 75 + (1.28)(13) = 91.64 $\approx$ 91.6.

237. Find the first quartile for the variable "final averages".

ANSWER:

$P(x < Q_1) = 0.25 \Rightarrow P\left(z < \dfrac{Q_1 - 75}{13}\right) = 0.25 \Rightarrow \dfrac{Q_1 - 75}{13} = -0.67$

$\Rightarrow Q_1 = 75 + (-0.67)(13) = 66.29 \approx 66.3$

238. The weights of ripe watermelons grown at Mr. Cooper's farm are normally distributed with a standard deviation of 2.5 lbs. Find the mean weight of Mr. Cooper's ripe watermelons if only 5% weigh less than 12 lbs.

ANSWER:

$P(x < 12) = 0.05 \Rightarrow P\left(z < \dfrac{12 - \mu}{2.5}\right) = 0.05 \Rightarrow \dfrac{12 - \mu}{2.5} = -1.645$

$\Rightarrow \mu = 12 - (2.5)(-1.645) = 16.1125 \approx 16.11$ lbs

239. A machine fills containers with a mean weight per container of 16.0 oz. If no more than 5% of the containers are to weigh less than 15.75 oz, what must the standard deviation of the weights equal: Assume the weights are normally distributed.

ANSWER:

$P(x < 15.75) = 0.05 \Rightarrow P\left(z < \dfrac{15.75 - 16}{\sigma}\right) = 0.05 \Rightarrow \dfrac{-0.25}{\sigma} = -1.645 \Rightarrow \sigma = 0.152$

240. Find the area under the normal curve for z between $z(0.90)$ and $z(0.05)$.

ANSWER:
The area to the right of $z(0.90)$ is 0.90; the area to the right of $z(0.05)$ is 0.05; therefore the area between $z(0.90)$ and $z(0.05)$ is found by 0.90 - 0.05, which is 0.85.

241. Find $z(0.025) - z(0.95)$.

ANSWER:
$z(0.025) - z(0.95) = 1.96 - (-1.645) = 3.605$

QUESTIONS 242 THROUGH 245 ARE BASED ON THE FOLLOWING INFORMATION:
The *z* notation, $z(\alpha)$, combines two related concepts, the *z*-score and the area to the right, into a mathematical symbol.

242. If z(A) = 0.15, identify the letter A as being a *z*-score or being an area.

ANSWER:
A is an area. *z* is 0.15 and the area to the right of $z = 0.15$ is 0.5000 - 0.0596 = 0.4404.

243. If *z*(0.15) = B, identify the letter B as being a *z*-score or being an area.

ANSWER:
B is a *z*-score. 0.15 is the area to the right of $z = B$. Use 0.35 to look up the *z*-score on the standard normal table. B = *z* =1.04.

244. If *z*(C) = -0.04, identify the letter C as being a *z*-score or being an area.

ANSWER:
C is an area. *z* is -0.04 and the area to the right of $z = -0.04$ is 0.5000 + 0.0160 = 0.516.

245. If –*z*(0.04) = D, identify the letter D as being a *z*-score or being an area.

ANSWER:
D is a *z*-score. D is to the left of zero (negative), use 0.46 to look it up; D = *z* = -1.75.

QUESTIONS 246 THROUGH 249 ARE BASED ON THE FOLLOWING INFORMATION:
Understanding the *z* notation, $z(\alpha)$, requires us to know whether we have a *z*-score or an area. Expressions use the *z* notation in a variety of ways, some typical and some not so typical.

246. Find *z*(0.09).

ANSWER:
z(0.09) = 1.34

247. Find the area between *z*(0.89) and *z*(0.11).

ANSWER:
Area between *z*(0.89) and *z*(0.11) = 0.89 – 0.11 = 0.78

248. Find *z*(1.00 – 0.03).

ANSWER:
z(1.00 – 0.03) = *z*(0.97) = -1.88

249. Find *z*(0.02) – *z*(0.98).

ANSWER:
z(0.02) – *z*(0.98) = 2.05 – (-2.05) = 4.1

Chapter 6

QUESTIONS 250 THROUGH 255 ARE BASED ON THE FOLLOWING INFORMATION:
The long-term record for weather shows that for Northeast States, the annual precipitation has a mean of 39.50 inches and a standard deviation of 4.30 inches. Assume the annual precipitation amount has a normal distribution.

250. What is the probability that next year the precipitation amount is more than 45.0 inches?

ANSWER:
$P(x > 45) = P(z > 1.28) = 0.5 - 0.3997 = 0.1003$

251. What is the probability that next year the precipitation amount is between 44.0 and 48.0 inches?

ANSWER:
$P(44 < x < 48) = P(1.05 < z < 1.98) = 0.4761 - 0.3531 = 0.123$

252. What is the probability that next year the precipitation amount is between 30.0 and 38.0 inches?

ANSWER:
$P(30 < x < 38) = P(-2.21 < z < -0.35) = 0.4864 - 0.1368 = 0.3496$

253. What is the probability that next year the precipitation amount is more than 36.0 inches?

ANSWER:
$P(x > 36) = P(z > -0.81) = 0.50 + 0.291 = 0.791$

254. What is the probability that next year the precipitation amount is less than 50.0 inches?

ANSWER:
$P(x < 50) = P(z < 2.44) = 0.50 + 0.4927 = 0.9927$

255. What is the probability that next year the precipitation amount is less than 33.0 inches?

ANSWER:
$P(x < 33) = P(z < -1.51) = .50 - 0.4345 = 0.0655$

QUESTIONS 256 AND 257 ARE BASED ON THE FOLLOWING INFORMATION:
The length of life of a certain type of DVD is approximately normally distributed with a mean of 5.0 years and a standard deviation of 1.5 years.

256. If this type of DVD is guaranteed for 2 years, what is the probability that the DVD you purchased will require replacement under the guarantee?

ANSWER:
$P(x < 2) = P(z < -2.0) = 0.50 - 0.4772 = 0.0228$

257. What period of time should the manufacturer of this type of DVD give as a guarantee if it is willing to replace only 0.5% of the DVDs?

ANSWER:

$$P(x < t) = 0.005 \Rightarrow P\left(z < \frac{t-5}{1.5}\right) = 0.005 \Rightarrow \frac{t-5}{1.5} = -2.575$$

$$\Rightarrow t = 5 + (-2.575)(1.5) = 1.138 \text{ years} \approx 1.12 \text{ years}$$

QUESTIONS 258 THROUGH 263 ARE BASED ON THE FOLLOWING INFORMATION:
The grades on an examination whose mean is 475 and whose standard deviation is 75 are normally distributed.

258. Anyone who scores below 315 will be retested. What percentage does this represent?

ANSWER:
$P(x < 315) = P(z < -2.13) = 0.50 - 0.4834 = 0.0166$

259. The top 10% are to receive a special commendation. What score must be surpassed to receive this special commendation?

ANSWER:
$$P(X > a) = 0.10 \Rightarrow P\left(z > \frac{a-475}{75}\right) = 0.10 \Rightarrow \frac{a-475}{75} = 1.28$$
$$\Rightarrow a = 47\ 5 + (1.28)(75) = 571$$

260. Find the first quartile for the grades on this examination.

ANSWER:
$$P(x < Q_1) = 0.25 \Rightarrow P\left(z < \frac{Q_1-475}{75}\right) = 0.25 \Rightarrow \frac{Q_1-475}{75} = -0.67$$
$$\Rightarrow Q_1 = 475 + (-0.67)(75) = 424.75$$

261. Find the third quartile for the grades on this examination.

ANSWER:
$$P(x < Q_3) = 0.75 \Rightarrow P\left(z < \frac{Q_3-475}{75}\right) = 0.75 \Rightarrow \frac{Q_3-475}{75} = 0.67$$
$$\Rightarrow Q_3 = 475 + (0.67)(75) = 525.25$$

262. Recall that the interquartile range of a distribution is the difference between the first and third quartiles. Find the interquartile range for the grades on this examination.

ANSWER:
Interquartile range = Q_3 - Q_1 = 525.25 – 424.75 = 100.5

263. Find the grade such that only 1% will score above it. What does this grade represent?

ANSWER:
$$P(x > a) = 0.01 \Rightarrow P\left(z > \frac{a-475}{75}\right) = 0.01 \Rightarrow \frac{a-475}{75} = 2.33$$
$$\Rightarrow a = 475 + (2.33)(75) = 649.75$$
This grade represents the 99th percentile for the grades on this examination.

Section 6.6

True-False Questions

264. For a binomial distribution with a fixed value of p, the binomial distribution begins to look like a normal distribution as n increases in size.
ANSWER: T

265. A binomial distribution has $n = 100$ and $p = 0.01$. The normal distribution provides a reasonable approximation to the probability of getting two or fewer successes in the 100 trials.
ANSWER: F

266. Every binomial distribution may be approximated reasonably by an appropriate normal distribution.
ANSWER: F

267. In a binomial distribution, if $np \geq 5$ then it must follow that $n(1 - p) \geq 5$.
ANSWER: F

268. Probabilities associated with a binomial distribution can be reasonably estimated by using the normal probability distribution.
ANSWER: T

269. The addition and subtraction of 0.5 to the z-value from a discrete variable is commonly called the continuity correction factor. It is a common method of converting a continuous variable into a discrete variable.
ANSWER: F

270. The binomial random variable is discrete, whereas the normal random variable is continuous.
ANSWER: T

271. The normal distribution provides a reasonable approximation to a binomial probability distribution whenever the values of np and nq both equal or exceed 5.
ANSWER: T

Multiple-Choice Questions

272. Consider the binomial random variable x with $n = 50$ and $p = 0.5$. Suppose we want to use a normal approximation to find the probability of at least 30 successes. A reasonable approximation would be obtained by computing:

A) $P(29.5 < x < 30.5)$.
B) $P(x < 30.5)$
C) $P(x > 29.5)$
D) $P(59.5 < x < 100.5)$
ANSWER: C

273. In which of the following binomial distributions is the normal approximation appropriate?

A) $n = 50$, $p = 0.01$
B) $n = 500$, $p = 0.001$
C) $n = 100$, $p = 0.05$
D) $n = 50$, $p = 0.02$
ANSWER: C

Applied and Computational Questions

274. In a southern state, 5% of all individuals who drive automobiles are not properly licensed. Use the normal approximation of the binomial distribution to find the probability that among 200 randomly selected individuals, between seven and nine, inclusive, are not properly licensed.

ANSWER:
0.3093

275. Find the 90th percentile for a binomial distribution having 400 identical trials, and probability of success of 0.1.

ANSWER:
48

276. If 15% of the population is left-handed, find the probability that in a class of 35 students that 3 or fewer are left-handed.

ANSWER:
0.2033

QUESTIONS 277 AND 278 ARE BASED ON THE FOLLOWING INFORMATION:
Consider a binomial distribution with 15 identical trials, and probability of success of 0.5.

277. Find the probability that $x = 2$ using the binomial tables.

ANSWER:
0.003

278. Use the normal approximation to find the probability that $x = 2$.

ANSWER:
0.004

279. A machine cuts circular filters from large rolls of material. If 7.3% of the filters fail to meet specifications, use the normal approximation to the binomial to compute the probability that a sample of 100 of the filters will contain 5 or fewer that fail to meet specifications.

ANSWER:
0.2451

Chapter 6

QUESTIONS 280 AND 281 ARE BASED ON THE FOLLOWING INFORMATION:
Consider a binomial distribution with 12 identical trials, probability of success of 0.3.

280. Find the probability that $x = 3$ using the binomial tables.

ANSWER:
0.023

281. Use the normal approximation to find the probability that $x = 3$.

ANSWER:
0.022

QUESTIONS 282 AND 283 ARE BASED ON THE FOLLOWING INFORMATION:
Consider a binomial distribution with 15 identical trials, and probability of success of 0.5.

282. Use the binomial tables to find $P(3 < x \leq 7)$.

ANSWER:
0.733

283. Use the normal approximation to find $P(3 < x \leq 7)$.

ANSWER:
0.734

284. Use the normal approximation of the binomial distribution to find the probability of obtaining at least 60 heads when a coin is flipped 100 times.

ANSWER:
0.0287

285. If 68% of all individuals who take a qualifying examination fail it on the first attempt, use the normal approximation of a binomial distribution to find the probability that in a group of 171 individuals taking the examination for the first time at least 60 will pass.

ANSWER:
0.2177

286. A drug manufacturer states that only 5% of the patients using a high blood pressure drug will experience side effects. Doctors at a large university hospital use the drug in treating 200 patients. What is the probability that 15 or fewer of the 200 patients experience side effects?

ANSWER:
Let x represent the number of patients in the 200 who will experience a side effect.
$\mu = np = (200)(0.05) = 10.0$

$\sigma = \sqrt{npq} = \sqrt{(200)(0.05)(0.95)} = 3.082$

The formula $z = (x - \mu)/\sigma$ reduces to $z = (x - 10.0) / 3.082$. Then, $P(x \leq 15) \approx P(x < 15.5) = P(z < 1.78) = 0.5000 + 0.4625 = 0.9625$.

287. Suppose we have a binomial distribution with n = 180 and p = 0.45. Furthermore, suppose we want to use a normal approximation to find the probability of at least 120 successes. Explain why we need only compute $P(x > 119.5)$ instead of $P(119.5 < x < 180.5)$.

ANSWER:
The value 180.5 converts to a z-score of 14.91. For all practical purposes, the area from $z = 0$ to 14.91 is 0.5. Therefore, $P(x > 119.5) \approx P(119.5 < x < 180.5)$.

288. Find the normal approximation for the binomial probability $P(x = 6)$, where n = 15 and p = 0.4. Compare this to the value of $P(x = 6)$ obtained from the binomial table.

ANSWER:
$\mu = np = (15)(0.4) = 6.0$, and $\sigma = \sqrt{npq} = \sqrt{(15)(0.4)(0.6)} = 1.897$
The formula $z = (x - \mu)/\sigma$ reduces to $z = (x - 6.0) / 1.897$. Then, $P(x = 6) = P(5.5 < x < 6.5) = P(-0.26 < z < 0.26) = 2(0.1026) = 0.2052$. Using the table of binomial probabilities, we have: $P[x = 6 \mid B(n = 15, p = 0.4)] = 0.207$.

289. If 25% of all students entering a certain university drop out during or at the end of their first year, what is the probability that more than 550 of this year's entering class of 2000 will drop out during or at the end of their first year?

ANSWER:
$\mu = np = (2000)(0.25) = 500.0$

$\sigma = \sqrt{npq} = \sqrt{(2000)(0.25)(0.75)} = 19.3649$

The formula $z = (x - \mu)/\sigma$ reduces to $z = (x - 500.0) / 19.3649$. Then,

$P(x > 550) = P(x \geq 551) = P(x > 550.5) = P(z > 2.61) = 0.5000 - 0.4955 = 0.0045$.

QUESTIONS 290 THROUGH 293 ARE BASED ON THE FOLLOWING INFORMATION:
A test-scoring machine is known to record an incorrect grade on 5% of the exams it grades.

290. Find by the appropriate method, the probability that the machine records 2 wrong grades in a set of 10 exams.

ANSWER:
$P(\text{2 wrong in 10}) = P[x = 2 \mid B(n = 10, p = 0.05)] = 0.075$

291. Find by the appropriate method, the probability that the machine records no more than 2 wrong grades in a set of 10 exams.

ANSWER:
$P(\text{no more than 2 wrong in 10}) = P[x = 0, 1, 2 \mid B(n = 10, p = 0.05)]$
$= 0.599 + 0.315 + 0.075 = 0.989$

292. Find by the appropriate method, the probability that the machine records no more than 2 wrong grades in a set of 15 exams.

ANSWER:
$P(\text{no more than 2 wrong in 15}) = P[x = 0,1, 2 \mid B(n = 15, p = 0.05)]$
$= 0.463 + 0.366 + 0.135 = 0.964$

293. Find by the appropriate method, the probability that the machine records no more than 2 wrong grades in a set of 150 exams.

ANSWER:
P(no more than 2 wrong in 150) = $P[x \le 2 \mid B(150, 0.05)]$

$\mu = np = (150)(0.05) = 7.5$

$\sigma = \sqrt{npq} = \sqrt{(150)(0.05)(0.95)} = 2.6693$

Then,

$P(x \le 2) = P(x < 2.5) = P[z < (2.5 - 7.5)/2.6693]$
$= P(z < -1.87) = 0.5000 - 0.4693 = 0.0307$

QUESTIONS 294 THROUGH 296 ARE BASED ON THE FOLLOWING INFORMATION:
It is believed that 60% of married couples with children agree on methods of disciplining their children. Assuming this to be the case, in a random survey of 200 married couples is conducted by a researcher.

294. What is the probability that exactly 115 couples who agree?

ANSWER:
P (exactly 115 of 200 agree) = $P[x = 115 \mid B(200, 0.60)]$
$\mu = np = (200)(0.60) = 120.0$, and

$\sigma = \sqrt{npq} = \sqrt{(200)(0.60)(0.40)} = 6.9282$

Then,

$P(x = 115) = P(114.5 < x < 115.5) = P[(114.5 - 120)/6.9282 < z < (115.5 - 120)/6.9282]$
$= P(-0.79 < z < -0.65) = 0.2852 - 0.2422 = 0.043$

295. What is the probability that fewer than 115 couples who agree?

ANSWER:
$P(x < 115) = P(x \le 114) = P(x < 114.5) = P[z < (114.5 - 120)/6.9282] = P(z < -0.79)$
$= 0.5000 - 0.2852 = 0.2148$

296. What is the probability that more than 110 couples who agree?

ANSWER:
$P(x > 110) = P(x \ge 111) = P(x > 110.5) = P[z > (110.5 - 120)/6.9282] = P(z > -1.37)$
$= 0.5000 + 0.4147 = 0.9147$

297. For a binomial distribution with $n = 10$, and $p = 0.2$, does the normal distribution provide a reasonable approximation? Use the "rule of thumb" for normal approximation to the binomial distribution.

ANSWER:
Since $np = 2 < 5$ and $nq = 8 > 5$, the normal approximation to the binomial distribution is not appropriate in this case.

298. For a binomial distribution with $n = 100$, and $p = 0.05$, does the normal distribution provide a reasonable approximation? Use the "rule of thumb" for normal approximation to the binomial distribution.

ANSWER:
Since $np = 5 \geq 5$ and $nq = 95 > 5$, the normal approximation to the binomial distribution is appropriate in this case.

299. For a binomial distribution with $n = 600$, and $p = 0.1$, does the normal distribution provide a reasonable approximation? Use the "rule of thumb" for normal approximation to the binomial distribution.

ANSWER:
Since $np = 60 > 5$ and $nq = 540 > 5$, the normal approximation to the binomial distribution is appropriate in this case.

300. For a binomial distribution with $n = 50$, and $p = 0.4$, does the normal distribution provide a reasonable approximation? Use the "rule of thumb" for normal approximation to the binomial distribution.

ANSWER:
Since $np = 20 > 5$ and $nq = 30 > 5$, the normal approximation to the binomial distribution is appropriate in this case.

301. For a binomial distribution with $n = 12$, and $p = 0.4$, does the normal distribution provide a reasonable approximation? Use the "rule of thumb" for normal approximation to the binomial distribution.

ANSWER:
Since $np = 4.8 < 5$ and $nq = 7.2 > 5$, the normal approximation to the binomial distribution is not appropriate in this case.

<u>**QUESTIONS 302 THROUGH 304 ARE BASED ON THE FOLLOWING INFORMATION:**</u>
In order to see what happens when the normal approximation is improperly used, consider the binomial distribution with $n = 10$ and $p = 0.7$. Since $np = 7$ and $nq = 3$, the rule of thumb ($np \geq 5$ and $nq \geq 5$) is not satisfied.

302. Find the probability of eight or more successes using the binomial tables.

ANSWER:
$P(x \geq 8) = 0.233 + 0.121 + 0.028 = 0.382$

303. Find the probability of eight or more successes using the normal approximation.

ANSWER:
Since $\mu = np = 7.0$ and $\sigma = \sqrt{npq} = \sqrt{(10)(0.70)(0.30)} = 1.449$, then

$P(x \geq 8)$ (for a discrete random variable x)
$= P(x \geq 7.5)$ (for a continuous random variable x)
$= P\left(z \geq \frac{7.5 - 7.0}{1.449}\right) = P(z \geq 0.35) = 0.50 - 0.1368 = 0.3632$

304. Compare your answers to questions 302 and 303.

ANSWER:
There is a difference of 0.0188 between the two answers.

QUESTIONS 305 AND 306 ARE BASED ON THE FOLLOWING INFORMATION:
Consider a binomial distribution with n = 14 and p = 0.6.

305. Find the probability of seven successes using the binomial tables.

ANSWER:
$P(x = 7) = 0.157$

306. Find the probability of seven successes using the normal approximation.

ANSWER:
Since $\mu = np = 8.40$ and $\sigma = \sqrt{npq} = \sqrt{(14)(0.60)(0.40)} = 1.833$, then

$P(x = 7)$ (for a discrete random variable x)
$= P(6.5 < x < 7.5)$ (for a continuous random variable x)

$$= P\left(\frac{6.5-8.4}{1.833} < z < \frac{7.5-8.4}{1.833}\right) = P(-1.04 < z < -0.49) = 0.3508 - 0.1879 = 0.1629$$

QUESTIONS 307 AND 308 ARE BASED ON THE FOLLOWING INFORMATION:
Consider a binomial distribution with n = 12 and p = 0.4.

307. Find the probability of five successes using the binomial tables.

ANSWER:
$P(x = 5) = 0.227$

308. Find the probability of five successes using the normal approximation.

ANSWER:
Since $\mu = np = 4.8$ and $\sigma = \sqrt{npq} = \sqrt{(12)(0.40)(0.60)} = 1.697$, then

$P(x = 5)$ (for a discrete random variable x)
$= P(4.5 < x < 5.5)$ (for a continuous random variable x)

$$= P\left(\frac{4.5-4.8}{1.697} < z < \frac{5.5-4.8}{1.697}\right) = P(-0.18 < z < 0.41) = 0.0714 + 0.1591 = 0.2305$$

QUESTIONS 309 AND 310 ARE BASED ON THE FOLLOWING INFORMATION:
Consider a binomial distribution with n = 12 and p = 0.05.

309. Find the probability of one or fewer successes using the normal approximation.

ANSWER:
Since $\mu = np = 0.60$ and $\sigma = \sqrt{npq} = \sqrt{(12)(0.05)(0.95)} = 0.755$, then

$P(x \leq 1)$ (for a discrete random variable x)
$= P(x < 1.5)$ (for a continuous random variable x)

$$= P\left(z < \frac{1.5-0.60}{0.755}\right) = P(z < 1.19) = 0.50 + 0.383 = 0.883$$

310. Find the probability of one or fewer successes using the binomial tables.

ANSWER:
$P(x \leq 1) = 0.540 + 0.341 = 0.881$

311. If 25% of all students entering a certain university drop out during or at the end of their first year, what is the probability that more than 420 of this year's entering class of 1500 will drop out during or at the end of their first year?

ANSWER:
Since np = (1500)(0.25) = 375.0 > 5 and nq = (1500)(0.75) = 1125 > 5, the normal approximation to the binomial is appropriate. Now,
$\mu = np = 375$ and $\sigma = \sqrt{npq} = \sqrt{(1500)(0.25)(0.75)} = 25.617$, then
$P(x > 420) = P(x \geq 421)$ (for a discrete random variable x)
$= P(x \geq 420.5)$ (for a continuous random variable x)
$= P\left(z \geq \frac{420.5 - 375.0}{25.617}\right) = P(z \geq 1.78) = 0.50 - 0.4625 = 0.0375$

__QUESTIONS 312 THROUGH 316 ARE BASED ON THE FOLLOWING INFORMATION:__
Suppose that x has a binomial distribution with $n = 25$ and $p = 0.4$.

312. Explain why the normal approximation to the binomial distribution is reasonable.

ANSWER:
Since np = (25)(0.40) = 10 > 5 and nq = (25)(0.60) = 15 > 5, the normal approximation to the binomial distribution is reasonable.

313. Find the mean and standard deviation of the normal distribution that is used in the approximation.

ANSWER:
Mean = $\mu = np = 10$ and standard deviation = $\sigma = \sqrt{npq} = \sqrt{(25)(0.40)(0.60)} = 2.449$

314. Approximate the probability of $x = 5$.

ANSWER:
$P(x = 5)$ (for a discrete random variable x)
$= P(4.5 < x < 5.5)$ (for a continuous random variable x)
$= P\left(\frac{4.5 - 10.0}{2.449} < z < \frac{5.5 - 10.0}{2.449}\right) = P(-2.25 < z < -1.84) = 0.4878 - 0.4671 = 0.0207$

315. Use the binomial formula to find the probability of $x = 5$

ANSWER:
$P(x = 5) = \binom{25}{5}(0.4)^5(0.6)^{20} = 0.0199$

316. Compare your answers to questions 314 and 315.

ANSWER:
In this situation, the normal approximation to the binomial is excellent. The difference between the two answers is 0.0207 – 0.0199 = 0.0008.

Chapter 6

QUESTIONS 317 THROUGH 321 ARE BASED ON THE FOLLOWING INFORMATION:
It is believed that 60% of married couples with children agree on methods of disciplining their children. Assume that a survey of 250 married couples is conducted.

317. Explain why the normal approximation to the binomial distribution is reasonable.

ANSWER:
Since np = (250)(0.60) = 150 > 5 and nq = (250)(0.40) = 100 > 5, the normal approximation to the binomial distribution is reasonable.

318. Find the mean and standard deviation of the normal distribution that is used in the approximation.

ANSWER:
Mean = $\mu = np = 150$ and standard deviation = $\sigma = \sqrt{npq} = \sqrt{(250)(0.60)(0.40)} = 7.746$

319. What is the probability we would find exactly 135 couples who agree?

ANSWER:
$P(x = 135)$ (for a discrete random variable x)
$= P(134.5 < x < 13\ 5.5)$ (for a continuous random variable x)
$= P\left(\frac{134.5-150.0}{7.746} < z < \frac{135.5-150.0}{7.746}\right) = P(-2.00 < z < -1.87) = 0.4772 - 0.4693 = 0.0079$

320. What is the probability we would find fewer than 135 couples who agree?

ANSWER:
$P(x < 135)$ (for a discrete random variable x)
$= P(x < 134.5)$ (for a continuous random variable x)
$= P\ z[(134.5-150.0)/7.746]$
$= P(z < -2.00) = 0.50 - 0.4772 = 0.0228$

321. What is the probability we would find more than 135 couples who agree?

ANSWER:
$P(x > 135)$ (for a discrete random variable x)
$= P(x > 135.5)$ (for a continuous random variable x)
$= P\ z[(135.5-150.0)/7.746]$
$= P(z > -1.87) = 0.50 + 0.4693 = 0.9693$

QUESTIONS 322 THROUGH 325 ARE BASED ON THE FOLLOWING INFORMATION:
A recent study showed that 75% of commercial airline flights in and out of the US airports were on-time arrivals and 19% were on late departures. Three hundred flights are to be randomly identified from all flights and their flight logs examined closely.

322. What is the mean and standard deviation of commercial airline flights in and out of the US airports that were on-time arrivals?

ANSWER:
Mean = $\mu = np = (300)(0.75) = 225$

Standard deviation = $\sigma = \sqrt{npq} = \sqrt{(300)(0.75)(0.25)} = 7.5$

323. What is the probability that more than 80% of the sample will be on-time arrival?

ANSWER:
Since 80% of 300 is 240, then
$P(x > 240)$ (for a discrete random variable x)
$= P(x > 240.5)$ (for a continuous random variable x)
$= P\, z[(240.5 - 225)/7.5]$
$= P(z > 2.07) = 0.50 - 0.4808 = 0.0192$

324. What is the mean and standard deviation of commercial airline flights in and out of the US airports that were on late departure?

ANSWER:
Mean $= \mu = np = (300)(0.19) = 57$
Standard deviation $= \sigma = \sqrt{npq} = \sqrt{(300)(0.19)(0.81)} = 6.795$

325. What is the probability that less than 15% of the sample will have departed late?

ANSWER:
Since 15% of 300 is 45, then
$P(x < 45)$ (for a discrete random variable x)
$= P(x < 44.5)$ (for a continuous random variable x)
$= P\, z[(44.5 - 57.0)/6.795] = P(z < -1.84) = 0.50 - 0.4671 = 0.0329.$

QUESTIONS 326 THROUGH 330 ARE BASED ON THE FOLLOWING INFORMATION:
A soda-filling machine is known to under fill an incorrect amount of soda on 5% of the cans it fills.

326. Find by the appropriate method, the probability that the machine under fills 1 can in a set of 15 cans.

ANSWER:
P(1 under filled can in 15 cans) $= P[x = 1 \mid B(n = 15, p = 0.05)] = 0.366$

327. Find by the appropriate method, the probability that the machine under fills no more than 3 cans in a set of 15 cans.

ANSWER:
P(no more than 3 under filled cans in 15 cans) $= P[x = 0, 1, 2, 3 \mid B(n = 15, p = 0.05)]$
$= 0.463 + 0.366 + 0.135 + 0.031$
$= 0.995$

328. Find by the appropriate method, the probability that the machine under fills no more than 3 cans in a set of 10 cans.

ANSWER:
P(no more than 3 under filled cans in 10 cans) $= P[x = 0, 1, 2, 3 \mid B(n = 10, p = 0.05)]$
$= 0.599 + 0.315 + 0.075 + 0.010$
$= 0.999$

329. Find by the appropriate method, the probability that the machine under fills no more than 3 cans in a set of 200 cans.

ANSWER:
P(no more than 3 under filled cans in 200 cans) = $P[x \leq 3 \mid B(200, 0.05)]$

$\mu = np = (200)(0.05) = 10.0$

$\sigma = \sqrt{npq} = \sqrt{(200)(0.05)(0.95)} = 3.3091$

Then,

$P(x \leq 3) = P(x < 3.5) = P[z < (3.5 - 10.0) / 3.3091]$
$= P(z < -1.96) = 0.5000 - 0.4750 = 0.025$

330. Find by the appropriate method, the probability that the machine under fills no less than 2 cans in a set of 12 cans.

ANSWER:
P(no less than 2 under filled cans in 12 cans) = $P[x = 2, 3, \cdots, 12 \mid B(n = 12, p = 0.05)]$
$= 1.0 - P[x = 0, 1 \mid B(n = 12, p = 0.05)]$
$= 1.0 - (0.540 + 0.341)$
$= 0.119$

Chapter 7

Sample Variability

Sections 7.1 Through 7.3

True-False Questions

1. For any population, the sampling distributions of sample means and the sampling distribution of sample ranges will have the same shape provided the same sample size is used in both distributions.
   ANSWER: F

2. In general, the term "*standard error*" is the name only used for the standard deviation of the sampling distribution of sample means.
   ANSWER: F

3. The histogram for a population and the histogram for a sampling distribution of a sample mean have the same shape.
   ANSWER: F

4. The sampling distribution of sample means will be approximately normally distributed for large samples when the parent population is *not* normally distributed.
   ANSWER: T

5. The term "standard error of the mean" has the same meaning as the "standard deviation of the sample mean."
   ANSWER: T

6. The standard error of the mean is the standard deviation of the population from which the samples have been taken.
   ANSWER: F

7. We do *not* need to repeatedly sample a population in order to use the concept of the sampling distribution.
   ANSWER: F

8. As the sample size increases, the sampling distribution of the sample means from a normal distribution has a normal curve that becomes more peaked.
ANSWER: T

9. The Central Limit Theorem provides us with a description of the three characteristics of a sampling distribution of sample medians.
ANSWER: F

10. The histograms of all sampling distributions are symmetrically shaped.
ANSWER: F

11. The standard error of the mean increases as the sample size increases.
ANSWER: F

12. A sample obtained in such a way that each possible sample of fixed size *n* has an equal probability of being selected is referred to as a random sample.
ANSWER: T

13. The Central Limit Theorem states that if all possible random samples of size *n* are taken from any population, the sampling distribution of sample means becomes approximately normal when the sample size *n* is large enough.
ANSWER: T

14. The sampling distribution of sample means is normal for samples of all sizes, provided that the parent sampled population has a normal distribution.
ANSWER: T

15. The fundamental goal of a survey is to come up with the same results that would have been obtained had every single member of a population been interviewed.
ANSWER: T

16. Central Limit Theorem states that the sampling distribution of sample means will more closely resemble the normal distribution regardless of the sample size.
ANSWER: F

17. The sampling distribution of a sample statistic is the distribution of values for a sample statistic obtained from repeated samples, all of the same size and all drawn from the same population.
ANSWER: T

18. A random sample is a sample obtained in such a way that each possible sample of fixed size *n* selected from the same population has a chance or probability of being selected.
ANSWER: F

19. If the sampled distribution is normal, then the sampling distribution of sample means (SDSM) is normal and the Central Limit Theorem does not apply.
ANSWER: T

20. The basic purpose for considering what happens when a population is repeatedly sampled is to form sampling distributions. The sampling distribution is then used to describe the variability that occurs from one sample to the next.
ANSWER: T

21. The standard error of the sample mean is the standard deviation of the population from which the samples have been selected.
ANSWER: F

22. Repeated samples are commonly used in the field of production control, in which samples are taken to determine whether a product is of the proper size or quantity. When the sample statistic does not fit the standards, a mechanical adjustment of the machinery is necessary.
ANSWER: T

23. The histograms of all sampling distributions are symmetric.
ANSWER: F

24. The mean of the sampling distribution of sample means $\bar{x}$ is equal to the mean of the population from which the samples have been selected.
ANSWER: T

Multiple-Choice Questions

25. Which of the following is *not* a characteristic of the sampling distribution of a sample statistic?

A) The distribution of values is obtained by means of repeated sampling.
B) The samples are all of size *n*.
C) The samples are all drawn from the same population.
D) The mean is zero and the standard deviation is one.
ANSWER: D

26. Assume that you have repeatedly taken samples of size 5 from a population of 30. What can be said about the individual sample means?

A) They will be the population mean.
B) They will vary, but be close to the population mean.
C) The mean of the means will equal zero.
D) The mean will equal 5.
ANSWER: B

27. As the sample size increases, what happens to the standard error of the mean ($\sigma_{\bar{x}}$)?

A) Increases
B) Decreases
C) Remains the same
D) Becomes negative
ANSWER: B

28. Given that all possible random samples of size *n* are taken from any population, which of the following would be true?

A) $\mu_{\bar{x}} = \mu$ and $\sigma_{\bar{x}} = \sigma$.
B) $\mu_{\bar{x}} < \mu$ and $\sigma_{\bar{x}} > \sigma$.
C) $\mu_{\bar{x}} = \mu$ and $\sigma_{\bar{x}} < \sigma$.
D) Need to see the raw data before can make any true statement.
ANSWER: C

29. If all possible random samples of size n are taken from a population, and the mean of each sample is determined, what can you say about the mean of the sample means?

A) It is larger than the population mean.
B) It is exactly the same as the population mean.
C) It is smaller than the population mean.
D) None of the above.
ANSWER: B

30. As the size of the sample increases, what happens to the shape of the sampling distribution of sample means?

A) Becomes positively skewed
B) Becomes negatively skewed
C) Becomes uniformly distributed
D) Becomes approximately normal
ANSWER: D

31. If all possible random samples of size n are taken from a population that is not normally distributed, and the mean of each sample is determined, what can you say about the sampling distribution of sample means?

A) It is positively skewed.
B) It is negatively skewed.
C) It is approximately normal provided that n is large enough.
D) None of the above.
ANSWER: C

32. If the standard deviation of the sampling distribution of sample means is 5.0 for samples of size 16, then the population standard deviation must be

A) 20.
B) 5.0.
C) 3.2.
D) 80.
ANSWER: A

33. Which of the following statements about the Central Limit Theorem is correct?

A) The sample mean $\overline{x}$ is always equal to the population mean μ.
B) The sampling distribution of sample means $\overline{x}$ is approximately normal for large sample sizes.
C) The sample mean $\overline{x}$ is equal to the population mean μ for large sample sizes.
D) The sampling distribution of the population mean μ is approximately normal, provided that sample size is large enough.
ANSWER: B

34. Consider a large population with a mean of 100 and a standard deviation of 21. A random sample of size 36 is taken from this population. The standard error of the sampling distribution of sample mean is equal to:

A) 16.67.
B) 3.50.
C) 12.25.
D) 1.71.
ANSWER: **B**

35. For a sampling distribution of sample means, $\sigma_{\bar{x}}$ is equal to

A) σ.
B) $\sigma / \sqrt{n}$.
C) s.
D) $\sqrt{\sigma} / n$.
ANSWER: **B**

36. If all possible samples of size *n* are taken from a large population with a mean of 30 and a standard deviation of 5, then the standard error of sample means equals 1.0 only for samples of size

A) 40.
B) 35.
C) 30.
D) 25.
ANSWER: **D**

37. Which of the following statements is false?

A) The standard error of the mean $(\sigma_{\bar{x}})$ is the standard deviation of the sampling distribution of sample means.
B) If the sampled population is not normal, the sampling distribution of sample means will still be approximately normally distributed under the right conditions.
C) The standard error of the mean $(\sigma_{\bar{x}})$ is the standard deviation of the sampling distribution of sample means.
D) None of the above
ANSWER: **A**

Short-Answer Questions

38. There are 50 possible samples of size two when selected with replacement from a total of 10 items. In order to be a random sample, each possible sample must have what probability of being selected?

ANSWER:
0.02

Chapter 7

39. Determine the number of ways that two letters can be selected from {*A, B, C, D*} if order in the sample is not to be considered. List the possible samples.

ANSWER:
6 possible ways: {A, B}, {A, C}, {A, D}, {B, C}, {B, D}, and {C, D}

40. Determine the number of ways that two letters can be selected from {*A, B, C, D*} if order in the sample is to be considered. List the possible samples.

ANSWER:
12 possible ways: {A, B}, {B, A}, {A, C}, {C, A}, {A, D}, {D, A}, {B, C}, C, B}, {B, D}, {D, B}, {D, C} and {C, D}

41. How many samples of size 5 are possible when selecting from a set of 10 distinct integers if the sampling is done with replacement?

ANSWER:
100,000 samples

42. Explain why the sample means become more variable as the sample size decreases.

ANSWER:
With a smaller sample size there will be more "gaps" between the values; as the sample size increases the "gaps" become filled in.

43. What name do we give to the standard deviation of the sampling distribution of sample means?

ANSWER:
Standard error of the mean

44. Suppose samples of size 50 are selected from the distributions listed in parts a through e below. What type of distribution will $\bar{x}$ have in each of the five cases?

(a) A uniform distribution
(b) A normal distribution
(c) A distribution that is skewed to the right
(d) A distribution that is skewed to the left
(e) A bimodal distribution

ANSWER:
The sample mean would have a normal distribution in part (b) since the parent population is normal. In all other parts, the distribution is approximately normal since n = 50 > 30, so Central Limit Theorem does apply.

45. Consider the integers {0, 1, 2, 3, 4}. If all samples of size 3 are taken, with replacement, and the sampling distribution of the sample mean is found, what would the mean of the sample mean equal?

ANSWER:
Mean of the sample means = 2

46. Consider the integers {10, 20, 30, 40, 50, 60}. If all samples of size 3 are taken, with replacement, and the sampling distribution of the sample mean is found, what would the mean of the sample mean equal?

ANSWER:
Mean of the sample means = 20

47. Discuss the effect on the standard error of the mean as the sample size increases.

ANSWER:
As the sample size increases, the standard error of the mean decreases.

48. How does the bell-shaped curve for the sampling distribution of sample means for samples of size n = 100 compare to the bell-shaped curve for the sampling distribution of sample means for samples of size n = 60?

ANSWER:
Both distributions are normally distributed. With n = 100 the distributions has a standard error of 0.1σ, while the distributions for n = 60 has a standard error of 0.129σ.

49. Abby stated that "a sampling distribution of the standard deviation tell you how the standard deviation varies from sample to sample." Debra argues that "a population distribution tells you that." Who is right? Justify your answer.

ANSWER:
Abby is right. A population distribution is a distribution formed for all x values that make up the entire population.

50. Lily says that it is the "size of each sample used" and Sue says that it is the "number of samples used" that determines the spread of an empirical sampling distribution. Who is right? Justify your answer.

ANSWER:
Lily is right. The standard error is found by dividing the standard deviation by the square root of the *sample size*.

51. If a population has a standard deviation σ of 25 units, what is the standard error of the mean if samples of size 80 are selected?

ANSWER:
$\sigma_{\bar{x}} = \sigma / \sqrt{n} = 25 / \sqrt{80} = 2.795$

52. In sampling, there is a fundamental principle called equal probability of selection. What does this principle say?

ANSWER:
The equal probability of selection principle states that if every member of a population has an equal probability of being selected in a sample, then that sample will be representative of the population.

53. If a population has a standard deviation σ of 25 units, what is the standard error of the mean if samples of size 20 are selected?

ANSWER:
$\sigma_{\bar{x}} = \sigma / \sqrt{n}$ = 25/ $\sqrt{20}$ = 5.59

54. What is the total measure of the area for any probability distribution?

ANSWER:
1.0

55. Is the statement "$\bar{x}$ becomes less variable as n increases" correct? Justify the statement

ANSWER:

Yes, the statement is correct, simply because the standard error of the sample mean $\bar{x}$ is given by $\sigma_{\bar{x}} = \sigma / \sqrt{n}$; and as n increases, the value of this fraction, the standard deviation of sample mean, gets smaller.

56. If a population has a standard deviation σ of 25 units, what is the standard error of the mean if samples of size 40 are selected?

ANSWER:
$\sigma_{\bar{x}} = \sigma / \sqrt{n}$ = 25/ $\sqrt{40}$ = 3.953

Applied and Computational Questions

QUESTIONS 57 THROUGH 59 ARE BASED ON THE FOLLOWING INFORMATION:
Consider the set of numbers {0, 1, 2, 3, 4}.

57. Make a list of all possible samples of size 2 that could be drawn with replacement from this set of numbers.

ANSWER:
All possible samples of size 2 using replacement are listed below:

(0,0)	(0,1)	(0,2)	(0,3)	(0,4)
(1,0)	(1,1)	(1,2)	(1,3)	(1,4)
(2,0)	(2,1)	(2,2)	(2,3)	(2,4)
(3,0)	(3,1)	(3,2)	(3,3)	(3,4)
(4,0)	(4,1)	(4,2)	(4,3)	(4,4)

58. Construct the sampling distribution of sample means for the samples in question 57.

ANSWER:

$\bar{x}$	0.0	0.5	1.0	1.5	2.0	2.5	3.0	3.5	4.0
$P(\bar{x})$	0.04	0.08	0.12	0.16	0.20	0.16	0.12	0.08	0.04

59. Construct the sampling distribution of sample ranges for the samples in question 57.

ANSWER:

Range	0.0	1.0	2.0	3.0	4.0
P(Range)	0.20	0.32	0.24	0.16	0.08

<u>QUESTIONS 60 THROUGH 62 ARE BASED ON THE FOLLOWING INFORMATION:</u>
Consider the set of numbers {1, 3, 5, 7}.

60. Make a list of all possible samples of size 2 that could be drawn with replacement from this set of numbers.

ANSWER:
All possible samples of size 2 using replacement are listed below:

(1,1) (1,3) (1,5) (1,7)
(3,1) (3,3) (3,5) (3,7)
(5,1) (5,3) (5,5) (5,7)
(7,1) (7,3) (7,5) (7,7)

61. Construct the sampling distribution of sample means for the samples in question 60.

ANSWER:

$\overline{x}$	1.0	2.0	3.0	4.0	5.0	6.0	7.0
$P(\overline{x})$	0.0625	0.125	0.1875	0.25	0.1875	0.125	0.0625

62. Construct the sampling distribution of sample ranges for the samples in question 60.

ANSWER:

Range	0.0	2.0	4.0	6.0
P(Range)	0.25	0.375	0.25	0.125

63. If a population has a mean equal to 25 and a standard deviation equal to 5, give the mean of the sample means and the standard error for each of the sample sizes 9, 100, 225, and 10,000, respectively. What trend do you notice for the mean and standard error?

ANSWER:

n	$\mu_{\overline{x}}$	$\sigma_{\overline{x}}$
9	25	1.667
100	25	0.500
225	25	0.333
10,000	25	0.050

The mean remains constant, but the standard error decreases as *n* increases.

QUESTIONS 64 THROUGH 69 ARE BASED ON THE FOLLOWING INFORMATION:
Let a very small population consist of five numbers: 10, 20, 30, 40, and 50, each having probability of being selected equal to 0.2. Consider all possible samples (selected with replacement) of size 2 that could be selected.

64. Find the mean of the population.

ANSWER:
μ =30.0

65. Find the standard deviation of the population.

ANSWER:
σ = 14.142

66. Find the sampling distribution of the sample mean.

ANSWER:

$\bar{x}$	10.0	15.0	20.0	25.0	30.0	35.0	40.0	45.0	50.0
$P(\bar{x})$	0.04	0.08	0.12	0.16	0.20	0.16	0.12	0.08	0.04

67. Find the mean of the sample mean using your answer to question 66.

ANSWER:
$\mu_{\bar{x}} = \sum[\bar{x} \cdot P(\bar{x})]$ = 30.0

68. Find the standard error of the mean using your answer to question 66.

ANSWER:
$\sigma_{\bar{x}} = \sqrt{\sum[\bar{x}^2 \cdot P(\bar{x})] - (\mu_{\bar{x}})^2} = \sqrt{1000 - (30)^2} = \sqrt{100} = 10$

69. Verify that $\mu_{\bar{x}} = \mu$, and $\sigma_{\bar{x}} = \sigma / \sqrt{n}$.

ANSWER:
$\mu_{\bar{x}} = 30 = \mu$, and $\sigma_{\bar{x}} = 10 \approx 14.142 / \sqrt{2} = 9.9999 = \sigma / \sqrt{n}$

QUESTIONS 70 THROUGH 72 ARE BASED ON THE FOLLOWING INFORMATION:
A certain population has a bimodal distribution with a mean of 58.5 and a standard deviation of 2.5. Many samples of size 25 are randomly selected and their means calculated.

70. What shape would you expect the distribution of all sample means to have?

ANSWER:
Approximately normal distribution

71. What value would you expect to find for the mean of the sample means?

ANSWER:
Approximately 58.5

72. What value would you expect to find for the standard deviation of the sample means?

ANSWER:
0.5

QUESTIONS 73 THROUGH 75 ARE BASED ON THE FOLLOWING INFORMATION:
Consider the set of numbers {1, 2, 3, 4}.

73. Make a list of all possible samples of size 2 that could be drawn with replacement from this set of numbers.

ANSWER:
All possible samples of size 2 using replacement are listed below:

(1,1)	(1,2)	(1,3)	(1,4)
(2,1)	(2,2)	(2,3)	(2,4)
(3,1)	(3,2)	(3,3)	(3,4)
(4,1)	(4,2)	(4,3)	(4,4)

74. Construct the sampling distribution of sample means for the samples in question 73.

ANSWER:

$\overline{x}$	1.0	1.5	2.0	2.5	3.0	3.5	4.0
$P(\overline{x})$	0.0625	0.125	0.1875	0.25	0.1875	0.125	0.0625

75. Construct the sampling distribution of sample ranges for the samples in question 73.

ANSWER:

Range	0.0	1.0	2.0	3.0
P(*Range*)	0.25	0.375	0.25	0.125

76. A pair of dice is rolled 25 times, the sum of the dice observed each time, and the mean of the 25 rolls is computed. This procedure is repeated 99 more times, and the 100 means are plotted on a histogram. The mean of the distribution will be close to what number?

ANSWER:
7

QUESTIONS 77 THROUGH 79 ARE BASED ON THE FOLLOWING INFORMATION:
Consider the set of numbers {3, 5}.

77. Make a list of all possible samples of size 3 that could be drawn with replacement from this set of numbers.

ANSWER:
All possible samples of size 2 using replacement are listed below:

(3,3,3)	(3,3,5)	(3,5,3)	(5,3,3)
(3,5,5)	(5,3,5)	(5,5,3)	(5,5,5)

78. Construct the sampling distribution of sample means for the samples in question 77.

ANSWER:

$\overline{x}$	3.0	3.67	4.33	5.0
$P(\overline{x})$	0.125	0.375	0.375	0.125

79. Construct the sampling distribution of sample ranges for the samples in question 77.

ANSWER:

Range	0.0	2.0
P(*Range*)	0.25	0.75

QUESTIONS 80 THROUGH 82 ARE BASED ON THE FOLLOWING INFORMATION:
Consider the set of numbers {10, 20, 30, 40, 50}.

80. Make a list of all possible samples of size 2 that could be drawn with replacement from this set of numbers.

ANSWER:
All possible samples of size 2 using replacement are listed below:

(10,10)	(10,20)	(10,30)	(10,40)	(10,50)
(20,10)	(20,20)	(20,30)	(20,40)	(20,50)
(30,10)	(30,20)	(30,30)	(30,40)	(30,50)
(40,10)	(40,20)	(40,30)	(40,40)	(40,50)
(50,10)	(50,20)	(50,30)	(50,40)	(50,50)

81. Construct the sampling distribution of sample means for the samples in question 80.

ANSWER:

$\overline{x}$	10.0	15.0	20.0	25.0	30.0	35.0	40.0	45.0	50.0
$P(\overline{x})$	0.04	0.08	0.12	0.16	0.20	0.16	0.12	0.08	0.04

82. Construct the sampling distribution of sample ranges for the samples in question 80.

ANSWER:

Range	0.0	10.0	20.0	30.0	40.0
P(*Range*)	0.20	0.32	0.24	0.16	0.08

QUESTIONS 83 THROUGH 85 ARE BASED ON THE FOLLOWING INFORMATION:
Consider the set of even single-digit integers {0, 2, 4, 6}.

83. Make a list of all the possible samples of size 3 that can be drawn from this set of integers. (Sample with replacement; that is, the first number is drawn, observed, then replaced before the next drawing.)

ANSWER:

0,0,0	0,2,0	0,4,0	0,6,0	6,0,0	6,2,0	6,4,0	6,6,0
0,0,2	0,2,2	0,4,2	0,6,2	6,0,2	6,2,2	6,4,2	6,6,2
0,0,4	0,2,4	0,4,4	0,6,4	6,0,4	6,2,4	6,4,4	6,6,4
0,0,6	0,2,6	0,4,6	0,6,6	6,0,6	6,2,6	6,4,6	6,6,6

2,0,0	2,2,0	2,4,0	2,6,0	4,0,0	4,2,0	4,4,0	4,6,0
2,0,2	2,2,2	2,4,2	2,6,2	4,0,2	4,2,2	4,4,2	4,6,2
2,0,4	2,2,4	2,4,4	2,6,4	4,0,4	4,2,4	4,4,4	4,6,4
2,0,6	2,2,6	2,4,6	2,6,6	4,0,6	4,2,6	4,4,6	4,6,6

84. Construct the sampling distribution of the sample medians for samples of size 3.

ANSWER:

$\tilde{x}$	$P(\tilde{x})$
0	10/64
2	22/64
4	22/64
6	10/64

85. Construct the sampling distribution of the sample means for samples of size 3.

ANSWER:

$\overline{x}$	$P(\overline{x})$
0/3	1/64
2/3	3/64
4/3	6/64
6/3	10/64
8/3	12/64
10/3	12/64
12/3	10/64
14/3	6/64
16/3	3/64
18/3	1/64

QUESTIONS 86 THROUGH 88 ARE BASED ON THE FOLLOWING INFORMATION:

Assume that the average amount spent per month for long-distance calls through the long-distance carrier is \$38.25, and that the standard deviation is \$11.75. If a sample of 100 customers is selected, the mean amount spent per month for long-distance calls of this sample belongs to a sampling distribution.

86. What is the shape of this sampling distribution? Why?

ANSWER:
The shape of the sampling distribution of sample means is approximately normal since n = 100 is large and Central Limit Theorem does apply in this case.

87. What is the mean of this sampling distribution?

ANSWER:
$\mu_{\overline{x}} = \mu = \$38.25$

88. What is the standard deviation of this sampling distribution?

ANSWER:
$\sigma_{\overline{x}} = \sigma / \sqrt{n} = 11.75 / \sqrt{100} = \$1.175$

Chapter 7

QUESTIONS 89 THROUGH 91 ARE BASED ON THE FOLLOWING INFORMATION:
Consider the set of odd single-digit integers {2, 4, 6, 8}.

89. Make a list of all samples of size 2 that can be drawn from this set of integers. (Sample with replacement; that is, the first number is drawn, observed, then replaced before the next drawing.)

ANSWER:

2,2	2,4	2,6	2,8
4,2	4,4	4,6	4,8
6,2	6,4	6,6	6,8
8,2	8,4	8,6	8,8

90. Construct the sampling distribution of sample means for samples of size 2 selected from this set.

ANSWER:

$\bar{x}$	2	3	4	5	6	7	8
$P(\bar{x})$	0.0625	0.125	0.1875	0.25	0.1875	0.125	0.0625

91. Construct the sampling distributions of sample ranges for samples of size 2.

ANSWER:

R	0	2	4	6
$P(R)$	0.25	0.375	0.25	0.125

QUESTIONS 92 THROUGH 97 ARE BASED ON THE FOLLOWING INFORMATION:
Consider a very small, finite population, consisting of the set of odd single-digit integers {3, 5, 7, 9, and 11}.

92. Make a list of all samples of size 2 that can be drawn with replacement from this set of integers. (Sample with replacement means that the first number is drawn, observed, then replaced before the next drawing.)

ANSWER:

(3,3)	(3,5)	(3,7)	(3,9)	(3,11)
(5,3)	(5,5)	(5,7)	(5,9)	(5,11)
(7,3)	(7,5)	(7,7)	(7,9)	(7,11)
(9,3)	(9,5)	(9,7)	(9,9)	(9,11)
(11,3)	(11,5)	(11,7)	(11,9)	(11,11)

93. Construct the sampling distribution of sample means for samples of size 2 selected from this small population.

ANSWER:

$\bar{x}$	3	4	5	6	7	8	9	10	11
$P(\bar{x})$	0.04	0.08	0.12	0.16	0.20	0.16	0.12	0.08	0.04

94. Calculate the mean of the sampling distribution of sample means in question 93.

ANSWER:

$\mu_{\bar{x}} = \sum \bar{x} \cdot P(\bar{x}) = 7.0$

95. Calculate the population mean.

ANSWER:
$\mu = \sum x / N = 35 / 5 = 7.0$

96. Compare your answers to questions 94 and 95. What did you notice? What is your conclusion?

ANSWER:
The two answers are the same. We may conclude that the mean of the sampling distribution of sample means, $\mu_{\bar{x}}$, is equal to the population mean, μ.

97. Construct the sampling distributions of sample ranges for samples of size 2 selected from this small population.

ANSWER:

R	0	2	4	6	8
$P(R)$	0.20	0.32	0.24	0.16	0.08

QUESTIONS 98 THROUGH 100 ARE BASED ON THE FOLLOWING INFORMATION:
Consider the set of even single-digit integers (0, 2, 4, 6, 8).

98. Make a list of all the possible samples of size 3 that can be drawn with replacement from this set of integers.

ANSWER:

000 020 040 060 080
002 022 042 062 082
004 024 044 064 084
006 026 046 066 086
008 028 048 068 088

200 220 240 260 280
202 222 242 262 282
204 224 244 264 284
206 226 246 266 286
208 228 248 268 288

400 420 440 460 480
402 422 442 462 482
404 424 444 464 484
406 426 446 466 486
408 428 448 468 488

600 620 640 660 680
602 622 642 662 682
604 624 644 664 684
606 626 646 666 686
608 628 648 668 688

800 820 840 860 880
802 822 842 862 882
804 824 844 864 884
806 826 846 866 886
808 828 848 868 888

99. Construct the sampling distribution of the sample medians for samples of size 3.

ANSWER:

$\tilde{x}$	0	2	4	6	8
$P(\tilde{x})$	0.104	0.248	0.296	0.248	0.104

100. Construct the sampling distribution of the sample means for samples of size 3.

ANSWER:

$\overline{x}$	$P(\overline{x})$
0/3	0.008
2/3	0.024
4/3	0.048
6/3	0.080
8/3	0.120
10/3	0.144
12/3	0.152
14/3	0.144
16/3	0.120
18/3	0.080
20/3	0.048
22/3	0.024
24/3	0.008

101. What does the sampling distribution of sample means (SDSM) say if all possible random samples, each of size *n*, are taken from any population with mean μ, and standard deviation σ ?

ANSWER:
The SDSM states that the sampling distribution of sample means will have a mean $\mu_{\overline{x}}$ equal to μ, and have a standard deviation $\sigma_{\overline{x}}$ equal to $\sigma/\sqrt{n}$. Furthermore, if the sampled population has a normal distribution, then the sampling distribution of $\overline{x}$ will also be normal for samples of all sizes.

QUESTIONS 102 THROUGH 104 ARE BASED ON THE FOLLOWING INFORMATION:
A certain population has a mean of 529 and a standard deviation of 29.7. Many samples of size 36 are randomly selected and means calculated.

102. What value would you expect to find for the mean of all these sample means? Why?

ANSWER:
529; since $\mu_{\overline{x}} = \mu$

103. What value would you expect to find for the standard deviation of all these sample means?

ANSWER:
$\sigma_{\overline{x}} = \sigma/\sqrt{n} = 29.7/\sqrt{36} = 4.95$

104. What shape would you expect the distribution of all these samples means to have? Why?

ANSWER:
According to Central Limit Theorem (*n* = 36 is large), we would expect the shape of the distribution of all these samples means to be approximately normal.

QUESTIONS 105 THROUGH 107 ARE BASED ON THE FOLLOWING INFORMATION:
Egyptians watch an average of 2.5 hours of television per person per day. If the standard deviation for the number of hours of television watched per day is 1.6 and a random sample of 225 Egyptians is selected, the mean of this sample belongs to a sampling distribution.

105. What is the shape of this sampling distribution? Why?

ANSWER:
According to Central Limit Theorem (n = 225 is large), we would expect the shape of this sampling distribution to be approximately normal.

106. What is the mean of this sampling distribution?

ANSWER:
$\mu_{\bar{x}} = \mu$ = 2.5

107. What is the standard deviation of this sampling distribution?

ANSWER:
$\sigma_{\bar{x}} = \sigma / \sqrt{n} = 1.6 / \sqrt{225} = 0.107$

QUESTIONS 108 THROUGH 110 ARE BASED ON THE FOLLOWING INFORMATION:
Suppose the annual consumption of chicken mean is 20.84 pounds per person, and that the standard deviation for the consumption of chicken per person is 9.193 pounds. The mean weight of chicken consumed for a sample of 200 randomly selected people is one value of many that form the sampling distribution of sample means.

108. Describe the shape of this sampling distribution. Justify your answer.

ANSWER:
According to Central Limit Theorem (n = 200 is large), we would expect the shape of this sampling distribution to be approximately normal.

109. What is the mean value for this sampling distribution?

ANSWER:
$\mu_{\bar{x}} = \mu$ = 20.84

110. What is the standard deviation of this sampling distribution?

ANSWER:
$\sigma_{\bar{x}} = \sigma / \sqrt{n} = 9.193 / \sqrt{200} = 0.65$

Section 7.4

True-False Questions

111. If we are sampling from a normal population with a mean of 100 and a standard deviation of 10, $P(90 < x < 100) < P(90 < \bar{x} < 100)$.
ANSWER: T

112. A sugar company packages sugar in 5-pound bags. The amount of sugar per bag varies according to a normal distribution and has a mean equal to 5.0 pounds and a standard deviation equal to 0.05 pounds. The computation of probabilities of events involving weights of individual bags of sugar will utilize the variable $z = (x - 5.0) / 0.05$ while the computation of probabilities of events involving the weights of sample means for samples of size $n = 25$ each will utilize the variable $z = (\bar{x} - 5.0) / 0.01$.
ANSWER: T

113. As the sample size *n* increases, the standard error of the sample means $\sigma_{\bar{x}}$ becomes smaller so that the distribution of sample means becomes much narrower.
ANSWER: T

114. The standard error of the sample mean increases as the sample size increases.
ANSWER: F

115. The shape of the distribution of sample means is always that of a normal distribution.
ANSWER: F

116. We need to take repeated samples in order to use the concept of the sampling distribution.
ANSWER: T

Multiple-Choice Questions

117. A soft drink bottling machine is set to dispense soft drink into containers labeled 16 ounces. While the actual quantities vary, they are normally distributed with a mean of 16.1 ounces and a standard deviation of 0.015 ounces. If a random sample of 25 bottles was selected, then 90% of the sample would have weights between

A) 15.275 and 16.925 ounces
B) 15.770 and 16.43 ounces
C) 15.935 and 16.265 ounces
D) 15.875 and 16.325 ounces
ANSWER: C

118. A normal distributed population has a mean of 250 pounds and a standard deviation of 10 pounds. Given $n = 20$, what is the probability that this sample will have a mean value between 245 and 255 pounds?

A) 0.9750
B) 0.4875
C) 0.3830
D) 0.0876
ANSWER: A

Short-Answer Questions

119. A manufacturer of light bulbs claims that the bulbs have a mean life of 800 hours with a standard deviation of 20 hours. You test a random sample of 100 of these bulbs and find a sample mean of 750 hours. Discuss the likelihood of the manufacturer's claim.

ANSWER:
If the manufacturer's claim is true, $\bar{x}$ = 750 has a z-score of –25.0, an extremely unlikely occurrence. Therefore, it seems unlikely the manufacturer's claim is true.

120. Consider a population with a mean μ of 51 and a standard deviation σ of 5.1. Calculate the z-score for an $\bar{x}$ of 48.5 from a sample of size 36.

ANSWER:
$$z = \frac{\bar{x} - \mu}{\sigma / \sqrt{n}} = \frac{48.5 - 51}{5.1 / \sqrt{36}} = -2.94$$

121. It is known that when the width of the normal curve narrows, the height of the curve has to increase. Why?

ANSWER:
Recall that the area (probability) under the normal curve is always exactly one. So as the width of the curve narrows, the height of the curve has to increase in order to maintain this area.

Applied and Computational Questions

QUESTIONS 122 AND 123 ARE BASED ON THE FOLLOWING INFORMATION:
Samples of size 50 are selected from a normal population with a mean of 78.1 and a standard deviation of 16.2.

122. Calculate $P(71.1 < x < 85.1)$.

ANSWER:
0.3328

123. Calculate $P(71.1 < \bar{x} < 85.1)$.

ANSWER:
0.9978

124. A sugar company packages sugar in 5-pound bags. The amount of sugar per bag varies according to a normal distribution. A sample of 15 bags is selected from the day's production, and if the total weight of the sample is less than 74.5 pounds, the fill per bag is increased. If the mean for the day is 5.00 pounds and the standard deviation is 0.05 pounds, what is the probability that the fill per bag will be increased?

ANSWER:
0.0049

125. If we are sampling from a normal population with a mean of 80 and a standard deviation of 12, what size sample must be taken so that the middle 90% of the sampling distribution of sample means falls between 78.35 and 81.65?

ANSWER:
144

126 If we are sampling from a normal population with a mean of 50 and a standard deviation of 5, what size sample must be taken so that the middle 90% of the sampling distribution of sample means falls between 48.5 and 51.5?

ANSWER:
30

QUESTIONS 127 AND 128 ARE BASED ON THE FOLLOWING INFORMATION:
Samples of size 10 are selected from a normal population with a mean of 35.5 and a standard deviation of 6.5.

127. Calculate $P(29.5 < x < 40.0)$.

ANSWER:
0.5761

128. Calculate $P(29.5 < \bar{x} < 40.0)$.

ANSWER:
0.9839

129. A sugar company packages sugar in 5-pound bags. The amount of sugar per bag varies according to a normal distribution. A sample of 25 bags is selected from the day's production, and if the mean of the sample is less than 4.98 pounds, the fill per bag is increased. If the mean for the day is 5.00 pounds per bag and the standard deviation is 0.05 pounds, what is the probability that the fill per bag will be increased?

ANSWER:
0.0228

130. The daily production of product parts has lengths that are normally distributed with a mean of 3.0 cm and a standard deviation of 0.05 cm. The daily production is 100% inspected if a sample of 25 has a mean length that exceeds 3.02 cm or is less than 2.98 cm. What is the probability that a daily production is 100% inspected?

ANSWER:
0.0456

131. A sample of size 50 is selected from a normal distribution having a mean equal to 95 and a standard deviation equal to 15. What is the probability of selecting a sample having a mean exceeding 100?

ANSWER:
0.0091

132. A normal population has a mean equal to 100 and a standard deviation equal to 5. If a sample of size 25 is selected, what is the probability that the sample mean will be between 98.04 and 101.96?

ANSWER:
0.95

133. A population has a mean equal to 50. To have only a 10% chance of getting a sample of size 36 whose mean exceeds 52.5, what must the standard deviation equal?

ANSWER:
Standard deviation = 11.7

134. A random sample of 100 times is selected from a population having a mean equal to 75. If there is a 20% probability that the sample mean will be at the most 70 and assuming $z = -0.84$, what would be the population standard deviation?

ANSWER:
5.95

135. A population has a mean equal to x and a standard deviation equal to y. Find the 90th percentile for the distribution of sample means based on samples of size 64.

ANSWER:
$x + 0.16y$

136. A normal population has a mean of 40 and a standard deviation of 10. If the probability that a sample of size n will have a mean greater than 45 is 0.0062, find n.

ANSWER:
$n = 5$

137. A normal population has a mean of 64 and a standard deviation of 10. If the probability that a sample of size $n = 25$ will have a sample mean less than $\bar{x}$ is 0.0062, find $\bar{x}$.

ANSWER:
Sample mean = 41

QUESTIONS 138 THROUGH 140 ARE BASED ON THE FOLLOWING INFORMATION:
A normal population has a mean of 75 and a standard deviation of 12.5.

138. If the probability that a sample of size n will have a mean greater than 77 is 0.2389, find n.

ANSWER:
$n = 20$

139. If the probability that a sample of size $n = 100$ will have a sample mean of at least $\bar{x}$ is 0.9452, find $\bar{x}$.

ANSWER:
$\bar{x} = 73$

140. If the probability that a sample of size $n = 100$ will have a sample mean greater than $\bar{x}$ is 0.0548, find $\bar{x}$.

ANSWER:
$\bar{x} = 77$

QUESTIONS 141 AND 142 ARE BASED ON THE FOLLOWING INFORMATION:
Individual scores of a placement examination are normally distributed with a mean of 84.2 and a standard deviation of 12.8.

141. If the score of an individual is randomly selected, find the probability that the score will be less than 90.0.

ANSWER:
0.6736

142. If a random sample of size $n = 20$ is selected, find the probability that the sample mean will be less than 90.0.

ANSWER:
0.9788

143. The mean of a population is 64, and its standard deviation is 12. Samples of size $n = 40$ are randomly selected. Find a value of k such that 90% of all such samples will have a mean $\bar{x}$ such that $64 - k < \bar{x} < 64 + k$.

ANSWER:
$k = 3.13$

QUESTIONS 144 THROUGH 149 ARE BASED ON THE FOLLOWING INFORMTION:
Assume that the population of heights of male college students is normally distributed with mean μ of 68 inches and standard deviation σ of 3.75 inches. A random sample of 16 heights is obtained.

144. Describe the distribution of x, height of male college students.

ANSWER:
Heights are normally distributed with mean $\mu = 68$ and standard deviation $\sigma = 3.75$.

145. Find the proportion of male college students whose height is greater than 70 inches.

ANSWER:
$P(x > 70) = P[z > (70 - 68)/3.75] = P(z > 0.53) = 0.5000 - 0.2091 = 0.2909$

146. Describe the distribution of $\bar{x}$, the mean of samples of size 16.

ANSWER:
The distribution of $\bar{x}$'s will be normally distributed, since the sampled population is normal.

147. Find the mean and standard error of the $\overline{x}$ distribution.

ANSWER:
$\mu_{\overline{x}} = \mu$ = 68; $\sigma_{\overline{x}} = \sigma / \sqrt{n} = 3.75 / \sqrt{16}$ = 0.9375

148. Find $P(\overline{x} > 70)$.

ANSWER:
$P(\overline{x} > 70) = P[z > (70 - 68)/0.9375] = P(z > 2.13) = 0.5000 - 0.4834 = 0.0166$

149. Find $P(\overline{x} < 67)$.

ANSWER:
$P(\overline{x} < 67) = P[z < (67 - 68)/0.9375] = P(z < -1.07) = 0.5000 - 0.3577 = 0.1423$

QUESTIONS 150 THROUGH 153 ARE BASED ON THE FOLLOWING INFORMATION:
Suppose that the average speed of winds in Honolulu, Hawaii, equals 12 miles per hour, and that wind speeds are approximately normally distributed with a standard deviation of 3.4 miles per hour.

150. Find the probability that the wind speed on any one reading will exceed 13.5 miles per hour.

ANSWER:
Since $\mu = 12$, and $\sigma = 3.4$, then
$P(x > 13.5) = P[z > (13.5 - 12)/3.4] = P(z > 0.44) = 0.5000 - 0.1700 = 0.33$

151. Find the probability that the mean of a random sample of 9 readings exceeds 13.5 miles per hour.

ANSWER:
Since $\mu = 12$, and $\sigma = 3.4$, then
$P(\overline{x} > 13.5) = P[z > (13.5 - 12)/(3.4/\sqrt{9})] = P(z > 1.32) = 0.5000 - 0.4066 = 0.0934$

152. Do you think the assumption of normality is reasonable? Explain.

ANSWER:
It is hard to tell if the assumption of normality is reasonable or not without studying wind speeds more extensively. However, it would not be surprising if wind speeds have a mounded distribution that could reasonably be approximated by the normal distribution. One might also expect the distribution to be skewed to the right since very high winds can occur. However, the assumption of normality seems reasonable.

153. What effect do you think the assumption of normality had on the answers to 150 and 151? Explain.

ANSWER:
The assumption of normality allowed the use of the normal probability distribution to estimate the probabilities.

Chapter 7

QUESTIONS 154 THROUGH 157 ARE BASED ON THE FOLLOWING INFORMATION:
Suppose that the average weekly earnings for employees in general automotive repair shops is \$450, and that the standard deviation for the weekly earnings for such employees is \$50. A sample of 100 such employees is selected at random.

154. Find the probability that the mean of the sample is less than \$445.

ANSWER:
$P(\bar{x} < 445) = P[z<(445 - 450)/(50/\sqrt{100})] = P(z< -1.0) = 0.5000 - 0.3413 = 0.1587$

155. Find the probability that the mean of the sample is between \$445 and \$455.

ANSWER:
$P(445< \bar{x} <455) = P[(445 - 450)/(50/\sqrt{100}) < z <(455 - 450)/(50/\sqrt{100})]$
$= P(-1.0 < z < 1.0) = 2(0.3413) = 0.6826$

156. Find the probability that the mean of the sample is greater than \$460.

ANSWER:
$P(\bar{x} > 460) = P[z > (460 - 450)/(50/\sqrt{100})] = P(z > 2.0) = 0.5000 - 0.4772 = 0.0228$

157. Explain why the assumption of normality about the x distribution was not involved in the answers to 154, 155, and 156.

ANSWER:
The sample size is large; n = 100 is greater than 30, so Central Limit Theorem does apply.

QUESTIONS 158 THROUGH 162 ARE BASED ON THE FOLLOWING INFORMATION:
The diameters of oranges in a certain orchard are normally distributed with a mean of 5.26 inches and a standard deviation of 0.50 inches.

158. What percentage of the oranges in this orchard has diameters less than 4.5 inches?

ANSWER:
$P(x < 4.5) = P[z < (4.5 - 5.26)/0.5] = P(z< -1.52) = 0.5000 - 0.4357 = 0.0643$ or 6.43%

159. What percentage of the oranges in this orchard is larger than 5.12 inches?

ANSWER:
$P(x >5.12) = P[z > (5.12 - 5.26)/0.5] = P(z >-0.28) = 0.5000 + 0.1103 = 0.6103$ or 61.03%

160. A random sample of 100 oranges is gathered and the mean diameter obtained was $\bar{x}$ = 5.12. If another sample of size 100 is taken, what is the probability that its sample mean will be greater than 5.12 inches?

ANSWER:
$P(\bar{x} > 5.12) = P[z >(5.12 - 5.26)/(0.5/\sqrt{100}] = P(z >-2.80) = 0.5000 + 0.4974 = 0.9974$

161. Why is the *z*-score used in answering questions 158, 159 and 160?

ANSWER:
z is used in questions 158 and 159 since the distribution of *x* is given to be normal, and it is also used in question 160 since the sampling distribution of $\overline{x}$ is normal. (The sampled population is normal).

162. Why the *z*- formula used in question 160 is different from that used in questions 158 and 159?

ANSWER:
Questions 158 and 159 are distributions of individual *x*-values, while question 160 is a sampling distribution of $\overline{x}$ values.

163. A manufacturer of light bulbs says that its light bulbs have a mean life of 800 hours and a standard deviation of 120 hours. You purchased 169 of these bulbs with the idea that you would purchase more if the mean life of your sample were more than 780 hours. What is the probability that you will not buy again from this manufacturer?

ANSWER:
Given information: $\mu = 800,\ \sigma = 120,\ \text{and}\ n = 169$

$P(\overline{x} < 780) = P[z < (780 - 800)/(120/\sqrt{169})] = P(z < -2.17) = 0.500 - 0.485 = 0.015$

164. A tire manufacturer claims (based on years of experience with its tires) that the mean mileage is 45,000 miles and the standard deviation is 6000 miles. A consumer agency randomly selects 100 of these tires and finds a sample mean of 41,000. Should the consumer agency doubt the manufacturer's claim?

ANSWER:
Given information: $\mu = 45000,\ \sigma = 6000,\ \text{and}\ n = 100$

$P(\overline{x} < 41{,}000) = P[z < (41{,}000 - 45{,}000)/(6{,}000/\sqrt{100})] = P(z < -6.67) = 0.0000+$
Yes, the consumer agency shows doubt the manufacturer's claim.

165. The baggage weights for passengers using a domestic airline are normally distributed with a mean of 22 lbs. and a standard deviation of 4 lbs. If the limit on total luggage weight is 2250 lbs., what is the probability that the limit will be exceeded for 100 passengers?

ANSWER:
Given information: $\mu = 22,\ \sigma = 4,\ \text{and}\ n = 100$. Let $\sum x$ represent the total baggage weight for the 100 passengers:

$P(\sum x > 2250) = P(\sum x / n > 2250/100) = P(\overline{x} > 22.5)$
$= P[z > (22.5 - 22) / (4/\sqrt{100})] = P(z > 1.25) = 0.5000 - 0.3944 = 0.1056$

Chapter 7

QUESTIONS 166 THROUGH 172 ARE BASED ON THE FOLLOWING INFORMATION:
A random sample of size 36 is to be selected from a population that has a mean μ of 75 and a standard deviation σ of 15.

166. This sample of 36 has a mean value of $\bar{x}$ which belongs to a sampling distribution. Find the shape of this sampling distribution.

ANSWER:
According to Central Limit Theorem (n = 36 is large), the shape of this sampling distribution would be approximately normal.

167. Find the mean of this sampling distribution.

ANSWER:
$\mu_{\bar{x}} = \mu = 75$

168. Find the standard error of this sampling distribution.

ANSWER:
$\sigma_{\bar{x}} = \sigma / \sqrt{n} = 15 / \sqrt{36} = 2.5$

169. What is the probability that this sample mean will be between 68 and 82?

ANSWER:
$P(68 < \bar{x} < 82) = P(-2.8 < z < 2.8) = 2\,(0.4974) = 0.9948$

170. What is the probability that the sample mean will have a value greater than 72?

ANSWER:
$P(\bar{x} > 72) = P(z > -1.2) = 0.50 + 0.3849 = 0.8849$

171. What is the probability that the sample mean will be within 2 units of the mean?

ANSWER:
$P(73 < \bar{x} < 77) = P(-0.8 < z < 0.8) = 2\,(0.2881) = 0.5762$

172. What is the probability that the sample mean will be within 3 units of the mean?

ANSWER:
$P(72 < \bar{x} < 78) = P(-1.2 < z < 1.2) = 2\,(0.3849) = 0.7698$

QUESTIONS 173 THROUGH 180 ARE BASED ON THE FOLLOWING INFORMATION:
Consider the approximately normal population of weights of female college students with mean μ of 118 pounds and standard deviation σ of 6.8 pounds. A random sample of 16 weights is obtained.

173. Describe the distribution of x, weight of female college student.

ANSWER:
Weights are approximately normally distributed with a μ = 118 and σ = 6.8.

174. Find the proportion of female college students whose weight is greater than 120 pounds.

ANSWER:
$P(x > 120) = P(z > 0.29) = 0.50 - 0.1141 = 0.3859$

175. Describe the distribution of $\bar{x}$, the mean of samples of size 16.

ANSWER:
The distribution of $\bar{x}$, the mean of samples of size 16, will be approximately normally distributed.

176. Find the mean and standard error of the $\bar{x}$ distribution.

ANSWER:
Mean = $\mu_{\bar{x}} = \mu = 118$ and standard deviation = $\sigma_{\bar{x}} = \sigma / \sqrt{n} = 6.8 / \sqrt{16} = 1.70$

177. Find the probability that the sample mean weight exceeds 121 pounds.

ANSWER:
$P(\bar{x} > 121) = P(z > 1.76) = 0.50 - 0.4608 = 0.0392$

178 Find the probability that the sample mean weight is less than 114 pounds.

ANSWER:
$P(\bar{x} < 114) = P(z < -2.35) = 0.50 - 0.4906 = 0.0094$

179. Find the probability that the sample mean weight is between 116 and 121 pounds.

ANSWER:
$P(116 < \bar{x} < 121) = P(-1.18 < z < 1.76) = 0.3810 + 0.4608 = 0.8418$

180. Within what limits does the middle 95% of the sampling distribution of sample means for samples of size 16 fall?

ANSWER:
The middle 95% of the sampling distribution of $\bar{x}$ is bounded by $z = \pm 1.96$. Therefore,

$$-1.96 = \frac{\bar{x} - 118}{1.7} \Rightarrow \bar{x} - 118 = -3.332 \Rightarrow \bar{x} = 114.668 \approx 114.7 \text{ pounds, and}$$

$$1.96 = \frac{\bar{x} - 118}{1.7} \Rightarrow \bar{x} - 118 = 3.332 \Rightarrow \bar{x} = 121.332 \approx 121.3 \text{ pounds}$$

Therefore, the middle 95% of the sampling distribution of sample mean weights of female college students is bounded by 114.7 pounds and 121.3 pounds.

QUESTIONS 181 THROUGH 184 ARE BASED ON THE FOLLOWING INFORMATION:
A recent study showed that the average amount that high school graduates in USA spend on their open house is \$932. Assume that amounts spent are normally distributed with a standard deviation of \$348, and that open houses for 36 high school graduates are randomly selected from Lansing, Michigan.

181. Describe the distribution of $\bar{x}$; the sample average amount spent on open houses of high school graduates.

ANSWER:
The distribution of $\bar{x}$ will be normally distributed, since the sampled population is normal.

182. Find the mean and standard error of the $\bar{x}$ distribution.

ANSWER:
Mean = $\mu_{\bar{x}} = \mu$ = \$932, and standard deviation = $\sigma_{\bar{x}} = \sigma / \sqrt{n} = 348 / \sqrt{36} =$ \$58

183. Find the probability that the sample mean cost to have an open house is between \$816 and \$874.

ANSWER:
$P(820 < \bar{x} < 874) = P(-2.0 < z < -1.0) = 0.4772 - 0.3413 = 0.1359$

184. Find the probability that the sample mean cost to have an open house is higher than \$1042.

ANSWER:
$P(\bar{x} > 1042) = P(z > 1.90) = 0.50 - 0.4713 = 0.0287$

QUESTIONS 185 THROUGH 189 ARE BASED ON THE FOLLOWING INFORMATION:
A recent report in a women magazine stated that the average age for women to marry in the United States is now 25 years of age, and that the standard deviation is assumed to be 3.2 years. A sample of 50 U.S. women is randomly selected.

185. Describe the distribution of $\bar{x}$; the sample average age for women to marry in the United States.

ANSWER:
The distribution of $\bar{x}$ will be approximately normal (n = 50 is large).

186. Find the mean and standard error of the $\bar{x}$ distribution.

ANSWER:
Mean = $\mu_{\bar{x}} = \mu$ = 25, and standard deviation = $\sigma_{\bar{x}} = \sigma / \sqrt{n} = 3.2 / \sqrt{50} = 0.4525$

187. Find the probability that the sample mean age for women to marry is at most 24 years.

ANSWER:
$P(\bar{x} \leq 24) = P(z \leq -2.21) = 0.50 - 0.4864 = 0.0136$

188. Find the probability that the sample mean age for women to marry is more than 25.5 years.

ANSWER:
$P(\bar{x} > 25.5) = P(z > 1.10) = 0.50 - 0.3643 = 0.1357$

189. Find the probability that the sample mean age for women to marry is between 24 and 25 years.

ANSWER:
$P(24 < \bar{x} < 25) = P(-2.21 < z < 0.0) = 0.4864$

Chapter 8

Inferential Statistics

Sections 8.1 Through 8.3

True-False Questions

1. A confidence interval estimate for μ will always contain the corresponding point estimate for μ.
   ANSWER: T

2. In real-world problems, the population standard deviation is often unknown.
   ANSWER: T

3. A sample statistic whose sampling distribution has a mean value equal to the value of the population parameter being estimated is referred to as unbiased statistic.
   ANSWER: T

4. The maximum error of estimate is controlled by three factors: level of confidence, sample size, and standard deviation.
   ANSWER: T

5. The objective of inferential statistics is to use the information contained in the sample data to increase our knowledge of the sample.
   ANSWER: F

6. If the maximum error E is expressed as a multiple of the standard deviation σ, then the actual value of σ is not needed in order to calculate the sample size.
   ANSWER: T

7. The sample mean, $\bar{x}$, is an unbiased statistic because the mean value of the sampling distribution of sample means, $\mu_{\bar{x}}$, is larger than the population mean, μ.
   ANSWER: F

8. The sample mean, $\bar{x}$, is the point estimate (single number value) for the mean μ of the sampled population.
   ANSWER: T

9. The variability of a statistic is measured by the standard deviation of its sampled population.
ANSWER: F

10. The quality of an estimation procedure (or method) is greatly enhanced if the sample statistic is both *less variable* and *unbiased.*
ANSWER: T

11. An unbiased statistic is defined as a sample statistic whose sampling distribution has a mean value equal to the value of the population parameter being estimated. A statistic that is not unbiased is a biased statistic.
ANSWER: T

12. The Central Limit Theorem can only be applied to large samples when the data provide a strong indication of a unimodal distribution that is approximately symmetric.
ANSWER: F

13. A point estimate for a parameter is a single number designed to estimate a quantitative parameter of a population, usually the value of the corresponding sample statistic.
ANSWER: T

14. The sampling distribution of sample means (SDSM) and the Central Limit Theorem provide the information needed to describe how close the point estimate, *s*, is expected to be to the population standard deviation, σ.
ANSWER: F

15. $z(\alpha/2)$ in the formula $\bar{x} \pm z(\alpha/2)\left(\frac{\sigma}{\sqrt{n}}\right)$ is the *confidence coefficient*. It is the number of multiples of the standard error needed to formulate an interval estimate of the correct width to have a level of confidence of $1-\alpha$.
ANSWER: T

Multiple-Choice Questions

16. When estimating a population mean with a confidence interval estimate, then *E* is:

A) equal to the level of confidence.
B) one-half the width of the confidence interval.
C) a multiple of the population mean.
D) a multiple of the population standard deviation.
ANSWER: B

17. Suppose you selected 200 different samples from a large population and used each sample to construct a 0.95 confidence interval estimate for the population mean. How many of the 200 confidence interval estimates should you expect to actually contain the population mean μ?

A) 200
B) 190
C) 100
D) 95
ANSWER: B

18. What value is always located at the center of a confidence interval for μ?

A) E
B) μ
C) $\bar{x}$
D) σ
ANSWER: C

19. You are constructing a 95% confidence interval using the following information: n = 60, $\bar{x}$ = 65.5, s = 2.5, and E = 0.7. What is the value of the middle of the interval?

A) 0.7
B) 2.5
C) 0.95
D) 65.5
ANSWER: D

20. Which of the following statements is false?

A) An interval estimate is a interval bounded by two values and used to estimate the value of a population parameter.
B) The values that bound a confidence interval are statistics calculated from the sample that is being used as the basis for the estimation.
C) Level of confidence, denoted by α , is the proportion of all interval estimates that do not include the parameter being estimated.
D) Confidence interval is an interval estimate with a specified level of confidence.
ANSWER: C

21. Which of the following statements is correct?

A) The sampling distribution of sample means $\bar{x}$ is distributed about a mean equal to μ with a standard error equal to $\sigma/\sqrt{n}$.
B) If the randomly sampled population is normally distributed, then $\bar{x}$ is normally distributed for all sample sizes.
C) If the randomly sampled population is not normally distributed, then $\bar{x}$ is approximately normally distributed for sufficiently large sample sizes.
D) All of the above
ANSWER: D

22. Which of the following statements is false?

A) $\sigma/\sqrt{n}$ is the standard error of the mean, or the standard deviation of the sampling distribution of sample means.
B) $z(\alpha/2)\left(\frac{\sigma}{\sqrt{n}}\right)$ is the width of the confidence interval (the product of the confidence coefficient and the standard error) and is called the maximum error of estimate, E.
C) The higher the level of confidence, the more likely the interval is to contain the parameter, and the narrower the interval, the precise the estimation.
D) None of the above.
ANSWER: B

23. Which of the following statements is false?

A) The confidence interval has two basic characteristics that determine its quality: its level of confidence and its width.
B) It is preferred that the confidence interval has a high level of confidence and be precise (narrow) at the same time.
C) When we solve for the sample size *n*, it is customary to round down to the next larger integer, no matter what fraction (or decimal) results.
D) None of the above

ANSWER: C

Short-Answer Questions

24. Discuss the difference between a point estimate for a parameter and an interval estimate for a parameter.

ANSWER:
Point estimate for a parameter is a single value, the value of the corresponding sample statistic. An interval estimate is an interval bounded by two values.

25. Five hundred confidence intervals, each having level of confidence 85%, were computed for population mean μ. Approximately how many of the confidence intervals would not capture μ?

ANSWER:
75

26. When a $(1-\alpha)$ 100% confidence interval is formed for μ, what is the probability that the interval will not contain μ within its limits?

ANSWER:
α

27. Explain why there needs to be a balance between *n*, $1-\alpha$, and *E*.

ANSWER:
There needs to be a balance between *n*, $1-\alpha$, and *E* to insure acceptable interval results (as high as possible level of confidence while minimizing error and keeping *n* as small as possible).

28. If the sample mean is used to estimate μ and a maximum error of estimate is specified, then *n* may be determined for a known standard deviation and a given level of confidence. If the maximum error of estimate is doubled, what is the affect on the required sample size?

ANSWER:
The sample size is divided by four.

29. Does decreasing the sample size increase or decrease the width of the confidence interval for a particular parameter (all other things remaining the same)?

ANSWER:
Increase the width of the confidence interval.

30. Consider the statement: "The variance among the test scores on last week's exam in your statistics class was 112". Identify each numeral value that appears above by name (mean, variance, etc.) and by symbol ($\bar{x}$, σ , etc.)

ANSWER:
Sample variance = s^2 = 112

31. Explain the difference between a *point estimate* and an *interval estimate*.

ANSWER:
A *point estimate* for a parameter is a single number designed to estimate a quantitative parameter of a population, usually the value of the corresponding sample statistic. An *interval estimate* is an interval bounded by two values and used to estimate the value of a population parameter. The values that bound the interval are statistics calculated from the sample that is being used as the basis for the estimation.

32. Consider the statement: "The mean height of a sample of 50 senior high school boys is 68 inches". Identify each numeral value that appears above by name (mean, variance, etc.) and by symbol ($\bar{x}$, σ , etc.)

ANSWER:
Sample size = n = 50, and sample mean = $\bar{x}$ = 68.

33. Explain the difference between an *interval estimate* and *confidence interval.*

ANSWER:
An *interval estimate* is an interval bounded by two values and used to estimate the value of a population parameter. The values that bound the interval are statistics calculated from the sample that is being used as the basis for the estimation. A *confidence interval* is an interval estimate with a specified level of confidence.

34. Consider the statement: "The standard deviation for I.Q. scores is 12.3". Identify each numeral value that appears above by name (mean, variance, etc.) and by symbol ($\bar{x}$, σ , etc.)

ANSWER:
Population standard deviation = σ = 12.3

35. The number 1.96 in the formula $\bar{x} \pm 1.96\left(\frac{\sigma}{\sqrt{n}}\right)$ is the *confidence coefficient.* What does this mean?

ANSWER:
It is the number of multiples of the standard error needed to formulate an interval estimate of the correct width to have a level of confidence 1-α = 1 - 0.05 = 0.95 or 95%.

36. Consider the statement: "The mean height of all cadets who have ever entered West Point is 69 inches". Identify each numeral value that appears above by name (mean, variance, etc.) and by symbol ($\bar{x}$, σ , etc.)

ANSWER:
Population mean = μ = 69

Applied and Computational Questions

37. What value would the standard deviation need to be in order for $\bar{x}$ (based on 150 observations) to estimate μ with a maximum error of estimate equal to 0.15 and with 95% confidence?

ANSWER:
0.94

38. A sample of size 40 is taken from a population having σ = 2.7. If the mean of the sample equals 48.5, then give a point estimate for μ and find an 85% confidence interval for μ .

ANSWER:
Point estimate of μ is $\bar{x}$ = 48.5, and (47.9 to 49.1) is an 85% confidence interval.

39. A study was conducted to estimate the mean amount spent on Christmas gifts for a typical family having two children. A sample of size 150 was taken, and the mean amount spent was \$225. Assuming a standard deviation equal to \$50, find a 95% confidence interval for μ , the mean for all such families.

ANSWER:
(217 to 233)

40. What sample size would be needed to estimate the population mean to within one-half standard deviation with 95% confidence?

ANSWER:
16

41. A machine is programmed to put 737 grams of salt in a container. Due to uncontrolled variation in the process, there is variation in content from container to container. To estimate the mean amount of salt per container, a sample of 50 boxes is selected and $\bar{x}$ = 739.5 grams. From experience with the machine, it is known that the σ = 7.5 grams. Find a 90% confidence interval for μ .

ANSWER:
(737.7 to 741.3)

42. A 95% confidence interval estimate for a population mean was computed to be (44.8 to 50.2). Determine the mean of the sample, which was used to determine the interval estimate.

ANSWER:
$\bar{x} = 47.5$

QUESTIONS 43 THROUGH 46 ARE BASED ON THE FOLLOWING QUESTIONS:
A sample was selected from a normal population with a standard deviation $\sigma = 6.1$. The sample values are 114, 120, 108, 118, 119, 123, 117, 124, 115, and 129.

43. Construct a confidence interval estimate of the population mean with 0.90 level of confidence.

ANSWER:
(115.52 to 121.88)

44. Construct a confidence interval estimate of the population mean with 0.95 level of confidence.

ANSWER:
(114.92 to 122.48)

45. Construct a confidence interval estimate of the population mean with 0.99 level of confidence.

ANSWER:
(113.72 to 123.68)

46. Based on your answers to questions 43, 44, and 45, what is the relationship between the level of confidence and the width of the confidence interval?

ANSWER:
The larger the level of confidence, the wider the width of the confidence interval.

QUESTIONS 47 THROUGH 49 ARE BASED ON THE FOLLOWING INFORMATION:
A random sample of the amount paid for taxi fare from downtown to the airport was obtained and produced the following summary statistics: $n = 15$, $\sum x = 301$, $\sum x^2 = 6159$.

47. Find a point estimate for the population mean.

ANSWER:
$\bar{x} = \sum x / n = 301/15 = 20.0667$

48. Find a point estimate for the population variance.

ANSWER:
$s^2 = [\sum x^2 - (\sum x)^2 / n]/(n-1) = [6159 - (301)^2 / 15]/14 = 8.4952$

49. Find a point estimate for the population standard deviation.

ANSWER:
$s = \sqrt{s^2} = \sqrt{8.4952} = 2.9146$

50. Find the level of confidence assigned to an interval estimate of the mean formed using the interval $\overline{x} - 1.28 \cdot \sigma_{\overline{x}}$ to $\overline{x} + 1.28 \cdot \sigma_{\overline{x}}$.

ANSWER:
0.3997 + 0.3997 = 0.7994 or 79.94%

51. Find the level of confidence assigned to an interval estimate of the mean formed using the interval $\overline{x} - 1.75 \cdot \sigma_{\overline{x}}$ to $\overline{x} + 1.75 \cdot \sigma_{\overline{x}}$.

ANSWER:
0.4599 + 0.4599 = 0.9198 or 91.98%

52. Find the level of confidence assigned to an interval estimate of the mean formed using the interval $\overline{x} - 1.96 \cdot \sigma_{\overline{x}}$ to $\overline{x} + 1.96 \cdot \sigma_{\overline{x}}$.

ANSWER:
0.4750 + 0.4750 = 0.9500 or 95.00%

53. Find the level of confidence assigned to an interval estimate of the mean formed using the interval $\overline{x} - 2.33 \cdot \sigma_{\overline{x}}$ to $\overline{x} + 2.33 \cdot \sigma_{\overline{x}}$.

ANSWER:
0.4901 + 0.4901 = 0.9802 or 98.02%

54. Find the level of confidence assigned to an interval estimate of the mean formed using the interval $\overline{x} - 2.75 \cdot \sigma_{\overline{x}}$ to $\overline{\mathrm{x}} + 2.75 \cdot \sigma_{\overline{\mathrm{x}}}$.

ANSWER:
0.4970 + 0.4970 = 0.9940 or 99.40%

QUESTIONS 55 THROUGH 57 ARE BASED ON THE FOLLOWING INFORMATION:
Consider the information: the sampled population is normally distributed, the population standard deviation σ = 10.4, the sample size n = 60, and the sample mean $\overline{x}$ = 81.2.

55. Find the 98% confidence interval for μ.

ANSWER:
The parameter of interest = μ, normality indicated, σ = 10.4, 1-α = 0.98, n = 60, and $\overline{x}$ = 81.2. Then, $\alpha/2$ = 0.01; z (0.01) = 2.33, and $E = z(\alpha/2) \cdot \sigma / \sqrt{n}$ = (2.33) (10.4 / $\sqrt{60}$) = (2.33)(1.3426) = 3.128. Hence, $\overline{x} \pm E = 81.2 \pm 3.128$, and the 98% confidence interval for μ is 78.072 to 84.328.

56. Interpret the confidence interval in question 55.

ANSWER:
With 98% confidence we can say the population mean μ is between 78.072 and 84.328.

57. Are the assumptions satisfied? Explain why.

ANSWER:
Yes; the sampled population normally distributed.

QUESTIONS 58 THROUGH 61 ARE BASED ON THE FOLLOWING INFORMATION:
In a recent article, it was reported that the mean percentile score on the California Achievement Test (CAT) for 20 students was 59.80. Assume the population of CAT scores is normally distributed and that $\sigma = 20.5$.

58. Find a point estimate for the mean of the population the sample represents.

ANSWER:
$\bar{x} = 59.80$

59. Find the maximum error of estimate for a level of confidence equal to 95%.

ANSWER:
$E = z(\alpha/2) \cdot \sigma / \sqrt{n}$ = (1.96) (20.5 / $\sqrt{20}$) = 8.985

60. Construct a 95% confidence interval for the population mean.

ANSWER:
$\bar{x} \pm E$ = 59.80 $\pm$ 8.985. Then, the 95% confidence interval for μ is 50.815 to 68.785.

61. Explain the meaning of the answers to questions 58, 59, and 60.

ANSWER:
The above answers are the main parts of the 95% confidence interval for the population mean μ .

QUESTIONS 62 AND 63 ARE BASED ON THE FOLLOWING INFORMATION:
Given the following information: the sample size n = 20, the sample mean $\bar{x}$ = 75.3, and the population standard deviation σ = 6.0.

62. Find the 0.99 confidence interval for μ .

ANSWER:
The parameter of interest = μ , normality cannot be assumed for x; with n = 20, the Central Limit Theorem does not assure us that $\bar{x}$ will be approximately normal either. It may be meaningless to complete the procedure. However, since σ = 6.0, 1-α = 0.99, n = 20, $\bar{x}$ =75.3, then, $\alpha/2$ = 0.005; z (0.005) = 2.58, and $E = z(\alpha/2) \cdot \sigma / \sqrt{n}$ = (2.58) (6.0/$\sqrt{20}$) = (2.58) $\cdot$ (1.3416) = 3.46. Hence,
$\bar{x} \pm E = 75.3 \pm 3.46$, and the 99% confidence interval for μ is 71.84 To 78.76.

63. Are the assumptions satisfied? Explain why.

ANSWER:
No; the distribution for the variable x is unknown, and n = 20 is not large enough to satisfy the Central Limit Theorem. The resulting interval is likely to have a level of confidence that is unknowingly less than 99%.

64. How large a sample should be taken if the population mean is to be estimated with 99% confidence to within $72? The population has a standard deviation of $800.

ANSWER:
$n = [z(\alpha/2) \cdot \sigma / E]^2 = [(2.58)(800)/72]^2$ = 821.78 or 822.

Chapter 8

QUESTIONS 65 AND 66 ARE BASED ON THE FOLLOWING INFORMATION:
By measuring the amount of time it takes a component of a product to move from one workstation to the next, an engineer has estimated that the standard deviation is 4.5 seconds.

65. How many measurements should be made in order to be 95% certain that the maximum error of estimation will not exceed 1 second?

ANSWER:
$n = [z(\alpha/2)\cdot\sigma/E]^2 = [(1.96)(4.5)/1]^2 = 77.79$ or 78

66. What sample size is required for a maximum error of 2 seconds?

ANSWER:
$n = [z(\alpha/2)\cdot\sigma/E]^2 = [(1.96)(4.5)/2]^2 = 19.44$ or 20

QUESTIONS 67 THROUGH 69 ARE BASED ON THE FOLLOWING INFORMATION:
Waiting times (in hours) at a popular restaurant are believed to be approximately normally distributed with a standard deviation of 1.5 hours during busy periods.

67. A sample of 20 customers revealed a mean waiting time of 1.58 hours. Construct the 95% confidence interval for the population mean.

ANSWER:
μ = The mean waiting time (in hours) at a popular restaurant. Normality indicated. Since $\sigma = 1.5$, $1-\alpha = 0.95$, $n = 20$, and $\bar{x} = 1.58$, then $\alpha/2 = 0.025$; and $z(0.025) = 1.96$. Hence, the maximum error of estimate. $E = z(\alpha/2)\cdot\sigma/\sqrt{n} = (1.96)(1.5/\sqrt{20}) = 0.657$. Then, $\bar{x} \pm E = 1.58 \pm 0.657$, and the 95% confidence interval for μ is 0.923 to 2.237.

68. Suppose that the mean of 1.58 hours had resulted from a sample of 32 customers. Find the 95% confidence interval.

ANSWER:
μ = The mean waiting time (in hours) at a popular restaurant. Normality indicated. Since $\sigma = 1.5$, $1-\alpha = 0.95$, $n = 20$, and $\bar{x} = 1.58$, then $\alpha/2 = 0.025$; and $z(0.025) = 1.96$. Hence, the maximum error of estimate. $E = z(\alpha/2)\cdot\sigma/\sqrt{n} = (1.96)(1.5/\sqrt{32}) = 0.52$. Then, $\bar{x} \pm E = 1.58 \pm 0.52$, and the 95% confidence interval for μ is 1.06 to 2.10.

69. What effect does a larger sample size have on the confidence interval?

ANSWER:
The larger sample size causes a narrower interval.

70. An automobile manufacturer wants to estimate the mean gasoline mileage that its customers will obtain with its new compact model. How many sample runs must be performed in order that the estimate be accurate to within 0.25 mpg at 95% confidence? (Assume that $\sigma = 2.0$.)

ANSWER:
$n = [z(\alpha/2)\cdot\sigma/E]^2 = [(1.96)(2.0)/0.25]^2 = 245.86$ or 246

71. Being unbiased and having a small variability are two desirable characteristics of a statistic if it is going to be used as an estimator. Describe how the Central Limit Theorem addresses both of these properties when estimating the mean of a population.

ANSWER:
$\bar{x}$ is an unbiased estimator of the population mean μ since $\mu_{\bar{x}} = \mu$; and the variability $\sigma_{\bar{x}} = \sigma / \sqrt{n}$ decreases as *n* increases.

QUESTIONS 72 THROUGH 74 ARE BASED ON THE FOLLOWING INFORMATION:
A random sample of taxi fares (in dollars) from Big Rapids to Ford International airport in Grand Rapids, Michigan, was obtained: 55, 59, 57, 63, 61, 57, 56, 58, 52, 58, 60, 62, 55, 58, and 60.

72. Find a point estimate for the population mean.

ANSWER:
$\bar{x} = \sum x / n =$ 871 / 15 = 58.067

73. Find a point estimate for the population variance.

ANSWER:

$$s^2 = \frac{\sum x^2 - \frac{(\sum x)^2}{n}}{n-1} = \frac{50{,}695 - \frac{(871)^2}{15}}{14} = 118.933 / 14 = 8.495$$

74. Find a point estimate for the population standard deviation.

ANSWER:
s = $\sqrt{s^2} = \sqrt{8.495}$ = 2.915

75. Find the level of confidence assigned to an interval estimate of μ formed using the interval $\bar{x}$ -1.15$\cdot\sigma_{\bar{x}}$ to $\bar{x}$ +1.15$\cdot\sigma_{\bar{x}}$.

ANSWER:
Level of confidence = 0.3749 + 0.3749 = 0.7498 or 74.98%

76. Find the level of confidence assigned to an interval estimate of μ formed using the interval $\bar{x}$ -1.65$\cdot\sigma_{\bar{x}}$ to $\bar{x}$ +1.65$\cdot\sigma_{\bar{x}}$.

ANSWER:
Level of confidence = 0.4505 + 0.4505 = 0.901 or 90.1%

77. Suppose there are two statistics that will serve as an estimator for the same parameter. One of them is biased and the other unbiased. Everything else being equal, explain why you usually would prefer an unbiased estimator to a biased estimator.

ANSWER:
An unbiased estimate has a sampling distribution with the mean of its statistic equal to the parameter in question.

78. If a statistic is unbiased, does that ensure it is a good estimator? Why or why not? What other considerations must be taken into account?

ANSWER:
No. The variability could be too large to be an effective estimator.

79. Find the level of confidence assigned to an interval estimate of μ formed using the interval $\bar{x}$ - 2.17$\cdot\sigma_{\bar{x}}$ to $\bar{x}$ +2.17$\cdot\sigma_{\bar{x}}$.

ANSWER:
Level of confidence = 0.4850 + 0.4850 = 0.97 of 97%

80. Describe a situation that might occur in which the biased statistic might be a better choice as an estimator than the unbiased statistic.

ANSWER:
Possibly when the biased statistic is biased by a predictable amount and the unbiased statistic is far more variable.

81. Find the level of confidence assigned to an interval estimate of μ formed using the interval $\bar{x}$ -2.58$\cdot\sigma_{\bar{x}}$ to $\bar{x}$ +2.58$\cdot\sigma_{\bar{x}}$.

ANSWER:
Level of confidence = 0.4951 + 0.4951 = 0.9902 of 99.02%

82. Find the level of confidence assigned to an interval estimate of μ formed using the interval $\bar{x}$ -1.96$\sigma_{\bar{x}}$ to $\bar{x}$ +1.96$\sigma_{\bar{x}}$.

ANSWER:
Level of confidence = 0.4750 + 04750 = 0.95 of 95%

83. Find the level of confidence assigned to an interval estimate of μ formed using the interval$\bar{x}$ -2.33$\sigma_{\bar{x}}$ to $\bar{x}$ -2.33$\sigma_{\bar{x}}$.

ANSWER:
Level of confidence = 0.4901 + 0.4901 = 0.9802 or 98.02%

84. Find the level of confidence assigned to an interval estimate of μ formed using the interval $\bar{x}$ -1.645$\cdot\sigma_{\bar{x}}$ to $\bar{x}$ +1.645$\cdot\sigma_{\bar{x}}$.

ANSWER:
Level of confidence = 0.4495 + 0.4505 = 0.90 or 90%

85. Determine the value of the confidence coefficient $z(\alpha/2)$ if 1-α = 0.90.

ANSWER:
α = 0.10 $\Rightarrow$ $z(\alpha/2)$= z(0.05) = 1.645

86. Determine the value of the confidence coefficient $z(\alpha/2)$ if 1-α = 0.95.

ANSWER:
α = 0.05 $\Rightarrow$ $z(\alpha/2)$= z(0.025) = 1.96

87. Determine the value of the confidence coefficient $z(\alpha/2)$ for 98% confidence.

ANSWER:
$\alpha = 0.02 \Rightarrow z(\alpha/2) = z(.01) = 2.33$

88. Determine the value of the confidence coefficient $z(\alpha/2)$ for 99% confidence.

ANSWER:
$\alpha = 0.01 \Rightarrow z(\alpha/2) = z(0.005) = 2.575$

89. Determine the level of the confidence given the confidence coefficient $z(\alpha/2)=1.645$.

ANSWER:
$z(\alpha/2)=1.645 \Rightarrow \alpha/2 = 0.05 \Rightarrow \alpha = .10 \Rightarrow 1 - \alpha = 0.90$

90. Determine the level of the confidence given the confidence coefficient $z(\alpha/2)=1.96$.

ANSWER:
$z(\alpha/2)=1.96 \Rightarrow \alpha/2 = 0.025 \Rightarrow \alpha = 0.05 \Rightarrow 1 - \alpha = 0.95$

91. Determine the level of the confidence given the confidence coefficient $z(\alpha/2)=2.575$.

ANSWER:
$z(\alpha/2)=2.575 \Rightarrow \alpha/2 = 0.005 \Rightarrow \alpha = 0.01 \Rightarrow 1 - \alpha = 0.99$

92. Determine the level of the confidence given the confidence coefficient $z(\alpha/2)=2.05$.

ANSWER:
$z(\alpha/2)=2.05 \Rightarrow \alpha/2 = 0.0202 \Rightarrow \alpha = .0404 \Rightarrow 1 - \alpha = 0.9899$

93. Determine the level of the confidence given the confidence coefficient $z(\alpha/2)=2.88$.

ANSWER:
$z(\alpha/2)=2.88 \Rightarrow \alpha/2 = 0.002 \Rightarrow \alpha = .004 \Rightarrow 1 - \alpha = 0.996$

QUESTIONS 94 AND 95 ARE BASED ON THE FOLLOWING INFORMATION:
Consider a random sample of size n = 100, and mean $\bar{x}=125$. Assume that the population standard deviation $\sigma=15$.

94. Find the 0.90 confidence interval for μ.

ANSWER:
$z(\alpha/2) = z(0.05) = 1.645$ and $E = z(\alpha/2)\cdot\sigma/\sqrt{n} = 1.645\cdot 15/\sqrt{100} = 2.4675$
$\bar{x} \pm E = 125 \pm 2.4675$. Hence the 90% confidence interval for μ. is 122.5325 to 127.4675.

95. Are the assumptions satisfied? Explain why.

ANSWER:
A sample of size 100 should be large enough for the Central Limit Theorem to apply and ensure that the sampling distribution of sample means will be normally distributed.

QUESTIONS 96 AND 97 ARE BASED ON THE FOLLOWING INFORMATION:
Consider a random sample of size n = 20, and mean $\bar{x}$ =70.3. Assume that the population standard deviation σ = 5.4.

96. Find the 0.99 confidence interval for μ .

ANSWER:
$z(\alpha/2)$ = z(0.005) = 2.575 and E = $z(\alpha/2)\cdot\sigma/\sqrt{n} = 2.575\cdot 5.4/\sqrt{20}$ = 3.109
$\bar{x} \pm E = 70.3 \pm 3.109$ .Hence the 99% confidence interval for μ. is 67.191 to 73.409

97. Are the assumptions satisfied? Explain why.

ANSWER:
The assumptions are not satisfied since the distribution for the variable x is unknown, and a sample of size n = 20 is not large enough to satisfy the Central Limit Theorem and assure us that $\bar{x}$ will be approximately normal. The resulting interval is likely to have a level of confidence that is unknowingly less than 99%.

98. Discuss the effect that the point estimate has on the confidence interval for μ .

ANSWER:
The point estimate is the center of the confidence interval; as it changes in value, the interval "slides" along the number line, but does not change in width.

99. Discuss the effect that the level of confidence has on the confidence interval for μ .

ANSWER:
When the level of confidence increases or decreases, $z(\alpha/2)$ also increases or decreases; thus the confidence interval increases or decreases, respectively, in width.

100. Discuss the effect that the sample size has on the confidence interval for μ.

ANSWER:
When the sample size increases, the denominator in the confidence interval formula increases causing the maximum error to decrease; thus the confidence interval decreases in width. Contrarily, if the sample size decreases, the denominator decreases and the maximum error increases and the width of the confidence interval increases

101. Discuss the effect that the variability of the characteristic being measured has on the confidence interval for μ .

ANSWER:
The variability of the characteristic being measured is the standard deviation. When the standard deviation is larger, the width of the confidence interval also increases. Likewise, if the standard deviation is smaller, the width of the confidence interval will decrease.

QUESTIONS 102 THROUGH 108 ARE BASED ON THE FOLLOWING INFORMATION:
The lengths of 225 fish caught in Lake Michigan had a mean of 15.0 inches. Assume that the population standard deviation is 2.5 inches.

102. Give a point estimate for μ .

ANSWER:
$\bar{x}$ = 15

103. Find the 90% confidence maximum error of estimate for μ.

ANSWER:
$z(\alpha/2)$ = z(0.05) = 1.645 and E = $z(\alpha/2)\cdot\sigma/\sqrt{n} = 1.645\cdot 2.5/\sqrt{225}$ = 0.274

104. Find the 90% confidence interval for the population mean length.

ANSWER:
$\bar{x} \pm E = 15 \pm 0.274$ .Hence the 90% confidence interval for μ is 14.726 to 15.274.

105. Find the 98% confidence maximum error of estimate for μ .

ANSWER:
$z(\alpha/2)$ = z(0.01) = 2.33 and E = $z(\alpha/2)\cdot\sigma/\sqrt{n} = 2.33\cdot 2.5/\sqrt{225}$ = 0.388

106. Find the 98% confidence interval for the population mean length.

ANSWER:
$\bar{x} \pm E = 15 \pm 0.388$.Hence the 90% confidence interval for μ is 14.612 to 15.388

107. What is the effect of increasing the level of confidence from 0.90 to 0.98 on the maximum error of estimate for μ ?

ANSWER:
When the level of confidence increases from 0.90 to 0.99, the confidence coefficient $z(\alpha/2)$ increases from 1.645 to 2.33; and thus the maximum error of estimate for μ increases from 0.274 to 0.388

108. What is the effect of increasing the level of confidence from 0.90 to 0.98 on the width of the confidence interval for μ ?

ANSWER:
When the level of confidence increases from 0.90 to 0.99, the confidence coefficient $z(\alpha/2)$ also increases from 1.645 to 2.33; and the maximum error of estimate *E* for μ increases from 0.274 to 0.388. As a result, the width of the confidence interval increases from 0.548 to 0.776.

Chapter 8

QUESTIONS 109 THROUGH 111 ARE BASED ON THE FOLLOWING INFORMATION:
A certain adjustment to a machine will change the length of the parts it is making but will not affect the standard deviation. The length of the parts is normally distributed, and the standard deviation is 0.5mm. After an adjustment is made, a random sample is taken to determine the mean length of parts now being produced. The resulting lengths are: 78.0, 78.7, 77.1, 79.0, 79.7, 77.6, 79.7, 79.2, 78.5, and 77.7.

109. What is the parameter of interest?

ANSWER:
The parameter of interest is the mean length of parts being produced after adjustment.

110. Find the point estimate for the mean length of all parts now being produced.

ANSWER:
$\bar{x}$ = 78.52

111. Find the 0.99 confidence interval for μ .

ANSWER:
$z(\alpha/2)$ = z(0.005) = 2.575 and E = $z(\alpha/2)\cdot\sigma/\sqrt{n} = 2.575\cdot 0.5/\sqrt{10}$ = 0.407
$\bar{x} \pm E = 78.52 \pm 0.407$ .Hence the 99% confidence interval for μ is 78.113 to 78.927.

QUESTIONS 112 AND 113 ARE BASED ON THE FOLLOWING INFORMATION:
By measuring the amount of time it takes a component of a product to move from one workstation to the next, an engineer has estimated that the standard deviation is 6 seconds.

112. How many measurements should be made in order to be 95% certain that the maximum error of estimation will not exceed 1.5 seconds?

ANSWER:

$z(\alpha/2)$ = z(0.025) = 1.96. Then, $n = \left(\frac{z(\alpha/2)\cdot\sigma}{E}\right)^2 = \left(\frac{(1.96)(6.0)}{1.5}\right)^2 = 61.47 \approx 62$

113. What sample size is required for a maximum error of 3.0 seconds?

ANSWER:

$z(\alpha/2)$ = z(0.025) = 1.96. Then, $n = \left(\frac{z(\alpha/2)\cdot\sigma}{E}\right)^2 = \left(\frac{(1.96)(6.0)}{3.0}\right)^2 = 15.37 \approx 16$

114. How large a sample would be needed to estimate the population mean weight of the new mini-laptop computers if the maximum error of estimate is to be 0.4 of one standard deviation with 95% confidence?

ANSWER:

$z(\alpha/2)$ = z(0.025) = 1.96. Then, $n = \left(\frac{z(\alpha/2)\cdot\sigma}{E}\right)^2 = \left(\frac{(1.96)(\sigma)}{0.4\sigma}\right)^2 = 24.01 \approx 25$

QUESTIONS 115 THROUGH 117 ARE BASED ON THE FOLLOWING INFORMATION:
In an effort to compare college costs in State of Michigan, a sample of 36 junior students is randomly selected statewide from the private colleges and 36 more from the public colleges. The private college sample resulted in a mean of \$27,650 and the public college sample mean was \$11,360.

115. Assume the annual college fees for private colleges have a mounded distribution and the standard deviation is \$1725. Find the 95% confidence interval for the mean costs for private colleges.

ANSWER:
$z(\alpha/2)$ = z(0.025) = 1.96 and E = $z(\alpha/2)\cdot\sigma/\sqrt{n} = 1.96\cdot 1725/\sqrt{36}$ = 563.5
$\bar{x} \pm E = 27{,}650 \pm 563.5$. Hence, the 95% confidence interval for μ is \$27,086.5 to \$28,213.5.

116. Assume the annual college fees for public colleges have a mounded distribution and the standard deviation is \$1125. Find the 95% confidence interval for the mean costs for public colleges.

ANSWER:
$z(\alpha/2)$ = z(0.025) = 1.96 and E = $z(\alpha/2)\cdot\sigma/\sqrt{n} = 1.96\cdot 1125/\sqrt{36}$ = 367.5
$\bar{x} \pm E = 11{,}360 \pm 367.5$. Hence, the 95% confidence interval for μ is \$10,992.5 to \$11,727.5.

117. Compare the confidence intervals found in questions 115 and 116 and describe the effect the two different sample standard deviations had on the resulting answers.

ANSWER:
When the standard deviation decreases from 1725 to 1125, the width of the confidence interval also decreases from \$1127 to \$735.

118. Find the sample size needed to estimate μ of a normal population with σ = 3.5 to within 1.0 unit at the 98% level of confidence.

ANSWER:
$z(\alpha/2)$ = z(0.01) = 2.33. Then, $n = \left(\frac{z(\alpha/2)\cdot\sigma}{E}\right)^2 = \left(\frac{(2.33)(3.5)}{1.0}\right)^2 = 66.50 \approx 67$

119. How large a sample should be taken if the population mean is to be estimated with 90% confidence to within \$75 if the population has a standard deviation of \$800?

ANSWER:
$z(\alpha/2)$ = z(0.05) = 1.645. Then, $n = \left(\frac{z(\alpha/2)\cdot\sigma}{E}\right)^2 = \left(\frac{(1.645)(800)}{75.0}\right)^2 = 307.89 \approx 308$

Chapter 8

QUESTIONS 120 THROUGH 122 ARE BASED ON THE FOLLOWING INFORMATION:
The weights of full boxes of Frosted Mini-Wheat cereal are normally distributed with a standard deviation of 0.52 oz. A sample of 18 randomly selected boxes produced a mean weight of 24.3 oz.

120. Find the 95% confidence interval for the true mean weight of a box of this cereal.

ANSWER:
$z(\alpha/2)$ = z(0.025) = 1.96 and E = $z(\alpha/2)\cdot\sigma/\sqrt{n}=1.96\cdot 0.52/\sqrt{18}$ = 0.2402
$\bar{x}\pm E=24.3\pm 0.2402$. Hence the 95% confidence interval for μ is 24.0598 to 24.5402

121. Find the 99% confidence interval for the true mean weight of a box of this cereal.

ANSWER:
$z(\alpha/2)$ = z(0.005) = 2.575 and E = $z(\alpha/2)\cdot\sigma/\sqrt{n}=2.575\cdot 0.52/\sqrt{18}$ = 0.3156
$\bar{x}\pm E=24.3\pm 0.3156$. Hence the 99% confidence interval for μ is 23.9844 to 24.6156

122. What effect did the increase in the level of confidence have on the width of the confidence interval?

ANSWER:
When the level of confidence increases from 0.95 to 0.99, the confidence coefficient $z(\alpha/2)$ also increases from 1.96 to 2.575. As a result, the width of the confidence interval increases from 0.4804 to 0.6312.

123. A pharmaceutical company wants to estimate the mean response time for Tenormin 50 mg tablets to reduce blood pressure. How large of a sample should they take in order to estimate the mean response time to within 0.80 week at 95% confidence? Assume that σ = 4.2 weeks.

ANSWER:

$z(\alpha/2)$ = z(0.025) = 1.96. Then, $n=\left(\frac{z(\alpha/2)\cdot\sigma}{E}\right)^2=\left(\frac{(1.96)(4.2)}{0.80}\right)^2=105.88\approx 106$.

124. We are interested in estimating the mean life of a new product. How large a sample do we need to take in order to estimate the mean to within 0.20 of a standard deviation with 90% confidence?

ANSWER:

$z(\alpha/2)$ = z(0.05) = 1.645. Then, $n=\left(\frac{z(\alpha/2)\cdot\sigma}{E}\right)^2=\left(\frac{(1.645)(\sigma)}{0.2\sigma}\right)^2=67.65\approx 68$.

Section 8.4

True-False Questions

125. When we reject the null hypothesis, we are certain that the null hypothesis is false.
ANSWER: F

126. If our decision in a hypothesis test is to fail to reject the null hypothesis, then we know that the null hypothesis must be true.
ANSWER: F

127. If α is reduced and β remains constant, then the sample size n must be increased.
ANSWER: T

128. The alternative hypothesis, sometimes referred to as the research hypothesis, is supported by using the sample evidence to contradict the null hypothesis.
ANSWER: T

129. If a hypothesis test concerning a population mean is conducted at a level of significance equal to 0.05, then the probability of a Type II error equals 0.95 for any value of μ associated with the alternative hypothesis.
ANSWER: F

130. If β is decreased and n remains constant, then α must also decrease.
ANSWER: F

131. Depending on the statement of the original problem, the equal sign may be in the null hypothesis or the alternative hypothesis.
ANSWER: F

132. Rejection of a null hypothesis that is false is a Type II error.
ANSWER: F

133. The risk of a Type I error is directly controlled in a hypothesis test by establishing a level for α .
ANSWER: T

134. β is the probability of a Type I error.
ANSWER: F

135. α is the measure of the area under the curve of the standard score that lies in the rejection region for the null hypothesis.
ANSWER: T

136. 1 - α is known as the level of significance of a hypothesis test.
ANSWER: F

137. Failing to reject the null hypothesis when it is false is a correct decision.
ANSWER: F

Chapter 8

138. To conclude that the mean is larger (or smaller) than a claimed value, the calculated value of the test statistic must fall in the rejection (critical) region.
ANSWER: T

139. The null hypothesis is sometimes referred to as the *research hypothesis* since it represents what the researcher hopes will be found to be true.
ANSWER: F

140. Failing to reject the null hypothesis when it is true is referred to as Type *A* correct decision.
ANSWER: T

141. Rejecting the null hypothesis when it is false is referred to as Type *B* correct decision.
ANSWER: T

142. Hypothesis is a statement that something is true.
ANSWER: T

143. A Type *A* correct decision occurs when the null hypothesis is false, and we decide in its favor.
ANSWER: F

144. A Type I error is committed when a true null hypothesis is rejected - that is, when the null hypothesis is true but we decide against it.
ANSWER: T

145. The Greek letter α is always the probability of rejecting the null hypothesis.
ANSWER: F

146. A Type *B* correct decision occurs when the null hypothesis is true, and the decision is in opposition to the null hypothesis.
ANSWER: F

147. A Type II error is committed when we decide in favor of a null hypothesis that is actually false.
ANSWER: T

148. The Type I error often results in what represents a "lost opportunity".
ANSWER: F

149. Test statistic is a random variable whose value is calculated from the sample data and is used in making the decision "fail to reject H_o" or "reject H_o".
ANSWER: T

150. When writing the decision and the conclusion, remember that the decision is about H_a and the conclusion is a statement about whether or not the contention of H_o was upheld.
ANSWER: F

Multiple-Choice Questions

151. You have rejected the null hypothesis when it is false, and therefore you have made a

A) Type *A* correct decision.
B) Type *B* correct decision.
C) Type I error.
D) Type II error.
ANSWER: B

152. Consider the situation: "A newly developed drug will not increase incidences of heart attacks among its users." Which of the following would be the most appropriate choices for α and β?

A) $\alpha = 0.001$ and $\beta = 0.10$
B) $\alpha = 0.01$ and $\beta = 0.05$
C) $\alpha = 0.025$ and $\beta = 0.01$
D) $\alpha = 0.10$ and $\beta = 0.001$
ANSWER: D

153. Which of the following is the name given to rejecting the null hypothesis when it is true?

A) Type *A* correct decision.
B) Type *B* correct decision.
C) Type I error.
D) Type II error.
ANSWER: C

154. Consider the following nonmathematical situation: "I do not have to study for my statistics test." Which of the following would be the most appropriate choices for α and β?

A) $\alpha = 0.001$ and $\beta = 0.10$
B) $\alpha = 0.01$ and $\beta = 0.05$
C) $\alpha = 0.025$ and $\beta = 0.01$
D) $\alpha = 0.10$ and $\beta = 0.001$
ANSWER: D

155. Consider the following nonmathematical situation: "The brakes on my automobile are in need of repair." Which of the following would be the most appropriate choices for α and β?

A) $\alpha = 0.001$ and $\beta = 0.10$
B) $\alpha = 0.01$ and $\beta = 0.05$
C) $\alpha = 0.025$ and $\beta = 0.01$
D) $\alpha = 0.10$ and $\beta = 0.001$
ANSWER: A

Chapter 8

156. Which of the following is the name given to failing to reject the null hypothesis when it is true?

A) Type *A* correct decision
B) Type *B* correct decision
C) Type I error
D) Type II error
ANSWER: A

157. Which of the following is the probability of making a Type I error?

A) α
B) $1 - \alpha$
C) β
D) $1 - \beta$
ANSWER: A

158. You have failed to reject the null hypothesis when it is false, and therefore you have made a

A) Type *A* correct decision.
B) Type *B* correct decision.
C) Type I error.
D) Type II error.
ANSWER: D

159. Which of the following is the probability of making a Type II error?

A) α
B) $1 - \alpha$
C) β
D) $1 - \beta$
ANSWER: C

160. Which of the following is the probability of making a Type *A* correct decision?

A) α
B) $1 - \alpha$
C) β
D) $1 - \beta$
ANSWER: B

161. Which of the following is the probability of making a Type *B* correct decision?

A) α
B) $1 - \alpha$
C) β
D) $1 - \beta$
ANSWER: D

162. Which of the following is the probability of having the computed value of the test statistic fall in the critical region when the null hypothesis is true?

A) α
B) $1 - \alpha$
C) β
D) $1 - \beta$
ANSWER: A

163. Which of the following is the probability of having the computed value of the test statistic fall in the non-critical region when the null hypothesis is true?

A) α
B) $1 - \alpha$
C) β
D) $1 - \beta$
ANSWER: B

164. Which of the following statements is false regarding the null hypothesis H_o?

A) It is the hypothesis we will test.
B) This is a statement that a sample statistics has a specific value.
C) It is so named because it is the "starting point" for the investigation. (The phrase "there is no difference" is often used in its interpretation.)
D) None of the above.
ANSWER: B

165. Which of the following statements is false regarding the alternative hypothesis H_a?

A) It is a statement about the same population parameter that is used in the null hypothesis.
B) It is a statement that specifies the population parameter has a value different, in some way, from the value given the null hypothesis.
C) The rejection of the null hypothesis will imply the likely truth of this alternative hypothesis.
D) None of the above.
ANSWER: D

166. Which of the following statements is false?

A) The *basic idea* of the hypothesis test is for the evidence to have a chance to "disprove" the alternative hypothesis.
B) The null hypothesis is the statement that the evidence might disprove.
C) Your *concern* (belief or desired outcome), as the person doing the testing, is expressed in the alternative hypothesis.
D) The alternative hypothesis is sometimes referred to as the *research hypothesis;* since it represents what the researcher hopes will be found to be "true".
ANSWER: A

Chapter 8

167. Which of the following statements is false?

A) The probability assigned to the Type I error is α (called "alpha"; α is the first letter of the Greek alphabet).
B) The probability of the Type II error is β (called "beta"; β is the second letter of the Greek alphabet).
C) The most frequently used probability values for α and β are 0.01 and 0.05.
D) 1-α is the probability of a correct decision when the null hypothesis is true (i.e., probability of Type *B* correct decision).

ANSWER: D

168. Which of the following statements is false?

A) 1-β is the probability of a correct decision when the null hypothesis is false (i.e., probability of Type *B* correct decision).
B) 1-β is called the *power of the statistical test,* since it is the measure of the ability of a hypothesis test to reject a false null hypothesis, a very important characteristic.
C) If α is decreased, then either β must decrease, or *n* must be decreased.
D) There is an interrelationship among the probability of the Type I error (α), the probability of the Type II error (β), and the sample size (*n*).

ANSWER: C

169. Which of the following statements is true?

A) The null hypothesis is the statement that is "on trial", and therefore the decision must be about it.
B) The contention of the alternative hypothesis is the thought that brought about the need for a decision.
C) The question that led to the alternative hypothesis must be answered when the conclusion is written.
D) All of the above

ANSWER: D

Short-Answer Questions

170. When considering error, explain the relationship between α, β, and *n*.

ANSWER:
If holding *n* constant, then as α increases β decreases (and visa versa). An increase in *n* will help decrease both Types of error.

171. If you do not reject the null hypothesis when some alternative hypothesis is correct, what Type error do you make?

ANSWER:
Type II error

172. What error could be made if the test statistic falls in the noncritical region?

ANSWER:
Type II

173. What proportion of the probability distribution is in the critical region, provided the null hypothesis is correct?

ANSWER:
α

174. What error could be made if the test statistic falls in the critical region?

ANSWER:
Type I

175. What proportion of the probability distribution is in the noncritical region, provided the null hypothesis is not correct?
ANSWER:
β

176. If the null hypothesis is false, the probability of a correct decision is identified by what symbol?

ANSWER:
$1-\beta$

177. You are investigating a complaint that "special computer brand takes too much time" to start. State the null and alternative hypotheses.

ANSWER:
H_o: Special computer brand does not take too much time to start
H_a: Special computer brand takes too much time to start

178. If the probability of Type II error, β, decreases, how does this affect the probability of Type I error, α, or the sample size *n*?

ANSWER:
If β decreases, then either α increases or *n* must be increased.

179. If the null hypothesis is false, the probability of a decision error is identified by what symbol?

ANSWER:
β

180. If the sample size *n* decreases, how does this affect the probability of Type I error, α, or the probability of Type II error, β?

ANSWER:
If *n* is decreased, then either α increases or β increases.

181. You are testing a new security system and you are concerned that the system is not reliable. State the null and alternative hypotheses.

ANSWER:
H_o: The security system is reliable
H_a: The security system is not reliable

Chapter 8

182. If the null hypothesis is true, what decision error could be made?

ANSWER:
Type I error

183. If the null hypothesis is false, what decision error could be made?

ANSWER:
Type II error

184. If the decision "reject H_o" is made, what decision error could have been made?

ANSWER:
Type I error

185. If the decision "fail to reject H_o" is made, what decision error could have been made?

ANSWER:
Type II error

186. Find the power of a test when the probability of committing Type II error is 0.01

ANSWER:
Power = $1 - \beta = 1 - 0.01 = 0.99$

187. If the null hypothesis is true, the probability of a decision error is identified by what symbol?

ANSWER:
α

188. If the null hypothesis is true, the probability of a correct decision is identified by what symbol?

ANSWER:
$1-\alpha$

189. Find the power of a test when the probability of making Type II error is 0.10

ANSWER:
Power = $1 - \beta = 1 - 0.10 = 0.90$

190. Explain why the probability of rejecting the null hypothesis is not always α

ANSWER:
The probability of rejecting the null hypothesis is α only if the null hypothesis is true.

191. Find the power of a test when the probability of committing Type II error is 0.05

ANSWER:
Power = $1 - \beta = 1 - 0.05 = 0.95$

Applied and Computational Questions

192. Describe the action that would result in a Type I error and a Type II error if the following null hypothesis was tested; " H_o : The majority of Americans favor laws against assault weapons."

ANSWER:
A Type I error occurs when it is determined that the majority of Americans do not favor laws against assault weapons when, in fact, the majority do favor such laws.
A Type II error occurs when it is determined that the majority of Americans do favor laws against assault weapons when, in fact, they do not favor such laws.

193. Describe the action that would result in a Type I error and a Type II error if the following null hypothesis was tested; " H_o : This fast-food menu is not low fat."

ANSWER:
A Type I error occurs when it is determined that the fast food is low fat when, in fact, it is not low fat.
A Type II error occurs when it is determined that the fast food is not low fat when, in fact, it is low fat.

194. Describe the action that would result in a Type I error and a Type II error if the following null hypothesis was tested; " H_o : This old book must not be thrown away"

ANSWER:
A Type I error occurs when it is determined that the old book must be thrown away when, in fact, it should not be thrown.
A Type II error occurs when it is determined that the old book must not be thrown away when, in fact, it should be thrown.

195. Describe the action that would result in a Type I error and a Type II error if the following null hypothesis was tested; " H_o : There is no waste in the US Defense Department spending."

ANSWER:
A Type I error occurs when it is determined that there is waste in the US Defense Department spending when, in fact, there is not waste.
A Type II error occurs when it is determined that there is no waste in the US Defense Department spending when, in fact, there is waste.

196. Describe the action that would result in a correct decision Type *A* and a correct decision Type *B*, if the following null hypothesis was tested; " H_o : The majority of Americans favor laws against assault weapons."

ANSWER:
Type *A* correct decision: The majority of Americans do favor laws against assault weapons and it is decided that they do favor the laws.
Type *B* correct decision: The majority of Americans do not favor laws against assault weapons and it is decided that they do not favor the laws.

Chapter 8

197. Describe the action that would result in a correct decision Type *A* and a correct decision Type *B*, if the following null hypothesis was tested; "H_o : This fast-food menu is not low fat."

ANSWER:
Type *A* correct decision: The fast food menu is not low fat and it is decided that it is not low fat.
Type *B* correct decision: The fast food menu is low fat and it is decided that it is low fat.

198. Describe the action that would result in a correct decision Type *A* and a correct decision Type *B*, if the following null hypothesis was tested; "H_o : This old book must not be thrown away"

ANSWER:
Type *A* correct decision: The old book must not be thrown away and it is decided that it should not be thrown.
Type *B* correct decision: The old book must be thrown away and it is decided that it should be thrown.

199. Describe the action that would result in a correct decision Type *A*, and a correct decision Type *B*, if the following null hypothesis was tested; "H_o : There is no waste in the US Defense Department spending."

ANSWER:
Type *A* correct decision: There is no waste in US Defense Department spending and it is decided that there is no waste.
Type *B* correct decision: There is waste in US Defense Department spending and it is decided that there is waste.

QUESTIONS 200 AND 201 ARE BASED ON THE FOLLOWING INFORMATION:
A normally distributed population is known to have a standard deviation of 5, but its mean is in question. It has been argued to be either $\mu = 70$ or $\mu = 80$, and the following hypothesis test has been devised to settle the argument. The null hypothesis, $H_o : \mu = 70$, will be tested using one randomly selected data and comparing it to the critical value 76. If the data is greater than or equal to 76, the null hypothesis will be rejected.

200. Find α, the probability of committing the Type I error.

ANSWER:
α = *P*(rejecting H_o when H_o is true)
$= P(x \geq 76 \mid \mu = 70) = P[z > (76 - 70)/5] = P(z > 1.20)$ = 0.5000 – 0.3849 = 0.1151

201. Find β, the probability of committing the Type II error.

ANSWER:
β = *P*(accepting H_o when H_o is false)
$= P(x < 76 \mid \mu = 80) = P(z < (76 - 80)/5) = P(z < -0.80)$ = 0.5000 – 0.2881 = 0.2119

202. In order to complete a hypothesis test, we will need to write a conclusion that carefully describes the meaning of the decision relative to the intent of the hypothesis test. What does this mean?

ANSWER:
If the decision is "reject H_a" then the conclusion should be worded something like, "There is sufficient evidence at the α level of significance to show that........(the meaning of the alternative hypothesis)". On the other hand, if the decision is "fail to reject H_a" then the conclusion should be worded something like, "There is not sufficient evidence at the α level of significance to show that...........(the meaning of the alternative hypothesis)".

203. You want to show that professors find the new method of teaching calculus is more effective than traditional method. State the null and alternative hypotheses.

ANSWER:
H_o: The new method of teaching calculus is not more effective than traditional method
H_a: The new method of teaching calculus is more effective than traditional method

204. You are trying to show that smoking has an effect on a person's health. State the null and alternative hypotheses.

ANSWER:
H_o: Smoking has no effect on a person's health
H_a: Smoking has an effect on a person's health

QUESTIONS 205 THROUGH 208 ARE BASED ON THE FOLLOWING INFORMATION:
A statistician is interested in testing the null hypothesis H_o :Iraq was not a threat to US national security vs. the alternative hypothesis H_a :Iraq was a threat to US national security.

205. Identify the following situation as Type *A* or *B* correct decision or Type I or II error:
Truth of situation: Null hypothesis was false
Conclusion: The null hypothesis was failed to be rejected

ANSWER:
Type II error

206. Identify the following situation as Type *A* or *B* correct decision or Type I or II error:
Truth of situation: Null hypothesis was false
Conclusion: the null hypothesis was rejected

ANSWER:
Type B correct decision

207. Identify the following situation as Type *A* or *B* correct decision or Type I or II error:
Truth of situation: Null hypothesis was true
Conclusion: the null hypothesis was rejected

ANSWER:
Type I error

208. Identify the following situation as Type *A* or *B* correct decision or Type I or II error:
Truth of situation: Null hypothesis was true
Conclusion: The null hypothesis was failed to be rejected

ANSWER:
Type *A* correct decision

QUESTIONS 209 THROUGH 214 ARE BASED ON THE FOLLOWING INFORMATION:
When an airplane is inspected, the inspector is looking for anything that might indicate the plane might not be safe to fly.

209. State the null and alternative hypotheses.

ANSWER:
H_o: The plane will be safe to fly
H_a: The plane will not be safe to fly

210. Depending on the truth of the null hypothesis and the decision reached, describe what is meant by Type *A* correct decision in this situation as a possible outcome.

ANSWER:
Type *A* correct decision:The plane will be safe to fly and the inspector okayed its use.

211. Depending on the truth of the null hypothesis and the decision reached, describe what is meant by Type *B* correct decision in this situation as a possible outcome.

ANSWER:
Type *B* correct decision: The plane will not be safe to fly and the inspector did not okay its use.

212 Depending on the truth of the null hypothesis and the decision reached, describe what is meant by Type I error in this situation as a possible outcome.

ANSWER:
Type I error: The plane will be safe to fly and the inspector did not okay its use.

213. Depending on the truth of the null hypothesis and the decision reached, describe what is meant by Type II error in this situation as a possible outcome.

ANSWER:
Type II error: The plane will not be safe to fly and the inspector Okayed its use.

214. Describe the seriousness of the two possible errors in questions 212 and 213.

ANSWER:
The Type I error is not at all serious. A plane that is safe to fly will not be allowed to fly. The Type II error is very serious. A plane that is not safe to fly will be allowed to fly and passengers may get hurt to the extent that all may die as a result of a crash.

215. You are testing a new formula for hand lotion hoping to show it is effective on "dry or damaged skin". State the null and alternative hypotheses.

ANSWER:

H_o: The new formula for hand lotion is not effective on dry or damaged skin

H_a: The new formula for hand lotion is effective on dry or damaged skin

216. You are trying to show that tennis lessons have a positive effect on a child's self esteem. State the null and alternative hypotheses.

ANSWER:

H_o: Tennis lessons have no positive effect on a child's self esteem

H_a: Tennis lessons have positive effect on a child's self esteem

QUESTIONS 217 THROUGH 222 ARE BASED ON THE FOLLOWING INFORMATION:
When a medic at the scene after the collapse of the World Trade Center In New York on September 11, 2002 inspects each victim, he administers the appropriate medical assistant to all victims, unless he is certain the victim is dead.

217. State the null and alternative hypotheses.

ANSWER:

H_o: The victim is alive

H_a: The victim is not alive

218. Depending on the truth of the null hypothesis and the decision reached, describe what is meant by Type *A* correct decision in this situation as a possible outcome.

ANSWER:

Type *A* correct decision: The victim is alive and is treated as though alive.

219. Depending on the truth of the null hypothesis and the decision reached, describe what is meant by Type *B* correct decision in this situation as a possible outcome.

ANSWER:

Type *B* correct decision: The victim is dead and treated as dead.

220. Depending on the truth of the null hypothesis and the decision reached, describe what is meant by Type I error in this situation as a possible outcome.

ANSWER:

Type I error: The victim is alive, but is treated as though dead.

221. Depending on the truth of the null hypothesis and the decision reached, describe what is meant by Type II error in this situation as a possible outcome.

ANSWER:

Type II error: The victim is dead, but treated as if alive.

222. Describe the seriousness of the two possible errors in questions 220 and 221.

ANSWER:
The Type I error is very serious. The victim may very well be dead shortly without the attention that is not being received.
The Type II error is not as serious. The victim is receiving attention that is of no value. This would be serious only if there were other victims that needed this attention.

223. Consider the null hypothesis:" H_o : The majority of Americans favor laws against abortion." Describe the action that would result in a Type I error and a Type II error if this hypothesis was tested.

ANSWER:
A Type I error occurs when it is determined that the majority of Americans do not favor laws against abortion when, in fact, the majority do favor such laws.
A Type II error occurs when it is determined that the majority of Americans do favor laws against abortion when, in fact, they do not favor such laws.

224. Consider the null hypothesis:" H_o : This fast-food menu is not low sodium." Describe the action that would result in a Type I error and a Type II error if this hypothesis was tested.

ANSWER:
A Type I error occurs when it is determined that the fast food is low sodium when, in fact, it is not low sodium.
A Type II error occurs when it is determined that the fast food is not low sodium when, in fact, it is low sodium.

225. Consider the null hypothesis:" H_o : This historical building must not be demolished." Describe the action that would result in a Type I error and a Type II error if this hypothesis was tested.

ANSWER:
A Type I error occurs when it is determined that the historical building must be demolished when, in fact, it should not be demolished.
A Type II error occurs when it is determined that the historical building must not be demolished when, in fact, it should be demolished.

226. Consider the null hypothesis:" H_o : there is no waste in Bush's government spending." Describe the action that would result in a Type I error and a Type II error if this hypothesis was tested.

ANSWER:
A Type I error occurs when it is determined that there is waste in Bush's government spending when, in fact, there is no waste.
A Type II error occurs when it is determined that there is no waste in Bush's government spending when, in fact, there is waste.

227. If α is assigned the value 0.001, what are we saying about the Type I error?

ANSWER:
The Type I error is very serious and, therefore, we are willing to allow it to occur with a probability of 0.001; that is, only 1 chance in 1000.

228. Consider the null hypothesis:" H_o : The majority of Americans favor laws against abortion." Describe the action that would result in a correct decision Type *A* and a correct decision Type *B* if this hypothesis was tested.

ANSWER:
Type *A* correct decision: The majority of Americans do favor laws against abortion and it is decided that they do favor the laws.
Type *B* correct decision: The majority of Americans do not favor laws against abortion and it is decided that they do not favor the laws.

229. Consider the null hypothesis:" H_o : This fast-food menu is not low sodium." Describe the action that would result in a correct decision Type *A* and a correct decision Type *B* if this hypothesis was tested.

ANSWER:
Type *A* correct decision: The fast food menu is not low sodium and it is decided that it is not low sodium.
Type *B* correct decision: The fast food menu is low sodium and it is decided that it is low sodium.

230. Consider the null hypothesis:" H_o : This historical building must not be demolished." Describe the action that would result in a correct decision Type *A* and a correct decision Type *B* if this hypothesis was tested.

ANSWER:
Type *A* correct decision: The historical building must not be demolished and it is decided that it should not be demolished.
Type *B* correct decision: The historical building must be demolished and it is decided that it should be demolished.

231. Consider the null hypothesis:" H_o : there is no waste in Bush's government spending." Describe the action that would result in a correct decision Type *A* and a correct decision Type *B* if this hypothesis was tested.

ANSWER:
Type *A* correct decision: There is no waste in Bush's government spending and it is decided that there is no waste.
Type *B* correct decision: There is waste in Bush's government spending and it is decided that there is waste.

232. If α is assigned the value 0.05, what are we saying about the Type I error?

ANSWER:
The Type I error is somewhat serious and, therefore, we are willing to allow it to occur with a probability of 0.05; that is, 1 chance in 20.

233. If α is assigned the value 0.10, what are we saying about the Type I error?

ANSWER:
The Type I error is not at all serious and, therefore, we are willing to allow it to occur with a probability of 0.10; that is, 1 chance in 10.

Chapter 8

234. If β is assigned the value 0.001, what are we saying about the Type II error?

ANSWER:
The Type II error is very serious and, therefore, we are willing to allow it to occur with a probability of 0.001; that is, only 1 chance in 1000.

235. If β is assigned the value 0.05, what are we saying about the Type II error?

ANSWER:
The Type II error is somewhat serious and, therefore, we are willing to allow it to occur with a probability of 0.05; that is, 1 chance in 20.

236. If β is assigned the value 0.10, what are we saying about the Type II error?

ANSWER:
The Type II error is not at all serious and, therefore, we are willing to allow it to occur with a probability of 0.10; that is, 1 chance in 10.

<u>QUESTIONS 237 AND 238 ARE BASED ON THE FOLLOWING INFORMATION:</u>
The owner of a life insurance company is concerned with the effectiveness of a television commercial to promote his company.

237. What null hypothesis is he testing if he commits a Type I error when he erroneously says that the commercial is effective?

ANSWER:
H_o: Commercial is not effective

238. What null hypothesis is he testing if he commits a Type II error when he erroneously says that the commercial is effective?

ANSWER:
H_o: Commercial is effective

<u>QUESTIONS 239 THROUGH 241 ARE BASED ON THE FOLLOWING INFORMATION:</u>
A normally distributed population is known to have standard deviation of 4, but its mean is in question. It has been argued to be either μ = 90 or μ = 100, and the following hypothesis test has been devised to settle the argument. The null hypothesis, $H_o : \mu$ = 90, will be tested using one randomly selected data and comparing it to the critical value 96. If the data is greater than or equal to 96, the null hypothesis will be rejected.

239. Calculate α ; the probability of the Type I error.

ANSWER:
α = P(Type I error)
= P(rejecting H_o when the H_o is true)
= $P(x \geq 96 \mid \mu = 90)$
= $P(z \geq (96 - 90) / 4)$
= $P(z > 1.50)$
= 0.5000 - 0.4332
= 0.0668

240. Calculate β ; the probability of the Type II error.

ANSWER:

β = *P*(Type II error)

= *P*(failing to reject H_o when the H_o is false)
= $P(x < 96 \mid \mu = 100)$
= $P(z < (96 - 100) / 4)$
= $P(z < -1.0) = 0.5000 - 0.3413 = 0.1587$

241. Find the power of the statistical test.

ANSWER:
Power = 1 - β = 1.0 – 0.1587 = 0.8413

Sections 8.5 and 8.6

True-False Questions

242. In a particular hypothesis test, if $\alpha = 0.01$ and *p*-value = 0.019, then the correct decision would be to fail to reject the null hypothesis.
ANSWER: T

243. The classical approach to hypothesis testing is completed using a five-step model.
ANSWER: T

244. In a particular hypothesis test, if $\alpha = 0.05$ and *p*-value = 0.042, then the correct decision would be to fail to reject the null hypothesis.
ANSWER: F

245. If the noncritical region in a hypothesis test is made wider (assuming σ and *n* remain fixed), then α becomes larger.
ANSWER: F

246. In testing $H_o : \mu = \mu_o$ vs. $H_a : \mu \neq \mu_o$, the term highly significant is commonly used in research findings if 0.01<*p*-value$\leq$0.05.
ANSWER: F

247. The *probability-value approach*, or simply *p-value approach*, is the hypothesis test process that has gained popularity in recent years, largely as a result of the convenience and the "number crunching" ability of the computer.
ANSWER: T

248. If the *p*-value is *less than or equal to* the level of significance, α , then the decision must be not to reject H_o.
ANSWER: F

249. In testing $H_o : \mu = \mu_o$ vs. $H_a : \mu \neq \mu_o$, non-statistically significant is commonly used in research findings if *p*-value>0.05.
ANSWER: T

250. The alternative hypothesis assigns a specific value to the parameter in question, and therefore "equals" will always be part of the alternative hypothesis.
ANSWER: F

251. Probability value, or *p*-value is the probability that the test statistic could be the value it is or a more extreme value (in the direction of the alternative hypothesis) when the null hypothesis is true.
ANSWER: T

252. In testing $H_o : \mu = \mu_o$ vs. $H_a : \mu \neq \mu_o$, statistically significant is commonly used in research findings if *p*-value$\leq$0.01.
ANSWER: F

253. The *fundamental idea of the p-value* is to express the degree of belief in the null hypothesis.
ANSWER: T

254. If the *p*-value is *greater than* the level of significance α , then the decision must be to reject H_o .
ANSWER: F

255. The alternative hypothesis is referred to as being "two-tailed" when H_a is "not equal".
ANSWER: T

256. After the null and alternative hypotheses are established, we always work under the assumption that the null hypothesis is a true statement until there is sufficient evidence to reject it.
ANSWER: T

Multiple-Choice Questions

257. Choose the pair of words that correctly completes the following statement: "The *p*-value of a hypothesis test is the ______ level of significance for which the observed sample information is _____ provided the null hypothesis is true."

A) smallest, not significant
B) smallest, significant
C) largest, not significant
D) largest, significant
ANSWER: B

258. In a particular hypothesis test, the *p*-value is 0.0211. What must be true of α in order to reject the null hypothesis?

A) $\alpha > 0.0211$
B) $\alpha \geq 0.0211$
C) $\alpha < 0.0211$
D) $\alpha \leq 0.0211$
ANSWER: B

259. You have conducted a hypothesis test and found that *p*-value = 0.04. Based on this information you know that you cannot reject the null hypothesis if

A) $\alpha < 0.04$.
B) $\alpha > 0.04$.
C) $\alpha \leq 0.04$.
D) $\alpha \geq 0.04$.
ANSWER: A

Chapter 8

260. In the classical approach to hypothesis testing, we use an asterisk" $*$ " to identify which of the following?

A) The level of significance
B) The value of the parameter stated in the null hypothesis
C) The critical value
D) The computed value of the test statistic
ANSWER: D

261. Which of the following would be the correct hypotheses for testing the claim that the mean lifetime of a cellular phone battery, while the phone is left on, is less than 24 hours?

A) $H_o: \mu = 24, \quad H_a: \mu \neq 24$
B) $H_o: \mu = 24(\geq), \quad H_a: \mu < 24$
C) $H_o: \mu = 24(\leq), \quad H_a: \mu > 24$
D) $H_o: \mu > 24, \quad H_a: \mu \leq 24$
ANSWER: B

262. Which of the following would be the null hypothesis in testing the claim that the mean GPA of all college graduates majoring in computer science in U.S. colleges is different from 3.14?

A) $H_o: \mu = 3.14$
B) $H_o: \mu = 3.14(\geq)$
C) $H_o: \mu = 3.14(\leq)$
D) $H_o: \mu \neq 3.14$
ANSWER: A

263. Which of the following would be the correct hypotheses for testing the claim that the mean monthly rainfall in Toledo daily during April is no less than 2.5 inches?

A) $H_o: \mu = 2.5, \quad H_a: \mu \neq 2.5$
B) $H_o: \mu = 2.5(\geq), \quad H_a: \mu < 2.5$
C) $H_o: \mu = 2.5(\leq), \quad H_a: \mu > 2.5$
D) $H_o: \mu > 2.5, \quad H_a: \mu = 2.5(\leq)$
ANSWER: B

264. Which of the following would be the alternative hypothesis in testing the claim that the mean distance students commute to campus is no more than 7.1 miles?

A) $H_a: \mu \neq 7.1$
B) $H_a: \mu < 7.1$
C) $H_a: \mu > 7.1$
D) $H_a: \mu = 7.1(\leq)$
ANSWER: C

265. Which of the following would be the correct hypotheses for testing the claim that the mean cost of a meal at a fast food restaurant is less than $3.79?

A) $H_o: \mu = 3.79, \quad H_a: \mu \neq 3.79$
B) $H_o: \mu = 3.79(\geq), \quad H_a: \mu < 3.79$
C) $H_o: \mu = 3.79(\leq), \quad H_a: \mu > 3.79$
D) $H_o: \mu > 3.79, \quad H_a: \mu = 3.79(\leq)$
ANSWER: B

266. Which of the following statements is true regarding the *p*-value?

A) When the *p*-value is miniscule (like 0.0003), the null hypothesis would be rejected by everybody because the sample results are very unlikely for a true H_o. However, when the *p*-value is fairly small (like 0.012), the evidence against H_o is quite strong and H_o will be rejected by many.
B) When the *p*-value begins to get larger (say, 0.02 to 0.08), there is too much probability that data like the sample involved could have occurred even if H_o were true, and the rejection of H_o is not an easy decision.
C) When the *p*-value gets large (like 0.15 or more), the data are not at all unlikely if the H_o is true, and no one will reject H_o.
D) All of the above
ANSWER: D

267. Which of the following statements is false?

A) Critical region is the set of values for the test statistic that will cause us to always reject the null hypothesis for specific level(s) of significance α.
B) Critical region is the set of values for the test statistic that will cause us to always reject the null hypothesis for any level of significance α.
C) The set of values that are not in the critical region is called the noncritical region (sometimes called the acceptance region).
D) None of the above
ANSWER: B

Short-Answer Questions

268. Suppose the null hypothesis is "the mean diameter of parts produced by a machine is 0.85" (μ = 0.85) and the alternative is μ > 0.85. If *n* items are tested and based on the results, it is concluded that μ > 0.85 when in fact μ = 0.85. What Type of error is made?

ANSWER:
Type I error

269. Suppose that we want to test the hypothesis that the mean fill by a bottling machine is less than 12 ounces. Explain the conditions that would exist if we make an error in decision by committing a Type I error.

ANSWER:
We reject the null hypothesis that $\mu \geq 12$ when, in fact, $\mu \geq 12$.

Chapter 8

270. Suppose that we want to test the hypothesis that the mean IQ of a large group of students is at least 105. Explain the conditions that would exist if we make an error in decision by committing a Type II error.

ANSWER:
We fail to reject the null hypothesis that $\mu \geq 105$ when, in fact, $\mu < 105$.

271. Do large or small values for the *p*-value help support the alternative hypothesis?

ANSWER:
The smaller the *p*-value, the stronger the support for the alternative hypothesis

272. An experimenter is testing the following hypothesis, $H_o : \mu = 14.8 (\geq)$ and $H_a : \mu < 14.8$, using the *p*-value approach and from his sample information computes a *p*-value of 0.0778. Then he sets the value of α = 0.10 so that he may reject the null hypothesis. Discuss the procedure described.

ANSWER:
An honest experimenter decides on the seriousness of Type I error and sets α before performing the test, not after the test is performed.

273. State the null and alternative hypotheses used to test the following claim: "The mean of ACT scores is 25."

ANSWER:
H_o : μ = 25 vs. H_a : $\mu \neq 25$

274. Briefly discuss the *advantages of the p-value approach*.

ANSWER:
(1) The results of the test procedure are expressed in terms of a continuous probability scale from 0.0 to 1.0, rather than simply on a "reject" or "fail to reject" basis.
(2) A *p*-value can be reported and the user of the information can decide on the strength of the evidence as it applies to his / her situation.
(3) Computers can do all the calculations and report the *p*-value, thus eliminating the need for tables.

275. For the following pair of values, *p*-value = 0.025 and α = 0.05, state the decision that will be reached and state why.

ANSWER:
Reject H_o since *p*-value = 0.025 < α = 0.05

276. State the null and alternative hypotheses used to test the following claim: "The mean lifetime of fluorescent light bulbs is at most 2000 hours."

ANSWER:
H_o : μ = 2000 ($\leq$) vs. H_a : μ > 2000

277. What decision is reached when the *p*-value is smaller than α ?

ANSWER:
The decision will be: reject H_o.

278. State the null and alternative hypotheses used to test the claim that "The mean score on that MCAT (Medical College Admission Test) is different from 27."

ANSWER:
$H_o : \mu = 27$ vs. $H_a : \mu \neq 27$

279. Assume that z is the test statistic and calculate the value of z^* for testing the null hypothesis $H_o : \mu = 150.0$ when $\sigma = 4.5,\ n = 15,\ \overline{x} = 147.8$

ANSWER:
$$z^* = \frac{\overline{x} - \mu}{\sigma / \sqrt{n}} = \frac{147.8 - 150}{4.5 / \sqrt{15}} = -1.89$$

280. What decision is reached when α is equal to the *p*-value?

ANSWER:
The decision will be: reject. H_o

281. State the null and alternative hypotheses used to test the claim that "The mean selling price of foreign made mini vans is no less than $38,000."

ANSWER:
$H_o : \mu = 38,000\ (\geq)$ vs. $H_a : \mu < 38,000$

282. For the following pair of values, *p*-value = 0.12 and α = 0.10, state the decision that will be reached and state why.

ANSWER:
Fail to reject H_o since *p*-value = 0.12 > α = 0.10

283. Assume that z is the test statistic and calculate the value of z^* for testing the null hypothesis $H_o : \mu = 415$ when $\sigma = 42.6,\ n = 50,\ \overline{x} = 430$

ANSWER:
$$z^* = \frac{\overline{x} - \mu}{\sigma / \sqrt{n}} = \frac{430 - 415}{42.6 / \sqrt{50}} = 2.49$$

Applied and Computational Questions

284. Use the *p*-value approach to test the hypotheses $H_o : \mu = 500(\geq)$ vs. $H_a : \mu < 500$ at the 0.05 level of significance, given that $\sigma = 30.2$, and that a sample of size 81 produced a sample mean of 508.2.

ANSWER:
p-value = 0.0073. Since *p*-value < α, reject the null hypothesis and conclude that the population mean is less than 500.

285. For the hypothesis test, $H_o: \mu = a(\geq)$ vs. $H_a: \mu < a$, *p*-value = 0.0013. Give the calculated value for the test statistic.

ANSWER:
$z^* = -3.0$

286. A statistician was testing the following hypotheses: $H_o: \mu = 500$ vs. $H_a: \mu \neq 500$. The *p*-value approach was to be used. A sample of size 49 gave a sample mean of 508. Given that $\sigma = 30.2$, and $\alpha = 0.01$, find the *p*-value, and write your conclusion.

ANSWER:
p-value = 0.008. Since *p*-value $< \alpha$, reject the null hypothesis and conclude that the population mean is not 500.

287. The mean cost for a home nationwide is reported to be \$80,000 with a standard deviation equal to \$9,500. To test that the mean in Omaha is less than the national mean, 35 homes for sale are randomly selected and the mean is found to be \$65,000. Assuming the variability is the same locally as nationally, write the appropriate null and alternative hypotheses for this situation, calculate the *p*-value for the test, and write your conclusion.

ANSWER:
$H_o: \mu = 80,000(\geq)$ vs. $H_a: \mu < 80,000$

p-value is practically zero, since $z^* = -9.34$. Therefore we reject the null hypothesis and conclude that the mean cost for homes in Omaha is less than the national mean of \$80,000.

288. For the hypothesis test, $H_o: \mu = a$ vs. $H_a: \mu \neq a$, *p*-value = 0.1260. Give the calculated value for the test statistic.

ANSWER:
$z^* = \pm 1.53$

289. For the hypothesis test, $H_o: \mu = a(\leq)$ vs. $H_a: \mu > a$, *p*-value = 0.2358. Give the calculated value for the test statistic.

ANSWER:
$z^* = 0.72$

290. For testing the hypothesis $H_o: \mu = a$ vs. $H_a: \mu \neq a$, give the absolute value of the calculated test statistic, which would correspond to *p*-value = 0.0672.

ANSWER:
$|z^*| = 1.83$

291. For testing the hypothesis $H_o: \mu = a$ vs. $H_a: \mu \neq a$, give the absolute value of the calculated test statistic, which would correspond to *p*-value = 0.0120.

ANSWER:
$|z^*| = 2.34$

292. For a national compliance test for diabetics, μ = 74 and σ = 4. To test that diabetic patients at a particular hospital have this mean versus a value different than the national mean, the test is administered to 50 diabetic patients at the hospital, and $\bar{x}$ = 78.5. Assuming the variability in test scores at the hospital is the same as that at the national level, find the *p*-value for this hypothesis test, and write your conclusion.

ANSWER:
p-value is practically zero, since $z^* = 7.95$. Therefore we reject the null hypothesis and conclude that the mean for diabetic patients at this hospital have is different than the national mean

293. For the hypothesis $H_o : \mu = a$ vs. $H_a : \mu \neq a$, give the absolute value of the calculated test statistic, which would correspond to *p*-value = 0.1336.

ANSWER:
$|z^*| = 1.50$

<u>QUESTIONS 294 THROUGH 296 ARE BASED ON THE FOLLOWING INFORMATION:</u>
A machine is programmed to put 737 grams of salt in each container that passes underneath its nozzle. In order to test $H_o : \mu = 737(\leq)$ vs. $H_a : \mu > 737$, a sample of 35 boxes of salt is selected. It is found that $\bar{x} = 740.5$, and it is known that $\sigma = 7.5$ grams.

294. Give the calculated test statistic.

ANSWER:
$z^* = 2.76$

295. Calculate the *p*-value.

ANSWER:
p-value = 0.0029

296. Test the hypothesis at α = 0.01, and write your conclusion.

ANSWER:
Since *p*-value $< \alpha$, reject the null hypothesis and conclude that the machine, on average, puts more than 737 grams of salt in each container that passes underneath its nozzle.

297. Calculate the *p*-value for testing $H_o : \mu = 25(\geq)$ vs. $H_a : \mu < 25$, if the value of the test statistic z^* = -2.84.

ANSWER:
p-value = 0.0023

298. Calculate the *p*-value for testing $H_o : \mu = 50$ vs. $H_a : \mu \neq 50$, if the value of the test statistic z^* = 1.98.

ANSWER:
p-value = 0.0478

299. Calculate the *p*-value for testing $H_o : \mu = c(\leq)$ vs. $H_a : \mu > c$, if the value of the test statistic z^* = 3.16.

ANSWER:
p-value = 0.0008

300. Consider the hypothesis test $H_o : \mu = 165(\leq)$ vs. $H_a : \mu > 165$, with σ = 15. Find the critical value of the test statistic $\bar{x}$ if samples of size 50 and α = 0.01 are utilized.

ANSWER:
169.94

QUESTIONS 301 THROUGH 303 ARE BASED ON THE FOLLOWING INFORMATION:
The following terms are commonly used in research findings: if 0.01< *p*-value ≤ 0.05, the result is said to be statistically significant. If *p*-value ≤ 0.01, the result is said to be highly significant. If *p*-value > 0.05, the result is said to be non-statistically significant, statistically significant, or highly significant.

301. Classify $H_o : \mu = 19$ vs. $H_a : \mu < 19$, as non-statistically significant, statistically significant, or highly significant if the value of the test statistic is $z^* = -1.73$.

ANSWER:
Statistically significant

302. Classify $H_o : \mu = 17$ vs. $H_a : \mu \neq 17$, as non-statistically significant, statistically significant, or highly significant if the value of the test statistic is $z^* = 3.21$.

ANSWER:
Highly significant

303. Classify $H_o : \mu = 13$ vs. $H_a : \mu > 13$, as non-statistically significant, statistically significant, or highly significant if the value of the test statistic is $z^* = 123$.

ANSWER:
Non-statistically significant

QUESTIONS 304 THROUGH 307 ARE BASED ON THE FOLLOWING INFORMATION:
A sample of size 35 is used to test $H_o : \mu = 65(\geq)$ vs. $H_a : \mu < 65$, and produced a sample mean $\bar{x}$ = 63.5. Assume that the population standard deviation is σ = 2.5.

304. What is the computed value of the test statistic?

ANSWER:
$z^* = -3.55$

305. What distribution does the test statistic have when the null hypothesis is true?

ANSWER:
Standard normal distribution

306. Is the alternative hypothesis one-tailed or two-tailed?

ANSWER:
One-tailed alternative

307. What is the *p*-value?

ANSWER:
p-value < 0.0002

QUESTIONS 308 THROUGH 311 ARE BASED ON THE FOLLOWING INFORMATION:
A sample of size 40 is used to test $H_o: \mu = 65$ vs. $H_a: \mu \neq 65$, and produced a sample mean $\bar{x}$ = 66.2. Assume that the population standard deviation is σ = 2.5.

308. What is the computed value of the test statistic?

ANSWER:
$z^* = 3.04$

309. What distribution does the test statistic have when the null hypothesis is true?

ANSWER:
Standard normal distribution

310. Is the alternative hypothesis one-tailed or two-tailed?

ANSWER:
Two-tailed alternative

311. What is the *p*-value?

ANSWER:
p-value = 0.0024

QUESTIONS 312 AND 313 ARE BASED ON THE FOLLOWING INFORMATION:
A random sample was selected from a normal population with a standard deviation σ = 5.70. The sample values were 236, 240, 229, 237, 241, 243, 239, 228, 231, and 225.

312. Compute the *p*-value for the hypothesis test: $H_o: \mu = 235.8 (\geq)$ vs. $H_a: \mu < 235.8$.

ANSWER:
p-value = 0.3085

313. What is your conclusion at the 0.10 level of significance?

ANSWER:
Since *p*-value $> \alpha$; we fail to reject the null hypothesis and conclude that the population mean is at least 235.8.

314. Suppose the hypothesis $H_o: \mu = a(\leq)$ be tested versus $H_a: \mu > a$ at α = 0.01. If σ = *b*, and *n* = 100, how large must $\bar{x}$ be before the null hypothesis can be rejected?

ANSWER:
$\bar{x} \geq a + 0.233b$

315. In testing $H_o: \mu = 28.7(\geq)$ vs. $H_a: \mu < 28.7$, using the *p*-value approach, a *p*-value of 0.0764 was obtained. If σ = 9.8, find the sample mean which produced this *p*-value given that a sample of size *n* = 40 was randomly selected.

ANSWER:
$\bar{x} = 26.48$

316. Suppose a sample of size 50 was taken to test the null hypothesis $H_o: \mu = 75$ versus the alternative hypothesis $H_a: \mu < 75$ at $\alpha = 0.05$. Determine the critical region for this test.

ANSWER:
$z \leq -1.65$

317. We wish to test the null hypothesis "the mean is no more than 20," versus the alternative hypothesis "the mean is more than 20." The test statistic *z* is to be used. Find the value of α that corresponds to the critical region: $z \geq 1.68$.

ANSWER:
α = 0.0465

318. Suppose a sample of size 50 was taken to test the null hypothesis $H_o: \mu = 85$ versus the alternative hypothesis $H_a: \mu \neq 85$ at $\alpha = 0.05$. Determine the critical region for this test.

ANSWER:
$z \leq -1.96$ or $z \geq 1.96$

319. We wish to test the null hypothesis "the mean is no more than 20," versus the alternative hypothesis "the mean is more than 20." The test statistic *z* is to be used. Find the value of α that corresponds to the critical region: $z \geq 1.75$.

ANSWER:
α = 0.0401

320. Suppose a sample of size 50 was taken to test the null hypothesis $H_o: \mu = 95$ versus the alternative hypothesis $H_a: \mu > 95$ at $\alpha = 0.10$. Determine the critical region for this test.

ANSWER:
$z \geq 1.28$

321. A machine is programmed to put 737 grams of salt in each container that passes underneath its nozzle. In order to test $H_o: \mu = 737(\leq)$ vs. $H_a: \mu > 737$, a sample of 100 boxes of salt is selected. How large must the sample mean be before the null hypothesis can be rejected for α = 0.01? It is known that σ = 7.5 grams.

ANSWER:
$\bar{x} \geq 738.75$ grams

322. To test the null hypothesis that the average lifetime for a particular brand of bulb is 750 hours versus the alternative that the average lifetime is different from 750 hours, a sample of 75 bulbs is used. If the standard deviation is known to equal 50 hours and if α is equal to 0.01, what values for $\bar{x}$ will result in rejection of the null hypothesis?

ANSWER:
$\bar{x} \leq 735.1$ hours or $\bar{x} \geq 764.9$ hours

323. Suppose we were testing the hypothesis $H_o : \mu = 82.4(\leq)$ vs. $H_a : \mu > 82.4$, using α = 0.10. Suppose further that σ = 16.7. What is the smallest sample mean that would cause us to reject the null hypothesis using samples of size n = 35?

ANSWER:
$\bar{x} = 86.0$

324. Suppose we were testing the hypothesis $H_o : \mu = 76.9$ vs. $H_a : \mu \neq 76.9$, using α = 0.05. Suppose further that σ = 14.6. What is the smallest sample size that would cause us to reject the null hypothesis if the sample mean is 74.8?

ANSWER:
n = 186

325. If the probability of making a Type I error in a right-tailed test decreases from α = 0.05 to α = 0.01, does the critical value increase or decrease? By what amount does it increase or decrease?

ANSWER:
The critical value increases by 0.68.

326. Calculate the p -value for testing $H_o : \mu = 12$ vs. $H_a : \mu > 12$, $z^* = 1.58$.

ANSWER:
p – value = $P(z > 1.58)$ = 0.5000 – 0.4429 = 0.0571

327. Calculate the p-value for testing $H_o : \mu = 100$ vs. $H_a : \mu < 100$, $z^* = -0.75$.

ANSWER:
p – value = $P(z < -0.75)$ =0.5000 – 0.2734 = 0.2266

328. Calculate the p-value for testing $H_o : \mu = 15.6$ vs. $H_a : \mu \neq 15.6$, $z^* = 1.37$.

ANSWER:
$p - \text{value} = 2 \cdot P(z > 1.37) = 2(0.5000 - 0.4147) = 0.1706$

329. Calculate the p-value for testing $H_o : \mu = 9.46$ vs. $H_a : \mu < 9.46$, $z^* = -2.19$.

ANSWER:
p – value = $P(z < -2.19)$ = 0.5000 – 0.4857 = 0.0143

Chapter 8

330. Calculate the *p*-value for testing $H_o : \mu = 115$ vs. $H_a : \mu \neq 115, \quad z^* = -0.99$.

ANSWER:
p – value = 2 · $P(z > 0.93) = 2(0.5000 - 0.3238) = 0.3524$

331. Find the value of z^* for testing $H_o : \mu = 40$ vs. $H_a : \mu > 40$ when *p*-value = 0.0582. Sketch a normal curve to display the results.

ANSWER:
P = *p*-value = $P(z > z^*) = 0.0582$

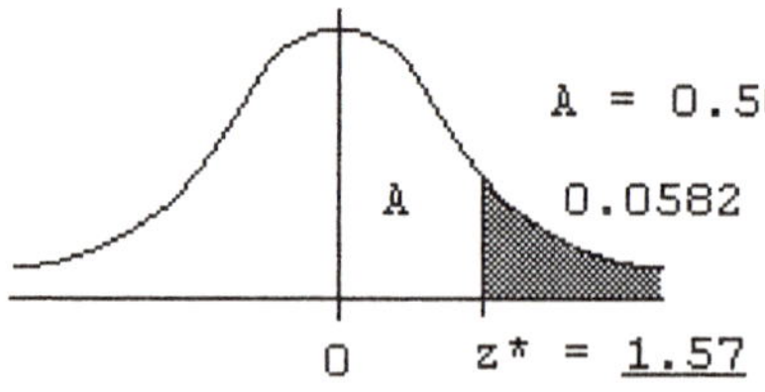

A = 0.5000 - 0.0582 = 0.4418

332. Find the value of z^* for testing $H_0 : \mu = 40$ versus $H_a : \mu < 40$ when *p*-value = 0.0166. Sketch a normal curve to display the results.

ANSWER:
P = $P(z < z^*) = 0.0166$

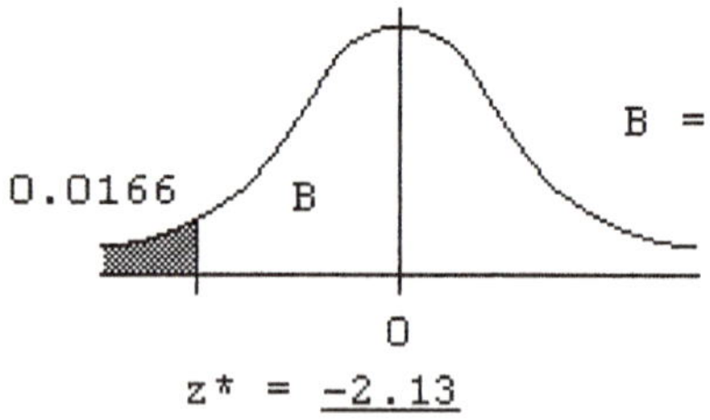

B = 0.5000 - 0.0166 = 0.4834

333. Find the value of z^* for testing $H_o : \mu = 40$ versus $H_a : \mu \neq 40$ when *p*-value = 0.0042. Sketch a normal cure to display the results.

ANSWER:
P = $P(z < -z^*) + P(z > +z^*) = 2 \cdot P(z > +z^*) = 0.0042$. Hence, $P(z > +z^*) = 0.0021$

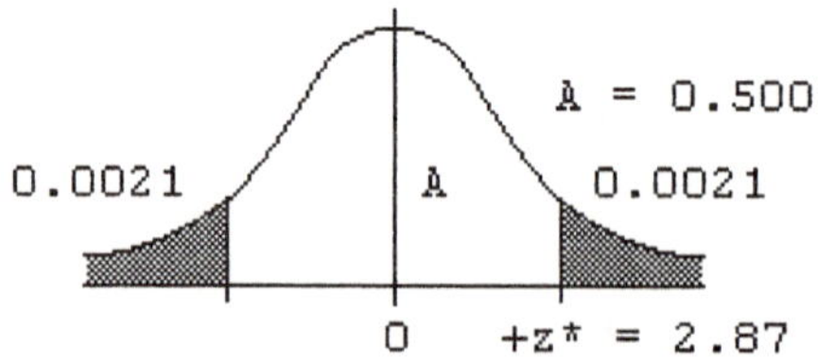

A = 0.5000 - 0.0021 = 0.4979

z* = -2.87 or 2.87 [or ±2.87]

334. The null hypothesis, $H_o: \mu = 50$, was tested against the alternative hypothesis, $H_a: \mu > 50$. A sample of 100 resulted in a calculated *p*-value of 0.102. If $\sigma = 4.0$, find the value of the sample mean, $\bar{x}$.

ANSWER:

P = $P(z > z^*) = 0.1020$

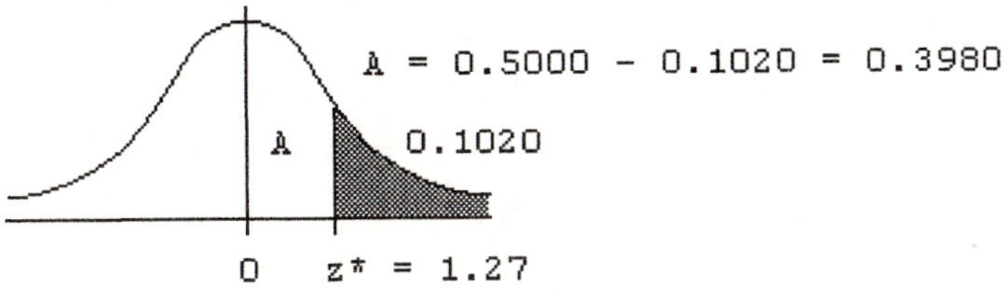

The formula $z = (\bar{x} - \mu)/(\sigma/\sqrt{n})$ reduces to $1.27 = (\bar{x} - 50)/(4.0/\sqrt{100})$. Solving for $\bar{x}$, we get $\bar{x} = 50 + (1.27)(4.0/\sqrt{100}) = 50.508$.

QUESTIONS 335 THROUGH 338 ARE BASED ON THE FOLLOWING INFORMATION:

The following computer output was used to complete a hypothesis test.

TEST OF MU = 7.25 VS MU N. E. 7.25
THE ASSUMED SIGMA = 1.25

N	MEAN	STDEV	SE MEAN	Z	P VALUE
80	7.654	1.152	0.1398	2.89	0.0038

335. State the null and alternative hypotheses.

ANSWER:

$H_o: \mu = 7.25$ vs. $H_a: \mu \neq 7.25$

336. If the test is completed using $\alpha = 0.05$, what decision and conclusion are reached? Explain.

ANSWER:

Since *p*-value = 0.0038 < α, we reject H_o; and conclude that the population mean is significantly different from 7.25.

337. Verify the value of the standard error of the mean.

ANSWER:

$\sigma_{\bar{x}} = \sigma/\sqrt{n} = 1.25/\sqrt{80} = 0.1398$

Chapter 8

338. Find the values for $\sum x$ and $\sum x^2$.

ANSWER:

$\bar{x} = \sum x / n \Rightarrow \sum x = n \cdot \bar{x} = (80)(7.654) = 612.32$

Since $s^2 = [\sum x^2 - (\sum x)^2 / n)]/(n-1)$; s = 1.152, n = 80, and $\sum x$ = 612.32, then $(1.152)^2 = [\sum x^2 - (612.32)^2 / 80]/(80-1)$. Hence, $\sum x^2$ =4791.5385.

339. Determine the critical region and critical values for z that would be used to test the null hypothesis $H_o: \mu = 25$ vs. $H_a: \mu \neq 25$, at the level of significance α = 0.10. Sketch a normal curve to display the results.

ANSWER:

$z \leq -1.65, \quad z \geq 1.65$

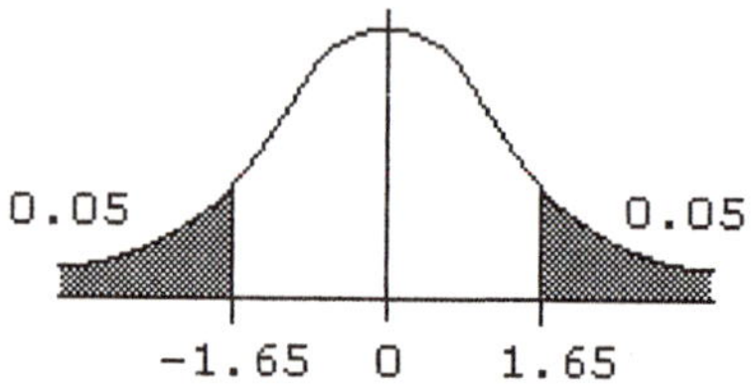

340. Determine the critical region and critical values for z that would be used to test the null hypothesis $H_o: \mu = 32(\leq)$ vs. $H_a: \mu > 32$, at the level of significance α =0.01. Sketch a normal curve to display the results.

ANSWER:

$z \geq 2.33$

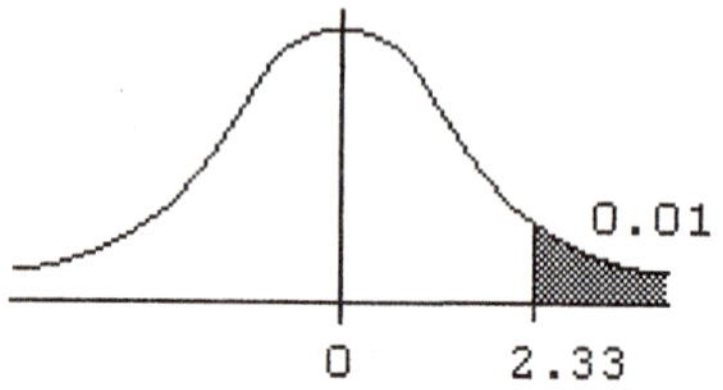

341. Determine the critical region and critical values for z that would be used to test the null hypothesis $H_o: \mu = 13(\geq)$ vs. $H_a: \mu < 13$, at the level of significance α =0.05. Sketch a normal curve to display the results.

ANSWER:

$z \leq -1.65$

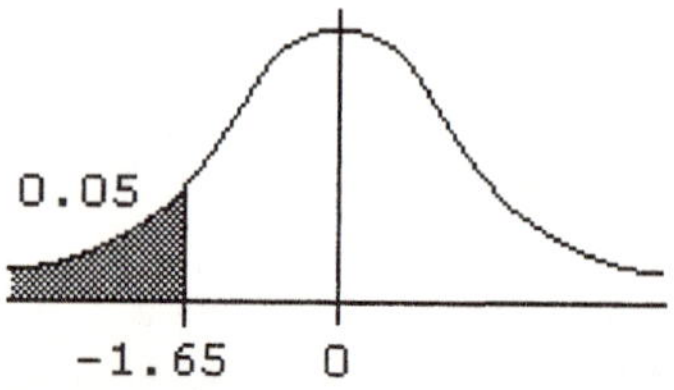

342. Determine the critical region and critical values for z that would be used to test the null hypothesis $H_o: \mu = 18$ vs. $H_a: \mu \neq 18$, at the level of significance α =0.01. Sketch a normal curve to display the results.

ANSWER:
$z \leq -2.58, \quad z \geq 2.58$

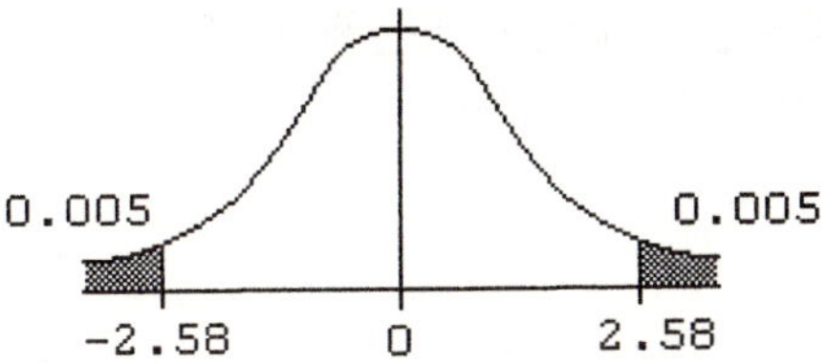

343. The manager at Fed Express feels that the weights of packages shipped recently are less than in the past. Records show that in the past packages have had a mean weight of 38.5 lb. and a standard deviation of 13.4 lb. A random sample of last month's shipping records yielded a mean weight of 34.2 lb. for 64 packages. Is this sufficient evidence to reject the null hypothesis in favor of the manager's claim? Use $\alpha = 0.01$.

ANSWER:
μ = The mean weight of packages shipped by Fed Express

$H_o: \mu = 38.5(\geq)$ vs. $H_a: \mu < 38.5$

Normality assumed. Since $\sigma = 13.4$, $n = 64, \bar{x} = 34.2$, then

$z^* = (\bar{x} - \mu)/(\sigma/\sqrt{n}) = (34.2 - 38.5)/(13.4/\sqrt{64}) = -2.57$

Critical value at $\alpha = 0.01$ is: $-z(0.01) = -2.33$

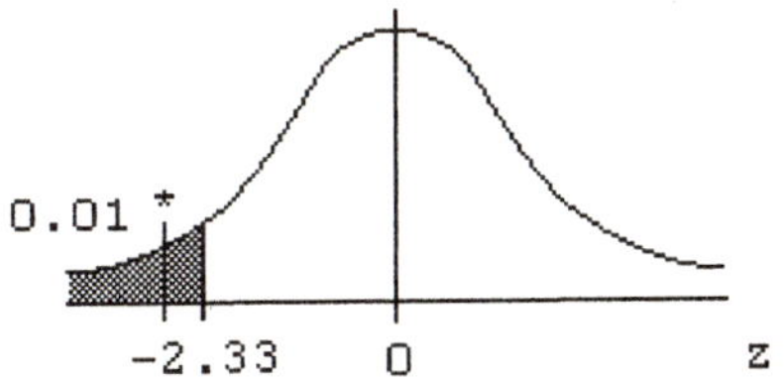

z^* falls in the critical region, therefore we reject H_o at the 0.01 level of significance in favor of the manager's claim, and conclude that the population mean is significantly less than the mean of 38.5.

344. Find the value of $\bar{x}$ for testing $H_o: \mu = 320$, given that z^* = 2.6, $\sigma = 21$, and n = 60.

ANSWER:
The formula $z = (\bar{x} - \mu)/(\sigma/\sqrt{n})$ reduces to $2.60 = (\bar{x} - 320)/(21/\sqrt{60})$.

Solving for $\bar{x}$, we get $\bar{x}$ = 320 + (2.60)(21/$\sqrt{60}$) = 327.0488.

Chapter 8

345. Find the value of $\bar{x}$ for testing $H_o: \mu = 80,$ given that z^* = -0.95, $\sigma = 6.75,$ and n = 36.

ANSWER:

The formula $z = (\bar{x} - \mu)/(\sigma / \sqrt{n})$ reduces to $-0.95 = (\bar{x} - 80)/(6.75/\sqrt{36})$.

Solving for $\bar{x}$, we get $\bar{x}$ = 80 + (-0.95)(6.75/$\sqrt{36}$) = 78.9313.

QUESTIONS 346 THROUGH 349 ARE BASED ON THE FOLLOWING INFORMATION:

From a population of unknown mean μ and a known standard deviation $\sigma = 6.0$, a sample of $n = 100$ is selected and the sample mean 43.5 is found.

346. Determine the 95% confidence interval for μ.

ANSWER:

Normality assumed. Since $\sigma = 6.0$, $1-\alpha = 0.95$, $n = 100,$ and $\bar{x} = 43.5$, then $\alpha/2 = 0.025;$ and $z(0.025) = 1.96$. Hence,

$E = z(\alpha/2)\cdot\sigma/\sqrt{n} = (1.96)(6/\sqrt{100})$ = 1.176. Then, $\bar{x} \pm E = 43.5 \pm 1.176$, and the 95% confidence interval for μ is 42.324 to 44.676.

347. Complete the hypothesis test involving $H_a: \mu \neq 42$ using the *p*-value approach and $\alpha = 0.05.$

ANSWER:

$H_o: \mu = 42$ vs. $H_a: \mu \neq 42$

Normality assumed. Since $\sigma = 6.0$, $n = 100,$ and $\bar{x} = 43.5$, then

$z^* = (\bar{x} - \mu)/(\sigma/\sqrt{n}) = (43.5 - 42)/(6/\sqrt{100}) = 2.5$

P = $2P(z > 2.5)$. Using the table of standard normal distribution, then we get:

P = 2(0.5000 – 0.4938) = 0.0124

Since **P** $< \alpha$, we reject H_o at the 0.05 level of significance, and conclude that there is sufficient evidence to support the contention that the mean is not equal to 42.

348. Complete the hypothesis test involving $H_a: \mu \neq 42$ using the classical approach and $\alpha = 0.05.$

ANSWER:

$H_o: \mu = 42$ vs. $H_a: \mu \neq 42$

Normality assumed. Since $\sigma = 6.0$, $n = 100,$ and $\bar{x} = 43.5$, then

$z^* = (\bar{x} - \mu)/(\sigma/\sqrt{n}) = (43.5 - 42)/(6/\sqrt{100}) = 2.5$

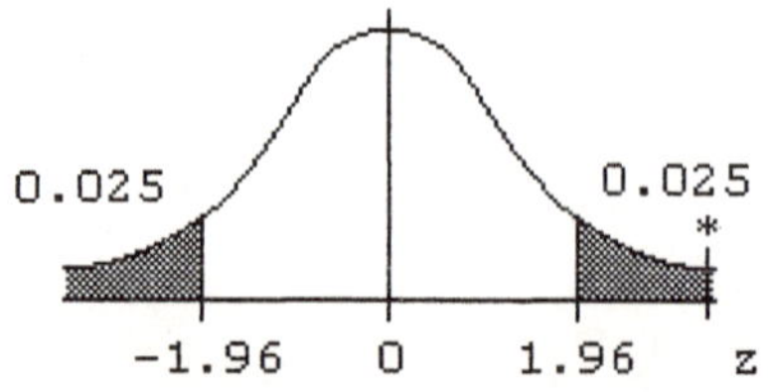

The critical values at α = 0.05 are: $\pm z(0.025) = \pm 1.96$

z^* falls in the critical region, therefore we reject H_o at the 0.05 level of significance, and conclude that there is sufficient evidence to support the contention that the mean is not equal to 42.

349. Describe the relationship between these three separate procedures performed in questions 346, 347 and 348.

ANSWER:
The null hypothesis is rejected at the 0.05 level of significance since z^* = 2.5 is in the critical region, or *p*-value is less than α, and μ = 42 is not within the 95% confidence interval estimate of 42.324 to 44.676.

QUESTIONS 350 AND 351 ARE BASED ON THE FOLLOWING INFORMATION:
In Meijer supermarket, the customer's waiting time to check out is approximately normally distributed with a standard deviation of 2.5 minutes. A sample of 25 customer waiting times produced a mean of 8.2 minutes. Is this evidence sufficient to reject the supermarket's claim that its customer checkout time averages no more than 7 minutes? Complete this hypothesis test using the 0.02 level of significance.

350. Solve using the *p*-value approach.

ANSWER:
μ = the mean customer checkout time at Meijer supermarket.

$H_o : \mu = 7 (\leq)$ vs. $H_a : \mu > 7$

Normality indicated. Since $\sigma = 2.5$, $n = 25$, $\bar{x} = 8.2$, then

$z^* = (\bar{x} - \mu)/(\sigma/\sqrt{n}) = (8.2 - 7.0)/(2.5/\sqrt{25}) = 2.40$

P = $P(z > 2.40) = 0.5000 - 0.4918 = 0.0082$. Since **P**$<\alpha$, we reject H_o at α = 0.02. The sample does provide sufficient evidence to conclude that the mean waiting time is more than the claimed 7 minutes.

351. Solve using the classical approach.

ANSWER:
μ = the mean customer checkout time at Meijer supermarket.

$H_o : \mu = 7 (\leq)$ vs. $H_a : \mu > 7$

Normality indicated. Since $\sigma = 2.5$, $n = 25$, $\bar{x} = 8.2$, then

$z^* = (\bar{x} - \mu)/(\sigma/\sqrt{n}) = (8.2 - 7.0)/(2.5/\sqrt{25}) = 2.40$

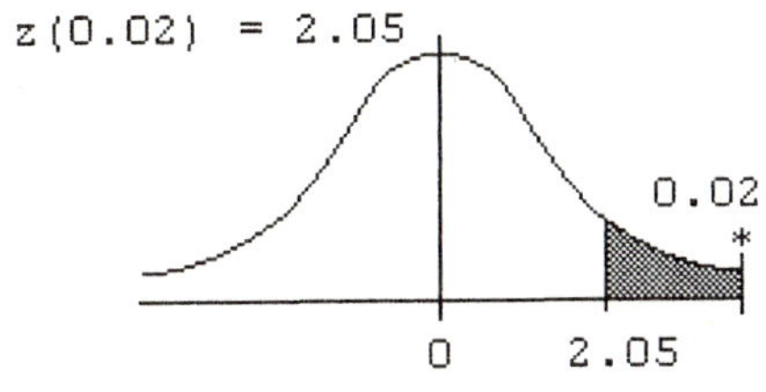

Since z^* falls in the critical region, we reject H_o at the 0.02 level of significance. The sample does provide sufficient evidence to conclude the mean waiting time is more than the claimed 9 minutes.

QUESTIONS 352 AND 353 ARE BASED ON THE FOLLOWING INFORMATION:
The Food and Drug Administration (FDA) must approve all drugs before they can be marketed by a drug company. The FDA must weigh the error of marketing an ineffective drug, with the usual risks of side effects, against the consequences of not allowing an effective drug to be sold. Suppose, using standard medical treatment, that the mortality rate (*r*) of a certain disease is known to be *C*. A manufacturer submits for approval a new drug that is supposed to treat this disease. The FDA sets up the hypothesis to test the mortality rate for the drug as follows:

(1) $H_o: r = C,\quad H_a: r < C,\quad \alpha = 0.005;$ or

(2) $H_o: r = C,\quad H_a: r > C,\quad \alpha = 0.005.$

352. If C = 0.95, which test do you think the FDA should use? Explain.

ANSWER:
$H_a: r > C$. Failure to reject H_o will result in the new drug being marketed. Because of the high current mortality rate (0.95), burden of proof is on the old ineffective drug.

353. If C = 0.05, which test do you think the FDA should use? Explain

ANSWER:
$H_a: r < C$. Failure to reject H_o will result in the new drug not being marketed. Because of the low current mortality rate (0.05), burden of proof is on the new drug.

354. State the null and alternative hypotheses used to test the following claim: "The mean weight of college female students is 120 pounds."

ANSWER:
$H_o: \mu = 120$
$H_a: \mu \neq 120$

355. For the following pair of values, *p*-value = 0.021 and α = 0.01, state the decision that will occur and state why.

ANSWER:
Fail to reject H_o since *p*-value = 0.021 > α = 0.01.

356. State the null and alternative hypotheses used to test the following claim: "The mean life of fluorescent light bulbs is at least 1650 hours."

ANSWER:
$H_o: \mu = 1650\ (\geq)$
$H_a: \mu < 1650$

357. What decision is reached when the *p*-value is greater than α ?

ANSWER:
The decision will be: fail to reject H_o.

358. State the null and alternative hypotheses used to test the claim that "The mean score on that ACT is different from 22."

ANSWER:
$H_o : \mu = 22$
$H_a : \mu \neq 22$

359. Assume that *z* is the test statistic and calculate the value of z^* for testing the null hypothesis $H_o : \mu = 140.0$ when $\sigma = 4.2,\ n = 15,\ \bar{x} = 143.8$

ANSWER:

$$z^* = \frac{\bar{x} - \mu}{\sigma / \sqrt{n}} = \frac{143.8 - 140}{4.2 / \sqrt{15}} = 3.50$$

360. What decision is reached when α is greater then the *p*-value?

ANSWER:
The decision will be: reject H_o.

361. State the null and alternative hypotheses used to test the claim that "The mean selling price of full-size cars is no more than $35,000."

ANSWER:
$H_o : \mu = 35{,}000\ (\leq)$
$H_a : \mu > 35{,}000$

362. For the following pair of values, *p*-value = 0.016 and α = 0.025, state the decision that will occur and state why.

ANSWER:
Reject H_o since *p*-value = 0.016 < α = 0.025.

363. Assume that *z* is the test statistic and calculate the value of z^* for testing the null hypothesis $H_o : \mu = 515$ when $\sigma = 38.3,\ n = 60,\ \bar{x} = 500$.

ANSWER:

$$z^* = \frac{\bar{x} - \mu}{\sigma / \sqrt{n}} = \frac{500 - 515}{38.3 / \sqrt{60}} = -3.03$$

364. For the following pair of values, *p*-value = 0.068 and α = 0.10, state the decision that will occur and state why.

ANSWER:
Reject H_o since *p*-value = 0.068 < α = 0.10.

Chapter 8

365. Assume that z is the test statistic and calculate the value of z^* for testing the null hypothesis $H_o: \mu = 11.6$ when $\sigma = 1.54,\ n = 16,\ \overline{x} = 12.3$

ANSWER:

$$z^* = \frac{\overline{x} - \mu}{\sigma / \sqrt{n}} = \frac{12.3 - 11.6}{1.54 / \sqrt{16}} = 1.82$$

Chapter 9

Inferences Involving One Population

Sections 9.1 and 9.2

True-False Questions

1. For a sample of size n = 31, the critical value of the t-distribution equals the corresponding critical value of the standard normal distribution.
ANSWER: F

2. In considering Student's t-distribution, we see that $t = (\bar{x} - \mu)/(s/\sqrt{n})$ is distributed with a variance less than 1.
ANSWER: F

3. The t-distribution approaches the normal distribution as the number of degrees of freedom decreases.
ANSWER: F

4. As n becomes larger, the value of $t(n-1, \alpha/2)$ becomes closer and closer in value to $z(\alpha/2)$.
ANSWER: T

5. If σ is unknown when completing a hypothesis test about the population mean, then the best estimate for the unknown standard deviation is the sample standard deviation s.
ANSWER: T

6. The Student's t-distributions have an approximately normal distribution but are more dispersed than the standard normal distribution.
ANSWER: T

7. In hypothesis testing about population mean μ, if the test statistic falls in the critical region, then the null hypothesis has been proven to be true.
ANSWER: F

8. When making inferences about one population mean when the value of the standard deviation σ is unknown, the *t*-score is the test statistic.
ANSWER: T

9. When the test statistic is *t* and the number of degrees of freedom gets very large, the critical value of *t* gets very close to that of the standard normal *z*.
ANSWER: T

10. The Student's *t*-distribution is distributed symmetrically about its mean, and approaches the standard normal distribution as the number of degrees of freedom increases.
ANSWER: T

11. Inferences about the population mean μ are based on the sample mean $\bar{x}$ and information obtained from the sampling distribution of sample means.
ANSWER: T

12. The test statistic $t = \frac{\bar{x} - \mu}{s/\sqrt{n}}$ is distributed so as to be less peaked at the mean and thicker at the tails than is the normal distribution.
ANSWER: T

13. The sampling distribution of sample means has a mean μ and a standard error of $\sigma/\sqrt{n}$ for all samples of size *n*.
ANSWER: F

14. The sampling distribution of sample means is normally distributed when the sampled population has a normal distribution or approximately normally distributed when the sample size is sufficiently large.
ANSWER: T

15. Samples as small as *n* =15 or 20 may be considered large enough for the Central Limit Theorem to hold if the sample data are unimodal, nearly symmetric, short-tailed, and without outliers.
ANSWER: T

16. The test statistic $t = \frac{\bar{x} - \mu}{s/\sqrt{n}}$ is distributed symmetrically about its mean $\mu\ (\mu \neq 0)$.
ANSWER: F

17. The *t*-distribution approaches the standard normal distribution as the number of degrees of freedom increases.
ANSWER: T

18. The test statistic $t = \frac{\bar{x} - \mu}{s/\sqrt{n}}$ is distributed with a variance greater than 1, but as the degrees of freedom increases, the variance approaches 1.
ANSWER: T

19. The number of degrees of freedom, *df*, is a statistic that identifies each different distribution of Student's *t*-distribution.
ANSWER: F

20. The number of degrees of freedom associated with s^2 is the divisor (*n*-1) used to calculate the sample variance s^2.
ANSWER: T

21. All the properties of the *t*-distribution hold only for degrees of freedom greater than or equal to 2.
ANSWER: F

22. The Central Limit Theorem indicates that the *t*-distribution can also be applied to nonnormal populations when the sample size is sufficiently large.
ANSWER: T

23. *t* (*df*, 0.95) is the same as *t* (*df*, 0.05) since the *t*-distribution is symmetric around its mean, zero.
ANSWER: F

24. Once *df* is "greater than 100," the critical values of the *t*-distribution are the same as the corresponding critical values of the standard normal distribution.
ANSWER: T

25. *t* (df, 0.90) is the same as -*t* (*df*, 0.10) since the *t*-distribution is symmetric around its mean, zero.
ANSWER: T

Multiple-Choice Questions

26. In a two-tailed test, with *n* = 20, the computed value of *t* is found to be $t^* = 1.85$. Assuming the sample is randomly selected from a normal population, then the *p*-value is given by:

A) 0.005 < *p*-value < 0.01.
B) 0.01 < *p*-value < 0.02.
C) 0.025 < *p*-value < 0.05.
D) 0.05 < p-value < 0.10.
ANSWER: D

27. You are testing the claim that the mean weight of a particular object is more than 4.0 ounces. Select the appropriate null hypothesis and alternative hypothesis for testing the claim.

A) $H_o: \mu = 4.0(\leq), H_a: \mu > 4.0$
B) $H_o: \mu > 4.0, H_a: \mu = 4.0$
C) $H_o: \mu = 4.0(\geq), H_a: \mu < 4.0$
D) $H_o: \mu < 4.0, H_a: \mu > 4.0$
ANSWER: A

28. When testing the claim that the printing speed for a certain inkjet printer is at least 6 pages per minute, which of the following would be the alternative hypothesis?

A) $H_a : \mu > 6.0$

B) $H_a : \mu = 6.0$

C) $H_a : \mu < 6.0$

D) $H_a : \mu \geq 6.0$

ANSWER: C

29. Which of the following would be the null hypothesis and alternative hypothesis in testing the claim that the mean gasoline consumption of a particular model of an automobile is no more than 19 miles per gallon?

A) $H_o : \mu = 19.0(\leq), H_a : \mu > 19.0$

B) $H_o : \mu > 19.0, H_a : \mu = 19.0$

C) $H_o : \mu = 19.0(\geq), H_a : \mu < 19.0$

D) $H_o : \mu < 19.0, H_a : \mu > 19.0$

ANSWER: A

30. Which of the following would be the null hypothesis and alternative in testing the claim that the mean waiting time to be served at a large post office is at least 6.5 minutes?

A) $H_o : \mu = 65.0(\leq), H_a : \mu > 65.0$

B) $H_o : \mu > 65.0, H_a : \mu = 65.0$

C) $H_o : \mu = 65.0(\geq), H_a : \mu < 65.0$

D) $H_o : \mu < 65.0, H_a : \mu > 65.0$

ANSWER: C

31. In comparing Student's *t*-distribution to the standard normal distribution, we see that Student's *t*-distribution is:

A) less peaked and thinner at the tails.
B) less peaked and thicker at the tails.
C) more peaked and thinner at the tails.
D) more peaked and thicker at the tails.
ANSWER: B

32. The measurement of a random sample of 30 female college students produced an average height of 66 inches and a standard deviation of 2.5 inches. The correct symbol for 2.5 inches is:

A) $\overline{x}$
B) s
C) σ
D) μ
ANSWER: B

33. A researcher wants to test the claim that the average female college student is at least 66 inches tall. A random sample of 25 female students produced a mean of 64.5 inches and a standard deviation of 1.23 inches. The correct symbol for 64.5 inches is:

A) $\bar{x}$.
B) s.
C) σ.
D) μ.
ANSWER: A

34. Which of the following is *not* a property of the Student's t - distribution?

A) Mean equals zero
B) Standard deviation is larger than one
C) Symmetrical about zero
D) Used in testing hypotheses about the population standard deviation σ.
ANSWER: D

35. Which of the following statements is false regarding the test statistic $t = \dfrac{\bar{x} - \mu}{s/\sqrt{n}}$?

A) It is distributed with a mean of zero.
B) It is distributed symmetrically about zero.
C) It is distributed so as to be more peaked at the mean and lighter at the tails than is the normal distribution.
D) None of the above
ANSWER: C

36. Which of the following statements is false?

A) Once *df* is greater than or equal to 10, the critical values of the *t*-distribution are the same as the corresponding critical values of the standard normal distribution.
B) *t*(df, 0.99) is the same as -*t*(*df*, 0.01) since the *t*-distribution is symmetric around its mean, zero.
C) *t*(10, 0.05) = 1.81
D) *t*(15, 0.95) = -1.75
ANSWER: A

37. Which of the following statements is false regarding a *t*-distribution with *df* = 15?

A) Its mean is zero.
B) Its 10th percentile is 1.34.
C) Its 95th percentile is 1.75.
D) Its third quartile is 0.691.
ANSWER: B

38. Which of the following statements is false regarding the test statistic $t = \dfrac{\bar{x} - \mu}{s/\sqrt{n}}$?

A) It is distributed so as to form a family of distributions, a separate distribution for each different number of degrees of freedom ($df \geq 1$).
B) It approaches the standard normal distribution as the number of degrees of freedom increases.
C) It is distributed with a variance greater than 1, but as the degrees of freedom increases, the variance approaches 1.
D) None of the above
ANSWER: D

Chapter 9

Short-Answer Questions

39. How is the standard error of the mean estimated?

ANSWER:
The sample standard deviation, s, is divided by the square root of the sample size.

40. Find the value of $t(10, 0.01)$.

ANSWER:
2.76

41. Find the value of $t(20, 0.025)$.

ANSWER:
2.09

42. Find the value of $t(13, 0.99)$.

ANSWER:
-2.65

43. Find the area under the t-distribution curve with $df = 15$ for $P(1.34 < t < 2.95)$.

ANSWER:
0.095

44. What distribution does the Student t-distribution approach as the degrees of freedom become larger?

ANSWER:
Standard normal distribution

45. Find the value of $t(28, 0.90)$.

ANSWER:
-1.31

46. Find the value of $t(18, 0.005)$.

ANSWER:
2.88

47. The alternative hypothesis is sometimes called the "research hypothesis." The conclusion is a statement written about the alternative hypothesis. Explain why these two statements are compatible.

ANSWER:
The alternative hypothesis expresses the concern; the conclusion answers the concern.

48. Find the value of $t(40, 0.10)$.

ANSWER:
1.30

49. Find the value of $t(9, 0.95)$.

ANSWER:
-1.83

50. Find the value of $t(25, 0.975)$.

ANSWER:
-2.06

Applied and Computational Questions

QUESTIONS 51 THROUGH 54 ARE BASED ON THE FOLLOWING INFORMATION:
To test the null hypothesis that the mean waist size for males under 40 years equals 34 inches versus the hypothesis that the mean differs from 34, the following data were collected: 33, 33, 30, 34, 34, 40, 35, 35, 32, 38, 34, 32, 35, 32, 32, 34, 36, 30.

51. Calculate the sample mean and sample standard deviation.

ANSWER:
$\bar{x}$ =33.833, and s = 2.526

52. Calculate the t^*-value of the test statistic.

ANSWER:
t^* = -0.25

53. Find the p-value.

ANSWER:
p-value > 0.50

54. Test the stated hypothesis at $\alpha = .05$ and write your conclusion.

ANSWER:
Since p-value $> \alpha$, we fail to reject H_o, and conclude that the mean waist size for males under 40 equals 34.

55. A new supervisor initiates procedures to reduce the mean time of 6.34 hours currently required to complete an assembly line procedure. In a random sample of 23 assembly line runs, the mean time required was 5.77 hours with a sample standard deviation of 1.82 hours. At the 0.05 level of significance, test the claim that the mean time has been reduced. Determine the critical region, the computed value of the test statistic, and the decision reached.

ANSWER:
Critical region: $t \le -1.72$, Computed value: -1.50, Decision: fail to reject H_o.

56. A machine produces 3-inch nails. A sample is obtained and the lengths determined. The results are as follows: 2.89, 2.95, 3.00, 3.05, 2.99, 2.96, 3.10, 3.06, 3.00, and 3.12. Find a 99% confidence interval for μ.

ANSWER:
(2.94 to 3.09)

57. In order to estimate the pulse rate for young males (less than 30 years), the following sample of pulse rates were obtained: 61, 73, 58, 64, 70, 64, 72, 60, 74, 65, 65, 80, 55, 72, 56, 56. Use these data to find a 95% confidence interval for μ, the mean for all such males.

ANSWER:
(61.3 to 69.3)

58. Solve the following equation for *x*: $t(10, 0.005) = x$.

ANSWER:
3.17

59. Solve the following equation for *x*: $t(x, 0.01) = 2.68$.

ANSWER:
12

60. Solve the following equation for *x*: $t(10, x) = 1.37$.

ANSWER:
0.10

QUESTIONS 61 THROUGH 64 ARE BASED ON THE FOLLOWING INFORMATION:
The program director for a medical assistants' program wishes to test the hypothesis that her students score higher than the national mean on the certified medical assistants' (CMA) exam. She randomly selects 15 recent graduates of the yearlong program and finds that $\bar{x} = 640$ and s = 25. Assume the national mean is 615.

61. State the null and alternative hypotheses.

ANSWER:
$H_o: \mu = 615(\leq), H_a: \mu > 615$

62. Calculate the value t^* of the test statistic.

ANSWER:
$t^* = 3.87$

63. Find the *p*-value of the test.

ANSWER:
p - value < 0.005

64. Test the hypothesis in question 61 at $\alpha = 0.01$ and write your conclusion.

ANSWER:
Since p -value $< \alpha$, we reject H_o and conclude that the program director for the medical assistants' program was right that her students score higher than the national mean (615) on the CMA exam.

65. Ten farms (randomly selected from a large agricultural region) were selected, and the yield per acre in wheat was determined for each. The summary data were as follows: $\bar{x} = 95.0$ and s = 85. Find a 95% confidence interval for the mean yield per acre for all such farms in this region.

ANSWER:
(88.9 to 101.1)

QUESTIONS 66 THROUGH 68 ARE BASED ON THE FOLLOWING INFORMATION:
A drug manufacturer produces 250-milligram capsules of a new antibiotic. A random sample is selected, and the amount of antibiotic in each capsule is determined. The results are as follows (in milligrams): 252, 246, 242, 250, 255, 258, 250, 252, 250, and 258.

66. Find a 95% confidence interval for μ, the mean amount of antibiotic per capsule.

ANSWER:
(247.7 to 254.9)

67. Give bound on the p-value, and test $H_o : \mu = 250$ vs. $H_a : \mu \neq 250$ at α =0.10.

ANSWER:
0.20 $\leq$ p-value $\leq$ 0.50. Since p-value $> \alpha$, we fail to reject the null hypothesis. We conclude that the average amount of antibiotic is 250-milligram.

68. Give the critical region, the computed test statistic, and your conclusion if you used these data to test the hypothesis in question 67 at the 0.05 level of significance.

ANSWER:
Critical region: $t \leq -2.26$ or $t \geq 2.26$, t^* = 0.82; Conclusion: unable to reject null hypothesis.

69. A machine produces 3-inch nails. A sample of 10 nails is obtained and the lengths determined. The results are as follows: 2.89, 2.95, 3.00, 3.05, 2.99, 2.96, 3.10, 3.06, 3.00, and 3.12. Use these results to test $H_o : \mu = 3.0$ vs. $H_a : \mu \neq 3.0$ at a level of significance equal to 0.01. Give the critical region, the computed test statistic, and the conclusion.

ANSWER:
Critical region: $t \leq -3.25$ or $t \geq 3.25$, t^* = 0.534; Conclusion: unable to reject the null hypothesis.

Chapter 9

QUESTIONS 70 THROUGH 73 ARE BASED ON THE FOLLOWING INFORMATION:
In order to test the claim that the mean of a particular normal population is greater than 4.8, the following random sample was selected: 5, 7, 3, 4, 5, 4, and 6. The test is to be completed using a level of significance $\alpha = 0.10$.

70. State the null and alternative hypotheses.

ANSWER:
$H_o : \mu = 4.8(\leq)$, vs. $H_a : \mu > 4.8$

71. State the test criteria.

ANSWER:
The test statistic is t^* and the level of significance is $\alpha = 0.10$. We reject H_o if $t^* > 1.44$.

72. Find the computed value of the test statistic.

ANSWER:
$t^* = 0.11$

73. State the decision and conclusion.

ANSWER:
Since $t^* = 0.11$, we fail to reject H_o. There is not sufficient evidence to suggest that the population mean is greater than 4.8.

QUESTIONS 74 THROUGH 76 ARE BASED ON THE FOLLOWING INFORMATION:
In order to test the claim that the mean of a particular normal population is greater than 7.6 the following random sample was selected: 11, 6, 8, 9, 7, 6, 5, 10, 9, and 8. The test is to be completed using $\alpha = 0.10$.

74. State the null and alternative hypotheses.

ANSWER:
$H_o : \mu = 7.6(\leq)$, vs. $H_a : \mu > 7.6$

75. Find the p -value.

ANSWER:
p -value > 0.25

76. State the decision and conclusion.

ANSWER:
Fail to reject H_o. There is not sufficient evidence to suggest that the population mean is greater than 7.6.

77. A sample of size $n = 14$ is selected from a normal population to construct a 95% confidence interval for a population mean. The following interval is obtained: (7.82 to 9.64). Find the sample standard deviation.

ANSWER:
$s = 1.576$

78. Find the first percentile of the Student's *t*-distribution with *df* = 20.

ANSWER:
–2.53

79. Find the 95th percentile of the Student's *t*-distribution with *df* = 20.

ANSWER:
1.72

QUESTIONS 80 THROUGH 82 ARE BASED ON THE FOLLOWING INFORMATION:
Suppose that the random variable *x* represents the cost of one book, and that the following sample summary values are given: $n = 40, \sum x = 540, \text{ and } \sum (x - \overline{x})^2 = 1620.$

80. Find the sample mean $\overline{x}$.

ANSWER:
$\overline{x} = \sum x / n = 540 / 40 = 13.5$

81. Find the sample standard deviation, *s*.

ANSWER:
$s = \sqrt{[\sum (x - \overline{x})^2 / (n-1)} = \sqrt{1620/39} = 6.445$

82. Find the 90% confidence interval to estimate the true mean textbook cost based on this sample.

ANSWER:
μ = The mean textbook cost. Normality assumed. Since $n = 40$, $\overline{x} = 13.50$, $s = 6.445$, and $1 - \alpha = 0.90$, then $\alpha / 2 = 0.05$; $df = n - 1 = 39$, and $t(39, 0.05) \approx 1.68$.
$E = t(df, \alpha/2) \cdot (s / \sqrt{n}) = (1.68)(6.445 / \sqrt{40}) = 1.71$. Hence,
$\overline{x} \pm E = 13.50 \pm 1.71$, and the 90% confidence interval of μ is 11.79 to 15.21.

83. Find the first quartile of the Student's *t*-distribution with *df* = 20.

ANSWER:
–0.687

84. Find the percent of the Student's *t*-distribution that lies between –1.37 and 2.76, when *df* = 10.

ANSWER:
1 – (0.10 + 0.01) = 0.89

85. Find the percent of the Student's *t*-distribution that lies between *t* ranges from –1.77 and 3.01, when *df* = 13.

ANSWER:
1 – (0.05 + 0.005) = 0.945

Chapter 9

QUESTIONS 86 THROUGH 88 ARE BASED ON THE FOLLOWING INFORMATION:
The pulse rates for 10 adult women were as follows: 60, 72, 58, 78, 66, 82, 78, 99, 70, and 80.

86. Find the sample mean.

ANSWER:
$\bar{x} = \sum x / n = 743 / 10 = 74.3$

87. Find the sample standard deviation.

ANSWER:
$s = \sqrt{[\sum x^2 - (\sum x)^2 / n]/(n-1)} = \sqrt{[56497 - (743)^2 / 10]/9} = 11.982$

88. Find 90% confidence interval to estimate the true mean pulse rate for women based on this sample.

ANSWER:
$\alpha / 2 = 0.05;\ df = n - 1 = 9,$ and $t(9, 0.05) = 1.83$.
$E = t(df, \alpha/2) \cdot (s/\sqrt{n}) = (1.83)(11.982/\sqrt{10}) = 6.93$. Hence,
$\bar{x} \pm E = 74.3 \pm 6.93$, and the 90% confidence interval of μ is 67.37 to 81.23.

89. Use the table of probability values for Student's *t*-distribution with 10 degrees of freedom to determine the *p*-value for testing $H_o: \mu = 13.5$ vs. $H_a: \mu > 13.5,$ when the test statistic $t^* = 1.94$.

ANSWER:
P = $P(t > +1.94 \mid df = 10)$; we have 0.037 < **P** < 0.043.

90. Use the table of probability values for Student's *t*-distribution with 10 degrees of freedom to determine the *p*-value for testing $H_o: \mu = 13.5$ vs. $H_a: \mu \neq 13.5,$ when the test statistic $t^* = 1.94$.

ANSWER:
P $= P(t < -1.94 \mid df = 10) + P(t > +1.94 \mid df = 10) = 2P(t > 1.94 \mid df = 10)$, we have 0.074 < **P** < 0.086.

91. Use the table of probability values for Student's *t*-distribution with 10 degrees of freedom to determine the *p*-value for testing $H_o: \mu = 13.5$ vs. $H_a: \mu \neq 13.5,$ when the test statistic $t^* = -1.94$.

ANSWER:
P = *p*-value $= P(t < -1.94 \mid df = 10) + (t > +1.94 \mid df = 10) = 2P(t > 1.94 \mid df = 10)$, we have 0.074 < **P** < 0.086.

92. Use the table of probability values for Student's *t*-distribution with 10 degrees of freedom to determine the *p*-value for testing $H_o: \mu = 13.5$ vs. $H_a: \mu < 13.5$, when the test statistic $t^* = -1.94$.

ANSWER:
P = $P(t < -1.94 \mid df = 10) = P(t > 1.94 \mid df = 10)$, we have 0.037 < **P** < 0.043.

QUESTIONS 93 THROUGH 96 ARE BASED ON THE FOLLOWING INFORMATION:
The *p*-value approach and classical approach, respectively, are two different approaches to hypothesis testing. The former approach requires finding the *p*-value of the test, and the later approach requires finding the critical value(s) and the rejection region(s). Both approaches lead to the same decision and conclusion.

93. Compare the *p*-value approach and classical approach to hypothesis testing by comparing the decision of the *p*-value approach to the decision of the classical approach, for testing $H_o: \mu = 100$ vs. $H_a: \mu \neq 100$, when n = 15, t^* = 1.60, and α = 0.05.

ANSWER:
The *p*-value approach:
P $= 2P(t > 1.60 \mid df = 14)$. Using the "Probability Values for Student's *t*-Distribution" table, we get 0.065 < ½ **P** < 0.068; hence 0.130 < **P** < 0.136. Since **P** > α, we fail to reject H_o.
The classical approach:

α = 0.05
$\alpha/2$ = 0.025
$\pm t(14, 0.025) = \pm 2.14$

$t^* = 1.60$
Fail to reject H_0

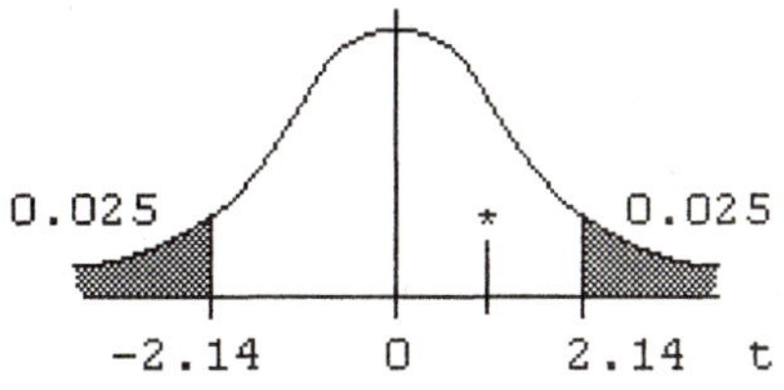

94. Compare the *p*-value approach and classical approach to hypothesis testing by comparing the decision of the *p*-value approach to the decision of the classical approach, for testing $H_o: \mu = 20$ vs. $H_a: \mu > 20$, when n = 25, t^* = 2.16, and α = 0.05.

ANSWER:
The *p*-value approach:
P $= P(t > 2.16 \mid df = 24)$. Using the "Probability Values for Student's *t*-Distribution" table, we get 0.019 < **P** < 0.024. Since **P** < α, we reject H_o.
The classical approach:

α = 0.05
$t(24, 0.05) = 1.71$

$t^* = 2.16$
Reject H_0

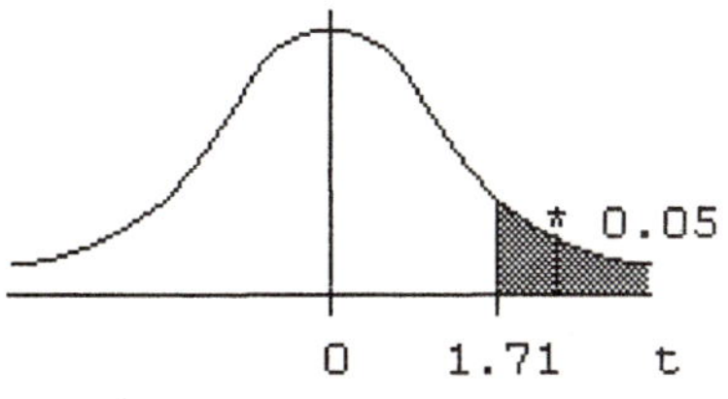

95. Compare the *p*-value approach and classical approach to hypothesis testing by comparing the decision of the *p*-value approach to the decision of the classical approach, for testing $H_o: \mu = 40$ vs. $H_a: \mu < 40$, when n = 45, t^* = -1.73, and α = 0.05.

ANSWER:

<u>The *p*-value approach:</u>

P= $P(t < -1.73 \mid df = 44) = P(t > 1.73 \mid df = 44)$ .Using the "Probability Values for Student's *t*-Distribution" table, we get 0.039 < **P** < 0.049. Since **P** < α , we reject H_o.

<u>The classical approach:</u>

```
α = 0.05
-t(44, 0.05) = -1.68

t* = -1.73
  Reject Ho
```

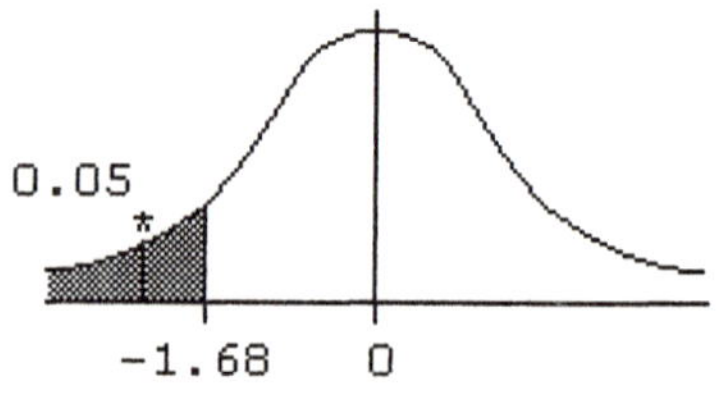

96. Compare the *p*-value approach and classical approach to hypothesis testing by comparing the decision of the *p*-value approach to the decision of the classical approach, for testing. Compare the results of the two techniques for questions 93, 94, and 95.

ANSWER:

The results of the two techniques for each of the decisions made to questions 93, 94, and 95 are identical.

97. State the null hypothesis, H_o, and the alternative hypothesis, H_a, that would be used to test: The mean weight of new born babies is at least 5 lbs.

ANSWER:

$H_o: \mu = 5(\geq)$ vs. $H_a: \mu < 5$

98. State the null hypothesis H_o, and the alternative hypothesis H_a, that would be used to test: The mean age of patients at Mecosta County General Hospital is no more than 56 years.

ANSWER:

$H_o: \mu = 56(\leq)$ vs. $H_a: \mu > 56$

99. State the null hypothesis, H_o, and the alternative hypothesis, H_a, that would be used to test: "The mean amount of fat in Healthy Choice meal is different from 15 mg."

ANSWER:

$H_o: \mu = 15$ vs. $H_a: \mu \neq 15$

QUESTIONS 100 THROUGH 102 ARE BASED ON THE FOLLOWING INFORMATION:
A large study involving over 20,000 individuals shows that the mean percentage intake of kilocalories from fat was 39% with a range from 6% to 72%. A small sample study was conducted at a university hospital to determine if the mean intake of patients at that hospital was different from 39%. A sample of 15 patients had a mean intake of 40.8% with a standard deviation equal to 6.5%. Assume that the sample is from a normally distributed population.

100. What evidence do you have that the assumption of normality is reasonable? Explain.

ANSWER:
The "population" data ranged from 6% to 72%, therefore the midrange is 39%. When the midrange is close in value to the mean, the distribution is approximately symmetrical; therefore, the assumption of normality is reasonable.

101. Test the hypothesis of "different from" at a level of significance equal to 0.05, using the *p*-value approach. Include t^*, *p*-value, and your conclusion.

ANSWER:
$\mu =$ The mean percentage intake of kilocalories from fat

$H_o: \mu = 39\%$ vs. $H_a: \mu \neq 39\%$. Normality indicated. Since $n = 15, \bar{x} = 40.8\%$, and $s = 6.5\%$, then $t^* = (\bar{x} - \mu)/(s/\sqrt{n}) = (40.8 - 39.0)/(6.5/\sqrt{15}) = 1.07$

P $= 2P(t > 1.07 \mid df = 14)$. Using the "Probability Values for Student's *t*-Distribution" table, we get 0.144 **<** ½ **P** < 0.169.Then 0.288 < **P** **<** 0.338. Since P> α, we fail to reject H_o.

102. Test the hypothesis of "different from" using the classical approach at a level of significance equal to 0.05. Include the critical values, t^*, and your conclusion.

ANSWER:
$\mu =$ The mean percentage intake of kilocalories from fat

$H_o: \mu = 39\%$ vs. $H_a: \mu \neq 39\%$

The critical values are $\pm t(14, 0.025) = \pm 2.14$;

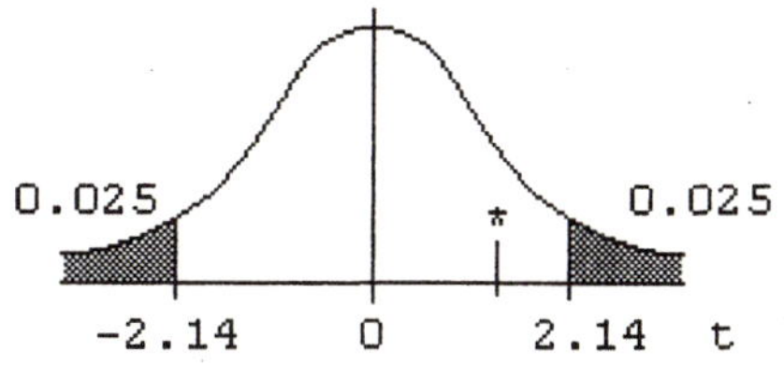

The test statistic t^* = 1.07 falls in the noncritical region, therefore we fail to reject H_o. We conclude that the sample does not provide sufficient evidence to justify the contention that the mean percentage is different than 39%, at the 0.05 level of significance.

Chapter 9

QUESTIONS 103 THROUGH 107 ARE BASED ON THE FOLLOWING INFORMATION:
It is claimed that the students at a certain university in Michigan will score an average of 85 on a given test. Is the claim reasonable if a random sample of test scores from this university yields 83, 92, 88, 87, 80, and 92? Assume test results are normally distributed.

103. Compute the sample mean and sample standard deviation.

ANSWER:
$\bar{x} = 87, \quad s = 4.817$

104. State the null and alternative hypotheses.

ANSWER:
$H_o : \mu = 85$ (reasonable) vs. $H_a : \mu \neq 85$ (not reasonable)

105. Calculate the value of the test statistic.

ANSWER:
$t^* = (\bar{x} - \mu)/(s/\sqrt{n}) = (87.0 - 85.0)/(4.817/\sqrt{6}) = 1.02$

106. Complete a hypothesis test at α = 0.05 using the *p*-value approach.

ANSWER:
P = *p*-value = $2P(t > 1.02 \mid df = 5)$; Using the "Probability Values for Student's *t*-Distribution" table, we have 0.161 < ½ ***P*** <0.182], then 0.322 < ***P*** < 0.364. Since ***P*** > α ; fail to reject H_o.

107. Complete a hypothesis test at α = 0.05 using the classical approach.

ANSWER:
The critical values are $\pm t(5, 0.025) = \pm 2.57$

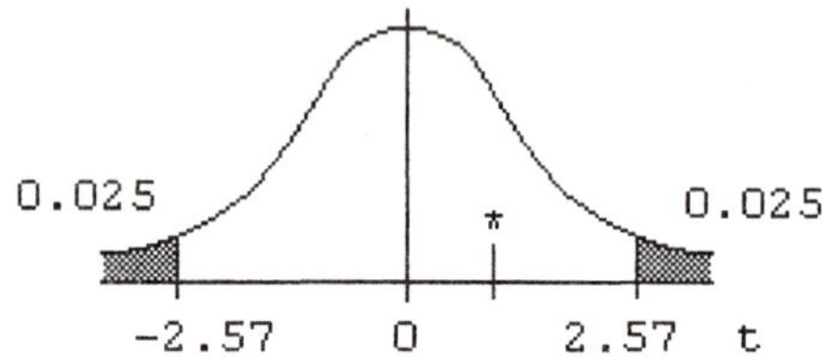

The test statistic t^* = 1.02 falls in the noncritical region, therefore we fail to reject H_o. The sample does not provide sufficient evidence to conclude at the 0.05 level of significance that the mean score is different from 85.

QUESTIONS 108 THROUGH 112 ARE BASED ON THE FOLLOWING INFORMATION:
Gasoline pumped from a supplier's pipeline is supposed to have an octane rating of 86.5. On 13 consecutive days a sample was taken and analyzed with the following results: 87.6, 85.4, 86.2, 87.4, 86.2, 86.6, 85.8, 85.1, 86.4, 86.3, 85.4, 85.6, and 86.1. Assume that the octane ratings have a normal distribution. We wish to determine at the 0.05 level of significance if there is sufficient evidence to show that these octane readings were taken from gasoline with a mean octane significantly less than 87.

108. Compute the sample mean and sample standard deviation.

ANSWER:
$\bar{x} = 86.162, \quad s = 0.742$

109. State the null and alternative hypotheses.

ANSWER:
$H_o: \mu = 86.5(\geq)$ vs. $H_a: \mu < 86.5$

110. Calculate the value of the test statistic.

ANSWER:
$t^* = (\bar{x} - \mu)/(s/\sqrt{n}) = (86.162 - 86.5)/(0.742/\sqrt{13}) = -1.64$

111. Complete the hypothesis test using the *p*-value approach.

ANSWER:
P = *p*-value = $P(t < -1.64 \mid df = 12) = P(t > 1.64 \mid df = 12)$; Using the "Probability Values for Student's *t*-Distribution" table, we have 0.057 < **P** < 0.068. Since **P** > α ; we fail to reject H_o. The sample does not provide sufficient evidence to conclude at the 0.05 level of significance that mean octane level is less than 87.5,

112. Complete the hypothesis test using the classical approach.

ANSWER:
The critical value is $-t(12, 0.05) = -1.78$

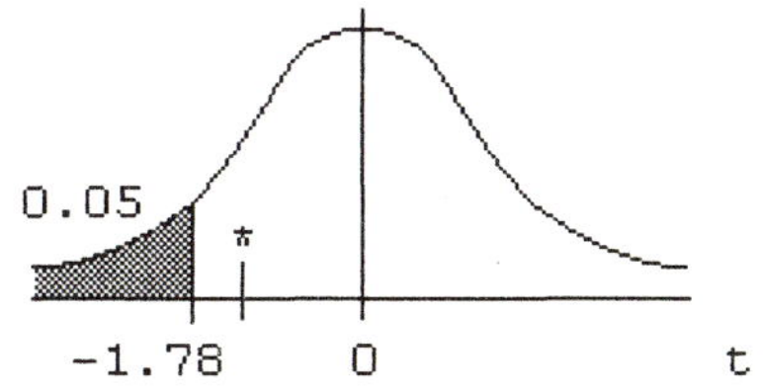

The test statistic t^* = -1.64 falls in the noncritical region, therefore we fail to reject H_o. The sample does not provide sufficient evidence to conclude at the 0.05 level of significance that mean octane level is less than 87.5,

QUESTIONS 113 AND 114 ARE BASED ON THE FOLLOWING INFORMATION:
A random sample of 20 weights is taken from babies born at the University of Iowa Hospital. A mean of 7.55 lb and a standard deviation of 1.85 lb were found for the sample. Based on past information, it is assumed that weights of newborns are normally distributed.

113. Estimate, with 95% confidence, the mean weight of all babies born in this hospital.

ANSWER:
$t(df, \alpha/2) = t(19, 0.025) = 2.09$
$E = t(df, \alpha/2)\cdot(s/\sqrt{n}) = 2.09\ (1.85/\sqrt{20}) = 0.865$
$\bar{x} \pm E = 7.55 \pm 0.865$. Thus, the 95% confidence interval for μ is 6.685 to 8.415.

114. Interpret the confidence interval in question 113.

ANSWER:
With 95% confidence, we estimate the mean weight of babies born at the University of Iowa Hospital to be between 6.685 to 8.415 lbs.

QUESTIONS 115 THROUGH 124 ARE BASED ON THE FOLLOWING INFORMATION:
Consider the Student's *t*-distribution with 20 degrees of freedom. Recall that the kth percentile, denoted by P_k, is a value such that at most k% of the ranked data are smaller in value than P_k and at most (100-k)% of the data are larger.

115. Find the first percentile.

ANSWER:
$P_1 = -2.53$

116. Find the 5th percentile.

ANSWER:
$P_5 = -1.72$

117. Find the 10th percentile.

ANSWER:
$P_{10} = -1.33$

118. Find the first quartile.

ANSWER:
$P_{25} = Q_1 = -0.687$

119. Find the median.

ANSWER:
$P_{50} = Q_2 = 0.0$

120. Find the third quartile.

ANSWER:
$P_{75} = Q_3$ = 0.687

121. Find the 90th percentile.

ANSWER:
P_{90} = 1.33

122. Find the 95th percentile.

ANSWER:
P_{95} = 1.75

123. Find the 99th percentile.

ANSWER:
P_{99} = 2.53

124. Find the interquartile range.

ANSWER:
$Q_3 - Q_1$ = 0.687 – (-0.687) = 1.374

125. Find the percent of the Student's *t*-distribution with *df* =10 that lies between –1.37 and 2.76.

ANSWER:
1 – (0.10 + 0.01) = 0.89

126. Find the percent of the Student's *t*-distribution with *df* =15 that lies between –1.75 and 2.60.

ANSWER:
1 – (0.05 + 0.01) = 0.94

127. Find the percent of the Student's *t*-distribution with *df* =20 that lies between – 0.687 and 2.09.

ANSWER:
1 – (0.25 + 0.025) = 0.725

128. Find the percent of the Student's *t*-distribution with *df* =25 that lies between 0.684 and 2.79.

ANSWER:
0.25 – 0.005 = 0.245

129. Ninety percent of Student's *t*-distribution lies between *t* = –1.81 and *t* =1.81 for how many degrees of freedom?

ANSWER:
df = 10

130. Ninety percent of Student's t-distribution lies to the right of t = –1.44 for how many degrees of freedom?

ANSWER:
$df = 6$

131. Eighty percent of Student's t-distribution lies between t = –1.40 and t =1.40 for how many degrees of freedom?

ANSWER:
$df = 8$

132. Ninety five percent of Student's t-distribution lies between t = –2.12 and t =2.12 for how many degrees of freedom?

ANSWER:
$df = 16$

133. Ninety eight percent of Student's t-distribution lies between t = –2.55 and t =2.55 for how many degrees of freedom?

ANSWER:
$df = 18$

134. Ninety nine percent of Student's t-distribution lies to the left of t = 2.68 for how many degrees of freedom?

ANSWER:
$df = 12$

135. Construct a 90% confidence interval estimate for the mean μ using the sample information n =21, $\bar{x}$ =13.6, and s =2.4.

ANSWER:
$t(df, \alpha/2) = t(20, 0.05) = 1.72$
$E = t(df, \alpha/2) \cdot (s/\sqrt{n}) = 1.72\ (2.4/\sqrt{21}) = 0.90$
$\bar{x} \pm E = 13.6 \pm 0.9$. The 90% confidence interval for μ is 12.78 to 14.5

QUESTIONS 136 THROUGH 139 ARE BASED ON THE FOLLOWING INFORMATION:
While doing an article on the high cost of college education, a reporter took a random sample of the cost of new textbooks for a semester. The random variable x is the cost of one book. Her sample data can be summarized by n = 51, $\sum x$ =4425.88, and $\sum(x-\bar{x})^2$ =12,280.12.

136. Find the sample mean $\bar{x}$.

ANSWER:
$\bar{x} = \sum x/n = 4425.88 / 51 = \$86.78$

137. Find the sample standard deviation, *s*.

ANSWER:
The sample variance is $s^2 = \Sigma(x-\bar{x})^2/(n-1) =$ 12,280.12 / 50 = 245.60. Hence the sample standard deviation *s* = \$15.67

138. Find the 90% confidence interval to estimate the true mean textbook cost for the semester based on this sample.

ANSWER:
$t(df, \alpha/2)$ = t(50, 0.05) = 1.68
$E = t(df, \alpha/2)\cdot(s/\sqrt{n})$ = 1.68 (15.67/$\sqrt{51}$) = 3.69
$\bar{x} \pm E = 86.78 \pm 3.69$. The 90% confidence interval for μ is \$83.09 to \$90.47

139. Interpret the confidence interval in question 138.

ANSWER:
With 90% confidence, we estimate the average cost of a college new textbook to be between \$83.09 and \$90.47.

QUESTIONS 140 THROUGH 144 ARE BASED ON THE FOLLOWING INFORMATION:
The pulse rates for 15 adult women were 95, 66, 76, 106, 84, 76, 81, 56, 68, 54, 74, 62, 78, 74, and 68.

140. Calculate the sample mean.

ANSWER:
$\bar{x} = \sum x/n =$ 1110 / 15 = 74

141. Calculate the sample standard deviation.

ANSWER:
The sample variance is $s^2 = \Sigma(x-\bar{x})^2/(n-1) =$ 2802 / 14 = 200.143. Hence the sample standard deviation *s* = 14.147.

142. Find the minimum error of estimate for 90% confidence interval for μ.

ANSWER:
$t(df, \alpha/2)$ = t(14, 0.05) = 1.76
$E = t(df, \alpha/2)\cdot(s/\sqrt{n})$ = 1.76 (14.147/$\sqrt{15}$) = 6.429

143. Find the lower and upper confidence limits for a 90% confidence interval.

ANSWER:
$\bar{x} \pm E = 74.000 \pm 6.429$. Hence, LCL = 67.571 $\approx$ 67.6 and UCL = 80.429 $\approx$ 80.4.

144. Interpret the confidence interval in question 143.

ANSWER:
With 90% confidence, we estimate the average pulse rate for adult women to be between 67.6 and 80.4.

Chapter 9

QUESTIONS 145 THROUGH 148 ARE BASED ON THE FOLLOWING INFORMATION:
The following data represent the scores for a sample of 20 high school students on a 25 points biology quiz: 20, 18, 15, 19, 17, 19, 19, 16, 15, 16, 17, 22, 19, 20, 16, 18, 18, 23, 15, and 16.

145. Use a computer to construct a 0.98 confidence interval for μ.

ANSWER:

Results for one-sample analysis for Quiz Score			
Summary measures			
	Sample size	20	
	Sample mean	17.900	
	Sample standard deviation	2.269	
Confidence interval for mean			
	Confidence level	98.0%	
	Sample mean	17.900	
	Std error of mean	0.507	
	Degrees of freedom	19	
	Lower limit	16.612	
	Upper limit	19.188	

146. What assumption is required to ensure the validity of the results to question 145?

ANSWER:
These results are based on the assumption that the variable Quiz Score is approximately normally distributed. If this is not the case, then these results might not be valid, especially a sample size of 20 is considered small.

147. Use a computer to construct a 0.98 confidence interval for μ.

ANSWER:

Results for one-sample analysis for Quiz Score		
Summary measures		
	Sample size	20
	Sample mean	17.900
	Sample standard deviation	2.269
Confidence interval for mean		
	Confidence level	90.0%
	Sample mean	17.900
	Std error of mean	0.507
	Degrees of freedom	19
	Lower limit	17.023
	Upper limit	18.777

148. What is the effect of decreasing the confidence level from 98% to 90%?

ANSWER:
The width of the confidence interval decreases from 2.576 to 1.754.

149. State null hypothesis, H_o, and the alternative hypothesis, H_a, that would be used to test the following claim "A chicken farmer claims that his chickens have a mean weight of 4 pounds."

ANSWER:
$H_o: \mu = 4$ vs. $H_a: \mu \neq 4$

150. State null hypothesis, H_o, and the alternative hypothesis, H_a, that would be used to test the following claim "The mean age of Egypt's commercial jets is less than 25 years."

ANSWER:
$H_o: \mu = 25\ (\geq)$ vs. $H_a: \mu < 25$

151. State null hypothesis, H_o, and the alternative hypothesis, H_a, that would be used to test the following claim "The mean monthly unpaid balance on Discover card accounts is more than $425."

ANSWER:
$H_o: \mu = 425\ (\leq)$ vs. $H_a: \mu > 425$

<u>QUESTIONS 152 THROUGH 155 ARE BASED ON THE FOLLOWING INFORMATION:</u>
Consider the Student's *t*-distribution with 10 degrees of freedom.

152. Determine the *p*-value for testing $H_o: \mu = 20$ vs. $H_a: \mu < 20$, if $t^* = -2.01$.

ANSWER:

P = *p*-value = *P*(*t* < -2.01 | *df* =10) = *P*(*t* > 2.01 | *df* =10)
Using the "Critical Values of Student's *t*-Distribution" Table, we get: 0.025 < **P** < 0.05
Using the "Probability Values for Student's *t*-Distribution" Table, we get: 0.031 < **P** < 0.037

153. Determine the *p*-value for testing $H_o: \mu = 20$ vs. $H_a: \mu > 20$, if $t^* = 2.01$.

ANSWER:
P = *p*-value = *P*(*t* > +2.01 | *df* =10);
Using the "Critical Values of Student's *t*-Distribution" Table, we get: 0.025 < **P** < 0.05
Using the "Probability Values for Student's *t*-Distribution" Table, we get: 0.031 < **P** < 0.037

154. Determine the *p*-value for testing $H_o: \mu = 20$ vs. $H_a: \mu \neq 20$, if $t^* = 2.01$.

ANSWER:
P = *p*-value = *P*(*t* < -2.01 | *df* =10) + *P*(*t* > +2.01 | *df* =10) = 2*P*(*t* > 2.01| *df* =10)
Using the "Critical Values of Student's *t*-Distribution" Table, we get: 0.05 < **P** < 0.10
Using the "Probability Values for Student's *t*-Distribution" Table, we get: 0.062 < **P** < 0.074

155. Determine the *p*-value for testing $H_o: \mu = 20$ vs. $H_a: \mu \neq 20$, if $t^* = -2.01$.

ANSWER:
P = *p*-value = *P*(*t* < -2.01 | *df* =10) + *P*(*t* > +2.01 | *df* =10) = 2*P*(*t* > 2.01| *df* =10)
Using the "Critical Values of Student's *t*-Distribution" Table, we get: 0.05 < **P** < 0.10
Using the "Probability Values for Student's *t*-Distribution" Table, we get: 0.062 < **P** < 0.074

Chapter 9

156. Draw an approximately normal distribution curve to determine the critical region and critical value(s) that would be used in the classical approach to test the hypothesis $H_o : \mu = 18$ vs. $H_a : \mu \neq 18$ given that $\alpha = 0.05$ and $n = 15$.

ANSWER:

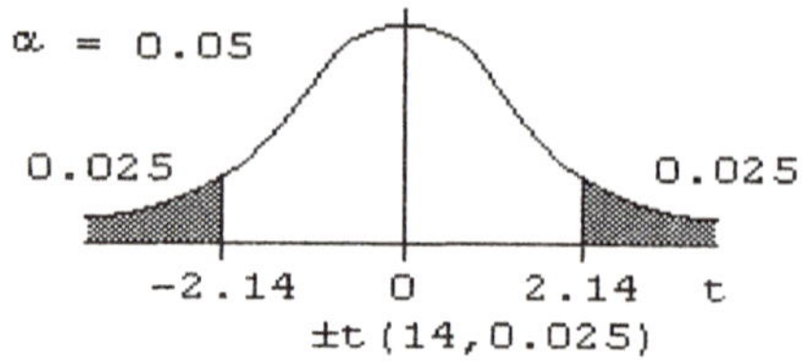

157. Draw an approximately normal distribution curve to determine the critical region and critical value(s) that would be used in the classical approach to test the hypothesis $H_o : \mu = 25$ vs. $H_a : \mu > 25$ given that $\alpha = 0.01$ and $n = 25$.

ANSWER:

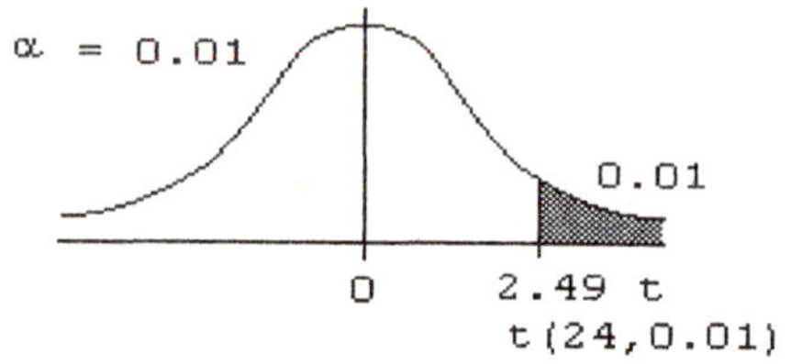

158. Draw an approximately normal distribution curve to determine the critical region and critical value(s) that would be used in the classical approach to test the hypothesis $H_o : \mu = -32$ vs. $H_a : \mu < -32$ given that α=0.05 and $n = 18$.

ANSWER:

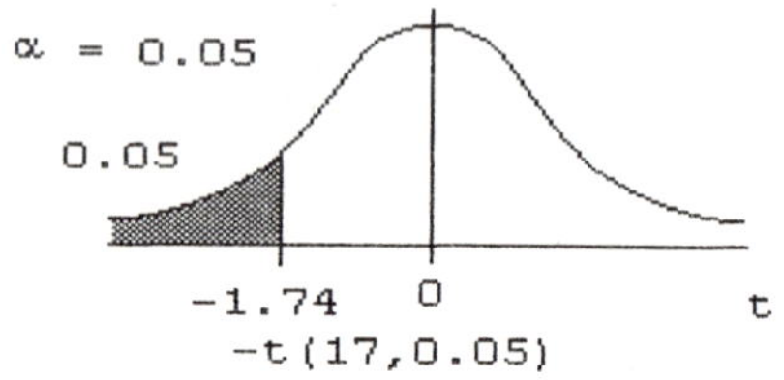

159. Draw an approximately normal distribution curve to determine the critical region and critical value(s) that would be used in the classical approach to test the hypothesis $H_o : \mu = 40$ vs. $H_a : \mu > 40$ given that α =0.01 and $n = 42$.

ANSWER:

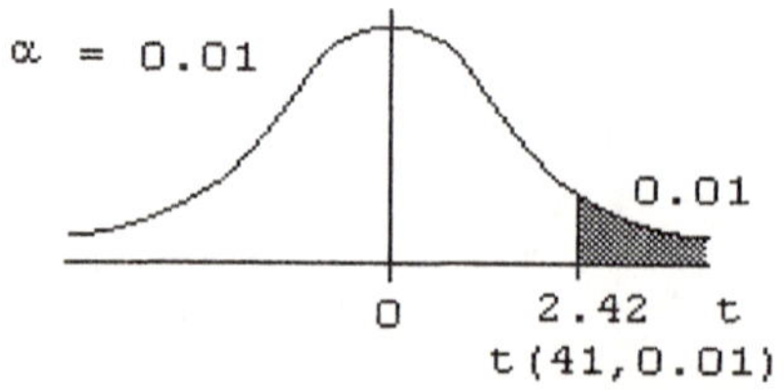

QUESTIONS 160 THROUGH 162 ARE BASED ON THE FOLLOWING INFORMATION:
Homes in nearby East Lansing, Michigan have a mean value of $178,750. It is assumed that homes in the vicinity of Michigan State University (MSU) have a higher value. To test this theory, a random sample of 12 homes is chosen from the MSU area. Their mean valuation is $182,210 and the standard deviation is $5,600. Assume prices are normally distributed, and that α =.05 is used in testing the appropriate hypothesis.

160. State the null and alternative hypotheses.

ANSWER:
$H_o: \mu = 178{,}750\ (\leq)$ vs. $H_a: \mu > 178{,}750$

161. Test the hypothesis in question 160 using the *p*-value approach.

ANSWER:
$t^* = (\bar{x} - \mu)/(s/\sqrt{n}) = (182{,}210 - 178{,}750)/(5{,}600/\sqrt{12}) = 2.14$
P = *p*-value = $P(t > 2.14 | df = 11)$.
Using the "Critical Values of Student's *t*-Distribution" Table, we get: 0.025 < **P** < 0.05.
Using the "Probability Values for Student's *t*-Distribution" Table, we get 0.024< **P** < 0.031.
Since *p*-value < α =.05, we reject H_o. The sample does provide sufficient evidence to justify the contention that the mean value is higher than $178,750 at the 0.05 level of significance.

162. Test the hypothesis in question 160 using the classical approach.

ANSWER:
$t(df, \alpha) = t(11, 0.05) = 1.80$
$t^* = (\bar{x} - \mu)/(s/\sqrt{n}) = (182{,}210 - 178{,}750)/(5{,}600/\sqrt{12}) = 2.14$
Since the value of the test statistic t^* = 2.14 falls in the rejection region, we reject H_o at α = 0.05, and reach the same conclusion as stated in question 161.

QUESTIONS 163 THROUGH 169 ARE BASED ON THE FOLLOWING INFORMATION:
The weights of 20 adult males were recorded as: 169, 174, 149, 152, 163, 175, 169, 133, 163, 170, 148, 167, 159, 166, 149, 155, 195, 127, 190, and 185. It is believed that the mean weight for adult males is at least 160 lb. Assume that the weights for adult males are normally distributed.

163. What are the null and alternative hypotheses?

ANSWER:
$H_o: \mu = 160\ (\leq)$ vs. $H_a: \mu > 160$

164. Use computer to calculate the sample mean and sample standard deviation.

ANSWER:

Summary measures		
	Sample size	20
	Sample mean	162.900
	Sample standard deviation	17.262

165. Calculate the appropriate value of the test statistic.

ANSWER:
$t^* = (\bar{x} - \mu)/(s/\sqrt{n}) = (162.9 - 160.0)/(17.262/\sqrt{20}) = 0.751$

166. Approximate the *p*-value of the test.

ANSWER:
P = *p*-value = $P(t > 0.751 | df = 19)$;
Using the "Probability Values for Student's *t*-Distribution" Table, we get: 0.216 < **P** < 0.246.

167. Find the exact *p*-value of the test.

ANSWER:
p-value = 0.231

168. Use computer to verify your answers to questions 165, 166, and 167.

ANSWER:

Test of mean<=160 versus one-tailed alternative		
	Hypothesized mean	160.000
	Sample mean	162.900
	Std error of mean	3.860
	Degrees of freedom	19
	t-test statistic	0.751
	p-value	0.231

169. Is there sufficient evidence to reject the null hypothesis? Test at α =.05.

ANSWER:
Since *p*-value > α =.05, we fail to reject H_o. The sample does not provide sufficient evidence to justify the contention that mean weight for adult males is higher than 160 lbs.

<u>QUESTIONS 170 THROUGH 172 ARE BASED ON THE FOLLOWING INFORMATION:</u>
The water pollution readings at Lake Michigan seem to be lower than last year. A sample of 15 readings was randomly selected from the records of this year's daily readings: 2.9, 3.2, 4.6, 3.1, 3.3, 3.7, 2.6, 2.9, 2.3, 3.3, 4.2, 2.9, 2.9, 3.1, and 2.6. A researcher claims that the mean of this year's pollution readings is significantly lower than last year's mean of 3.60. Assume that all such readings have a normal distribution.

170. State the null and alternative hypotheses.

ANSWER:
$H_o : \mu = 3.6\ (\geq)$ vs. $H_a : \mu < 3.6$

171. Use computer to calculate the sample mean and sample standard deviation.

ANSWER:

Summary measures		
	Sample size	15
	Sample mean	3.173
	Sample standard deviation	0.605

172. Does this sample provide sufficient evidence to support the researcher's claim at the 0.05 level? Use computer to complete the hypothesis test.

ANSWER:

Test of mean>=3.6 versus one-tailed alternative			
	Hypothesized mean	3.600	
	Sample mean	3.173	
	Std error of mean	0.156	
	Degrees of freedom	14	
	t-test statistic	-2.730	
	p-value	0.008	

Since *p*-value = 0.008 < α = 0.05, we reject H_o. Yes, the sample does provide sufficient evidence to support the researcher's claim that the mean of this year's pollution readings is significantly lower than last year's mean of 3.6, at the 0.05 level of significance.

173. The recommended number of hours of sleep per night is 8 hours, but everybody "knows" that the average college student sleeps less than 7 hours. The number of hours slept last night by15 randomly selected college students are: 5.0, 6.6, 6.0, 5.3, 7.6, 5.6, 6.9, 7.9, 6.7, 5.4, 6.5, 7.2, 5.9, 6.8, and 7.0. Assume that the variable sleeping hours is approximately normally distributed. Use a computer to test the hypothesis $H_o : \mu = 7$ vs. $H_a : \mu < 7$ at $\alpha = 0.02$.

ANSWER:

Summary measures		
	Sample size	15
	Sample mean	6.427
	Sample standard deviation	0.864
Test of mean>=7 versus one-tailed alternative		
	Hypothesized mean	7.000
	Sample mean	6.427
	Std error of mean	0.223
	Degrees of freedom	14
	t-test statistic	-2.570
	p-value	0.011

Since *p*-value = 0.011 < α = 0.02, we reject H_o.The sample does provide sufficient evidence to justify the belief that college student sleeps on average less than 7 hours per night.

Chapter 9

QUESTIONS 174 THROUGH 180 ARE BASED ON THE FOLLOWING INFORMATION:
It is claimed that medical students at the University of Michigan (U of M) score an average of 35 on MCAT. A random sample of test scores for ten students from U of M yields 30, 33, 35, 34, 29, 39, 30, 28, 32, and 38. Assume test results are normally distributed.

174. State the null and alternative hypotheses.

ANSWER:
$H_o: \mu = 35$ vs. $H_a: \mu \neq 35$

175. Use computer to calculate the sample mean and sample standard deviation.

ANSWER:

Summary measures	
Sample size	10
Sample mean	32.800
Sample standard deviation	3.736

176. Use computer to complete the hypothesis test using the *p*-value approach at $\alpha = 0.05$.

ANSWER:

Test of mean=35 versus two-tailed alternative	
Hypothesized mean	35.000
Sample mean	32.800
Std error of mean	1.181
Degrees of freedom	9
t-test statistic	-1.862
p-value	0.095

Since *p*-value = 0.096 $> \alpha = 0.05$, we fail to reject H_o. The sample does not provide sufficient evidence to justify that average MCAT scores for medical students at the University of Michigan is different from 35.

177. Complete the hypothesis test using the classical approach at $\alpha = 0.05$.

ANSWER:
$t(df, \alpha/2) = t(9, 0.025) = 2.26$. The critical values are ± 2.26.
The value of the test statistic $t^* = -1.862$ does not fall in the rejection region; therefore we fail to reject H_o at $\alpha = 0.05$. We reach the same conclusion as stated in question 176.

178. Use a computer to construct 95% confidence interval for the MCAT average score.

ANSWER:

Confidence interval for mean	
Confidence level	95.0%
Sample mean	32.800
Std error of mean	1.181
Degrees of freedom	9
Lower limit	30.128
Upper limit	35.472

179. Verify the lower and upper 95% confidence limits for μ shown on the computer output in question 178.

ANSWER:
$t(df, \alpha/2) = t(9, 0.025) = 2.26$
$E = t(df, \alpha/2)\cdot(s/\sqrt{n}) = 2.26\,(3.736/\sqrt{10}) = 2.67$
$\bar{x} \pm E = 32.8 \pm 2.67 \Rightarrow$ Lower limit = 30.13 and Upper limit = 35.47

180. Explain how to use the 95% confidence interval in question 179 to test the hypotheses in question 174 at $\alpha = 0.05$.

ANSWER:
Since the hypothesized value $\mu = 35$ falls in the 95% confidence interval, we fail to reject H_o at $\alpha = 0.05$.

QUESTIONS 181 THROUGH 184 ARE BASED ON THE FOLLOWING INFORMATION:
It has been suggested that abnormal male children tend to occur more in children born to older-than-average mothers. Case histories of 25 abnormal males were obtained, the ages of the 25 mothers were

21	39	31	21	29	28	34	45	21	41
31	38	40	38	32	28	37	28	16	39
35	29	43	27	42					

The mean age at which mothers in the general population give birth is 28.0 years. Assume ages have a normal distribution.

181. State the null and alternative hypotheses.

ANSWER:
$H_o: \mu = 28\ (\le)$ vs. $H_a: \mu > 28$

182. Use computer to calculate the sample mean and standard deviation.

ANSWER:

Summary measures		
	Sample size	25
	Sample mean	32.520
	Sample standard deviation	7.768

183. Does the sample give sufficient evidence to support the claim that abnormal male children have older-than-average mothers? Use computer and the *p*-value approach at $\alpha = 0.05$.

ANSWER:

Test of mean<=28 versus one-tailed alternative		
	Hypothesized mean	28.000
	Sample mean	32.520
	Std error of mean	1.554
	Degrees of freedom	24
	t-test statistic	2.909
	p-value	0.004

Since *p*-value = 0.004 < α = 0.05, we reject H_o. Yes, the sample provides sufficient evidence to support the claim that the mean age of mothers of abnormal male children is significantly greater than the mean age of mothers with normal male children, at the 0.05 level.

184. Does the sample give sufficient evidence to support the claim that abnormal male children have older-than-average mothers? Use computer and the classical approach at α = 0.05.

ANSWER:

$t(df, \alpha) = t(24, 0.05) = 1.71$

The value of the test statistic t^* = 2.909 does fall in the rejection region; therefore we reject H_o at $\alpha = 0.05$. We reach the same conclusion stated in question 183.

Section 9.3

True-False Questions

185. The maximum error of estimate for a proportion is a multiple of the standard error of proportion.
ANSWER: T

186. The best point estimate of the population proportion *p* is the observed proportion p'.
ANSWER: T

187. In determining the sample size required to estimate a population proportion, the size of sample needed may need to be reduced if a reasonably good estimate for *p* exists from previous studies or perhaps from a small pilot study.
ANSWER: T

188. The sampling distribution of sample proportions p' is approximately distributed as a Student's *t*-distribution.
ANSWER: F

189. The standard error of the sampling distribution of sample proportions p' is equal to $\sqrt{pq} / n$.
ANSWER: F

190. In practice, the sampling distribution of sample proportions p' is an approximately normal distribution if the sample size $n > 20$, the products *np* and *nq* are both greater than 5, and the sample consists of less than 10% of the population.
ANSWER: T

191. The maximum error of estimate for a proportion is given by $E = z(\alpha) \cdot \sqrt{pq} / n$.
ANSWER: F

192. If a random sample of size *n* is selected from a large population with *p* =*P*(success), then the sampling distribution of p' has a mean equal to p',
ANSWER: F

193. If a random sample of size *n* is selected from a large population with *p* =*P*(success), then the sampling distribution of p' has a standard error $\sigma_{p'}$ equal to $\sqrt{pq / n}$.
ANSWER: T

194. When we construct confidence interval for the population proportion *p*, we will base our estimation on the biased sample statistic p', where p' is the center of the confidence interval.
ANSWER: F

195. If a random sample of size n is selected from a large population with p =P(success), then the sampling distribution of p' has an approximately normal distribution if n is sufficiently large.
ANSWER: T

196. The sample size required for 1-α confidence interval of p is given by $n=[z(\alpha/2)]^2 \cdot p^* q^* / E^2$ where p^* and q^* are provisional values of p and q used for planning. If no provisional values for p and q are available, then use $p^* = 1.0$ and $q^* = 0.0$.
ANSWER: F

197. When the binomial parameter p is to be tested using a hypothesis-testing procedure, the test statistic is assumed to be normally distributed when the null hypothesis is true, when the assumptions for the test have been satisfied, and when n is sufficiently large (n>20, np > 5, and nq > 5).
ANSWER: T

Multiple-Choice Questions

198. Which of the following would be the hypotheses in testing the claim that the percentage of students who have part-time jobs is at least 82%?

A) $H_o : p = 0.82(\le), H_a : p > 0.82$
B) $H_o : p > 0.82, H_a : p = 0.82$
C) $H_o : p = 0.82(\ge), H_a : p < 0.82$
D) $H_o : p < 0.82, H_a : p > 0.82$
ANSWER: C

199. Which of the following would be the hypothesis for testing the claim that the proportion of students at a large university who smoke is significantly different from 0.15?

A) $H_o : p = 1.5(\le), H_a : p > 1.5$
B) $H_o : p = 1.5, H_a : p \ne 1.5$
C) $H_o : p > 1.5, H_a : p = 1.5$
D) $H_o : p < 1.5, H_a : p > 1.5$
ANSWER: B

200. Select the correct pair of hypothesis for: "Testing the claim that at most one-half of students at a large university favor an amendment to the student government bylaws."

A) $H_o : p = 0.5(\le), H_a : p > 0.5$
B) $H_o : p = 0.5, H_a : p \ne 0.5$
C) $H_o : p > 0.5, H_a : p = 0.5$
D) $H_o : p < 0.5, H_a : p > 0.5$
ANSWER: A

201. In references about the binomial probability of success, the largest possible value of *pq* where $p = P(\text{success})$ and $q = P(\text{failure})$ is:

A) 1.00.
B) 0.75.
C) 0.50.
D) 0.25.
ANSWER: D

202. When testing the claim that bags of M&M candies will have less than 3% broken pieces, which of the following would be the null hypothesis and alternative hypothesis?

A) $H_o : p = 0.03(\leq), H_a : p > 0.03$
B) $H_o : p = 0.03, H_a : p \neq 0.03$
C) $H_o : p = 0.03(\geq), H_a : p < 0.03$
D) $H_o : p < 0.03, H_a : p > 0.03$
ANSWER: C

203. As the binomial parameter *p* gets larger, then *q*

A) gets smaller.
B) also gets larger.
C) stays the same.
D) size depends on *n*.
ANSWER: A

204. If we do not know the value of the theoretical probability of a success on a single trial in a binomial experiment, then the best replacement available of the standard error of proportion is:

A) $\sqrt{npq}$
B) $\sqrt{np'q'}$
C) $\sqrt{pq / n}$
D) $\sqrt{p'q' / n}$
ANSWER: D

205. The standard deviation of the sampling distribution of the sample binomial probability p' is:

A) *p*
B) *np*
C) $\sqrt{npq}$
D) $\sqrt{pq / n}$
ANSWER: D

Chapter 9

206. The mean of the sampling distribution of the sample binomial probability p' is:

A) p
B) np
C) $\sqrt{npq}$
D) $\sqrt{pq}$
ANSWER: A

207. Which of the following is *not* true about the binomial parameter p?

A) It is the theoretical probability of success on a single trial in a binomial experiment.
B) It is estimated using p'.
C) It is the median of a population possessing a particular characteristic.
D) It is the proportion of a population possessing a particular characteristic.
ANSWER: **C**

208. Which of the following statements is false?

A) When finding the sample size n, always round up to the next larger integer, no matter how small the decimal.
B) When the binomial parameter p is to be tested using a hypothesis-testing procedure, the test statistic is assumed to be normally distributed when the null hypothesis is true, when the assumptions for the test have been satisfied, and when n is sufficiently large ($n > 20$, $np > 5$, and $nq > 5$).
C) There is a relationship between confidence intervals and one-tailed hypothesis tests when the level of confidence and the level of significance add up to one.
D) None of the above
ANSWER: **C**

209. Which of the following statements is false?

A) The point estimate is the center of the confidence interval, and the hypothesized mean is the center of the noncritical region.
B) If the hypothesized value of p is contained in the confidence interval, then the null hypothesis will be rejected.
C) If the hypothesized value of p does not fall within the confidence interval, then the test statistic will be in the critical region.
D) If the hypothesized value of p is contained in the confidence interval, then the test statistic will be in the noncritical region.
ANSWER: **B**

Short-Answer Questions

210. You may have noticed that the confidence coefficients and the critical values are the same when constructing $1-\alpha$ confidence interval or conducting a two-tailed hypothesis test at α level of significance. What does this mean?

ANSWER:
This means the width of the confidence interval and the width of the non-critical region are the same.

211. If the claim "65% of all new cars bought in 1991 were compacts" were tested, what distribution would be used to determine the *p*-value for the test?

ANSWER:
Standard normal distribution

212. Assume a random sample of size *n* is selected from a large population with *p* =*P*(success). Briefly discuss the practical guidelines that will ensure normality for the sample binomial probabilities p'.

ANSWER:
The sample size is greater than 20.
The products *np* and *nq* are both greater than 5.
The sample consists of less than 10% of the population.

Applied and Computational Questions

QUESTIONS 213 THROUGH 216 ARE BASED ON THE FOLLOWING INFORMATION:
A particular candidate claims she has the support of at least 60% of the voters in her district. A random sample of 150 voters yields 87 who support her. The candidate wishes to test her claim at the 0.05 level of significance.

213. State the null and alternative hypotheses.

ANSWER:
$H_o : p = 0.60(\geq), H_a : p < 0.60$

214. Determine the critical region.

ANSWER:
Critical region: $z < -1.65$

215. Compute the value of test statistic.

ANSWER:
Computed value: $z^* = -0.50$

216. State the decision and conclusion.

ANSWER:
Fail to reject H_o. There is not sufficient evidence to conclude that the candidate has the support of less than 60% of the voters.

217. In a random sample of 400 voters interviewed in a large city, 228 believed that the president was doing a good job. Construct a 99% confidence interval estimate for the true proportion of the voters in the city who thought the same way.

ANSWER:
(0.506, 0.634)

218. To test $H_o: p = 0.7(\leq)$ vs. $H_a: p > 0.7$, a sample of size 75 is selected at random. What is the minimum value of the binomial random variable *x* that would result in rejection of H_o if α = 0.05?

ANSWER:
Minimum value = 60

219. Determine the sample size that is required to estimate the true proportion of homes with a DVD if you want your estimate to be within 0.03 with 90% confidence.

ANSWER:
757 homes

220. A machine produces 3-inch nails. A sample of 100 nails is selected, and it is found that 25 are shorter than 3.00 inches. Find a 95% confidence interval of the proportion of all such nails that are shorter than 3.00 inches.

ANSWER:
(0.17 to 0.33)

QUESTIONS 221 THROUGH 223 ARE BASED ON THE FOLLOWING INFORMATION:
An insurance company reports that 75% of its claims are settled within two months of being filed. In order to test that the percent is less than seventy-five, a state insurance commission randomly selects 35 claims and determines that 23 of the 35 were settled within two months.

221. State the null and alternative hypotheses.

ANSWER:
$H_o: p = 0.75(\geq), H_a: p < 0.75$

222. Calculate the value of test statistic and *p*-value.

ANSWER:
z^* = -1.27, *p*-value = 0.102

223. State the decision and conclusion at the 0.05 level of significance.

ANSWER:
Since *p*-value > α, we fail to reject H_o. There is not sufficient evidence to conclude that the percentage of insurance company claims that are settled within two months of being filed is less than 75%.

224. A marketing research firm wishes to conduct a poll in a certain region to estimate the proportion of residents who would oppose the construction of a pipeline. Determine the sample size needed in order to be 90% confident that the sample proportion will be within 0.05 of the true proportion.

ANSWER:
n = 273

225. A company states that 80% of its seed will germinate. A consumer group plants 75 seeds produced by the company in order to test the hypothesis that less than 80% will germinate. ($H_o : p = 0.8(\geq), H_a : p < 0.8$). Find the *p*-value for the test if 52 of the 75 seeds germinate, and test the hypothesis at the 0.05 level of significance.

ANSWER:
p -value = 0.0104. Since *p* – value $< \alpha$, reject the null hypothesis. We conclude that less than 80% of the seeds germinate.

QUESTIONS 226 AND 227 ARE BASED ON THE FOLLOWING INFORMATION:
A sample is to be selected in order to estimate the proportion defective produced by a machine. The true proportion is to be estimated within 0.05 with 95% confidence.

226. Determine *n* if *p* is known to be close to 0.10.

ANSWER:
n = 139

227. Determine *n* if nothing is known about *p*.

ANSWER:
n = 385

QUESTIONS 228 AND 229 ARE BASED ON THE FOLLOWING INFORMATION:
In order to estimate the proportion of universities that provide some dental coverage for their employees, a survey was conducted. Thirty-eight out of 75 universities responded yes to the survey.

228. Give a point estimate for the proportion of all universities that provide some dental coverage.

ANSWER:
Point estimate = 0.51

229. Estimate the proportion of all universities that provide some dental coverage by constructing a 98% confidence interval for *p*.

ANSWER:
(0.37 to 0.64)

230. The null hypothesis being tested is "a coin is fair" and the alternative hypothesis is "the coin favors heads." Let *p* be the probability of a head occurring. The null hypothesis is $H_o : p = 0.5$, and the alternative is $H_a : p > 0.5$. The test statistic, *x*, is the number of heads to occur in a set of 12 tosses of this coin. Determine the largest critical region for which α does not exceed 0.05, by using a discrete variable. (Determine what values of *x* form the critical region, and state the corresponding value of α).

ANSWER:
Critical region: $x \geq 10$, $\alpha = 0.019$

Chapter 9

QUESTIONS 231 THROUGH 233 ARE BASED ON THE FOLLOWING INFORMATION:
On a test of 12 True/False questions we wish to test the null hypothesis that "a student guessed at the answers" versus "studied and performed better than would if simply guessed." The test statistic is *x*, the number of correct answers the student has in the 12.

231. Find α using a discrete variable if the critical region is $x > 8$.

ANSWER:
0.073

232. Find α using a discrete variable if the critical region is $x > 9$.

ANSWER:
0.073

233. Find α using a discrete variable if the critical region is $x > 10$.

ANSWER:
0.003

QUESTIONS 234 THROUGH 237 ARE BASED ON THE FOLLOWING INFORMATION:
In order to test at the 0.10 level of significance the claim that at least 60% of a large student population is in favor of an administrative proposal, a random sample of 150 students is selected. Of this number, 88 are in favor of the proposal.

234. State the null and alternative hypotheses.

ANSWER:
H_o: $p = 0.60$ (≥); H_a: $p < 0.60$

235. State the test criteria.

ANSWER:
The test statistic is z^*. The level of significance is $\alpha = 0.10$. The critical value is $z = -1.28$. Reject H_o if $z^* < -1.28$

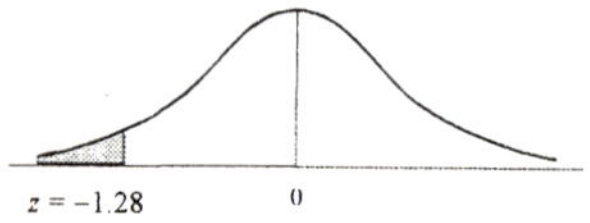

236. Find the computed value of the test statistic.

ANSWER:
$z^* = -0.33$

237. State the decision and conclusion.

ANSWER:
Fail to reject the null hypothesis. There is not sufficient evidence to indicate that the proportion of student population who are in favor of the administrative proposal is less than 0.60.

238. A sample study was randomly selected to construct a 95% confidence interval for *p*. The interval estimate was (0.078, 0.142). Find the value of p', the observed binomial probability.

ANSWER:
p' = 0.11

239. Find the best estimate of the standard error of p' if a sample of size 53 yields 16 successes.

ANSWER:
σ_p = 0.063

QUESTIONS 240 THROUGH 242 ARE BASED ON THE FOLLOWING INFORMTION:
A telephone survey was conducted to estimate the proportion of households with an answering machine. Of the 400 households surveyed, 85 had an answering machine.

240. Give a point estimate for the population proportion of households who have an answering machine.

ANSWER:
$p' = x / n =$ 85 / 400 = 0.2125

241. Give the maximum error of estimate with 95% confidence.

ANSWER:
$E = z(\alpha/2) \cdot \sqrt{p'q'/n} = 1.96 \cdot \sqrt{(0.2125)(0.7875)/400} =$ 0.0402

242. Construct a 95% confidence interval for the true proportion of households who have an answering machine.

ANSWER:
$p' \pm E = 0.2125 \pm 0.0402$. Then, the 95% interval for *p* is 0.1723 to 0.2527.

243. Independent bank randomly selected 400 checking-account customers and found that 150 of them also had savings accounts at this same bank. Construct a 95% confidence interval for the true proportion of checking-account customers who also have savings accounts.

ANSWER:
p = the proportion of checking account customers who also have savings accounts.
The sample was randomly selected and each subject's response was independent of those of the others surveyed.
$n = 400; \quad np = (400)(0.375) = 150 > 5, \quad nq = (400)(0.625) = 250 > 5$
$1 - \alpha = 0.95$; $z(\alpha/2) = z(0.025) = 1.96$
Since $n = 400, x = 150,$ then $p' = x/n = 150/400 = 0.375$.
$E = z(\alpha/2) \cdot \sqrt{p'q'/n} = 1.96 \cdot \sqrt{(0.375)(0.625)/400}$ = (1.96)(0.0242) = 0.0474
Then $p' \pm E = 0.375 \pm 0.0474$, and the 95% interval for *p* is 0.3276 to 0.4224.

Chapter 9

QUESTIONS 244 THROUGH 248 ARE BASED ON THE FOLLOWING INFORMATION:
A policeman wishes to conduct a survey in his city to determine what percent of the bicyclists own helmets. He decided to use the known national figure of 18% for his initial estimate of p.

244. Find the sample size if he wants his estimate to be within 0.02 with 90% confidence.

ANSWER:
$1-\alpha=0.90$; then $z(\alpha/2)=z(0.05)=1.65$. Since $E=0.02$, $p^*=0.18$, $q^*=0.82$,
then $n=\{[z(\alpha/2)]^2\cdot p^*\cdot q^*\}/E^2=(1.65)^2(0.18)(0.82)/(0.02)^2=1004.6$ or 1005

245. Find the sample size if he wants his estimate to be within 0.03 with 90% confidence.

ANSWER:
$1-\alpha=0.90$; then $z(\alpha/2)=z(0.05)=1.65$. Since $E = 0.03$, $p^*=0.18$, $q^*=0.82$,
then $n=\{[z(\alpha/2)]^2\cdot p^*\cdot q^*\}/E^2=(1.65)^2(0.18)(0.82)/(0.03)^2=446.49$ or 447

246. Find the sample size if he wants his estimate to be within 0.02 with 98% confidence.

ANSWER:
$1-\alpha=0.98$; then $z(\alpha/2)=z(0.01)=2.33$. Since $E = 0.02$, $p^*=0.18$, $q^*=0.82$,
then $n=\{[z(\alpha/2)]^2\cdot p^*\cdot q^*\}/E^2=(2.33)^2(0.18)(0.82)/(0.02)^2=2003.26$ or 2004

247. What effect does changing the level of confidence have on the sample size? Explain.

ANSWER:
Increasing the level of confidence increases the sample size.

248. What effect does changing the maximum error have on the sample size? Explain.

ANSWER:
Increasing the maximum error decreases the sample size.

249. It is known that about 15% of lung cancer patients survive for five years after diagnosis. Suppose a physician wants to see if this survival rate is accurate. How large a sample would he need to take to estimate the true proportion surviving for five years after diagnosis to within 1% with 95% confidence?

ANSWER:
$1-\alpha=0.95$; then $z(\alpha/2)=z(0.025)=1.96$. Since $E = 0.01$, $p^*=0.15$, $q^*=0.85$,
then $n=\{[z(\alpha/2)]^2\cdot p^*\cdot q^*\}/E^2=(1.96)^2(0.15)(0.85)/(0.01)^2=4898.04$ or 4899

250. Determine the *p*-value testing $H_o: p=0.25$ vs. $H_a: p\neq 0.25$, if the value of the test statistic $z^* = 1.84$.

ANSWER:
P = *p*-value = $2P(z>1.84)=2(0.5000-0.4671)=0.0658$

251. Determine the *p*-value testing $H_o: p = 0.75$ vs. $H_a: p \neq 0.75$, if the value of the test statistic $z^* = -2.05$.

ANSWER:
P = $2P(z < -2.05) = 2P(z > 2.05) = 2(0.5000 - 0.4798) = 0.0404$

252. Determine the *p*-value testing $H_o: p = 0.46$ vs. $H_a: p > 0.46$, if the value of the test statistic $z^* = 0.89$.

ANSWER:
P = $P(z > 0.89) = (0.5000 - 0.3133) = 0.1867$

253. Determine the *p*-value testing $H_o: p = 0.12$ vs. $H_a: p < 0.12$, if the value of the test statistic $z^* = -1.69$.

ANSWER:
P = $P(z < -1.69) = P(z > 1.69) = (0.5000 - 0.4545) = 0.0455$

254. The binomial random variable, *x*, may be used as the test statistic when testing hypotheses about the binomial parameter, *p*, when *n* is small (say, 15 or less). Use the table of binomial probabilities and determine the *p*-value for testing $H_o: p = 0.4$ vs. $H_a: p \neq 0.4$, where $n = 13$ and $x = 10$.

ANSWER:
P = $2P[x = 10, 11, 12, 13 \mid B(n = 13, p = 0.4)]$
= 2(0.006 + 0.001 + 2(0+)) = 2(0.007) = 0.014

255. The binomial random variable, *x*, may be used as the test statistic when testing hypotheses about the binomial parameter, *p*, when *n* is small (say, 15 or less). Use the table of binomial probabilities and determine the *p*-value for testing $H_o: p = 0.3$ vs. $H_a: p \neq 0.3$, where $n = 15$ and $x = 10$.

ANSWER:
P = $2P[x = 10, 11, 12, 13, 14, 15 \mid B(n = 15, p = 0.3)]$
= 2(0.003 + 0.001 + 4(0+)) = 2(0.004) = 0.008

256. The binomial random variable, *x*, may be used as the test statistic when testing hypotheses about the binomial parameter, *p*, when *n* is small (say, 15 or less). Use the table of binomial probabilities and determine the *p*-value for testing $H_o: p = 0.2$ vs. $H_a: p > 0.2$, where $n = 14$ and $x = 5$.

ANSWER:
P = $P[x = 5, 6, 7, 8, 9, \ldots, 14 \mid B(n = 14, p = 0.2)]$
= (0.086 + 0.032 + 0.009 + 0.002 + 6(0+) = 0.129

Chapter 9

257. The binomial random variable, *x,* may be used as the test statistic when testing hypotheses about the binomial parameter, *p*, when *n* is small (say, 15 or less). Use the table of binomial probabilities and determine the *p*-value for testing $H_o: p = 0.9$ vs. $H_a: p < 0.9$, where $n = 13$ and $x = 9$.

ANSWER:
P = $P[x = 0,1,2,...,9 \mid B(n = 13, p = 0.9)] = [7(0+) + 0.001 + 0.006 + 0.028] = 0.035$

258. Use the table of binomial probabilities to determine the critical region used in testing each of the hypothesis $H_o: p = 0.4$ vs. $H_a: p > 0.4$, where $n = 15$ and $\alpha = 0.05$. (*Note:* Since *x* is discrete, choose critical regions that do not exceed the value of α given.)

ANSWER:
Critical region is $x \geq 10, \alpha = 0.033$

259. Use the table of binomial probabilities to determine the critical region used in testing each of the hypothesis $H_o: p = 0.5$ vs. $H_a: p \neq 0.5$, where $n = 14$ and $\alpha = 0.05$. (*Note:* Since *x* is discrete, choose critical regions that do not exceed the value of α given.)

ANSWER:
Critical region is $x \leq 3$ or $x \geq 12, \alpha = 0.034$.

260. Use the table of binomial probabilities to determine the critical region used in testing each of the hypothesis $H_o: p = 0.6$ vs. $H_a: p < 0.6$, where $n = 10$ and $\alpha = 0.10$. (*Note:* Since *x* is discrete, choose critical regions that do not exceed the value of α given.)

ANSWER:
Critical region is $x \leq 3, \alpha = 0.055$

261. Use the table of binomial probabilities to determine the critical region used in testing each of the hypothesis $H_o: p = 0.7$ vs. $H_a: p > 0.7$, where $n = 13$ and $\alpha = 0.01$. (*Note:* Since *x* is discrete, choose critical regions that do not exceed the value of α given.)

ANSWER:
Critical region is $x = 13, \alpha = 0.010$

QUESTIONS 262 THROUGH 265 ARE BASED ON THE FOLLOWING INFORMATION:
State Farm insurance company states that 90% of its claims are settled within 5 weeks. A consumer group selected a random sample of 100 of the company's claims to test this statement. If the consumer group found that 75 of the claims were settled within 5 weeks, do they have sufficient reason to support their contention that fewer than 90% of the claims are settled within 5 weeks?

262. State the null and alternative hypotheses.

ANSWER:
$H_o: p = 0.90 (\geq)$ vs. $H_a: p < 0.90$

263. Identify the probability distribution to be used and calculate the test statistic.

ANSWER:
Since $n = 100 > 20$, $np = (100)(0.90) = 90 > 5$, and $nq = (100)(0.10) = 10 > 5$, then p' is expected to be approximately normally distributed.
$x = 75, p' = x/n = 75/100 = 0.75$. Then, the test statistic is
$z* = (p' - p)/\sqrt{pq/n} = (0.75 - 0.90)/\sqrt{(0.9)(0.1)/100} = -5.0$

264. Complete the test at the 0.05 level of significance using the *p*-value approach.

ANSWER:
P = *p*-value = $P(z < -5.0) = P(z > 5.0)$; Using the table of standard normal distribution, we have **P** = 0.5000 – 0.4999997 = 0.0000003. Since **P** $< \alpha$; we reject H_o.

265. Complete the test at the 0.05 level of significance using the classical approach.

ANSWER:
The critical value is: $-z(0.05) = -1.65$

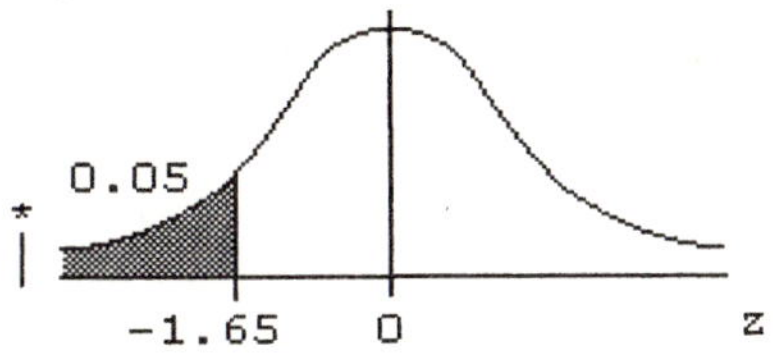

The test statistic z^* falls in the critical region, therefore we reject H_o, and conclude that the sample provides sufficient evidence that *p* is significantly less than 0.90; it appears that less than 90% are settled within 30 days as claimed, at the 0.05 level of significance.

266. The marketing research department of an automobile company conducted a survey to determine the proportion of unmarried women who prefer their model of sport cars. Thirty-five of the 100 unmarried women in the random sample preferred the company's model. Use a 95% confidence interval to estimate the proportion of all unmarried women who prefer this company's model of sport cars. Interpret your answer.

ANSWER:
P = The proportion of unmarried women who prefer a sport model car.
The sample was randomly selected and each subject's response was independent of those of the others surveyed.
$n = 100; \quad n > 20, \quad np = (100)(0.35) = 35 > 5, \quad nq = (100)(0.65) = 65 > 5$
Since $\alpha = 0.05$; then $z(\alpha/2) = z(0.025) = 1.96$.
$E = z(\alpha/2) \cdot \sqrt{p'q'/n} = 1.96\sqrt{(0.35)(0.65)/100}$ = (1.96)(0.0477) = 0.0935
Then, $p' \pm E = 0.35 \pm 0.0935$, and the 95% interval for *p* is 0.2565 to 0.4435.

Chapter 9

QUESTIONS 267 THROUGH 270 ARE BASED ON THE FOLLOWING INFORMATION:

The full-time student body of Big Rapids high school is composed of 50% males and 50% females. Does a random sample of students consisting of 25 male and 15 female from calculus course show sufficient evidence to reject the hypothesis that the proportion of male and female students who take this course is the same as that of the whole student body?

267. State the null and alternative hypotheses.

ANSWER:
p = The proportion of male students in calculus course.
$H_o: p = 0.50$ vs. $H_a: p \neq 0.50$

268. Identify the probability distribution to be used and calculate the value of the test statistic.

ANSWER:
p = The proportion of male students in calculus course.
Since $n = 40$; $n > 20$, $np = (40)(.50) = 20 > 5$, and $nq = (40)(0.50) = 20 > 5$, then p' is expected to be approximately normally distributed. $x = 25, p' = x/n = 25/40 = 0.625$.
Then, $z* = (p' - p)/\sqrt{pq/n} = (0.625 - 0.50)/\sqrt{(0.5)(0.5)/40} = 1.58$

269. Complete the test at the 0.05 level of significance using the p-value approach.

ANSWER:
p = The proportion of male students in calculus course
The p-value approach: **P** $= 2 \cdot P(z > 1.58)$; Using the table of standard normal distribution, we have **P** = 2(0.5000 – 0.4429) = 0.1142. Since **P** $> \alpha$; fail to reject H_o.
The sample provides sufficient evidence that the proportion is not significantly different than 0.50, at the 0.05 level; that is, the sample evidence does not indicate the proportion of males taking chemistry to be different than 50%.

270. Complete the test at the 0.05 level of significance using the classical approach.

ANSWER:
p = The proportion of male students in calculus course
The critical values are: $\pm z(0.025) = \pm 1.96$

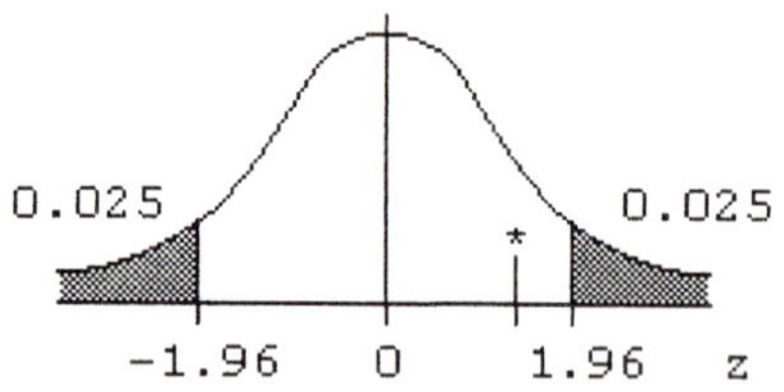

The test statistic z^* =1.58 falls in the noncritical region, therefore we fail to reject H_o. We reach the same conclusion as stated in question 269.

Section 9.4

True-False Questions

271. It is possible that a particular chi-square distribution has a sample size of 21 and the mean is also 21.
ANSWER: F

272. The chi-square distribution is used for inferences about the population mean μ when the standard deviation σ is unknown.
ANSWER: F

273. Often the concern with testing the variance (or standard deviation) is to keep its size under control or relatively small. Therefore, many of the hypotheses tests with chi-square will be one-tailed.
ANSWER: T

274. The Student's *t*-distribution is used for all inferences about a population's variance.
ANSWER: F

275. The chi-square distribution is a skewed distribution whose mean value is *n* for degrees of freedom larger than two.
ANSWER: F

276. When random samples are drawn from a normal population of a known variance σ^2, the quantity $(n-1)s^2/\sigma^2$ possesses a probability distribution that is known as the chi-square distribution, with $(n - 1)$ degrees of freedom.
ANSWER: T

277. The chi-square distributions, like the Student's *t*-distributions, are a family of probability distributions, with each member of the family being identified by the number of degrees of freedom.
ANSWER: T

278. The symbol $\chi^2(df, \alpha)$ is used to identify the critical value of chi-square with *df* degrees of freedom, and with α area to the left.
ANSWER: F

279. Inferences about the variance of a normally distributed population use the chi-square, χ^2, distributions.
ANSWER: T

280. When random samples are drawn from a normal population with a known variance σ^2, the quantity $(n-1)s^2/\sigma^2$ possesses a probability distribution that is known as the chi-square distribution with *n* -1 degrees of freedom.
ANSWER: T

Chapter 9

Multiple-Choice Questions

281. The mean age of 25 randomly selected college seniors was found to be 23.5 years, and the standard deviation of all college seniors was 1.3 years. The correct symbol for the 1.3 years is which of the following?

A) μ
B) s
C) σ
D) $\overline{x}$
ANSWER: C

282. Which of the following is a property of the chi-square distribution?

A) It can be positive or negative in value.
B) It is bell shaped.
C) It does not utilize degrees of freedom.
D) There is a separate distribution for each different sample size.
ANSWER: D

283. In a chi-square distribution, the mean is equal to the

A) degrees of freedom.
B) median.
C) mode.
D) standard deviation.
ANSWER: A

284. Which of the following statements is false as a property of the chi-square distribution?

A) χ^2 is nonnegative in value; it is zero or positively valued.
B) χ^2 is not symmetrical; it is skewed to the left
C) χ^2 is distributed so as to form a family of distributions, a separate distribution for each different number of degrees of freedom.
D) None of the above
ANSWER: B

285. Which of the following statements is false?

A) Inferences about the variance of a normally distributed population use the chi-square, χ^2, distributions.
B) $\chi^2(df,\alpha)$ (read "*chi-square of df, alpha*") is the symbol used to identify the critical value of chi-square with *df* degrees of freedom and with α area to the right.
C) When *df* >2, the mean value of the chi-square distribution is the square root of the *df*. Itself.
D) None of the above
ANSWER: C

286. Which of the following statements is false?

A) The t procedures for inferences about the mean were based on the assumption of normality, but they are generally useful even when the sampled population is nonnormal, especially for larger samples.
B) The statistical procedures for the standard deviation are very sensitive to nonnormal distributions (skewness, in particular), and this makes it difficult to determine whether an apparent significant result is the result of the sample evidence or a violation of the assumptions.
C) The test statistic that will be used in testing hypotheses about the population variance or standard deviation is obtained by using the formula $\chi^{2}* = (n-1)s^{2}/\sigma^{2}$ with $df = n - 1$.
D) None of the above

ANSWER: D

287. Which of the following critical values of the chi-square distribution is the smallest?

A) $\chi^{2}(15, 0.95)$
B) $\chi^{2}(18, 0.95)$
C) $\chi^{2}(32, 0.95)$
D) $\chi^{2}(40, 0.95)$

ANSWER: A

288. Which of the following critical values of the chi-square distribution is the largest?

A) $\chi^{2}(20, 0.95)$
B) $\chi^{2}(20, 0.75)$
C) $\chi^{2}(20, 0.50)$
D) $\chi^{2}(20, 0.25)$

ANSWER: D

289. Which of the following critical values of the chi-square distribution is the smallest?

A) $\chi^{2}(16, 0.01)$
B) $\chi^{2}(10, 0.10)$
C) $\chi^{2}(24, 0.50)$
D) $\chi^{2}(28, 0.95)$

ANSWER: B

290. Which of the following critical values of the chi-square distribution is the largest?

A) $\chi^{2}(20, 0.025)$
B) $\chi^{2}(12, 0.95)$
C) $\chi^{2}(8, 0.005)$
D) $\chi^{2}(15, 0.90)$

ANSWER: A

Chapter 9

Short-Answer Questions

291. For a chi-square distribution with a mean value of 30, find the area under the curve to the right of 34.8.

ANSWER:
0.25

292. State the null and alternative hypotheses for the claim: "the variance is greater than 16 ounces."

ANSWER:
$H_o: \sigma^2 = 16(\leq)$ vs. $H_a: \sigma^2 > 16$

293. If we correctly reject the claim that a population variance is at least 25.0, then can we also reject the claim that the population standard deviation is at least 5.0? Explain.

ANSWER:
The techniques employ the sample variance rather than the sample standard deviation. Since the standard deviation is the positive square root of the variance, talking about the variance is comparable to talking about the standard deviation. Thus, we could also reject the claim that the standard deviation is at least 5.0.

294. State the null hypothesis, H_o, and the alternative hypothesis, H_a, that would be used to test the claim: The standard deviation has increased from its previous value of 15.

ANSWER:
$H_o: \sigma = 15(\leq)$ vs. $H_a: \sigma > 15$

295. State the null hypothesis, H_o, and the alternative hypothesis, H_a, that would be used to test the claim: The standard deviation is no larger than 0.4 oz.

ANSWER:
$H_o: \sigma = 0.4(\leq)$ vs. $H_a: \sigma > 0.4$

296. State the null hypothesis, H_o, and the alternative hypothesis, H_a, that would be used to test the claim: The standard deviation is not equal to 5.2.

ANSWER:
$H_o: \sigma = 5.2$ vs. $H_a: \sigma \neq 5.2$

297. State the null hypothesis, H_o, and the alternative hypothesis, H_a, that would be used to test the claim: The variance is no less than 10.

ANSWER:
$H_o: \sigma^2 = 10(\geq)$ vs. $H_a: \sigma^2 < 10$

298. State the null hypothesis, H_o, and the alternative hypothesis, H_a, that would be used to test the claim: The variance is different from the value of 0.025.

ANSWER:
$H_o: \sigma^2 = 0.025$ vs. $H_a: \sigma^2 \neq 0.025$

299. State the null hypothesis, H_o, and the alternative hypothesis, H_a, that would be used to test the claim: The variance has increased from 13.2.

ANSWER:
$H_o: \sigma^2 = 13.2 (\leq)$ vs. $H_a: \sigma^2 > 13.2$

300. Find $\chi^2(12, 0.01)$.

ANSWER:
26.2

301. Find $\chi^2(15, 0.025)$.

ANSWER:
27.5

302. Find $\chi^2(20, 0.95)$.

ANSWER:
10.9

303. Find $\chi^2(25, 0.995)$.

ANSWER:
10.5

304. Find the critical value $\chi^2(16, 0.01)$.

ANSWER:
32.0

305. Find the critical value $\chi^2(18, 0.025)$.

ANSWER:
31.5

306. Find the critical value $\chi^2(10, 0.10)$.

ANSWER:
16.0

307. Find the critical value $\chi^2(24, 0.01)$.

ANSWER:
43.0

308. Find the critical value $\chi^2(28, 0.95)$.

ANSWER:
16.9

309. Find the critical value $\chi^2(13, 0.975)$.

ANSWER:
5.01

310. Find the critical value $\chi^2(40, 0.90)$.

ANSWER:
29.1

311. Find the critical value $\chi^2(50, 0.99)$.

ANSWER:
29.7

Applied and Computational Questions

QUESTIONS 312 THROUGH 315 ARE BASED ON THE FOLLOWING INFORMATION:
In order to test the claim that the variance of a particular normal population equals 9.0, the following random sample was selected: 58, 64, 57, 63, 62, 61, and 55.

The test is to be completed using α = 0.05.

312. State the null and alternative hypotheses.

ANSWER:
$H_o: \sigma^2 = 9.0, H_a: \sigma^2 \neq 9.0$

313. State the test criteria.

ANSWER:
Reject H_o if $\chi^2 < 1.24$ or $\chi^2 > 14.5$

314. Find the computed value of the test statistic.

ANSWER:
$\chi^2* = 7.56$

315. State the decision and conclusion.

ANSWER:
Fail to reject null. There is not sufficient evidence to suggest that the variance is not equal to 9.0.

316. Give a bound on the *p*-value for testing: $H_o : \sigma^2 = a$ vs. $H_a : \sigma^2 > a$ given that the computed test statistic = 25.2 and *n* = 15.

ANSWER:
0.025 < *P* < 0.050

317. Give a bound on the *p*-value for testing: $H_o : \sigma^2 = b$ vs. $H_a : \sigma^2 > b$ given that the computed test statistic = 6.10 and *n* = 15.

ANSWER:
0.025 < *P* < 0.050

318. The null hypothesis $H_o : \sigma^2 = 150$ is tested against $H_a : \sigma^2 > 150$. For a sample of size 20, the *p*-value has the bound 0.05 < *p* < 0.10. What is the range of s^2?

ANSWER:
$143.15 < s^2 < 158.42$

319. The hypothesis $H_o : \sigma^2 = 15$ is to be tested against $H_a : \sigma^2 > 15$ at α = 0.05. For a sample size of 20, what value for *s* would result in the rejection of H_o?

ANSWER:
Any value of *s* less than 2.824

320. A one-tailed hypothesis test for the standard deviation is to be performed. The null hypothesis is that $H_o : \sigma = 10$ and the alternative is $H_a : \sigma < 10$. A sample of size 15 and a level of significance equal to 0.05 is to be used. Give the critical region for this test.

ANSWER:
Critical region: $\chi^2 \leq 6.57$

QUESTIONS 321 THROUGH 324 ARE BASED ON THE FOLLOWING INFORMATION:
A drug manufacturer produces 250-milligram capsules of a new antibiotic. A random sample of ten such capsules is selected and the amount of antibiotic in each capsule is determined. The results are as follows (in milligrams): 252, 246, 242, 250, 255, 258, 250, 252, 250 and 258. This data is used to test $H_o : \sigma = 2.5$ vs. $H_a : \sigma > 2.5$.

321. Calculate the sample variance.

ANSWER:
s^2 = 5.1552

322. Calculate the value of the test statistic.

ANSWER:
x^2 =35.556

323. Give a bound on the *p*-value.

ANSWER:
p -value < 0.005

324. Test the hypothesis at the 0.01 level of significance.

ANSWER:
Since *p*-value $<\alpha$, we reject H_o. There is sufficient evidence to conclude that the population standard deviation is greater than 2.5.

QUESTIONS 325 THROUGH 328 ARE BASED ON THE FOLLOWING INFORMATION:
A machine produces 3-inch nails. A sample of ten nails is obtained and their lengths determined. The results are as follows: 2.89, 2.95, 3.00, 3.05, 2.99, 2.96, 3.10, 3.06, 3.00 and 3.12. This data is used test H_o: σ = 0.03 vs. H_a: $\sigma \neq$ 0.03.

325. Calculate the sample variance.

ANSWER:
s^2 = 0.00504

326. Calculate the value of the test statistic.

ANSWER:
X^2 =50.396

327. Give a bound on the *p*-value for the test.

ANSWER:
p-value < 0.005

328. Test the hypothesis at the 0.01 level of significance.

ANSWER:
Since *p*-value < α, we reject H_o. There is sufficient evidence to conclude that the population standard deviation is different from 0.03.

329. Give a bound on the *p*-value for the testing $H_o: \sigma^2 = 27 (\leq)$ vs. $H_a: \sigma^2 > 27$, with *df* = 16 and χ^2 = 28.4.

ANSWER:
0.025 < **P** < 0.050

330. Give a bound on the *p*-value for testing $H_o: \sigma^2 = 46.1$ vs. $H_a: \sigma^2 \neq 46.1$, with *df* = 20 and $\chi^2 = 9.01$.

ANSWER:
$0.02 < p < 0.05$

331. In testing the hypothesis $H_o: \sigma^2 = 30.0 (\leq)$ vs. $H_a: \sigma^2 > 30.0$, a sample of size *n* = 21 yielded $\chi^2 = 24.0$. Find the sample variance.

ANSWER:
$s^2 = 36.0$

332. Calculate the *p*-value for testing the alternative hypothesis $H_a: \sigma^2 \neq 18$, when *n* = 15, and $\chi^{2*} = 28.2$.

ANSWER:
P = $2P(\chi^{2*} > 28.2 \mid df = 14)$; Since 0.01 < ½ **P** < 0.025; then 0.02 < **P** <0.05

333. Calculate the *p*-value for testing the alternative hypothesis $H_a: \sigma^2 > 25$, when *n* = 16, and $\chi^{2*} = 30.6$.

ANSWER:
P = $P(\chi^{2*} > 30.6 \mid df = 15) = 0.01$

334. Calculate the *p*-value for testing the alternative hypothesis $H_a: \sigma^2 \neq 32$, when *df* = 20, and $\chi^{2*} = 33.1$.

ANSWER:
P = $2P(\chi^{2*} > 33.1 \mid df = 20)$; Since 0.025 < ½ **P** <0.05; then 0.05 < **P** < 0.10.

335. Calculate the *p*-value for testing the alternative hypothesis $H_a: \sigma^2 < 13$, when *df* = 30, and $\chi^{2*} = 17.4$.

ANSWER:
P = $P(\chi^{2*} < 17.4 \mid df = 30)$; then 0.025 < **P** < 0.05

QUESTIONS 336 THROUGH 339 ARE BASED ON THE FOLLOWING INFORMATION:
A random sample of 51 observations was selected from a normally distributed population. The sample mean was $\bar{x} = 88.6$, and the sample variance was $s^2 = 38.2$. We wish to determine if there is sufficient reason to conclude that the population standard deviation is not equal to 8 at the 0.05 level of significance.

336. State the null and alternative hypotheses.

ANSWER:
$H_o: \sigma = 8$ vs. $H_a: \sigma \neq 8$

Chapter 9

337. Calculate the value of the test statistic.

ANSWER:
$\chi^{2*} = (n-1)s^2/\sigma^2 = (50)(38.2)/(8)^2 = 29.84$

338. Complete the test using the *p*-value approach.

ANSWER:
P = *p*-value = $2P(\chi^2 < 29.84 \mid df = 50)$;

Since 0.01 < ½ **P** < 0.025; then 0.02 < **P** < 0.05 **P** < $\alpha = 0.5$; reject H_o. There is sufficient reason to conclude that the population standard deviation is not equal to 8, at the 0.05 level of significance.

339. Complete the test using the classical approach.

ANSWER:
The critical values are $\chi^2(50, 0.975) = 32.4$ and $\chi^2(50, 0.025) = 71.4$

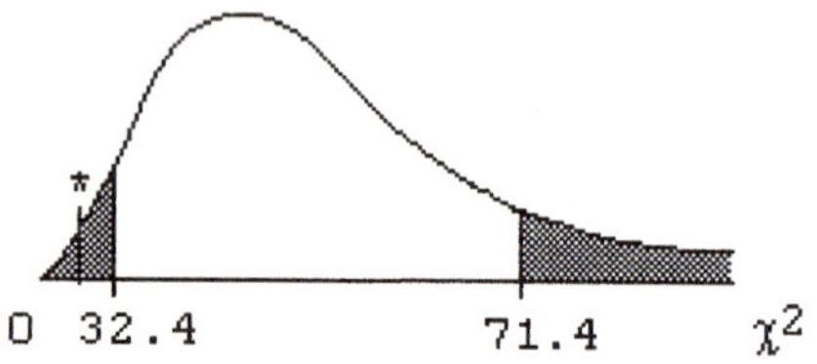

The test statistic χ^2* = 29.84 falls in the critical region, therefore we reject H_o. We reach the same conclusion as stated in question 338.

<u>QUESTIONS 340 THROUGH 344 ARE BASED ON THE FOLLOWING INFORMATION:</u>
A foreign car manufacturer claims that the miles per gallon for a certain model of their cars are normally distributed with a mean equal to 41.5 miles with a standard deviation equal to 3.5 miles. The following data are obtained from a random sample of 15 such cars; 39.0, 43.5, 41.0, 43.5, 37.0, 31.0, 38.5, 38.0, 39.0, 43.5, 46.0, 35.0, 33.0, 37.0, and 37.5. We wish to test the hypothesis that the standard deviation differs from 3.5.

340. Calculate the sample variance.

ANSWER:
$s^2 = 17.2024$

341. State the null and alternative hypotheses.

ANSWER:
$H_o: \sigma = 3.5$ vs. $H_a: \sigma \neq 3.5$

342. Calculate the value of the test statistic.

ANSWER:
$\chi^{2*} = (n-1)s^2/\sigma^2 = (14)(17.2024)/(3.50)^2 = 19.66$

343. Complete the test at α =0.05 using the *p*-value approach.

ANSWER:

P = *p*-value = $2 \cdot P(\chi^2 > 19.66 \mid df = 14)$; Since 0.10 < ½ **P** < 0.25, then 0.20 < **P** < 0.50

P $> \alpha = .05$; fail to reject H_o. There is not sufficient reason at the 0.05 level of significance to contradict the manufacturer's claim about the standard deviation, and conclude that it is different from 3.5.

344. Complete the test at α = 0.05 using the classical approach.

ANSWER:

The critical values are $\chi^2(14, 0.975) = 5.63$ and $\chi^2(14, 0.025) = 26.1$.

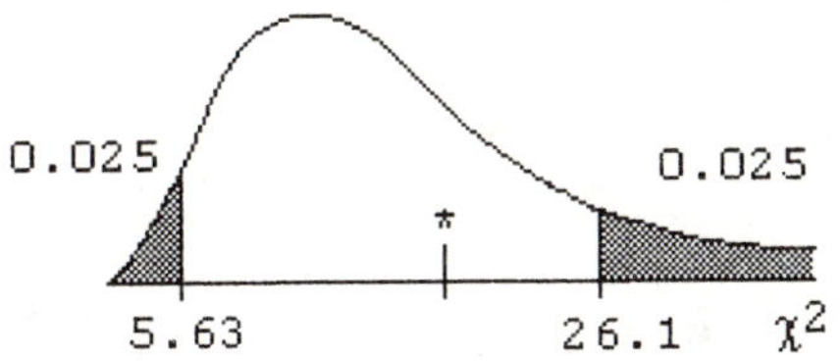

The test statistic χ^2* = 19.66 falls in the noncritical region, therefore we fail to reject H_o.
We reach the same conclusion as stated in question 343.

345. For a chi-square distribution having 25 degrees of freedom, find the area under the curve between $\chi^2(25, 0.94)$ and $\chi^2(25, 0.18)$.

ANSWER:

Area = $\chi^2(25, 0.94) - \chi^2(25, 0.18)$ = 0.94 – 0.18 = 0.76

QUESTIONS 346 THROUGH 349 ARE BASED ON THE FOLLOWING INFORMATION:
Consider a chi-square distribution with 15 degrees of freedom.

346. The central 80% of the distribution lies between what values?

ANSWER:

$\chi^2(15, 0.90)$ = 8.55 and $\chi^2(15, 0.10)$ = 22.3
Therefore the central 80% of the distribution lies between 8.55 and 22.3.

347. The central 90% of the distribution lies between what values?

ANSWER:

$\chi^2(15, 0.95)$ = 7.26 and $\chi^2(15, 0.05)$ = 25.0
Therefore the central 80% of the distribution lies between 7.26 and 25.0.

348. The central 95% of the distribution lies between what values?

ANSWER:
$\chi^2(15,0.975)$ = 6.26 and $\chi^2(15,0.025)$ = 27.5
Therefore the central 80% of the distribution lies between 6.26 and 27.5.

349. The central 99% of the distribution lies between what values?

ANSWER:
$\chi^2(15,0.995)$ = 4.60 and $\chi^2(15,0.005)$ = 32.8
Therefore the central 80% of the distribution lies between 4.60 and 32.8

350. For a chi-square distribution having 45 degrees of freedom, find the area under the curve between $\chi^2(45,0.98)$ and $\chi^2(45,0.13)$.

ANSWER:
Area = $\chi^2(45,0.98)$ - $\chi^2(45,0.13)$. = 0.98 – 0.13 = 0.85

QUESTIONS 351 THROUGH 358 ARE BASED ON THE FOLLOWING INFORMATION:
Problems often arise that require us to make inferences about variability (the spread of data). This is accomplished by performing hypotheses testing about the population variance σ^2 or the population standard deviation σ. This requires us to carefully state the null and alternative hypotheses based on the information provided to us.

351. State the null hypothesis, H_o, and the alternative hypothesis, H_a, that would be used to test the claim "The standard deviation has increased from its previous value of 20."

ANSWER:
$H_o: \sigma = 20\ (\leq)$ and $H_a: \sigma > 20$

352. State the null hypothesis, H_o, and the alternative hypothesis, H_a, that would be used to test the claim "The standard deviation is no larger than 0.2 oz".

ANSWER:
$H_o: \sigma = 0.2\ (\leq)$ and $H_a: \sigma > 0.2$

353. State the null hypothesis, H_o, and the alternative hypothesis, H_a, that would be used to test the claim "The standard deviation is not equal to 15."

ANSWER:
$H_o: \sigma = 15$ and $H_a: \sigma \neq 15$

354. State the null hypothesis, H_o, and the alternative hypothesis, H_a, that would be used to test the claim "The variance is no less than 24."

ANSWER:
$H_o: \sigma^2 = 24\ (\geq)$ and $H_a: \sigma^2 < 24$

355. State the null hypothesis, H_o, and the alternative hypothesis, H_a, that would be used to test the claim "The variance is different from the value of 0.01, the value called for in the specs."

ANSWER:
$H_o: \sigma^2 = 0.01$ and $H_a: \sigma^2 \neq 0.01$

356. State the null hypothesis, H_o, and the alternative hypothesis, H_a, that would be used to test the claim "The variance has decreased from its previous value of 32.25."

ANSWER:
$H_o: \sigma^2 = 32.25\ (\geq)$ and $H_a: \sigma^2 < 32.25$

357. State the null hypothesis, H_o, and the alternative hypothesis, H_a, that would be used to test the claim "The variance is at most 28."

ANSWER:
$H_o: \sigma^2 = 28\ (\leq)$ and $H_a: \sigma^2 > 28$

358. State the null hypothesis, H_o, and the alternative hypothesis, H_a, that would be used to test the claim "The standard deviation is at least 4.25."

ANSWER:
$H_o: \sigma = 4.25\ (\geq)$ and $H_a: \sigma < 4.25$

359. Find the value of the test statistic for testing $H_o: \sigma^2 = 500$ vs $H_a: \sigma^2 > 500$ using the sample information $n = 20$ and $s^2 = 682$.

ANSWER:
$\chi^{2*} = (n-1)s^2/\sigma^2$ = (19)(682) / 500 = 25.92

360. Find the value of the test statistic for testing $H_o: \sigma^2 = 55$ vs $H_a: \sigma^2 \neq 55$ using the sample information $n = 26$ and $s^2 = 75$.

ANSWER:
$\chi^{2*} = (n-1)s^2/\sigma^2$ = (25)(75) / 55 = 34.09

361. Place bounds on the *p*-value for testing $H_a: \sigma^2 \neq 24$, given that $n = 12$, and $\chi^2* = 20.8$

ANSWER:
P = *p*-value = 2 · *P*(χ^2 > 20.8 | *df* = 11). Since 0.025 < 1/2 **P** < 0.05; then, 0.05 < **P** < 0.10

362. Place bounds on the *p*-value for testing $H_a: \sigma^2 > 32$, given that $n = 16$, and $\chi^2* = 28.6$.

ANSWER:
P = *p*-value = *P*(χ^2 > 28.6 | *df* = 15). Then, 0.01 < **P** < 0.025

363. Place bounds on the *p*-value for testing $H_a: \sigma^2 \neq 40$, given that $df = 30$, and $\chi^{2*} = 48.9$

ANSWER:

P = *p*-value = $2 \cdot P(\chi^2 > 48.9 \mid df = 30)$. Since $0.01 < 1/2$ **P** < 0.025; then, $0.02 <$ **P** < 0.05

364. Place bounds on the *p*-value for testing $H_a: \sigma^2 < 16$, given that $df = 50$, and $\chi^{2*} = 30.4$

ANSWER:

P = *p*-value = $P(\chi^2 < 30.4 \mid df = 50) \Rightarrow 0.01 <$ **P** < 0.025.

365. Draw an approximate chi-square distribution and determine the critical region and critical value(s) that would be used to test $H_o: \sigma = 0.4$ vs. $H_a: \sigma > 0.4$, given that n =18 and $\alpha = 0.05$, using the classical approach:

ANSWER:

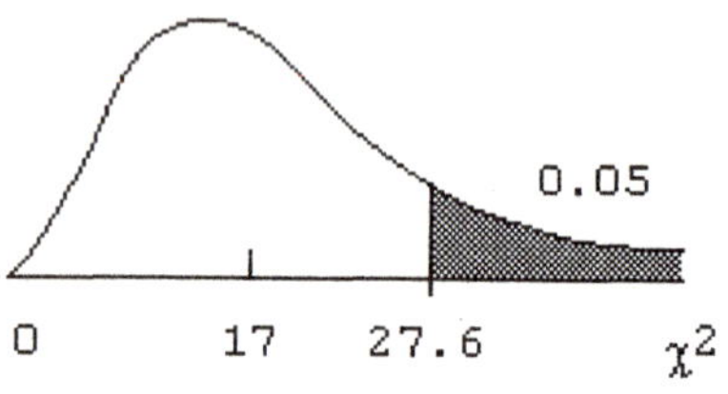

366. Draw an approximate chi-square distribution and determine the critical region and critical value(s) that would be used to test $H_o: \sigma^2 = 10$ and $H_a: \sigma^2 < 10$, with n =15 and $\alpha = 0.01$, using the classical approach:

ANSWER:

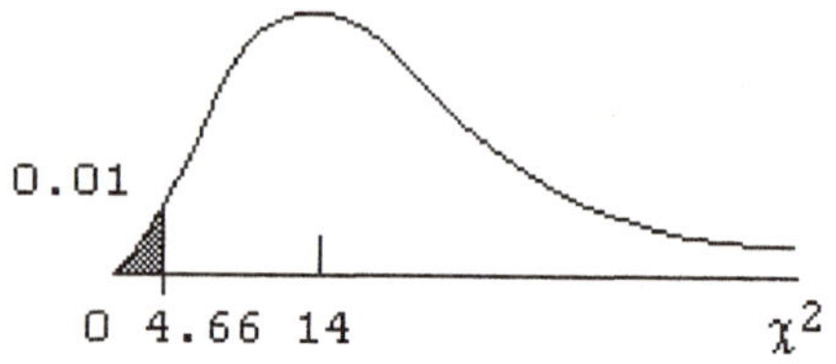

367. Draw an approximate chi-square distribution and determine the critical region and critical value(s) that would be used to test $H_o: \sigma = 12.4$ and $H_a: \sigma \neq 12.4$, with n =10 and $\alpha = 0.10$, using the classical approach:

ANSWER:

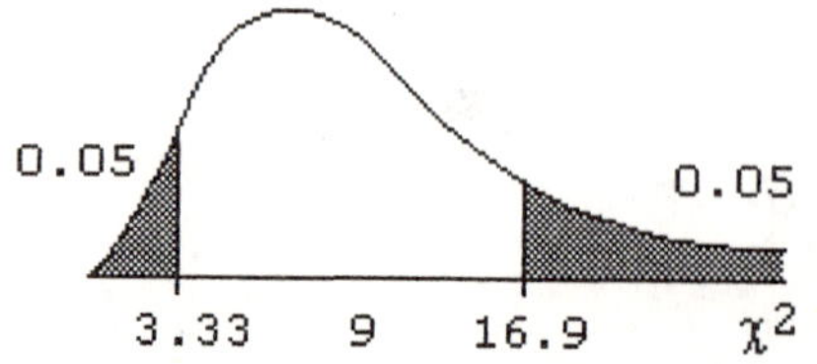

368. Place bounds on the *p*-value for testing $H_a: \sigma^2 < 44$, given that $n = 30$, and $\chi^{2*} = 18.9$

ANSWER:
P = *p*-value = $P(\chi^2 < 18.9 \mid df = 29) \Rightarrow 0.05 <$ **P** < 0.10.

369. Draw an approximate chi-square distribution and determine the critical region and critical value(s) that would be used to test $H_o: \sigma^2 = 0.09$ and $H_a: \sigma^2 \neq 0.09$, with $n = 8$ and $\alpha = 0.02$, using the classical approach:

ANSWER:

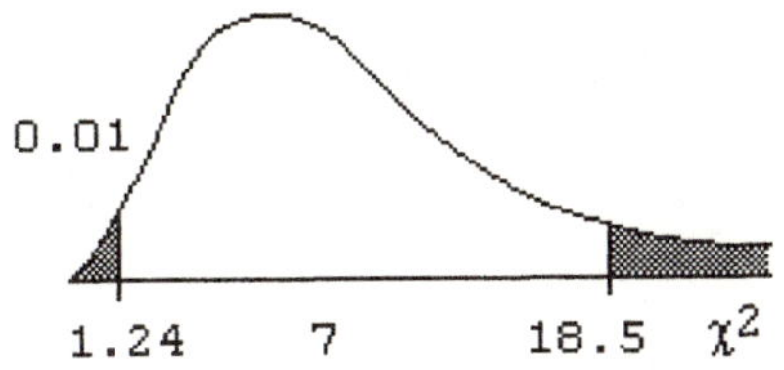

370. Draw an approximate chi-square distribution and determine the critical region and critical value(s) that would be used to test $H_o: \sigma = 0.6$ and $H_a: \sigma < 0.6$, with n =12 and $\alpha = 0.10$, using the classical approach:

ANSWER:

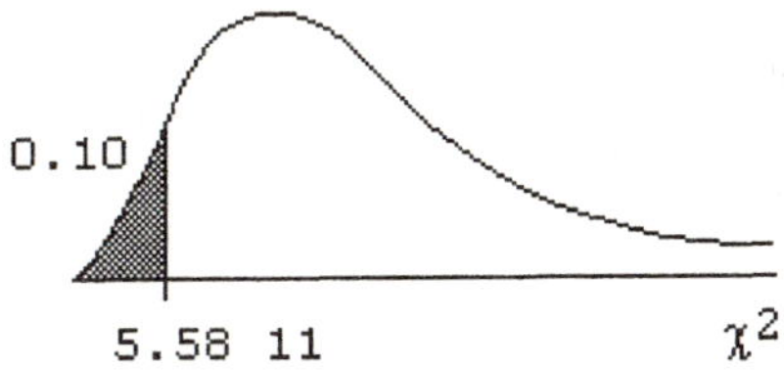

QUESTIONS 371 THROUGH 374 ARE BASED ON THE FOLLOWING INFORMATION:
A random sample of 51 observations was selected from a normally distributed population. The sample mean was $\bar{x}$ = 88.2, and the sample variance was s^2 =38.5. Suppose you will use this sample to determine whether there is sufficient reason to conclude that the population standard deviation is not equal to 8.2 at the 0.05 level of significance.

371. State the null and alternative hypotheses.

ANSWER:
$H_o: \sigma = 8.2$ and $H_a: \sigma \neq 8.2$

372. Calculate the value of the test statistic.

ANSWER:
$\chi^{2*} = (n-1)s^2/\sigma^2 = (50)(38.5)/(8.2)^2$ = 28.63

373. Complete the hypothesis test using the *p*-value approach.

ANSWER:

P = *p*-value = $2 \cdot P(\chi^2 < 28.63 \mid df = 50)$. Since $0.005 < 1/2\,\mathbf{P} < 0.01$; then, $0.01 < \mathbf{P} < 0.02$
Since *p*-value < α = 0.05, reject H_o. There is sufficient reason to conclude that the population standard deviation is not equal to 8.2, at the 0.05 level of significance

374. Complete the hypothesis test using the classical approach.

ANSWER:
The critical values are χ^2 (50, 0.975) = 32.4, and χ^2 (50, 0.025) = 71.4 as shown below.

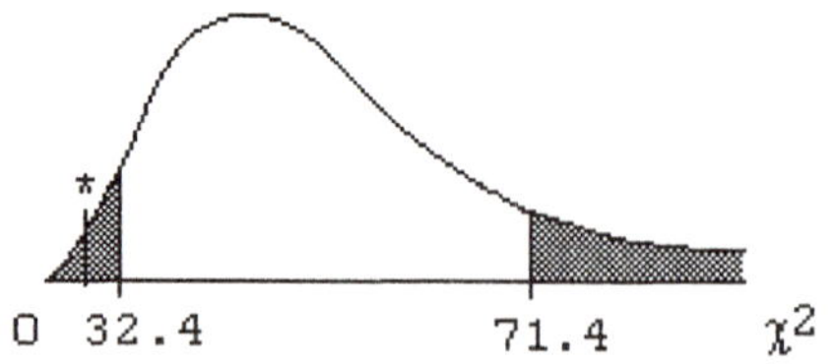

Since the test statistic χ^{2*} = 28.63 < 32.4, it falls in the rejection region and H_o is rejected. We reach the same conclusion as stated in question 373.

QUESTIONS 375 THROUGH 378 ARE BASED ON THE FOLLOWING INFORMATION:
The standard deviation of weights of certain 64.0-oz cans of tomato soup filled by a machine was 0.28 oz. A random sample of 20 cans showed a standard deviation of 0.38 oz. Suppose you will use this sample to determine whether there is an apparent increase in variability at the 0.10 level of significance. Assume can weight is normally distributed.

375. State the null and alternative hypotheses.

ANSWER:
$H_o: \sigma = 0.28\ (\leq 0)$ vs. $H_a: \sigma > 0.28$

376. Calculate the value of the test statistic.

ANSWER:
$\chi^{2*} = (n-1)s^2/\sigma^2 = (19)(0.38)^2/(0.28)^2 = 34.99$

377. Complete the hypothesis test using the *p*-value approach.

ANSWER:

P = *p*-value = $P(\chi^2 > 34.99 \mid df = 19)$. Then, $0.01 < \mathbf{P} < 0.025$
Since *p*-value < α = 0.10, reject H_o. There is sufficient reason to conclude that the apparent increase in variability is significant at the 0.10 level of significance

378. Complete the hypothesis test using the classical approach.

ANSWER:
The critical value is χ^2 (19, 0.10) = 27.2 as shown below.

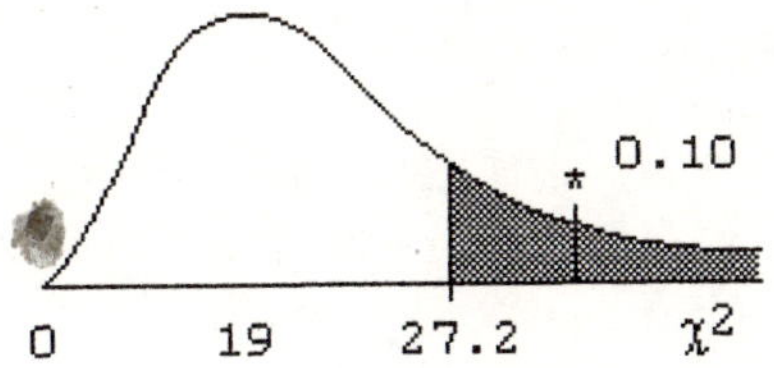

Since the test statistic $\chi^{2*} = 34.99 > 27.2$, it falls in the rejection region and H_o is rejected. We reach the same conclusion as stated in question 377.

QUESTIONS 379 THROUGH 382 ARE BASED ON THE FOLLOWING INFORMATION:
General Motors claims that their Malibu 2005 model has mean miles per gallon equal to 38 with a standard deviation equal to 4.0 mi. A random sample of 15 such cars and produced the following miles per gallon: 36.0, 37.0, 41.5, 44.0, 33.0, 31.0, 35.0, 34.5, 37.0, 41.5, 39.0, 41.5, 34.0, 29.0, and 36.5. Assume normality. Suppose you wish to use this sample to test the hypothesis that the standard deviation differs from 3.8 at level of significance $\alpha = 0.05$.

379. State the null and alternative hypotheses.

ANSWER:
$H_o: \sigma = 3.8$ and $H_a: \sigma \neq 3.8$

380. Use computer to provide summary statistics.

ANSWER:

Summary measures		
	Sample size	14
	Sample mean	36.714
	Sample standard deviation	4.375

381. Use computer to complete the hypothesis test using the *p*-value approach.

ANSWER:

Test of stdev=3.8 versus two-tailed alternative		
	Hypothesized standard deviation	3.800
	Sample standard deviation	4.375
	Ratio of standard deviations	1.151
	Degrees of freedom	13
	Chi-square test statistic	17.234
	p-value	0.479

Since *p*-value = 0.479 > $\alpha = 0.05$, we fail to reject H_o. There is not sufficient evidence to conclude that the population standard deviation is significantly different from 3.8 at the 0.05 level of significance. In other words, there is not sufficient reason to contradict the manufacturer's claim about the standard deviation, at the 0.05 level of significance.

382. Complete the hypothesis test using the classical approach.

ANSWER:

The critical values are χ^2 (14, 0.975) = 5.63 and χ^2(14, 0.025) = 26.1 as shown below.

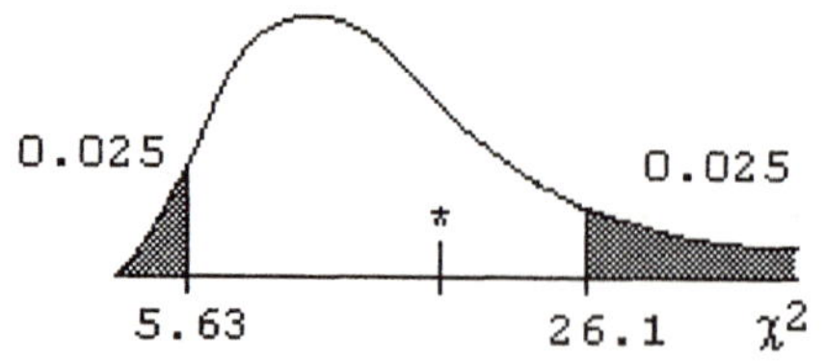

Since the test statistic χ^{2*} = 17.234 does not fall in the rejection region, we fail to reject H_o at α = 0.05. We reach the same conclusion as stated in question 381.

Chapter 10

INFERENCES INVOLVING TWO POPULATIONS

Sections 10.1 and 10.2

True-False Questions

1. Pretest versus posttest (before versus after) studies are usually independent samples.
ANSWER: F

2. In some experiments it is possible to collect data using either independent samples or dependent samples.
ANSWER: T

3. In independent sampling, a source can be a person, an object, or anything that yields a piece of data. However, in dependent sampling, a source must be a person.
ANSWER: F

4. Dependent samples result from using paired subjects.
ANSWER: T

5. If two samples have the same size, the samples may or may not be independent.
ANSWER: T

6. Two dependent samples may have different sample sizes.
ANSWER: F

7. Independent samples are obtained by using unrelated sets of subjects.
ANSWER: T

8. If two samples have the same size, the samples must be dependent.
ANSWER: F

Chapter 10

Multiple-Choice Questions

9. Which of the following statements is false?

A) When comparing two populations, we need two samples, one from each population.
B) If the same set of sources or related sets are used to obtain the data representing two populations, we have dependent samples.
C) If two unrelated sets of sources are used to obtain the data representing two populations, one set from each population, we have independent samples.
D) None of the above
ANSWER: **D**

10. Which of the following statements is false?

A) Comparing the final test scores of male and female students in your statistics class is an example of two dependent samples.
B) Pretest versus posttest (before versus after) studies usually use dependent samples.
C) Studies involving identical twins result in dependent samples of data.
D) None of the above
ANSWER: **A**

11. A political analyst in Michigan surveys a random sample of registered Democrats and compares the results with those obtained from a random sample of registered Republicans. This would be an example of:

A) dependent samples.
B) independent samples.
C) independent samples only if the sample sizes are equal.
D) dependent samples only if the sample sizes are equal.
ANSWER: **B**

12. Studies that involve paired subjects deal with

A) dating service samples.
B) independent samples.
C) dependent samples.
D) None of the above.
ANSWER: **C**

Applied and Computational Questions

13. Describe how one could select two dependent samples from among his/her co-workers in General Motors to compare their starting salaries after graduation from high school to their salaries when they continue working at GM and reach the age of 40.

ANSWER:
Randomly select a set of co-workers, obtain their two salaries (starting salaries after graduation from high school and salaries at the age of 40) from each of the selected co-workers.

14. Explain why studies involving identical twins result in dependent samples of data.

ANSWER:
Identical twins are so much alike that the information obtained from one would not be independent from the information obtained from the other twin.

15. Describe how one could select two independent samples from among his/her co-workers to compare the salaries of female and male workers.

ANSWER:
Divide the co-workers into two groups, males and females. Randomly select a sample from each of the two groups.

16. Twenty people were selected to participate in a psychology experiment. They answered a short multiple-choice quiz about their attitudes on abortion and then viewed a 50-minute film. The following day the same 20 people were asked to answer a follow-up questionnaire about their attitudes. At the completion of the experiment, the experimenter will have two sets of scores. Do these two samples represent dependent or independent samples? Explain.

ANSWER:
These two samples represent dependent samples. The two sets of data were obtained from the same set of 20 people, each person providing one piece of data for each sample.

17. An experiment is designed to study the effect diet has on the uric acid level. Thirty people are used for the study. Fifteen are randomly selected and given a junk-food diet. The other fifteen received a high-fiber, low-fat diet. Uric acid levels of the two groups are determined. Do the resulting sets of data represent dependent or independent samples? Explain.

ANSWER:
The resulting sets of data represent independent samples. The two samples are from two separate unrelated sets of fifteen people.

<u>**QUESTIONS 18 AND 19 ARE BASED ON THE FOLLOWING INFORMATION:**</u>
An auto insurance company is concerned that body shop "A" charges more for repair work than body shop "B" charges. It plans to send 20 cars to each body shop and obtain separate estimates for the repairs needed for each car.

18. How can the company do this and obtain independent samples? Explain in detail.

ANSWER:
Independent samples will result if the company sent a set of 20 cars to body shop "A", and another set of 20 cars to body shop "B". This means the company sent 40 cars, received 40 estimates (one estimate for each car).

19. How can the company do this and obtain dependent samples? Explain in detail.

ANSWER:
Dependent samples will result if the company sent the same set of 20 cars to both body shops "A" and "B". This means the company sent 20 cars, received 40 estimates (two estimates for each car, one from each body shop).

QUESTIONS 20 AND 21 ARE BASED ON THE FOLLOWING INFORMATION:
Suppose that 800 students in Michigan State University are taking elementary statistics this semester. Two samples of size 50 are needed in order to test some pre-course skill against the same skill after the students complete the course.

20. Describe how you would obtain your samples if you were to use dependent samples.

ANSWER:
Randomly select 50 students from the 800 students and take a measure of this skill from each of these 50 both before and after the course. This leads to 100 measurements from 50 students (two from each student).

21. Describe how you would obtain your samples if you were to use independent samples.

ANSWER:
Obtain a measurement of this skill from 50 randomly selected students before the course begins. Then obtain another sample of 50 randomly selected from those completing the course. This leads to 100 measurements from 100 students (one from each student).

Section 10.3

True-False Questions

22. In constructing a confidence interval for the mean difference in paired data we see that as the sample size increases the width of the interval also increases.
ANSWER: F

23. Suppose we were testing the hypothesis $H_o : \mu_d = 0(\geq)$, vs. $H_a : \mu_d < 0$, where $d = x_1 - x_2$. If we reject H_o, then this would indicate that the mean of population 2 is less than the mean of population 1.
ANSWER: F

24. In dependent sampling, two sets of data are combined into one set using $d = x_1 - x_2$. In this case, $\sum d / n = \bar{x}_1 - \bar{x}_2$.
ANSWER: T

25. Consider a right-tail hypothesis test concerning the mean difference between two dependent samples where $d = x_1 - x_2$. If we were to interchange the two populations, then the test would change to a left-tail hypothesis test.
ANSWER: T

26. In dependent sampling, the two data values, one from each set, that come from the same source are called paired data.
ANSWER: T

27. When the means of two unrelated samples are used to compare two populations, we are dealing with two dependent means.
ANSWER: F

28. The use of paired data often allows for the control of immeasurable or confounding variables because each pair is subjected to these confounding effects equally.
ANSWER: T

29. The *z*-distribution is used when two dependent means are to be compared.
ANSWER: F

30. In constructing a confidence interval for the mean difference in paired data, the interval increases in width when the sample size is increased.
ANSWER: F

31. When paired observations are randomly selected from normal populations, the paired difference, $d = x_1 - x_2$, will be normally distributed about a mean μ_d with a standard deviation σ_d.
ANSWER: T

32. The difference between two population means, when dependent samples are used, is equivalent to the mean of the paired differences.
ANSWER: T

33. The procedures for comparing two population means are based on the relationship between two sets of sample data, one sample from each population. When dependent samples are involved, the data are thought of as "paired data", where the pairs of data values are compared directly to each other by using the difference in their numerical values.
ANSWER: T

34. When paired observations are randomly selected from normal populations, the paired difference, $d = x_1 - x_2$, will be approximately normally distributed about a mean μ_d with a standard deviation of σ_d. In this situation, the *z*-test for one mean is applied.
ANSWER: F

35. In a confidence interval for the mean difference in paired data, the interval increases in width when the sample size is increased.
ANSWER: F

36. When paired observations are randomly selected from normal populations, the paired difference, $d = x_1 - x_2$, will be approximately normally distributed about a mean μ_d with a standard deviation of σ_d. In this situation, the *z*-test for one mean is applied with *df* =*n*-1, where *n* is the number of matched pairs of data.
ANSWER: T

37. When the means of two unrelated samples are used to compare two populations, we are dealing with two dependent means.
ANSWER: F

38. The *z*-distribution is used when two dependent means are to be compared.
ANSWER: F

Multiple-Choice Questions

39. When constructing a confidence interval for the mean difference in paired data, which of the following symbols indicates the middle point of the interval?

A) μ_d
B) σ_d
C) $\bar{d}$
D) s_d

ANSWER: C

40. A statistics professor is testing the claim that the use of computers will help students to better understand elementary statistics concepts. Based on this claim, if $d = X_{comp.} - X_{no\ comp.}$, which of the following would be the correct null and alternative hypotheses?

A) $H_o : \mu_d = 0, H_a : \mu_d \neq 0$
B) $H_o : \mu_d = 0(\leq), H_a : \mu_d > 0$
C) $H_o : \mu_d > 0, H_a : \mu_d = 0(\leq)$
D) $H_o : \mu_d < 0, H_a : \mu_d = 0(\geq)$

ANSWER: B

41. A research laboratory interested in the medicinal effect of herbs is testing the claim that a particular herb will reduce stress-related symptoms in adults. Based on this claim and assuming $d = X_{\text{after}} - X_{\text{before}}$, which of the following would be the correct null and alternative hypotheses?

A) $H_o: \mu_d = 0, H_a: \mu_d \neq 0$
B) $H_o: \mu_d > 0, H_a: \mu_d = 0(\leq)$
C) $H_o: \mu_d = 0(\leq), H_a: \mu_d > 0$
D) $H_o: \mu_d < 0, H_a: \mu_d = 0(\geq)$
ANSWER: C

42. You plan to test the dependent sampling claim: "a particular weight loss program is effective in weight reduction." What would be the null hypothesis, if $d = X_{\text{after}} - X_{\text{before}}$?

A) $H_o: \mu_d = 0$
B) $H_o: \mu_d = 0(\geq)$
C) $H_o: \mu_d \neq 0$
D) $H_o: \mu_d = 0(\leq)$
ANSWER: B

43. When using paired differences to test the mean difference between two dependent samples, which of the following is the point estimate of μ_d?

A) $\bar{d}$
B) $\mu_1 - \mu_2$
C) $x_1 - x_2$
D) $\sum d$
ANSWER: A

44. Which of the following statements is false?

A) When we test a null hypothesis about the mean difference, μ_d of two population means using two dependent samples, the test statistic used will be the difference between the sample mean *d* and the hypothesized value of μ_d, divided by the estimated standard error.
B) The assumption for inferences about the mean of paired differences μ_d is that the paired data are randomly selected from normally distributed populations.
C) The assumption for inferences about the mean of paired differences μ_d is that the paired data are randomly selected from *t*- distributed populations.
D) None of the above
ANSWER: C

Chapter 10

Short-Answer Questions

45. What is the assumption for inferences about the mean of paired differences μ_d?

ANSWER:
The paired data are randomly selected from normally distributed populations.

46. Consider *n* pairs represented by (x_i, y_i) for *i* = 1,2,...,*n*. Let $d_i = x_i - y_i$. If $\overline{x}$ is the mean of *x*-values and $\overline{y}$ is the mean of the *y*-values, express $\overline{d}$ in terms of $\overline{x}$ and $\overline{y}$.

ANSWER:
$\overline{d} = \overline{x} - \overline{y}$

47. In order to compare two scales, 30 objects are weighed on both scales. Each object would then have two weight values (one from scale 1 and one from scale 2). Based on the nature of the differences in the two weight measurements for the 30 objects, the two scales may be compared. Do these samples represent dependent or independent samples?

ANSWER:
Dependent samples

48. State the null and alternative hypotheses that would be used to test each of the following claims:
    a. The mean weight loss due to a special diet is at least 5 pounds. Assume dependent sampling was used.
    b. The mean adult body temperature is not 98.6°F.

ANSWER:
a. $H_o: \mu_d = 5 (\geq), H_a: \mu_d < 5$
b. $H_o: \mu_d = 98.6, H_a: \mu_d \neq 98.6$

49. Define the paired difference *d*, and then state the null hypothesis, H_o, and the alternative hypothesis, H_a, that would be used to test the claim "There is an increase in the mean difference between posttest and pretest scores for an introduction to macroeconomics course."

ANSWER:
Let *d* = posttest score – pretest score. Then, $H_o: \mu_d = 0\ (\leq)$ and $H_a: \mu_d > 0$

50. Define the paired difference *d*, and then state the null hypothesis, H_o, and the alternative hypothesis, H_a, that would be used to test the claim "As a result of a computer training session in Microsoft Office 2003, it is believed that the mean of the difference in performance scores will not be zero."

ANSWER:
Let *d* = scores after computer training session - scores before computer training session. Then, $H_o: \mu_d = 0$ and $H_a: \mu_d \neq 0$.

51. Define the paired difference *d*, and then state the null hypothesis, H_o, and the alternative hypothesis, H_a, that would be used to test the claim "The mean of the differences between pre and post self-esteem scores showed improvement after involvement in a community service project to build a playground for children."

ANSWER:
Let *d* = post self-esteem scores – pre self-esteem scores.
Then, $H_o: \mu_d = 0\ (\leq)$ and $H_a: \mu_d > 0$

52. Define the paired difference *d*, and then state the null hypothesis, H_o, and the alternative hypothesis, H_a, that would be used to test the claim "The mean of the differences between the posttest and the pretest scores is greater than 10."

ANSWER:
Let *d* = posttest score – pretest score. Then, $H_o: \mu_d = 10\ (\leq)$ and $H_a: \mu_d > 10$.

53. Define the paired difference *d*, and then state the null hypothesis, H_o, and the alternative hypothesis, H_a, that would be used to test the claim "The mean weight loss experienced by people on a new diet plan was more than 25 lb."

ANSWER:
Let *d* = weight before diet plan – weight after diet plan. Then,
$H_o: \mu_d = 25\ (\leq)$ and $H_a: \mu_d > 25$.

54. Define the paired difference *d*, and then state the null hypothesis, H_o, and the alternative hypothesis, H_a, that would be used to test the claim "The mean difference in the home reassessments from the two town assessors was no more than $800."

ANSWER:
Let *d* = home reassessment from first assessor - home reassessment from second assessor. Then, $H_o: \mu_d = 800\ (\leq)$ and $H_a: \mu_d > 800$

Applied and Computational Questions

55. Two different makes of stopwatches were used to time 12 different runners over a particular course. Using the times in seconds shown in the table below, find a 95% confidence interval for the mean time difference where *d* = Type 1 – Type 2.

	Runner											
Stopwatch	1	2	3	4	5	6	7	8	9	10	11	12
Type 1	59	49	64	60	54	47	49	58	66	76	70	66
Type 2	57	46	63	60	50	48	54	54	60	72	72	66

ANSWER:
(-0.65 to 3.31)

QUESTIONS 56 AND 57 ARE BASED ON THE FOLLOWING INFORMATION:
The exercise capacity of an individual is measured by the number of minutes the individual can exercise before certain medical criteria are met. The exercise capacity before and after basic training were measured for 20 marines. A summary of the data was provided as follows: $\sum d = 65$, $\sum d^2 = 1076$, where d = after capacity – before capacity. Assume that you wish to test $H_o : \mu_d = 0(\leq)$ vs. $H_a : \mu_d > 0$.

56. Calculate the test statistic.

ANSWER:
Value of the test statistic: $t^* = 2.15$.

57. Give a bound on the p-value.

ANSWER:
Using the Table of critical values of Student's t-distribution, we have: $0.01 < p < 0.025$.
Using the Table of probability values for Student's t-distribution, we have: $0.02 < P < 0.025$.

QUESTIONS 58 AND 59 ARE BASED ON THE FOLLOWING INFORMATION:
Consider the following paired data.

A	5	4	3	4	1
B	2	1	5	4	3

58. Calculate the differences $d = A - B$.

ANSWER:
The differences $d = A - B$ are 3, 3, -2, 0, and -2, respectively.

59. Calculate $\sum d$, $\sum d^2$, $\left(\sum d\right)^2$, $\bar{d}$, and s_d.

ANSWER:
$\sum d = 2$, $\sum d^2 = 26$, $\left(\sum d\right)^2 = 4$, $\bar{d} = 0.4$, $s_d = 2.5$

QUESTIONS 60 THROUGH 62 ARE BASED ON THE FOLLOWING INFORMATION:
Ten men compared two brands of razors. One side of the face was shaved by brand *A*, and the other was shaved by brand *B*. A "smoothness score" (from 1 to 10) was given by each person for each side. The side on which a given shaver was used was assigned by the flip of a coin and the smoothness scores are shown below.

					Man					
Razors	1	2	3	4	5	6	7	8	9	10
Brand A score	7	8	3	5	4	4	9	8	7	4
Brand B score	5	6	3	4	6	5	6	7	3	4

60. Calculate the differences d = *A* score – *B* score.

ANSWER:
D = 2, 2, 0, 1, -2, -1, 3, 1, 4, and 0.

61. Calculate $\sum d = 10, \sum d^2, (\sum d)^2, \bar{d}, s_d$.

ANSWER:

$\sum d = 10, \sum d^2 = 40, (\sum d)^2 = 100, \bar{d} = 1, s_d = 1.8257$

62. Test $H_o : \mu_d = 0$ vs. $H_a : \mu_d \neq 0$ by giving the critical region, t^*, and your conclusion. (Use α = 0.01).

ANSWER:

Critical region: $t < -3.25$ or $t > 3.25$; Value of the test statistic: $t^* = 1.732$; Conclusion: Unable to reject the null hypothesis.

QUESTIONS 63 THROUGH 66 ARE BASED ON THE FOLLOWING INFORMATION:

Two different testing agencies develop their own achievement tests for the same subject. Both tests are given to the same random sample of 10 students. The results are given below:

	Student									
Tests	1	2	3	4	5	6	7	8	9	10
Test A	83	79	96	87	93	90	77	73	85	84
Test B	90	88	98	83	97	94	82	80	92	88

Suppose we were to test the claim that there is no difference in the mean score for the two tests at the 0.01 level of significance.

63. Calculate the differences *d* = Test *A* – Test *B*.

ANSWER:

d = -7, -9, -2, 4, -4, -4, -5, -7, -7, and -4.Critical region: $t < -3.25$ or $t > 3.25$

64. Calculate $\sum d, \sum d^2, (\sum d)^2, \bar{d}$, and s_d.

ANSWER:

$\sum d = -45, \sum d^2 = 321, (\sum d)^2 = 2025, \bar{d} = -4.5, s_d = 3.6286$

65. State the null and alternative hypotheses.

ANSWER:

$H_o : \mu_d = 0$ vs. $H_a : \mu_d \neq 0$

66. Determine the critical region, the computed value of the test statistic, and the decision reached.

ANSWER:

Critical region: $t < -3.25$ or $t > 3.25$; Value of the test statistic: $t^* = -3.922$; Decision: reject null.

67. State the null hypothesis, H_o, and the alternative hypothesis, H_a, that would be used to test the following claims: The mean difference between the posttest and pretest scores is greater than 12.

ANSWER:

$H_o : \mu_d = 12(\leq);\quad H_a : \mu_d > 12;$ d = posttest score – pretest score

68. State the null hypothesis, H_o, and the alternative hypothesis, H_a, that would be used to test the following claims: The mean weight gain, due to the change in diet for the laboratory animals, is at least 8 oz.

ANSWER:

$H_o : \mu_d = 8(\geq);\quad H_a : \mu_d < 8;$ d = weight after – weight before

QUESTIONS 69 AND 70 ARE BASED ON THE FOLLOWING INFORMATION:
The following data were obtained from an experiment designed to estimate the reduction in diastolic blood pressure using a sample of 8 people, as a result of following a salt-free diet for two weeks. Assume diastolic readings to be normally distributed, and let d = Before – After.

Before	94	107	88	93	103	96	89	111
After	93	103	90	93	102	97	89	106

69. What is the point estimate for the mean reduction in the diastolic reading after two weeks on this diet?

ANSWER:

$n = 8, \sum d = 8, \sum d^2 = 48,$ Point estimate: $\bar{d} = \sum d / n$ = 8 / 8 = 1.0

70. Find the 98% confidence interval for the mean reduction in the diastolic reading.

ANSWER:

Normality indicated. Since $n = 8, \bar{d} = 1, s_d = 2.39, 1 - \alpha = 0.98,$ then df = 7, $\alpha / 2$ = 0.01, and $t(7, 0.01) = 3.00$.

Hence, $E = t(df, \alpha/2) \cdot (s_d / \sqrt{n}) = (3.00)(2.39 / \sqrt{8})$ = (3.00)(0.845) = 2.535. Then we get $\bar{d} \pm E = 1 \pm 2.535$, and the 98% confidence interval for μ_d is -1.535 to 3.535.

QUESTIONS 71 AND 72 ARE BASED ON THE FOLLOWING INFORMATION:
A sociologist is studying the effects of a certain motion picture film on the attitudes of white men toward black men. Twelve white men were randomly selected and asked to fill out a questionnaire before and after viewing the film. The scores received by the 12 men are shown in the table below. Assume the questionnaire scores are normally distributed.

Before	11	13	19	13	9	8	14	13	18	21	7	12
After	6	9	12	16	4	5	10	14	13	17	7	11

71. Construct a 95% confidence interval for the mean shift in score that takes place when this film is viewed.

ANSWER:
Sample statistics: $n = 12, \sum d = -34, \sum d^2 = 192, \bar{d} = -2.833, s_d = 2.949$, where *d* = after – before; the mean difference shift in score that takes place when a certain film is viewed. Normality indicated. Since *n* = 12, $1 - \alpha$ = 0.95, then *df* = 11, $\alpha/2$ = 0.025, and $t(11, 0.025) = 2.20$ .Hence,

$E = t(df, \alpha/2) \cdot (s_d / \sqrt{n}) = (2.20)(2.949/\sqrt{12})$ = (2.20)(0.8513) = 1.873. Then, we get $\bar{d} \pm E = -2.833 \pm 1.873$, and the 0.95 interval for μ_d is –4.706 to –0.96.

72. Use the confidence interval in question 71 to test $H_o : \mu_d = 0$ vs. $H_a : \mu_d \neq 0$ at α =0.05.

ANSWER:
Since the 95% confidence interval for μ_d does not include the hypothesized value 0, we reject H_o at α =0.05 and conclude that there is a difference in the mean score of the attitude of white men toward black men after viewing this motion picture film.

73. State the null hypothesis, H_o, and the alternative hypothesis, H_a, that would be used to test the following claims: The mean weight loss experienced by people on a new diet plan was no less than 15 lbs.

ANSWER:
$H_o : \mu_d = 15(\geq)$; $H_a : \mu_d < 15$; *d* = weight before – weight after

74. State the null hypothesis, H_o, and the alternative hypothesis, H_a, that would be used to test the following claims: The mean of the difference in performance scores due to special training session will not be zero.

ANSWER:
$H_o : \mu_d = 0$; $H_a : \mu_d \neq 0$; *d* = score after – score before

<u>QUESTIONS 75 THROUGH 78 ARE BASED ON THE FOLLOWING INFORMATION:</u>
The number of sit-ups that a person could do in one minute, both before and after a physical fitness course was recorded as shown below for ten randomly selected participants. Suppose you wish to determine whether a significant amount of improvement took place after the physical fitness course.

Before	30	23	25	29	27	25	30	45	34	26
After	31	27	25	35	34	35	32	52	50	43

75. State the null and alternative hypotheses.

ANSWER:
Let *d* = after – before; the mean difference between number of sit-ups a person can do before and after a physical fitness course. Then the null and alternative hypotheses are $H_o : \mu_d = 0$ vs. $H_a : \mu_d > 0$ (improvement).

76. Calculate the value of the test statistic.

ANSWER:
Normality assumed. Sample statistics: $n = 10,\ \bar{d} = 7.0,\ s_d = 5.8689$;
$t^* = (\bar{d} - \mu_d)/(s_d / \sqrt{n}) = (7.0 - 0.0)/(5.8689/\sqrt{10}) = 3.77.$

77. Test the hypotheses in question 75 at the 0.01 level of significance using the p - value approach.

ANSWER:
The p - value approach: **P** = $P(t > 3.77 \mid df = 9)$; Using the table of probability values for Student's t -distribution, we get 0.002 < **P** < 0.003. Since **P**<α , reject H_o and conclude that there is an improvement after the course.

78. Test the hypotheses in question 75 at the 0.01 level of significance Solve using the classical approach.

ANSWER:
The critical region is $t \geq 2.82$. Since t^* = 3.77 falls in the critical region, we reject H_o at α = 0.01 and conclude that there is an improvement after the course.

QUESTIONS 79 THROUGH 82 ARE BASED ON THE FOLLOWING INFORMATION:
Ten individuals with high cholesterol levels participated in a nutrition education session. The participants' cholesterol levels before and after the session were recorded as shown in the table below.

	Subject									
The session	1	2	3	4	5	6	7	8	9	10
Pre-session	300	284	255	240	260	295	315	265	280	245
Post-session	262	263	248	243	233	233	238	253	253	218

Let d = presession cholesterol level – postsession cholesterol level.
Suppose you wish to test the hypothesis that if participation in the nutrition education session lowers the cholesterol level. Assume normality.

79. State the null and alternative hypotheses.

ANSWER:
Let d = pre – post; the mean difference in cholesterol levels in pre and post education sessions. Then the null and alternative hypotheses are $H_o : \mu_d = 0 (\leq)$ vs. $H_a : \mu_d > 0$ (improvement).

80. Calculate the value of the test statistic.

ANSWER:
Normality indicated. Sample statistics: n = 10, $\bar{d} = 29.5,\ s_d = 24.369$;
$t^* = (\bar{d} - \mu_d)/(s_d / \sqrt{n}) = (29.5 - 0.0)/(24.369/\sqrt{10}) = 3.83.$

81. Test the hypotheses in question 79 at α = 0.05. Solve using the *p*-value approach.

ANSWER:
P = *p*-value = $P(t > 3.83 \mid df = 9)$; Using the table of probability values for Student's *t*-distribution, we have: 0.001 < **P** <0.003. Since **P** $< \alpha = 0.05$; reject H_o.

82. Test the hypotheses in question 79 at α = 0.05. Solve using the classical approach.

ANSWER:
The critical region is: $t \geq 1.83$. Since t^* = 3.83 falls in the critical region, we reject H_o at the 0.05 level of significance, and conclude that there is sufficient evidence that the education session does help to lower cholesterol levels.

83. Find the 95% confidence interval for μ_d given: *n* =25, $\bar{d}$ = 5.2, and s_d = 3.9.

ANSWER:
$t(df, \alpha/2)$ = $t(24, 0.025)$ = 2.06
E = $t(df, \alpha/2) \cdot s_d / \sqrt{n} = 2.06\,(3.9)/\sqrt{25}$ = 1.6068 $\approx$ 1.61
$\bar{d} \pm E$ = 5.2 $\pm$1.61. The lower and upper confidence limits are 3.59 and 6.81, respectively.

QUESTIONS 84 THROUGH 90 ARE BASED ON THE FOLLOWING INFORMATION:
Ten subjects with borderline-high cholesterol levels were recruited for a study. The study involved taking a nutrition education class. Cholesterol readings were taken before the class and three months after the class.

	Subject									
Ed. Class	1	2	3	4	5	6	7	8	9	10
Pre-class	238	293	253	298	258	282	243	263	278	313
Post-class	243	233	248	268	233	269	218	253	253	238

Let *d* = pre-class cholesterol- post-class cholesterol. Assume cholesterol readings to be normally distributed.

84. Use computer to provide summary measure for *d* = pre-class cholesterol- post-class cholesterol

ANSWER:

Summary measures for Pre-Class-Post Class		
	Sample size	10
	Sample mean	26.300
	Sample standard deviation	24.500

85. What is the mean $\bar{d}$ of the paired differences?

ANSWER:
$\bar{d}$ = 26.3

86. What is the standard deviation s_d of the paired differences?

ANSWER:
$s_d = 24.5$

87. Use computer to develop the 95% confidence interval for the mean amount of reduction in cholesterol readings resulting from taking the nutrition education class.

ANSWER:

Confidence interval for mean		
	Confidence level	95.0%
	Sample mean	26.300
	Std error of mean	7.747
	Degrees of freedom	9
	Lower limit	8.774
	Upper limit	43.826

88. The researcher who conducted the study believes that taking the nutrition education class is effective in reducing the cholesterol level. What are the appropriate null and alternative hypotheses?

ANSWER:
$H_o: \mu_d = 0\,(\le)$ and $H_a: \mu_d > 0$ (recall d = pre-class cholesterol- post-class cholesterol)

89. Use computer to test the hypotheses in question 88 at the 0.05 level of significance using the *p*-value approach.

ANSWER:

Test of mean<=0 versus one-tailed alternative		
	Hypothesized mean	0.000
	Sample mean	26.300
	Std error of mean	7.747
	Degrees of freedom	9
	t-test statistic	3.395
	p-value	0.004

Since *p*-value = 0.004 < α = 0.05, we reject H_o. There is sufficient evidence to support the researcher's claim that taking the nutrition education class is effective in reducing the cholesterol level.

90. Test the hypotheses in question 88 at the 0.05 level of significance using the classical approach.

ANSWER:
The critical value is $t(df, \alpha) = t(9, 0.05) = 1.83$. Since the value of the test statistic t^* = 3.395, we reject H_o at the 0.05 level of significance. We reach the same conclusion as stated in question 89.

QUESTIONS 91 THROUGH 97 ARE BASED ON THE FOLLOWING INFORMATION:
Salt-free diets are often prescribed to people with high blood pressure. The data shown below were obtained from an experiment designed to estimate the reduction in diastolic blood pressure as a result of following a salt-free diet for two weeks.

Before:	112	108	97	90	104	89	94	95	100	98
After:	107	104	98	90	103	91	94	94	102	96

Let d = diastolic blood pressure before diet – diastolic blood pressure after diet. Assume diastolic readings to be normally distributed.

91. Use computer to provide summary measure for d = before diet – after diet.

ANSWER:

Summary measures for Before-After		
	Sample size	10
	Sample mean	0.800
	Sample standard deviation	2.348

92. What is the point estimate for the mean reduction in the diastolic reading after two weeks on this diet?

ANSWER:
$\bar{d} = 0.80$

93. What is the standard deviation s_d of the paired differences?

ANSWER:
$s_d = 2.348$

94. Use computer to develop the 98% confidence interval for the mean reduction in the diastolic reading after two weeks on this diet.

ANSWER:

Confidence interval for mean		
	Confidence level	98.0%
	Sample mean	0.800
	Std error of mean	0.742
	Degrees of freedom	9
	Lower limit	-1.295
	Upper limit	2.895

95. If you are interested in determining whether a salt-free diet for two weeks is effective in reducing the diastolic blood pressure, state the appropriate null and alternative hypotheses.

ANSWER:
$H_o: \mu_d = 0\,(\leq)$ and $H_a: \mu_d > 0$ (recall d = before diet – after diet)

96. Use computer to test the hypotheses in question 95 at the 0.05 level of significance using the p-value approach.

ANSWER:

Test of mean<=0 versus one-tailed alternative		
	Hypothesized mean	0.000
	Sample mean	0.800
	Std error of mean	0.742
	Degrees of freedom	9
	t-test statistic	1.078
	p-value	0.155

Since *p*-value = 0.155 > α = 0.05, we fail to reject H_o. There is no sufficient evidence to indicate that a salt-free diet for two weeks is effective in reducing the diastolic blood pressure.

97. Test the hypotheses in question 95 at the 0.05 level of significance using the classical approach.

ANSWER:
The critical value is *t*(*df*, α) = *t*(9, 0.05) = 1.83 . Since the value of the test statistic t^* = 1.078, we fail to reject H_o at the 0.05 level of significance. We reach the same conclusion as stated in question 96.

<u>QUESTIONS 98 AND 99 ARE BASED ON THE FOLLOWING INFORMATION:</u>
Consider testing $H_o : \mu_d = 0 (\leq)$ vs. $H_a : \mu_d > 0$, with *n* =20 and t^* =1.95.

98. Place bounds on the *p*-value using the table of "critical values of Student's *t*-distribution" available in your textbook.

ANSWER:
P = *p*-value = *P*(*t* > 1.95 | *df* = 19) $\Rightarrow$0.025 < **P** < 0.05

99. Place bounds on the *p*-value using the table of "probability values for Student's *t*-distribution" available in your textbook.

ANSWER:
P = *p*-value = *P*(*t* > 1.95 | *df* = 19) $\Rightarrow$0.029 < **P** < 0.037

<u>QUESTIONS 100 AND 101 ARE BASED ON THE FOLLOWING INFORMATION:</u>
Consider testing $H_o : \mu_d = 0$ vs. $H_a : \mu_d \neq 0$, with *n* =25 and t^* = -2.27.

100. Place bounds on the *p*-value using the table of "critical values of Student's *t*-distribution" available in your textbook.

ANSWER:
P = *p*-value = *P*(*t* < -2.27 | *df* = 24) + *P*(*t* > 2.27 | *df* = 24) = 2. *P*(*t* > 2.27 | *df* = 24)
$\Rightarrow$ 0.01 <1/2 **P** < 0.025 $\Rightarrow$0.02 < **P** < 0.05

101. Place bounds on the *p*-value using the table of "probability values for Student's *t*-distribution" available in your textbook.

ANSWER:
P = *p*-value = $P(t < -2.27 \mid df = 24) + P(t > 2.27 \mid df = 24) = 2.\ P(t > 2.27 \mid df = 24)$
$\Rightarrow 0.015 < 1/2\ \mathbf{P} < 0.020 \Rightarrow 0.03 < \mathbf{P} < 0.04$

QUESTIONS 102 AND 103 ARE BASED ON THE FOLLOWING INFORMATION:

Consider testing $H_o : \mu_d = 0(\geq)$ vs. $H_a : \mu_d < 0$, with n =30 and t^* = -2.59.

102. Place bounds on the *p*-value for, using the table of "critical values of Student's *t*-distribution" available in your textbook.

ANSWER:
P = *p*-value = $P(t < -2.59 \mid df = 29) = P(t > 2.59 \mid df = 29) \Rightarrow 0.005 < \mathbf{P} < 0.01$

103. Place bounds on the *p*-value using the table of "probability values for Student's *t*-distribution" available in your textbook.

ANSWER:
P = *p*-value = $P(t < -2.59 \mid df = 29) = P(t > 2.59 \mid df = 29) \Rightarrow 0.007 < \mathbf{P} < 0.009$

QUESTIONS 104 AND 105 ARE BASED ON THE FOLLOWING INFORMATION:

Consider testing $H_o : \mu_d = 1.0(\leq)$ vs. $H_a : \mu_d > 1.0$, with n =10 and t^* =3.63.

104. Place bounds on the *p*-value for, using the table of "critical values of Student's *t*-distribution critical values of Student's *t*-distribution" available in your textbook.

ANSWER:
P = *p*-value = $P(t > 3.63 \mid df = 9) \Rightarrow \mathbf{P} < 0.005$

105. Place bounds on the *p*-value using the table of "probability values for Student's *t*-distribution" available in your textbook.

ANSWER:
P = *p*-value = $P(t > 3.63 \mid df = 9) \Rightarrow 0.002 < \mathbf{P} < 0.004$

106. Determine the test criteria that would be used to test $H_o : \mu_d = 0(\leq)$ vs. $H_a : \mu_d > 0$, with n =15 and $\alpha = 0.05$, using the classical approach.

ANSWER
df = 14

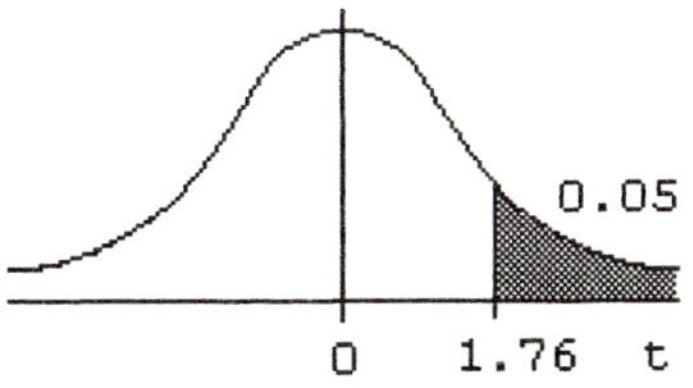

107. Determine the test criteria that would be used to test $H_o : \mu_d = 0$ vs. $H_a : \mu_d \neq 0$, with n =25 and $\alpha = 0.05$, using the classical approach

ANSWER:
df = 24

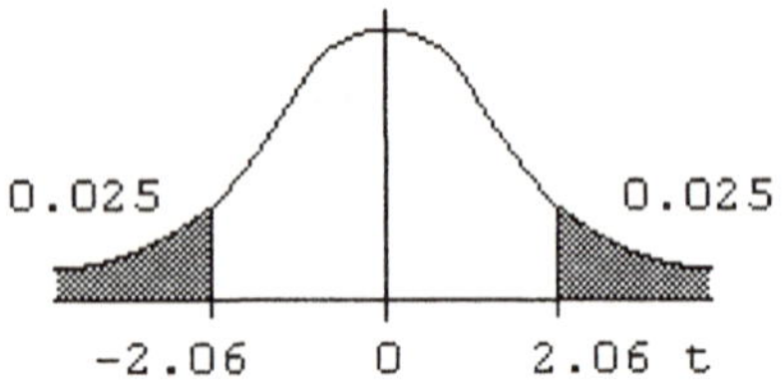

108. Determine the test criteria that would be used $H_o : \mu_d = 0(\geq)$ vs. $H_a : \mu_d < 0$, with n =12 and $\alpha = 0.10$, using the classical approach

ANSWER:
df = 11

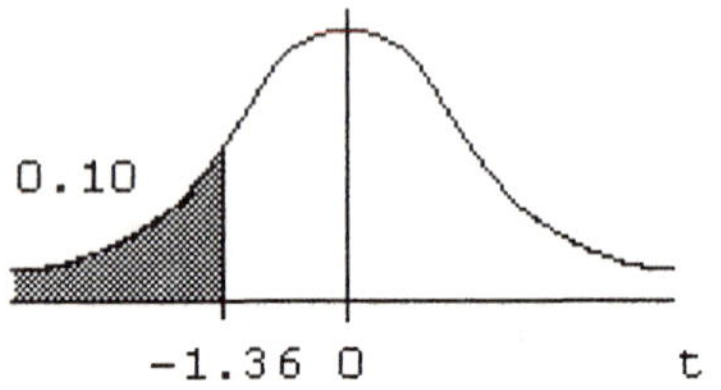

109. Determine the test criteria that would be used to test $H_o : \mu_d = 1.0(\leq)$ vs. $H_a : \mu_d > 1.0$, with n =18 and $\alpha = 0.01$, using the classical approach.

ANSWER:
df = 17

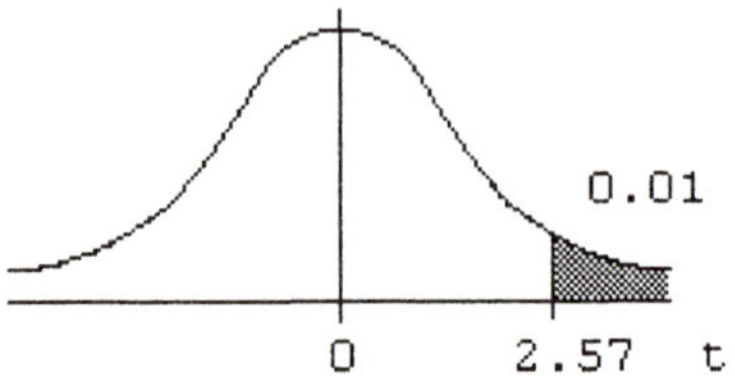

Section 10.4

True-False Questions

110. Confidence interval for the difference between the means of two populations using independent sampling may contain negative values.
ANSWER: T

111. With independent sampling, the sampling distribution of $\bar{x}_1 - \bar{x}_2$ is always normal.
ANSWER: F

112. If independent samples are drawn from two large populations, then the sampling distribution of $\bar{x}_1 - \bar{x}_2$ will be normally distributed.
ANSWER: F

113. In comparing two independent means when the σ's are unknown, we may use the standard normal distribution.
ANSWER: F

114. When making inferences about the difference between two independent means for the case when the number of degrees of freedom is estimated, the number of degrees of freedom for the critical value of *t* is equal to the smaller of $n_1 - 1$ or $n_2 - 1$.
ANSWER: T

115. If we are testing for the difference between two independent population means, it is assumed that the two populations are approximately normal and have equal variances.
ANSWER: T

116. A hypothesized difference between two population means, $\mu_1 - \mu_2$, must be zero in order to be able to make inferences about that difference.
ANSWER: F

117. When we test a null hypothesis about the difference between two population means, using two independent samples, the test statistic used will be the difference between the observed difference of the sample means and the hypothesized difference of the population means, divided by the estimated standard error.
ANSWER: T

118. A hypothesized difference between two population means, $\mu_1 - \mu_2$, can be any specified value. The most common value specified is zero; however, the difference can be nonzero.
ANSWER: T

119. In comparing two independent means when the σ's are unknown, we need to use the standard normal distribution.
ANSWER: F

Chapter 10

Multiple-Choice Questions

120. If two independent samples are used in a hypothesis test concerning the difference between population means for which the combined degrees of freedom is 20, which of the following could *not* be true about the sample sizes n_1 and n_2?

A) $n_1 = 12$ and $n_2 = 8$

B) $n_1 = 12$ and $n_2 = 10$

C) $n_1 = 13$ and $n_2 = 9$

D) Cannot be determined from the given information

ANSWER: A

121. If two independent samples are used in a hypothesis test concerning the difference between population means for which the combined degrees of freedom is 25, which of the following is true about the sample sizes n_1 and n_2?

A) $n_1 = 12$ and $n_2 = 13$

B) $n_1 = 13$ and $n_2 = 14$

C) $n_1 = 15$ and $n_2 = 12$

D) Cannot be determined from the given information

ANSWER: B

122. The director of student services for a large urban university is interested in testing the claim that evening college students have a higher grade point average than that of day students. Based on this claim, which of the following would be the correct null and alternative hypotheses?

A) $H_o: \mu_e = \mu_d (\geq), H_a: \mu_e < \mu_d$

B) $H_o: \mu_e > \mu_d, H_a: \mu_e \leq \mu_d$

C) $H_o: \mu_e = \mu_d, H_a: \mu_e \neq \mu_d$

D) $H_o: \mu_e - \mu_d = 0 (\leq), H_a: \mu_e - \mu_d > 0$

ANSWER: D

123. Which of the following are the null and alternative hypotheses that would be used to test the following claim using independent sampling: the mean gasoline consumption of automobile model *A* is no more than the mean gasoline consumption of automobile model *B*?

A) $H_o: \mu_A - \mu_B = 0 (\geq), H_a: \mu_A - \mu_B \neq 0$

B) $H_o: \mu_A - \mu_B = 0 (\geq), H_a: \mu_A - \mu_B < 0$

C) $H_o: \mu_A - \mu_B = 0 (\leq), H_a: \mu_A - \mu_B > 0$

D) $H_o: \mu_A - \mu_B = 0 (\leq), H_a: \mu_A - \mu_B \neq 0$

ANSWER: C

124. Which of the following would be the alternative hypothesis that would be used to test the claim that the mean IQ of individuals in population *A* is significantly different from the mean IQ of individuals in population *B*, assuming independent sampling?

A) $H_a : \mu_A - \mu_B = 0$
B) $H_a : \mu_A - \mu_B > 0$
C) $H_a : \mu_A - \mu_B < 0$
D) $H_a : \mu_A - \mu_B \neq 0$
ANSWER: D

125. Which of the following is not one of the required assumptions stated in your textbook for inferences about the difference between two population means, $\mu_1 - \mu_2$, using two samples?

A) The samples are randomly selected from their respective populations.
B) The population variances are equal.
C) The populations are normally distributed populations
D) The samples are selected in an independent manner.
ANSWER: B

126. Which of the following statements is false if independent samples of sizes n_1 and n_2 are drawn randomly from large populations with means μ_1 and μ_2 and variances σ_1^2 and σ_2^2, respectively?

A) The sampling distribution of $\bar{x}_1 - \bar{x}_2$, has mean $\mu_{\bar{x}_1 - \bar{x}_2} = \mu_1 - \mu_2$.
B) The sampling distribution of $\bar{x}_1 - \bar{x}_2$, has standard error $\sigma_{\bar{x}_1 - \bar{x}_2} = \sqrt{\frac{\sigma_1^2}{n_1} + \frac{\sigma_2^2}{n_2}}$.
C) The sampling distribution of $\bar{x}_1 - \bar{x}_2$, will be normally distributed, regardless of the sample sizes, If both populations have normal distributions.
D) None of the above
ANSWER: D

127. Which of the following is not one of the required assumptions stated in your textbook for inferences about the difference between two population means, $\mu_1 - \mu_2$, using two samples?

A) The samples are randomly selected from their respective populations.
B) The populations are normally distributed
C) The sample sizes are equal
D) The samples are selected in an independent manner.
ANSWER: C

Chapter 10

Short-Answer Questions

128. What is the assumption for inferences about the difference between two means, $\mu_1 - \mu_2$?

ANSWER:
The samples are randomly selected from normally distributed populations, and the samples are selected in an independent manner.

129. A group of sheep, infested with tapeworms, are randomly divided into two groups as follows. Each sheep is assigned a number (1 through 20) and then 10 numbers are selected by drawing 10 slips of paper from a box having the numbers 1 through 20 written on them. The drawing divides the sheep into two groups. One group is given a placebo and the other is given an experimental drug. After six weeks the sheep are sacrificed and tapeworm counts are made. Do these samples represent dependent or independent samples?

ANSWER:
Independent samples

130. State the null and alternative hypotheses that would be used to test each of the following claims:
   a. The difference between two population means is at most 8, assuming independent sampling.
   b. The proportion of male students (*M*) who ride bicycles on campus is no different than the proportion of female students (*F*) who ride bicycles on campus.

ANSWER:
a. $H_o: \mu_A - \mu_B = 8(\le), H_a: \mu_A - \mu_B > 8$
b. $H_o: p_M - p_F = 0, H_a: p_M - p_F \ne 0$

131. State the null and alternative hypotheses that would be used to test the claim "There is a difference between the mean salary of professors at two Michigan universities, say, A and B."

ANSWER:
$H_o: \mu_A - \mu_B = 0$ and $H_a: \mu_A - \mu_B \ne 0$

132. State the null and alternative hypotheses that would be used to test the claim "The mean of population A is greater than the mean of population B."

ANSWER:
$H_o: \mu_A - \mu_B = 0\ (\le)$ and $H_a: \mu_A - \mu_B > 0$

133. State the null and alternative hypotheses that would be used to test the claim "The mean age of workers at General Motors is less than the mean age of workers at Ford."

ANSWER:
$H_o: \mu_{GM} - \mu_F = 0\ (\ge)$ and $H_a: \mu_{GM} - \mu_F < 0$

134. State the null and alternative hypotheses that would be used to test the claim "There is no difference in the mean number of hours spent studying per week between male and female college students."

ANSWER:
$H_o: \mu_M - \mu_F = 0$ and $H_a: \mu_M - \mu_F \neq 0$

Applied and Computational Questions

135. A survey was conducted to compare the mean cost of a meal at fast food restaurants in two different cities. With the data below, set a 90% confidence interval on $\mu_1 - \mu_2$.

City	n	$\bar{x}$	s
A	40	\$4.05	\$0.55
B	35	\$4.85	\$0.85

ANSWER:
(0.75 to 0.85)

136. Suppose two independent samples of equal size are selected from two populations and both having standard deviation σ = 10. What common sample size is needed so that $\bar{x}_1 - \bar{x}_2$ has a standard error equal to 2?

ANSWER:
n = 50

137. If two independent random samples (each of size 50) are selected from a standard normal distribution, find the probability that the sample means are within 0.5 units of one another.

ANSWER:
$P(-0.5 \leq \bar{x}_1 - \bar{x}_2 \leq 0.5) = P(-2.5 \leq z \leq 2.5) = 0.9876$

138. An experiment was designed to test the effectiveness of a short course that teaches diabetic self-care. Fifty diabetics were enrolled in the course, and 50 others served as a control group. Six months after the course, blood tests were made to determine the hemoglobin A1C levels. This test measures the blood sugar control over the past few months. Based on the results, give the p-value for testing the hypothesis $H_o: \mu_1 - \mu_2 = 0$ vs. $H_a: \mu_1 - \mu_2 < 0$, at α = 0.05.

Diabetic Course Group	$\bar{x}_1 = 5.9$, $s_1 = 0.5$
Control Group	$\bar{x}_2 = 7.0$, $s_2 = 0.7$

ANSWER:
The value of the test statistic is: $t^* = -9.04$, p – value < 0.005
Since p-value $< \alpha$, we reject the null hypothesis. There is no sufficient evidence to indicate that the short course was not effective.

Chapter 10

QUESTIONS 139 THROUGH 142 ARE BASED ON THE FOLLOWING INFORMATION:

Attitude toward mathematics was measured for two different groups. The attitude scores range from 0 to 80 with the higher scores indicating a more positive attitude. One group consisted of Elementary Education majors, and the other group consisted of majors from several other areas. The data are shown below:

Group (major)	*n*	$\overline{x}$	*s*
Elementary Education (1)	75	42.7	15.5
Non-Elementary Education (2)	110	49.3	17.0

139. Calculate the value of the test statistic.

ANSWER:
Value of the test statistic: t^* =-2.73

140. Give the *p*-value when testing $H_o: \mu_1 - \mu_2 = 0$ vs. $H_a: \mu_1 - \mu_2 < 0$.

ANSWER:
p -value = 0.0039

141. Give the critical region, and the conclusion for testing the hypotheses in question 140.

ANSWER:
Critical region: $t < -1.67$; Conclusion: Reject the null hypothesis.

142. Set a 95% confidence interval for $\mu_1 - \mu_2$.

ANSWER:
(-10.63 to -2.57)

QUESTIONS 143 AND 144 ARE BASED ON THE FOLLOWING INFORMATION:

A sample of size 60 is selected from population 1, with $\overline{x}_1$ = 15.4 and s_1 = 1.7. A sample of size 40 is selected from population 2, with $\overline{x}_2$ = 16.8 and s_2 = 2.0. Suppose we were to test the claim that there is no difference in the population means at the 0.05 level of significance.

143. State the null and alternative hypotheses.

ANSWER:
$H_o: \mu_1 - \mu_2 = 0$ vs. $H_a: \mu_1 - \mu_2 \neq 0$

144. Determine the critical region, the computed value of the test statistic, the decision reached, and conclusion.

ANSWER:
Critical region: $t \leq -2.03$ or $t \geq 2.03$; Value of the test statistic: t^* = –3.64; Decision: Reject H_o. Conclusion: There is a difference in the population means at the 0.05 level of significance

<u>**QUESTIONS 145 AND 146 ARE BASED ON THE FOLLOWING INFORMATION:**</u>
An experiment was conducted to compare the mean absorptions of two drugs in specimens of muscle tissue. Eighty tissue specimens were randomly divided into two equal groups. Each group was tested with one of the two drugs. The sample results were as follows: $\bar{x}_A = 8.2, \bar{x}_B = 8.8, \quad s_A = 0.12$ and $s_B = 0.11$. Assume both populations are normal.

145. Construct the 98% confidence interval for the difference in the mean absorption rates.

ANSWER:
The difference between the mean absorption rates for two drugs is $\mu_B - \mu_A$. Normality indicated. $n_A = 40, \bar{x}_A = 8.2, s_A = 0.12$, $n_B = 40, \bar{x}_B = 8.8, s_B = 0.11$ .Then, $\bar{x}_B - \bar{x}_A$ =0.6. Since 1-α = 0.98, then α /2 = 0.01, and t(39, 0.01)$\approx$ 2.42. [We used the conservative approach in calculating the degrees of freedom; $df = \min(df_1 = n_1 - 1, df_2 = n_2 - 1)$ =39]
$E = t(df, \alpha/2) \cdot \sqrt{(s_B^2/n_B) + (s_A^2/n_A)} = (2.42) \cdot \sqrt{(0.11^2/40) + (0.12^2/40)}$ =0.062. Then we have $(\bar{x}_B - \bar{x}_A) \pm E$ = 0.60 $\pm$ 0.062, and the 98% confidence interval for $\mu_B - \mu_A$ is 0.538 to 0.662.

146. Use the confidence interval in question 145 to test the hypothesis that there is a difference in the mean absorptions of the two drugs at α = 0.02.

ANSWER:
$H_o: \mu_A - \mu_B = 0$ vs. $H_a: \mu_A - \mu_B \neq 0$. Since the 98% confidence interval for $\mu_A - \mu_B$ does not include the hypothesized value 0, we reject H_o and conclude that there is a difference in the mean absorption of the two drugs.

147. The two independent samples shown in the following table were obtained in order to estimate the difference between the two population means. Construct the 98% confidence interval.

Sample A	9	10	10	9	9	8	9	11	8	7
Sample B	9	4	6	5	5	7	6	8	6	4

ANSWER:
Sample statistics:
A: $n = 10, \quad \bar{x} = 9.0, \quad s^2 = 1.333$
B: $n = 10, \quad \bar{x} = 6.0, \quad s^2 = 2.667$
The difference between two means is $\mu_A - \mu_B$.
Normality assumed. Using sample information given above; $\bar{x}_A - \bar{x}_B$ =3
Since 1-α = 0.98, then α /2 = 0.01, df = min($n_A - 1, n_B - 1$) = 9, and t(9, 0.01) = 2.82.
Hence,
$E = t(df, \alpha/2) \cdot \sqrt{(s_A^2/n_A) + (s_B^2/n_B)} = (2.82)\sqrt{(1.333/10) + (2.667/10)}$ = 1.78. Then,
$(\bar{x}_A - \bar{x}_B) \pm E$ = 3 $\pm$ 1.78, and the 98% confidence interval for $\mu_A - \mu_B$ is 1.22 to 4.78.

Chapter 10

148. State the null and alternative hypotheses that would be used to test the following claims. There is a difference between the mean ages of students at two different colleges.

ANSWER:

$H_o: \mu_1 - \mu_2 = 0$ vs. $H_a: \mu_1 - \mu_2 \neq 0$

149. State the null and alternative hypotheses that would be used to test the following claims. The mean of population 1 is greater than the mean of population 2.

ANSWER:

$H_o: \mu_1 - \mu_2 = 0(\leq)$ vs. $H_a: \mu_1 - \mu_2 > 0$

150. Determine the *p*-value for the hypothesis test of the difference between two means with unknown population variances given $H_a: \mu_1 - \mu_2 > 0$, with $n_1 = 8$, $n_2 = 12, t^* = 1.4$

ANSWER:

We will use the conservative approach to determine the degrees of freedom; namely, *df* = min($n_1 - 1, n_2 - 1$), and use the table of probability values for Student's *t* –distribution. Then, **P** = $P(t > 1.4 \mid df = 7)$ = 0.102

151. State the null and alternative hypotheses that would be used to test the following claims. The difference between the mean weights of two populations is less than 50 pounds.

ANSWER:

$H_o: \mu_1 - \mu_2 = 25(\geq)$ vs. $H_a: \mu_1 - \mu_2 < 25$

152. Determine the *p*-value for the hypothesis test of the difference between two means with unknown population variances given $H_a: \mu_1 - \mu_2 < 0$, with $n_1 = 18$, $n_2 = 11$, $t^* = -2.9$

ANSWER:

We will use the conservative approach to determine the degrees of freedom; namely, *df* = min($n_1 - 1, n_2 - 1$), and use the table of probability values for Student's *t* –distribution. Then **P** = $P(t < -2.9 \mid df = 10) = P(t > 2.9 \mid df = 10)$ = 0.008

153. Determine the *p*-value for the hypothesis test of the difference between two means with unknown population variances given $H_a: \mu_1 - \mu_2 \neq 0$, with $n_1 = 30$, $n_2 = 13$, $t^* = 1.6$

ANSWER:

We will use the conservative approach to determine the degrees of freedom; namely, *df* = min($n_1 - 1, n_2 - 1$), and use the table of probability values for Student's *t* –distribution. Then **P** = $2P(t > 1.6 \mid df = 12)$ = 2(0.068) = 0.136

154. Determine the *p*-value for the following hypothesis test for the difference between two means with unknown population variances.

$H_a: \mu_1 - \mu_2 \neq 5, \text{ with } n_1 = 26, \quad n_2 = 38, \quad t^* = -2.1$

ANSWER:
We will use the conservative approach to determine the degrees of freedom; namely, *df* = min($n_1 - 1, n_2 - 1$), and use the table of probability values for Student's *t* –distribution. Then **P** = $2P(t^* > 2.1 \mid df = 25)$ = 2(0.023) = 0.046.

QUESTIONS 155 THROUGH 158 ARE BASED ON THE FOLLOWING INFORMATION:
Suppose a random sample of 20 homes east of State Street in Big Rapids, Michigan has a mean selling price of $128,000 and a standard deviation of $4500, and a random sample of 20 homes west of State Street has a mean selling price of $125,000 and a standard deviation of $2500. Suppose that you wish to test that there is a significant difference between the selling prices of homes in these two areas of Big Rapids at the 0.05 level.

155. State the null and alternative hypotheses.

ANSWER:
The difference between the mean selling prices of homes in two areas of Big Rapids is $\mu_E - \mu_W$. Therefore, $H_o: \mu_E - \mu_W = 0$ vs. $H_a: \mu_E - \mu_W \neq 0$.

156. Calculate the value of the test statistic.

ANSWER:
Normality assumed. Since
$n_E = 20, \bar{x}_E = 128,000, s_E = 4,500, n_W = 20, \bar{x}_W = 125,000, s_W = 2,500$, then,

$$t^* = [(\bar{x}_E - \bar{x}_W) - (\mu_E - \mu_W)] / \sqrt{(s_E^2 / n_E) + (s_W^2 / n_W)}$$

$$= [(128,000 - 125,000) - 0] / [\sqrt{(4500^2 / 20) + (2500^2 / 20)}] = 2.61.$$

157. Test the hypotheses in question 155 using the *p*-value approach.

ANSWER:
P = *p*-value = $2P(t > 2.61 \mid df = 19)$; Using the table of probability values for Student's *t*-distribution, we get 0.007 + < ½ **P** < 0.009; then 0.014 < **P** < 0.018. Since **P** $< \alpha$; reject H_o and conclude that there is not sufficient evidence at the 0.05 level of significance, to show that the mean home prices are different.

158. Test the hypotheses in question 155 using the classical approach.

ANSWER:
The critical regions are $t \leq -2.09$ and $t \geq 2.09$; t^* falls in the critical region, therefore we reject H_o, and conclude that there is not sufficient evidence at the 0.05 level of significance, to show that the mean home prices are different.

QUESTIONS 159 THROUGH 162 ARE BASED ON THE FOLLOWING INFORMATION:
The purchasing department for Meijer supermarket chain is considering two sources from which to purchase 10-lb bags of potatoes. A random sample taken from each source shows the following results.

	Idaho Supers	Idaho Best
Number of Bags Weighted	100	100
Mean Weight	10.3 lbs	10.5 lbs
Sample Variance	0.35	0.25

Suppose you wish to determine whether there is a difference between the mean weights of the 10-lb bags of potatoes.

159. State the null and alternative hypotheses.

ANSWER:
The difference in mean weights of 10-lb bags of potatoes is $\mu_b - \mu_s$. Therefore the null and alternative hypotheses are $H_o : \mu_b - \mu_s = 0$ vs. $H_a : \mu_b - \mu_s \neq 0$.

160. Calculate the value of the test statistic.

ANSWER:
Normality assumed. Sample information: $n_b = 100$ and $n_s = 100$,
$\overline{x}_b = 10.5, \overline{x}_s = 10.3, s_b^2 = 0.25,$ and $s_s^2 = 0.35$. Then

$$t^* = [(\overline{x}_b - \overline{x}_s) - (\mu_b - \mu_s)] / \sqrt{(s_b^2 / n_b) + (s_s^2 / n_s)}$$
$$= [(10.5 - 10.3) - 0] / [\sqrt{(0.25/100) + (0.35/100)}] = 2.58.$$

161. Test the hypotheses in question 159 at the 0.05 level of significance using the *p*-value approach.

ANSWER:
P = 2.*P*(t > 2.58 | df = 99); Using the table of probability values for the Student's *t* – distribution, 0.006 < ½ **P** <0.008; then 0.012 < **P** < 0.016. Since **P** < α = 0.05; reject H_o. There is sufficient evidence to indicate that there is a difference between the mean weights of the 10-lb bags of potatoes.

162. Test the hypotheses in question 159 at the 0.05 level of significance using the classical approach.

ANSWER:
The critical regions are: $t \leq -1.99$ and $t \geq 1.99$; *t** falls in the critical region, therefore we reject H_o. There is sufficient evidence to indicate that there is a difference between the mean weights of the 10-lb bags of potatoes.

QUESTIONS 163 THROUGH 166 ARE BASED ON THE FOLLOWING INFORMATION:
A test concerning some of the fundamental facts about AIDS was administered to two groups, one consisting of college graduates and the other consisting of high school graduates. A summary of test results follows:

College graduates: $n = 100$ $\bar{x} = 80.5$ $s = 6.5$

High school graduates: $n = 100$ $\bar{x} = 53.4$ $s = 10.7$

A professor wishes to determine whether these data show that the college graduates, on the average, score significantly higher on the test.

163. State the null and alternative hypotheses.

ANSWER:
The difference between mean scores college graduates and high school graduates is $\mu_c - \mu_h$. Then the hypotheses of interest are $H_o : \mu_c - \mu_h = 0$ vs. $H_a : \mu_c - \mu_h > 0$.

164. Calculate the value of the test statistic.

ANSWER:
Normality assumed. Since $\bar{x}_c = 80.5,\ \bar{x}_h = 53.4,\ s_c^2 = 42.25,\ s_h^2 = 114.49$, then

$$t^* = [(\bar{x}_c - \bar{x}_h) - (\mu_c - \mu_h)] / \sqrt{(s_c^2 / n_c) + (s_h^2 / n_h)}$$

$$= [(80.5 - 53.4) - 0] / [\sqrt{(42.25/100) + (114.49/100)}] = 21.65$$

165. Test the hypotheses in question 163 at α = 0.05 using the *p*-value approach.

ANSWER:
P = *p*-value = $P(t > 21.65 \mid df = 99)$; Using the table of probability values for Student's *t*-distribution, **P** = 0+. Since **P** $< \alpha = .05$; reject H_o.

166. Test the hypotheses in question 163 at α = 0.05 using the classical approach.

ANSWER:
The critical region is $t \geq 1.66$. The value of the test statistic t^* = 21.65 falls in the critical region, therefore we reject H_o. There is sufficient evidence to conclude that the college graduates did score significantly higher on the test.

QUESTIONS 167 AND 168 ARE BASED ON THE FOLLOWING INFORMATION:
Two independent random samples of sizes 16 and 20 were obtained to make inferences about the difference between two means.

167. If you're completing the inference with the aid of a computer and its statistical software, what is the number of degrees of freedom?

ANSWER:
Smaller of ($n_1 - 1, n_2 - 1$) $\leq$ *df* $\leq$ $n_1 + n_2 - 2$ = smaller of (15, 19) $\leq$ *df* $\leq$ 16+ 20 - 2
$\Rightarrow$ 15 $\leq$ df $\leq$ 34

Chapter 10

168. If you're completing the inference without the aid of a computer and its statistical software, what is the number of degrees of freedom?

ANSWER:
df = smaller of ($n_1 - 1, n_2 - 1$) = smaller of (15, 19) = 15

QUESTIONS 169 THROUGH 171 ARE BASED ON THE FOLLOWING INFORMATION:
The confidence coefficient $t(df, \alpha/2)$, is used to find the maximum error when estimating the difference between two means, $\mu_1 - \mu_2$. Assume you are completing the estimation without the aid of a computer and its statistical software.

169. Find the confidence coefficient when $1-\alpha = 0.95,\ n_1 = 20,\ n_2 = 15$.

ANSWER:
df = smaller of ($n_1 - 1, n_2 - 1$) = smaller of (19, 14) = 14 $\Rightarrow t(df, \alpha/2)$ = *t*(14, 0.025) = 2.14

170. Find the confidence coefficient when $1-\alpha = 0.98,\ n_1 = 45,\ n_2 = 31$.

ANSWER:
df = smaller of ($n_1 - 1, n_2 - 1$) = smaller of (44, 30) = 30 $\Rightarrow t(df, \alpha/2)$ = *t*(30, 0.01) = 2.46

171. Find the confidence coefficient when $1-\alpha = 0.99,\ n_1 = 18,\ n_2 = 40$.

ANSWER:
df = smaller of ($n_1 - 1, n_2 - 1$) = smaller of (17, 39) = 17 $\Rightarrow t(df, \alpha/2)$ = *t*(17, 0.005) = 2.9

172. Two independent random samples resulted in the following: Sample A: $n_A = 25,\ s_A = 8.7$, and Sample B: $n_B = 20,\ s_B = 10.5$. Find the estimate for the standard error for the difference between two means.

ANSWER:

Estimate standard error = $\sqrt{\frac{s_1^2}{n_1} + \frac{s_2^2}{n_2}} = \sqrt{\frac{(8.7)^2}{25} + \frac{(10.5)^2}{20}} = \sqrt{8.5401} = 2.92$

QUESTIONS 173 THROUGH 176 ARE BASED ON THE FOLLOWING INFORMATION:
A study comparing attitudes toward death was conducted in which organ donors (individuals who had signed organ donor cards) were compared with nondonors. Templer's Death Anxiety Scale (DAS) was administered to both groups. On this scale, high scores indicate high anxiety concerning death. The researcher who conducted the study believes that nondonors have mean anxiety scores higher than the mean anxiety scores of donors. The results were reported as follows?

	n	Mean	St. Dev.
Nonorgan Donors	65	7.80	3.65
Organ Donors	25	5.45	2.98

Define the population parameter of interest as $\mu_{non} - \mu_{donor}$; the difference between the mean anxiety scores of nondonors and the mean anxiety scores of donors.

173. Construct the 95% confidence interval for $\mu_{non} - \mu_{donor}$.

ANSWER:
df = smaller of ($n_1 - 1, n_2 - 1$) = smaller of (64, 24) = 24 $\Rightarrow t(df, \alpha/2)$ = *t*(24, 0.025) = 2.06

$$E = t(df,\alpha/2)\cdot\sqrt{\frac{s_1^2}{n_1}+\frac{s_2^2}{n_2}} = (2.06)\cdot\sqrt{\frac{(3.65)^2}{65}+\frac{(2.98)^2}{25}} = \text{(2.06)0.7485)} = 1.54$$

$((\bar{x}_1 - \bar{x}_2) \pm E = (7.80 - 5.45) \pm 1.54 = 2.35 \pm 1.54 \Rightarrow$ LCL = 0.81 and UCL = 3.89.

174. State the null and alternative hypotheses.

ANSWER:
$H_o: \mu_{non} - \mu_{donor} = 0\ (\leq)$ vs. $H_a: \mu_{non} - \mu_{donor} > 0$

175. Do the sample results support the researcher's belief? Test at the 0.05 level of significance using the *p*-value approach.

ANSWER:

$$t^* = \frac{(\bar{x}_1 - \bar{x}_2) - (\mu_1 - \mu_2)}{\sqrt{\frac{s_1^2}{n_1}+\frac{s_2^2}{n_2}}} = \frac{2.35 - 0}{0.7485} = 3.14$$

P = *p*-value = *P*(*t* > 3.14 | *df* = 24) $\Rightarrow$ **P** < 0.005. Since *p*-value < α = 0.05, we reject H_o. Yes, there is sufficient evidence to support the researcher's belief that nondonors have mean anxiety scores higher than the mean anxiety scores of donors.

176. Do the sample results support the researcher's belief? Test at the 0.05 level of significance using the classical approach.

ANSWER:
The critical value is $t(df, \alpha)$ = *t*(24, 0.05) = 1.71. Since t^* = 3.14 falls in the rejection region, we reject H_o. We reach the same conclusion as stated in question 175.

QUESTIONS 177 THROUGH 187 ARE BASED ON THE FOLLOWING INFORMATION:
At Ohio State University, a mathematics placement exam is administered to all students. Samples of 36 male and 30 female students are randomly selected from this year's student body and the following scores recorded. Assume the scores are approximately normally distributed.

Male	70	66	73	80	79	58	73	83	78	68
	69	82	66	83	80	78	52	79	84	77
	97	89	66	80	58	61	65	70	75	49
	59	69	79	72	77	74				

Female	79	74	92	87	81	76	83	89	81	81
	82	78	82	86	75	72	61	67	78	80
	87	67	72	95	71	77	53	74	76	79

Chapter 10

177. Use computer to find the mean and standard deviation, for each set of data.

ANSWER:

Summary stats for two samples			
		Male	Female
	Sample sizes	36	30
	Sample means	72.722	77.833
	Sample standard deviations	10.236	8.805

178. Use computer to construct 95% confidence interval for mean score for all male students.

ANSWER:

Confidence interval for mean score for male		
	Confidence level	95.0%
	Sample mean	72.722
	Std error of mean	1.706
	Degrees of freedom	35
	Lower limit	69.259
	Upper limit	76.186

179. Use computer to construct 95% confidence interval for mean score for all female students.

ANSWER:

Confidence interval for meanscore for female		
	Confidence level	95.0%
	Sample mean	77.833
	Std error of mean	1.608
	Degrees of freedom	29
	Lower limit	74.546
	Upper limit	81.121

180. Do the above results to questions 178 and 179 show the mean scores for males and females could be the same? Justify your answer. Be careful!!

ANSWER:
Yes, the mean scores for males and females could be the same since the two confidence intervals (69.259 to 76.186) and (74.546 and 81.121) do overlap.

181. Use computer to construct 95% confidence interval for the difference between the mean scores for male and female students.

ANSWER:

Confidence interval for difference between means			
	Confidence level	95.0%	
	Sample mean difference	-5.111	
	Pooled standard deviation	NA	
	Std error of difference	2.344	
	Degrees of freedom	64	
	Lower limit	-9.794	
	Upper limit	-0.428	

182. Do the results found in question 181 show that the mean scores for males and females could be the same? Explain.

ANSWER:
No, the results found in question 181 show the mean scores for males and females could not be the same since "zero" is not included in the interval (-0.794 to -0428).

183. Explain why the results to questions 178 and 179 can not be used to draw conclusions about the difference between the two means.

ANSWER:
The questions are asking for different information. In questions 178 and 179, two intervals are constructed that are each centered on separate sample means. In this case, the two sample means are a distance apart, but their intervals overlap allowing for the possibility of coming from populations with a common mean. Yet the two sample means are themselves far enough apart to be significantly different.

184. If you are interested in testing whether there is a difference for male and female students, state the appropriate null and alternative hypotheses.

ANSWER:
$H_o: \mu_{male} - \mu_{female} = 0$ and $H_a: \mu_{male} - \mu_{female} \neq 0$

185. Use computer to test the hypothesis in question 184 using the *p*-value approach at α = 0.05.

ANSWER:

	Male	*Female*
Mean	72.722	77.833
Variance	104.778	77.523
Observations	36	30
Hypothesized Mean Difference	0	
df	64	
t Stat	-2.180	
P(T<=t) one-tail	0.0165	
t Critical one-tail	1.669	
P(T<=t) two-tail	0.0329	
t Critical two-tail	1.998	

Since *p*-value = 0.0329 < α = 0.05, we reject H_o. There is significant evidence to indicate that the mean scores for male and female students are different.

186. Test the hypothesis in question 184 using the classical approach at α = 0.05.

ANSWER:
The critical values are $\pm t(df, \alpha/2) = \pm$ t(29, 0.025) = $\pm$2.05 [*df* = smaller (35,29) = 29].

Since the value of the test statistic t^* = -2.18 falls in the rejection region, we reject H_o. We reach the same conclusion stated in question 185.

187. Did you reach the same conclusion in questions 182, 184, and 185?

ANSWER:
Yes, we reached the same conclusion of rejecting H_o at the 0.05 level of significance.

Section 10.5

True-False Questions

188. Confidence interval estimates for the difference between the proportions of two populations always have values between −1 and 1.
ANSWER: T

189. In the hypothesis test, $H_o: p_1 - p_2 = 0$ and $H_a: p_1 - p_2 \neq 0$, concerning the difference between proportions of two independent samples, we are able to compute a pooled observed probability because p_1 and p_2 are unknown but assumed equal.
ANSWER: T

190. The standard normal score is used for all inferences concerning population proportions.
ANSWER: T

191. A pooled estimate for any statistic in a problem dealing with two populations is a value arrived at by combining the two separate sample statistics so as to achieve the best possible point estimate.
ANSWER: T

192. For right-hand tail test of the difference between proportions using two independent samples at the 5% level of significance, the critical value for the *z*-test is 1.65, but it is 1.96 for the *t*-test.
ANSWER: F

193. When we estimate the difference between two proportions, $p_1 - p_2$, we will base our estimate on the unbiased sample statistic $p'_1 - p'_2$.
ANSWER: T

194. When we estimate the difference between two proportions, $p_1 - p_2$, we will base our estimates on the unbiased sample statistic $x_1 - x_2$; the difference between number of successes in the two samples.
ANSWER: F

195. When the null hypothesis "there is no difference between two population proportions" is being tested, the test statistic will be the difference between the two population proportions, divided by the standard error.
ANSWER: F

Multiple-Choice Questions

196. Which of the following should be used as a point estimate of $p_1 - p_2$ when constructing confidence interval for estimating the difference between the proportions of two populations?

A) 0
B) $(x_1/n_1)-(x_2/n_2)$
C) $n_1 p_1' - n_2 p_2'$
D) $x_1 - x_2$

ANSWER: B

197. Which of the following would be the null hypothesis used to test the claim that the proportion of male students (*M*) who smoke at a particular college is greater than the proportion of female students (*F*) who smoke?

A) $H_o: p_M - p_F = 0(\geq)$
B) $H_o: p_M - p_F = 0(\leq)$
C) $H_o: p_M - p_F > 0$
D) $H_o: p_M - p_F < 0$

ANSWER: B

198. Select the correct hypotheses to test the claim that the proportion of female voters in Washington State (W) who favor a particular presidential candidate is the same as the proportion of voters in Connecticut (*C*) who favor the same candidate.

A) $H_o: p_W - p_C = 0(\leq), H_a: p_W - p_C > 0$
B) $H_o: p_W - p_C = 0(\geq), H_a: p_W - p_C < 0$
C) $H_o: p_W - p_C = 0, H_a: p_W - p_C \neq 0$
D) $H_o: p_W - p_C > 0, H_a: p_W - p_C < 0$

ANSWER: C

199. Select the correct hypotheses for testing the claim that the proportion of male voters (*M*) that support gun control is at least as large as the proportion of female voters (*F*) that support gun control.

A) $H_o: p_M - p_F = 0, H_a: p_M - p_F \neq 0$
B) $H_o: p_M - p_F = 0(\geq), H_a: p_M - p_F < 0$
C) $H_o: p_M - p_F < 0, H_a: p_M - p_F > 0$
D) $H_o: p_M - p_F = 0(\leq), H_a: p_M - p_F > 0$

ANSWER: B

200. The sampling distribution of $p_1' - p_2'$ is approximately normally distributed with a mean equal to:

A) $p_1 - p_2$
B) $n_1 p_1 - n_2 p_2$
C) $\sqrt{(p_1 q_1 / n_1) + (p_2 q_2 / n_2)}$
D) 0
ANSWER: A

201. Assume that two independent samples of sizes n_1 and n_2 are drawn randomly from large populations with $p_1 = P_1$(success) and $p_2 = P_2$ (success), respectively, and that $p_1' - p_2'$ is the difference between the observed proportions of the samples. Which of the following statements is false regarding the sampling distribution of $p_1' - p_2'$?

A) Its mean $\mu_{p_1' - p_2'} = p_1 - p_2$.
B) Its standard error $\sigma_{p_1' - p_2'} = \sqrt{\frac{p_1 q_1}{n_1} + \frac{p_2 q_2}{n_2}}$.
C) It has an approximately normal distribution if n_1 and n_2 are significantly larger.
D) None of the above
ANSWER: D

Short-Answer Questions

202. When estimating the difference between the proportions of two populations using a confidence interval estimate, why do we not use a pooled sample proportion?

ANSWER:
We do not use a pooled sample proportion because we do not know whether $p_1 = p_2$.

203. Only 48 of the 200 people interviewed were able to name the Secretary of State of the United States. Find the value for x, *n*, p', and q'.

ANSWER:
x = 48, *n* = 200, p' = x / *n* = 0.24, and q' =1- p' =0.76

204. Briefly discuss the practical *guidelines to ensure normality*, when comparing two population proportions.

ANSWER:
1) The sample sizes are both larger than 20.
2) The products $n_1 p_1$, $n_1 q_1$, $n_2 p_2$, and $n_2 q_2$ are larger than 5.
3) The samples consist of less than 10% of their respective populations.

NOTE p_1 and p_2 are unknown; therefore, the products mentioned in guideline 2 will be estimated by $n_1 p_1'$, $n_1 q_1'$ $n_2 p_2'$, and $n_2 q_2'$.

205. If $n_1 = 50,\ p_1' = 0.8,\ n_2 = 40,$ and $p_2' = 0.9$, would this satisfy the guidelines for approximately normal? Explain.

ANSWER:
$n_1 p_1'$ = (50)(0.8) = 40, $n_1 q_1'$ = (50)(0.2) = 10, $n_2 p_2'$ = (40)(0.9)= 36, and $n_2 q_2'$ = (40)(0.1) = 4 are not all greater than 5, therefore this situation does not satisfy the guidelines for approximately normal.

Applied and Computational Questions

206. Two different methods for teaching human anatomy were compared. One method is traditional lecture, and the other method utilizes computer-assisted instruction (CAI). Ninety out of 130 in the traditional method passed the course, and ninety-eight out of 125 in the CAI method passed the course. Let p_1 be the proportion of all students taking this course by the CAI method who would pass it, and let p_2 be a similar proportion for the traditional method. Find a 90% confidence interval for $p_1 - p_2$.

ANSWER:
(0 to 0.18)

207. In a survey of 150 men and 150 women, 36% of the men and 28% of the women listed the evening news as their primary source of information concerning world affairs. Set a 99% confidence interval on $p_1 - p_2$, where p_1 is the proportion of men and p_2 is the proportion of women who use the evening news as their primary source of information concerning world affairs.

ANSWER:
(-0.06 to 0.22)

208. Forty percent of 500 males were smokers and 30% of 600 females were smokers in a survey. Find the pooled observed probability for these two samples.

ANSWER:
Pooled observed probability = 0.345

QUESTIONS 209 AND 210 ARE BASED ON THE FOLLOWING INFORMATION:
A random sample of 500 persons was questioned regarding political affiliation and attitude toward government-sponsored mandatory testing of AIDS as shown in the table below.

	Favor	*Undecided*	*Opposed*	**Total**
Democrats	135	80	65	280
Republicans	95	60	65	220
Total	230	140	130	

A statistics student wants to determine if there is a difference in the proportions of Democrats and Republicans who are undecided regarding mandatory testing for AIDS.

209. State the null and alternative hypotheses.

ANSWER:
$H_o: P_1 - P_2 = 0$ vs. $H_a: P_1 - P_2 \neq 0$

210. Test the hypotheses at α = 0.05, by giving the critical region, test statistic z^*, and the conclusion.

ANSWER:
Critical regions: $z \le -1.96$ or $z \ge 1.96$; Value of the test statistic: = 0.32; Conclusion: unable to reject the null hypothesis. That is, there is no sufficient evidence to indicate that there is a difference in the proportions of Democrats and Republicans who are undecided regarding mandatory testing for AIDS.

211. Two different display types were compared to determine their effect upon sales for a new product. The results shown below were found regarding the number who looked at the product and the number who purchased the product. Give the *p*-value when $H_o: p_1 = p_2$ vs. $H_a: p_1 \neq p_2$ is tested. What is your conclusion?

Display Type	Number Who Looked	Number Who Purchased
1	850	75
2	700	70

ANSWER:
The value of the test statistic $z^* = -0.81$, and p –value = 0.418. Since *p*-value is relatively large, we fail to reject the null hypothesis and conclude that there is no difference between the proportion of customers who looked at the product and the proportion of customers who purchased the product.

QUESTIONS 212 AND 213 ARE BASED ON THE FOLLOWING INFORMATION:
A survey of 100 male and 100 female high school seniors showed that 35% of the males and 29% of the females had used marijuana previously. One wishes to determine if the results of this survey indicate a difference in proportions for the population of high school seniors?

212. State the null and alternative hypotheses.

ANSWER:
$H_o: P_1 - P_2 = 0$ vs. $\mathrm{H_a}: P_1 - P_2 \neq 0$

213. Test the hypotheses at α = 0.05, giving the critical region, the test statistic z^*, and your conclusion.

ANSWER:
Critical region: $z \le 1.96$ or $z \ge 1.96$; Value of the test statistic: $z^* = 0.91$ Conclusion: do not reject the null hypothesis. There is no sufficient evidence to indicate.

QUESTIONS 214 AND 215 ARE BASED ON THE FOLLOWING INFORMATION:
A marketing researcher analyst, interested in who purchased new computers, compared the buying average by men and women as shown below.

Gender	*Number Surveyed*	*Number Who Purchased*
Male	500	70
Female	450	100

214. If z^* = 3.1, calculate the *p*-value when testing $H_o: p_M = p_F$ vs. $H_a: p_M \neq p_F$.

ANSWER:
p-value = 0.002

215. If the level of significance is $\alpha = 0.05$, what would be your conclusion?

ANSWER:
Since p –value $< \alpha$, reject the null hypothesis and conclude that the proportion of male and female customers who purchased new computers are not the same.

216. In a random sample of 50 brown-haired individuals, 28 indicated that they used hair coloring. In another random sample of 50 blonde individuals, 34 indicated that they used hair coloring. Use a 95% confidence interval to estimate the difference in the proportion of these groups that use hair coloring.

ANSWER:
The difference in proportions of brown-haired and blonde individuals that use hair coloring is $p_{bl} - p_{br}$. Note that $n_1 > 20, n_2 > 20, n_1 p_1 > 5, n_2 p_2 > 5, n_1 q_1 > 5,$ and $n_2 q_2 > 5$, therefore the assumption of normality is met.
Sample information:

$n_{br} = 50, \quad x_{br} = 28, \quad p'_{br} = 28/50 = 0.56, \quad q'_{br} = 1 - 0.56 = 0.44$

$n_{bl} = 50, \quad x_{bl} = 34, \quad p'_{bl} = 34/50 = 0.68, \quad q'_{bl} = 1 - 0.68 = 0.32.$

Now,

$p'_{bl} - p'_{br} = 0.68 - 0.56 = 0.12,$ and $1 - \alpha = 0.95,$ then $\alpha/2 = 0.025;$ $z(0.025) = 1.96$, and

$$E = z(\alpha/2).\sqrt{(p'_{bl}.q'_{bl}/n_{bl}) + (p'_{br}.q'_{br}/n_{br})} = 1.96\sqrt{(0.68 \cdot 0.32/50) + (0.56 \cdot 0.44/50)}$$

$= (1.96)\ (0.096) = 0.188.$ Hence $(p'_{bl} - p'_{br}) \pm E = 0.12 \pm 0.188$,

and the 95% interval for $p_{bl} - p_{br}$ is –0.068 to 0.308.

217. State the null hypothesis, H_o, and the alternative hypothesis, H_a, that would be used to test the following claim: There is no difference between the proportions of men and women who will vote for the incumbent governor in the next election.

ANSWER:
$H_o: p_m - p_w = 0$ vs. $H_a: p_m - p_w \neq 0$

218. State the null hypothesis, H_o, and the alternative hypothesis, H_a, that would be used to test the following claim: The percentage of boys who play soccer is greater than the percentage of girls who play soccer.

ANSWER:
$H_o: p_b - p_g = 0\ (\leq)$ vs. $H_a: p_b - p_g > 0$

219. State the null hypothesis, H_o, and the alternative hypothesis, H_a, that would be used to test the following claim: The percentage of nurses who drive new cars is lower than the percentage of doctors of the same age who drive new cars.

ANSWER:
$H_o: p_n - p_d = 0\ (\geq)$ vs. $H_a: p_n - p_d < 0$

QUESTIONS 220 THROUGH 223 ARE BASED ON THE FOLLOWING INFORMATION:
In a survey of college students, one of the questions asked was "Have you ever cheated in a test?" Two hundred male and 200 female students were asked this question. Thirty percent of the male and 25% of the female responded "yes." Based on this survey, one wishes to determine whether there is a difference in the proportion of male and female responding "yes" to the above question at the 0.05 level of significance.

220. State the null and alternative hypotheses.

ANSWER:
The difference in the proportion of male and female responding "yes" to the survey question is $p_m - p_f$. Therefore the null and alternative hypotheses are:

$H_o: p_m - p_f = 0$ vs. $H_a: p_m - p_f \neq 0$.

221. Calculate the value of the test statistic.

ANSWER:
Since *n*'s >20, *np*'s and *nq*'s all > 5, $n_m = 200,\ p'_m = 0.30,\ n_f = 200,\ p'_f = 0.25$, then $p'_p = (x_m + x_f)/(n_m + n_f)$ = (60 + 50) / (200 + 200) = 0.275, and $q'_p = 1 - p'_p$ = 1.0 – 0.275 = 0.725. Hence, the value of the test statistic is

$$z^* = [(p'_m - p'_f) - (p_m - p_f)] / \sqrt{(p'_p)(q'_p)[(1/n_m) + (1/n_f)]}$$

$$= (0.30 - 0.25) / \sqrt{(0.275)(0.725)[(1/200) + (1/200)]} = 1.12.$$

222. Test the hypotheses in question 220 using the *p*-value approach.

ANSWER:
P = *p*-value = 2 · P(z > 1.12) = 2 (0.5000 – 0.3686) = 0.2628
Since P > α = 0.05, we fail to reject H_o. There is no sufficient evidence to indicate that there is a difference in the proportion of male and female responding "yes" to the above question.

223. Test the hypotheses in question 220 using the classical approach.

ANSWER:
The critical regions are: $z \leq -1.96$ and $z \geq 1.96$. Since $z^* = 1.12$ falls in the noncritical region, we fail to reject H_o. We reach the same conclusion as stated in question 222.

QUESTIONS 224 THROUGH 226 ARE BASED ON THE FOLLOWING INFORMATION:
It is believed that smoking boosts death risk for diabetics. A scientist investigated the smoking rates for male and female diabetics and obtained the following data.

Gender	*n*	Number Who Smoke
Male	400	172
Female	400	136

A researcher wants to test the hypothesis that smoking rate (proportion of smokers) is higher for males than females.

224. State the null and alternative hypotheses.

ANSWER:
$H_o: p_m - p_w = 0\ (\leq)$ vs. $H_a: p_m - p_w > 0$

225. Calculate the value of the test statistic.

ANSWER:
$p'_p = (x_m + x_w)/(n_m + n_w)$ = (172 +136) / (400 + 400) = 0.385, and $q'_p = 1 - p'_p$ = 1.0 0.385 = 0.615; Hence, the value of the test statistic is

$$z^* = [(p'_m - p'_w) - (p_m - p_w)] / \sqrt{(p'_p)(q'_p)[(1/n_m) + (1/n_w)]}$$

$$= (0.43 - 0.34) / \sqrt{(0.385)(0.615)[(1/400) + (1/400)]} = 2.62.$$

226. Calculate the *p*-value. What decision and conclusion would be reached at the 0.05 level of significance?

ANSWER:
P = $P(z > 2.62)$; Using the table of standard normal distribution, **P** = (0.5000 – 0.4956) = 0.0088. Since **P** < α = 0.05; we reject H_o . There is sufficient evidence to indicate that the smoking rate for male diabetics is significantly higher than for female diabetics, at the 0.05 level.

QUESTIONS 227 AND 228 ARE BASED ON THE FOLLOWING INFORMATION:
Of a random sample of 100 stocks on the New York Stock Exchange, 36 made a gain today. A random of 100 stocks on the American Stock Exchange showed 30 stocks making a gain.

227. Construct a 99% confidence interval estimate of the difference in the proportion of stocks making a gain.

ANSWER:
The difference in proportions of stocks making a gain is $p_n - p_a$.
n's > 20, *np*'s and *nq*'s all > 5.

$n_n = 100,\quad x_n = 36,\quad p'_n = 36/100 = 0.36,\quad q'_n = 1 - 0.36 = 0.64$

$n_a = 100,\quad x_a = 30,\quad p'_a = 30/100 = 0.30,\quad q'_n = 1 - 0.30 = 0.70$

Since $p'_n - p'_a = 0.36 - 0.30 = 0.06$, and $1 - \alpha = 0.99$, then $\alpha/2$ = 0.005; and *z*(0.005) = 2.58. Hence,

$$E = z(\alpha/2).\sqrt{(p'_n.q'_n / n_n) + (p'_a.q'_a / n_a)}$$

$$= 2.58\sqrt{(0.36 \cdot 0.64/100) + (0.30 \cdot 0.70/100)} = 0.171$$

$(p'_n - p'_a) \pm E = 0.06 \pm 0.171$. The 99% confidence interval for $p_n - p_a$ is –0.111 to 0.231

228. Does the answer to question 227 suggest that there is a significant difference between the proportions of stocks making gains on the two stock exchanges?

ANSWER:
No, there is no significant difference at the 0.01 level because the confidence interval estimate contains the value 0.

229. Calculate the estimate for the standard error of the difference between two proportions given that $n_1 = 50,\ p_1' = 0.9,\ n_2 = 40,\ \text{and}\ p_2' = 0.9$.

ANSWER:

$$\text{Standard error estimate} = \sqrt{\frac{p_1'q_1'}{n_1} + \frac{p_2'q_2'}{n_2}} = \sqrt{\frac{(0.9)(0.1)}{50} + \frac{(0.9)(0.1)}{40}} = 0.0636$$

230. Calculate the maximum error of estimate for a 95% confidence interval for the difference between two proportions if $n_1 = 32,\ p_1' = 0.32,\ n_2 = 38,\ \text{and}\ p_2' = 0.38$

ANSWER:
$z(\alpha/2)$ = z(0.025) = 1.96. Then,

$$E = z(\alpha/2)\cdot\sqrt{\frac{p_1'q_1'}{n_1} + \frac{p_2'q_2'}{n_2}} = (1.96)\cdot\sqrt{\frac{(0.32)(0.68)}{32} + \frac{(0.38)(0.62)}{38}}$$
$$= (1.96)(0.114) = 0.223$$

231. Calculate the maximum error of estimate for a 90% confidence interval for the difference between two proportions $n_1 = 33,\ p_1' = 0.35,\ n_2 = 37,\ \text{and}\ p_2' = 0.42$

ANSWER:
$z(\alpha/2)$ = z(0.05) = 1.645. Then,

$$E = z(\alpha/2)\cdot\sqrt{\frac{p_1'q_1'}{n_1} + \frac{p_2'q_2'}{n_2}} = (1.645)\cdot\sqrt{\frac{(0.35)(0.65)}{33} + \frac{(0.42)(0.58)}{37}}$$
$$= (1.645)(0.1161) = 0.191$$

QUESTIONS 232 THROUGH 236 ARE BASED ON THE FOLLOWING INFORMATION:
The proportions of defective parts produced by two machines were compared, and the following data were collected:
Machine 1: *n* = 200; number of defective parts =10
Machine 2: *n* = 200; number of defective parts = 6

232. Calculate the maximum error of estimate for a 90% confidence interval for the difference between the proportions of defective parts produced by the two machines.

ANSWER:
$n_1 = 200,\ p_1' = 10/200 = 0.05,\ n_2 = 200,\ \text{and}\ p_2' = 6/200 = 0.03$. $z(\alpha/2)$ = z(0.05) = 1.645. Then,

$$E = z(\alpha/2)\cdot\sqrt{\frac{p_1'q_1'}{n_1} + \frac{p_2'q_2'}{n_2}} = (1.645)\cdot\sqrt{\frac{(0.05)(0.95)}{200} + \frac{(0.03)(0.97)}{200}}$$
$$= (1.645)(0.0196) = 0.032$$

233. Determine a 90% confidence interval for $p_1 - p_2$.

ANSWER:
$(p_1' - p_2') \pm E = 0.02 \pm 0.032 \Rightarrow$ LCL = -0.012 and UCL = 0.052

234. If you wish to test there is no difference in the proportion of defective parts produced by both machines, state the appropriate null and alternative hypotheses.

ANSWER:
$H_o : p_1 - p_2 = 0$ and $H_a : p_1 - p_2 \neq 0$

235. Can you use the confidence interval in question 233 to test the hypotheses in question 234 at the 0.10 level of significance? Explain in detail.

ANSWER:
Yes, we can use the confidence interval in question 233 to test the hypotheses in question 234. Since the hypothesized value of zero falls in the 90% confidence interval, we fail to reject H_o at the 0.10 level of significance.

236. Based on your answer to question 235, what is your conclusion?

ANSWER:
There is no sufficient evidence to indicate a difference in the proportion of defective parts produced by both machines.

237. State null hypothesis, H_o, and the alternative hypothesis, H_a, that would be used to test the claim "There is no difference between the proportions of male and female students who will vote for the president of student government at Iowa State University."

ANSWER:
$H_o : p_M - p_F = 0$ and $H_a : p_M - p_F \neq 0$

238. State null hypothesis, H_o, and the alternative hypothesis, H_a, that would be used to test the claim "The percentage of boys who missed statistics classes is greater than the percentage of girls who missed the same classes."

ANSWER:
$H_o : p_B - p_G = 0 \, (\leq)$ and $H_a : p_B - p_G > 0$

239. State null hypothesis, H_o, and the alternative hypothesis, H_a, that would be used to test the claim "The percentage of college students who drive old cars is lower than the percentage of non-college people of the same age who drive old cars."

ANSWER:
Let p_1 = percentage of college students who drive old cars, and p_2 = percentage of non-college students who drive old cars. Then, $H_o : p_1 - p_2 = 0 \, (\geq)$ and $H_a : p_1 - p_2 < 0$

240. Determine the *p*-value that would be used to test $H_o: p_1 = p_2$ vs. $H_a: p_1 > p_2$, if the value of the test statistic $z^* = 2.12$

ANSWER:
P = *p*-value = *P*(*z* > 2.12) = 0.500 – 0.483 = 0.017

241. Determine the *p*-value that would be used to test $H_o: p_a = p_B$ vs. $H_a: p_A \neq p_B$, if the value of the test statistic $z^* = -2.28$.

ANSWER:
P = *p*-value = *P*(*z* < -2.28) + *P*(*z* > 2.28) = 2 *P*(*z* > 2.28) = 2 (0.5000 – 0.4887) = 0.0226

242. Determine the *p*-value that would be used to test $H_o: p_1 - p_2 = 0$ vs. $H_a: p_1 - p_2 < 0$, if the value of the test statistic $z^* = -0.75$.

ANSWER:
P = *p*-value = *P*(*z* < -0.75) = 0.5000 – 0.2734 = 0.2266

243. Determine the *p*-value that would be used to test $H_o: p_m - p_f = 0$ vs. $H_a: p_m - p_f > 0$, if the value of the test statistic $z^* = 3.09$.

ANSWER:
P = *p*-value = *P*(*z* > 3.09) = 0.50 – 0.499 = 0.001

244. Draw an approximate normal curve and determine the critical region and critical value(s) that would be used to test $H_o: p_1 = p_2$ vs. $H_a: p_1 > p_2$, with $\alpha = 0.05$.

ANSWER:

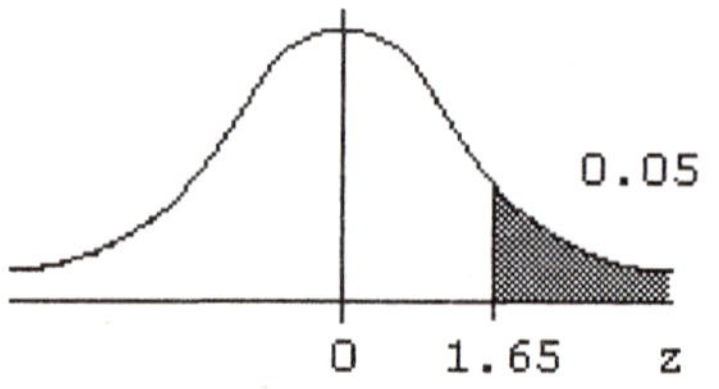

245. Draw an approximate normal curve and determine the critical region and critical value(s) that would be used to test $H_o: p_1 = p_2$ vs. $H_a: p_1 = p_2$, with $\alpha = 0.05$.

ANSWER:

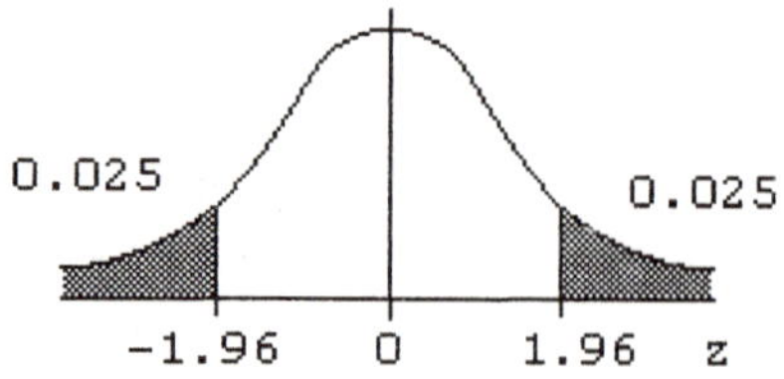

246. Draw an approximate normal curve and determine the critical region and critical value(s) that would be used to test $H_o: p_1 - p_2 = 0$ vs. $H_a: p_1 - p_2 = 0$, with $\alpha = 0.04$.

ANSWER:

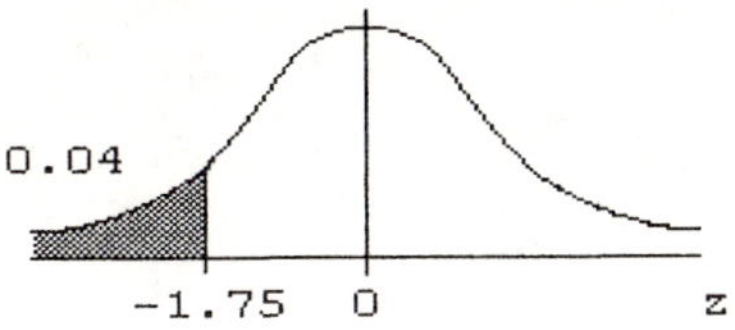

247. Draw an approximate normal curve and determine the critical region and critical value(s) that would be used to test $H_o: p_1 - p_2 = 0$ vs. $H_a: p_1 - p_2 = 0$, with $\alpha = 0.01$

ANSWER:

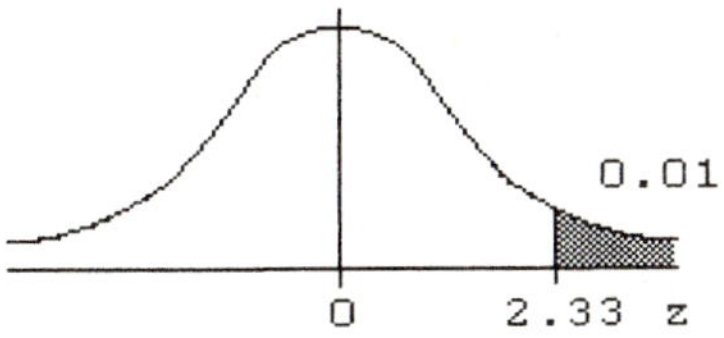

<u>**QUESTIONS 248 THROUGH 255 ARE BASED ON THE FOLLOWING INFORMATION:**</u>
Two randomly selected groups of citizens were exposed to different media campaigns that dealt with the image of a presidential candidate. One week later, the citizen groups were surveyed to see whether they would vote for the candidate. The results were as follows:

	Exposed to Conservative Image	Exposed to Moderate Image
Number in Sample	100	100
Proportion for the Candidate	0.42	0.46

A political analyst believes that there is no difference in the effectiveness of the two image campaigns.

248. Would this situation satisfy the guidelines for approximately normal? Explain.

ANSWER:
Let 1 = conservative and 2 = moderate.
$n_1 p_1' = (100)(0.42) = 42$, $n_1 q_1' = (100)(0.58) = 58$, $n_2 p_2' = (100)(0.46) = 46$, and $n_2 q_2' = (100)(0.54) = 54$ are all greater than 5, therefore this situation would satisfy the guidelines for approximately normal.

249. Calculate the maximum error of estimate for a 95% confidence interval for the difference between the two proportions of those who would vote for the presidential candidate.

ANSWER:
$z(\alpha/2) = z(0.025) = 1.96$. Then,

$$E = z(\alpha/2) \cdot \sqrt{\frac{p_1' q_1'}{n_1} + \frac{p_2' q_2'}{n_2}} = (1.96) \cdot \sqrt{\frac{(0.42)(0.58)}{100} + \frac{(0.46)(0.54)}{100}} = (1.96)(0.07) = 0.137.$$

250. Construct a 95% confidence interval for the difference between the two proportions of those who would vote for the presidential candidate.

ANSWER:
$(p_1' - p_2') \pm E = -0.04 \pm 0.137 \Rightarrow$ LCL = -0.177 and UCL = 0.097.

251. State the null and alternative hypotheses for this situation.

ANSWER:
$H_o : p_1 - p_2 = 0$ and $H_a : p_1 - p_2 \neq 0$

252. Calculate the value of the test statistics for testing the hypotheses in question 251.

ANSWER:
$p_p' = (x_1 + x_2)/(n_1 + n_2)$ = (42 + 46) / (100 + 100) = 0.44

$$z^* = \frac{p_1' - p_2'}{\sqrt{p_p' q_p' [(1/n_1) + (1/n_2)]}} = \frac{0.42 - 0.46}{\sqrt{(0.44)(0.56)[(1/100) + (1/100)]}}$$ = (-0.04) / (0.07) = -0.57

253. Test the hypotheses in question 251 at the 5% level of significance using the *p*-value approach.

ANSWER:
P = *p*-value = *P*(z < -0.57) + *P*(z > 0.57) = 2 *P*(z > 0.57) = 2 (0.50- 0.2157) = 0.5686. Since *p*-value = 0.5686 > α = 0.05, we fail to reject H_o. There is sufficient evidence to support the political analyst belief that there is no difference in the effectiveness of the two image campaigns.

254. Test the hypotheses in question 251 at the 5% level of significance using the classical approach.

ANSWER:
The critical values are $\pm$ z(α /2) = $\pm$z(0.025) = $\pm$1.96.

Since the value of the test statistic z^* = -0.57 does not fall in the rejection region, we fail to reject H_o. We reach the same conclusion as stated in question 253.

255. Can you use the confidence interval in question 250 to test the hypotheses in question 251? Explain in detail.

ANSWER:
Yes, we can use the confidence interval in question 250 to test the hypotheses in question 251. Since the hypothesized value of zero falls in the confidence interval, we fail to reject H_o.

Section 10.6

True-False Questions

256. The chi-square distribution is used for making inferences about the ratio of the variances of two populations.
ANSWER: F

257. The *F*-distribution is a symmetric distribution.
ANSWER: F

258. Inferences about the ratio of variances for two normally distributed populations use the Student's *t*-distribution with $n_1 + n_2 - 2$ degrees of freedom.
ANSWER: F

259. Inferences about the ratio of two variances require that the samples are randomly selected from *F*-distributed populations, and that the two samples are selected in an independent manner.
ANSWER: F

260. The critical *F*-value for samples of size 8 and 10 with 5% of the area in the right-hand tail is determined by the value *F*(8, 10, 0.05).
ANSWER: F

261. Inferences about the ratio of variances for two normally distributed populations use the *F*-distribution.
ANSWER: T

262. Each *F*-distribution is identified by two numbers of degrees of freedom, one for each of the two samples involved.
ANSWER: T

263. The tables of critical values for the *F*-distribution give only the right-hand critical values.
ANSWER: T

264. The chi-square distribution is used for making inferences about the ratio of the variances of two populations.
ANSWER: F

Multiple-Choice Questions

265. Which of the following is not one of the properties of the *F*- distribution?

A) *F* is nonnegative; it is zero or positive.
B) *F* is nonsymmetrical; it is skewed to the left
C) *F* is distributed so as to form a family of distributions
D) There is a separate *F*-distribution for each pair of numbers of degrees of freedom.
ANSWER: B

Chapter 10

266. In comparing the variances of two normally distributed populations using two independent samples, which of the following statements is false?

A) The test procedure uses the ratio of variances.
B) The null and alternative hypotheses are expressed as a ratio of the population variances.
C) It is recommended that the "larger" of the two sample variances be the numerator of the calculated *F*-statistic.
D) Is recommended that the "smaller" or "expected to be smaller" population variance be the numerator of the ratio in the null and alternative hypotheses.

ANSWER: D

267. Which of the following is not needed for calculating the critical values for the *F*-distribution?

A) The degrees of freedom associated with the sample whose variance is in the numerator of the calculated *F*.
B) The degrees of freedom associated with the sample whose variance is in the denominator of the calculated *F*.
C) The values of the two samples variances.
D) The area under the distribution curve to the right of the critical value being sought.

ANSWER: C

268. How many values are needed to identify a single critical value of the *F*-distribution?

A) 5
B) 4
C) 3
D) 2

ANSWER: C

Short-Answer Questions

269. Suppose we were to test the hypotheses, $H_o: \sigma_1^2/\sigma_2^2 = 1(\leq)$ vs. $H_a: \sigma_1^2/\sigma_2^2 > 1$, and then reject the null hypothesis, what would this suggest about which population is more variable? Why?

ANSWER:

This would suggest that population 1 is more variable since $\sigma_1^2/\sigma_2^2 > 1$ is equivalent to $\sigma_1^2 > \sigma_2^2$. If the variance of population 1 is greater than that of population 2, population 1 is more variable.

270. In using the *F*- test to test equality of variances in a two-tailed test, what can we do to insure that we will not need a left-tail critical value of *F*?

ANSWER:

Always use the sample with the largest variance for the "numerator." This will make F^* larger than 1 and place it in the right tail of the distribution.

271. What assumption must be met about two populations if we use the *F* test for equality of variances?

ANSWER:
The two populations must be normally distributed.

272. If a two-tailed test with n_1 = 10, n_2 = 18, and α = 0.05, find the right-tail critical value, assuming that $F^* = s_1^2 / s_2^2$.

ANSWER:
F(9, 17, 0.025) = 2.98

273. In a particular *F* test for the ratio of two variances, the test statistic $F^* = s_1^2 / s_2^2$ =331. If n_1 = 10 and n_2 = 12, find bounds for the *p*-value.

ANSWER:
0.025 < *p*-value < 0.05

274. Discuss properties of the *F*-distribution in regard to possible values of *F* and symmetry.

ANSWER:
Value of *F* is zero or positive. The *F*-distribution is not symmetric; distribution skewed to the right.

275. To conclude statistically at the 0.05 level of significance that population 1 is more variable than population 2, s_1^2 / s_2^2 must exceed what value if $n_1 = 10$ and $n_2 = 5$?

ANSWER:
Value of the test statistic F^* must exceed 6.00.

276. Testing the hypotheses, $H_o : \sigma_1^2 / \sigma_2^2 = 1 (\geq)$ vs. $H_a : \sigma_1^2 / \sigma_2^2 < 1$, given *F*(10,15,0.05) and s_2^2 = 10.1, what is the largest possible value of s_1 which would allow us to reject H_o?

ANSWER:
The largest possible value of s_1 is 5.06.

277. Suppose we were to test the hypotheses, $H_o : \sigma_1^2 / \sigma_2^2 = 1 (\geq)$ vs. $H_a : \sigma_1^2 / \sigma_2^2 < 1$, using the 0.05 level of significance. If n_1 = 31 and n_2 = 16, what is the smallest possible value of the ratio of s_1 / s_2 which causes us to reject H_o?

ANSWER:
The smallest possible ratio of s_1 / s_2 is 1.5.

278. Briefly discuss the assumptions for inferences about the ratio of two variances.

ANSWER:
The samples are randomly selected from normally distributed populations, and the two samples are selected in a independent manner.

Chapter 10

Applied and Computational Questions

QUESTIONS 279 AND 280 ARE BASED ON THE FOLLOWING INFORMATION:

An experiment was designed to compare two brands of fertilizers. Twenty plots on an experimental farm were randomly divided into two groups of 10 plots each. Brand *A* was applied to ten plots, and Brand *B* was applied to the other ten plots.

Brand	*n*	$\overline{x}$	*s*
A	10	17.5	1.2
B	10	20.2	4.7

Suppose you wish to test for unequal variability in yield at level of significance equal to 0.05. The results were as follows (in bushels of corn per plot).

279. State the null and alternative hypotheses.

ANSWER:
$H_o : \sigma_1^2 / \sigma_2^2 = 1$ vs. $H_a : \sigma_1^2 / \sigma_2^2 \neq 1$

280. Give the critical region, computed test statistic, and conclusion.

ANSWER:
Critical region: $F \geq 4.03$; Value of the test statistic: F^* = 15.3; Conclusion: Reject the null hypothesis of equal variances.

281. Find the following critical value for *F*: *F*(10, 30, 0.05).

ANSWER:
2.71

282. Find the following critical value for *F*: *F*(60, 150, 0.05).

ANSWER:
2.97

283. Find the following critical value for *F*: *F*(10, 15, 0.025).

ANSWER:
3.56

284. Find the following critical value for *F*: *F*(5, 20, 0.01).

ANSWER:
2.01

QUESTIONS 285 AND 286 ARE BASED ON THE FOLLOWING INFORMATION:

A study was designed to compare the self-care knowledge of two different groups of cardiac patients. A standard test was administered to the two groups. One group was selected from patients having only a high school education and the other was selected from college graduates who were cardiac patients. The results were as follows:

High School Graduates: $n_1 = 11, \quad \overline{x}_1 = 64.5, \quad s_1 = 2.7$

College Graduates: $n_2 = 11, \quad \overline{x}_2 = 74.3, \quad s_2 = 6.4$

One wishes to test for unequal variances at α = 0.05.

285. State the null and alternative hypotheses.

ANSWER:
$H_o: \sigma_1^2/\sigma_2^2 = 1$ vs. $H_a: \sigma_1^2/\sigma_2^2 \neq 1$

286. Give the critical region, computed test statistic, and conclusion.

ANSWER:
Critical region: $F \geq 3.72$; Value of the test statistic: $F^* = 5.62$; Conclusion: Reject the null hypothesis.

QUESTIONS 287 AND 288 ARE BASED ON THE FOLLOWING INFORMATION:
A researcher wishes to compare two different groups of students with respect to their mean time to complete a particular task. The time required is determined for each independent group as shown in the following summary: Suppose you wish to test the claim of unequal variances at α = 0.05., that there is no variance.

Technique	*n*	$\bar{x}$	s
1	10	23.5	2.7
2	8	20.4	5.2

287. State the null and alternative hypotheses.

ANSWER:
$H_o: \sigma_1^2/\sigma_2^2 = 1$ vs. $\mathrm{H_a}: \sigma_1^2/\sigma_2^2 \neq 1$

288. Give the critical region, test statistic value, and conclusion for the *F*-test.

ANSWER:
Critical region: $F \geq 4.20$; Value of the test statistic: $F^* = 3.71$; Do not reject the null hypothesis of equal variances.

289. Twenty individuals with cholesterol readings in the range from 250 to 275 were randomly divided into two groups of ten each. The two groups were put on two different diets and after 6 months, the change in cholesterol was determined for each individual. Using the summarized results shown below , give the critical region, the test statistic, and the conclusion for testing the null hypothesis of equal variances versus the alternative hypothesis of unequal variances at a level of significance equal to 0.05.

Diet	*n*	Mean change	SD
1	10	20.5	5.5
2	10	14.8	6.5

ANSWER:
Critical region: $F \geq 4.03$
Value of the test statistic: $F^* = 1.40$
Conclusion: Unable to reject the null hypothesis of equal variances.

290. A study was designed to compare the variability of male and female diastolic blood pressures. The null hypothesis was that the population standard deviations were equal versus the alternative that they were not equal. State the critical region for α = 0.05, F^*, and the conclusion if the following sample results were observed. Males: n = 25, s = 9.9 and Females: n = 25, s = 8.7

ANSWER:
Critical region: $F \geq 2.27$
Value of the test statistic: F^* = 1.29
Conclusion: Unable to reject the null, hypothesis of equal standard deviation.

291. State the null hypothesis H_o, and the alternative hypothesis H_a, that would be used to test the following claim: The variances of populations A and B are not equal.

ANSWER:
$H_o: \sigma_A^2 = \sigma_B^2$ vs. $H_a: \sigma_A^2 \neq \sigma_B^2$

292. State the null hypothesis H_o, and the alternative hypothesis, H_a, that would be used to test the following claim: The standard deviation of population 1 is larger than the standard deviation of population 2.

ANSWER:
$H_o: \sigma_1 = \sigma_2 (\leq 0)$ vs. $H_a: \sigma_1 > \sigma_2$

293. State the null hypothesis H_o, and the alternative hypothesis, H_a, that would be used to test the following claim: The ratio of the variances for populations A and B is different from 1.

ANSWER:
$H_o: \sigma_A^2 / \sigma_B^2 = 1$ vs. $H_a: \sigma_A^2 / \sigma_B^2 \neq 1$

294. State the null hypothesis H_o, and the alternative hypothesis, H_a, that would be used to test the following claim: The variability within population A is less than the variability within population B.

ANSWER:
$H_o: \sigma_A^2 / \sigma_B^2 = 1$ vs. $H_a: \sigma_A^2 / \sigma_B^2 < 1$ or equivalently, $H_o: \sigma_B^2 / \sigma_A^2 = 1$ vs. $H_a: \sigma_B^2 / \sigma_A^2 > 1$.

QUESTIONS 295 AND 296 ARE BASED ON THE FOLLOWING INFORMATION:
Two independent samples are drawn from a normally distributed population.

295. If each sample has a size of 3, find the probability that one of the sample variances is at least 39 times larger than the other one.

ANSWER:
$P(s_1^2 \geq 39 s_2^2 \text{ or } s_2^2 \geq 39 s_1^2) = P(s_1^2 / s_2^2 \geq 39) + P(s_2^2 / s_1^2 \geq 39)$
$= 2\ P(F \geq 39 \mid df = 2, 2)$
$= 2(0.025) = 0.05$ (since $F(2, 2, 0.025) = 39$)

296. If each sample has a size of 6, find the probability that one of the sample variances is no more than 11 times larger than the other one.

ANSWER:

$P(s_1^2 \geq 11s_2^2 \text{ or } s_2^2 \geq 11s_1^2) = P(s_1^2/s_2^2 \geq 11) + P(s_2^2/s_1^2 \geq 11)$

$= 2\,P[F \geq 11 \mid df = 5, 5]$

$= 2(0.01) = 0.02$ (since $F(5, 5, 0.01) = 11$)

QUESTIONS 297 THROUGH 300 ARE BASED ON THE FOLLOWING INFORMATION:

The standard deviation of Injury Severity Scores (ISS) for 40 children ten years or younger was 24.5, and the standard deviation for 40 children older than ten years was 7.5. Assume that ISS scores are normally distributed for both age groups. One wishes to determine whether there is sufficient evidence to conclude that the standard deviation of ISS scores for younger children is larger than the standard deviation of ISS scores for older children.

297. State the null and alternative hypotheses.

ANSWER:

The ratio of the standard deviations for scores of younger children and older children is σ_y/σ_o. Therefore the null and alternative hypotheses are given by $H_o: \sigma_y = \sigma_o (\leq 0)$ and $H_a: \sigma_y > \sigma_o$.

298. Calculate the value of the test statistic.

ANSWER:

Normality assumed, and independence exists. Since, $n_y = 40$, $s_y = 24.5$, $n_o = 40$, and $s_o = 7.5$, then $F^* = s_y^2/s_o^2 = (24.5)^2/(7.5)^2 = 10.67$.

299. Test the hypotheses in question 297 at $\alpha = 0.01$ using the *p*-value approach.

ANSWER:

P = *p*-value = $P(F > 10.67 \mid df = 39, 39)$. Using the *F*-distribution table, we get **P** < 0.01. Since **P** < α = 0.01, reject H_o.

300. Test the hypotheses in question 297 at $\alpha = 0.01$ using the classical approach.

ANSWER:

The critical region is F $\geq$2.11. Since the value of the test statistic F^* falls in the critical region, we reject H_o. There is sufficient evidence at the 0.01 level of significance that the standard deviation of scores for younger children is larger than the standard deviation for older children.

301. Reorganize the alternative hypothesis shown below so that the critical region will be the right-hand tail: $H_a: \sigma_2^2 < \sigma_1^2 \text{ or } \sigma_2^2/\sigma_1^2 < 1$ (population 2 is less variable)

ANSWER:

Reverse the direction of the inequality, and reverse the roles of the numerator and denominator. Therefore, $H_a: \sigma_1^2 > \sigma_2^2 \text{ or } \sigma_1^2/\sigma_2^2 > 1$ (Population 1 is less variable), and the calculated test statistic F^* will be s_1^2/s_2^2.

Chapter 10

302. State the null hypothesis, H_o, and the alternative hypothesis, H_a, that would be used to test the claim "The variances of populations A and B are not equal."

ANSWER:
$H_o: \sigma_A^2 / \sigma_B^2 = 1$ and $\sigma_A^2 / \sigma_B^2 \neq 1$

303. State the null hypothesis, H_o, and the alternative hypothesis, H_a, that would be used to test the claim "The standard deviation of population 1 is larger than the standard deviation of population 2".

ANSWER:
$H_o: \sigma_1 / \sigma_2 = 1\ (\leq)$ and $\sigma_1 / \sigma_2 > 1$

304. State the null hypothesis, H_o, and the alternative hypothesis, H_a, that would be used to test the claim "The ratio of the variances for populations C and D is different from 1."

ANSWER:
$H_o: \sigma_C^2 / \sigma_D^2 = 1$ and $\sigma_C^2 / \sigma_D^2 \neq 1$

305. State the null hypothesis, H_o, and the alternative hypothesis, H_a, that would be used to test the claim "The variability within population X is less than the variability within population Y."

ANSWER:
$H_o: \sigma_Y^2 / \sigma_X^2 = 1\ (\leq)$ and $\sigma_Y^2 / \sigma_X^2 > 1$

306. Use the table of critical values of the *F*-distribution to find *F*(24, 12, 0.01).

ANSWER:
3.78

307. Use the table of critical values of the *F*-distribution to find *F*(30, 40, 0.05).

ANSWER:
1.74

308. Use the table of critical values of the *F*-distribution to find *F*(12, 10, 0.025).

ANSWER:
3.62

309. Use the table of critical values of the *F*-distribution to find *F*(5, 20, 0.05).

ANSWER:
2.71

310. Use the table of critical values of the *F*-distribution to find *F*(15, 18, 0.05).

ANSWER:
2.27

311. Use the table of critical values of the *F*-distribution to find *F*(15, 9, 0.025).

ANSWER:
3.77

312. Use the table of critical values of the *F*-distribution to find *F*(40, 30, 0.01).

ANSWER:
2.30

313. Determine the *p*-value that would be used to test $H_o: \sigma_1 = \sigma_2$ vs. $H_a: \sigma_1 > \sigma_2$ with $n_1 = 8,\ n_2 = 15$ and $F^* = 2.96$.

ANSWER:
P = *p*-value = *P*(*F* > 2.96 | 7, 14) $\Rightarrow$0.025 < **P** < 0.05

314. Determine the *p*-value that would be used to test $H_o: \sigma_1^2 = \sigma_2^2$ vs. $H_a: \sigma_1^2 > \sigma_2^2$ with $n_1 = 21,\ n_2 = 21,$ and $F^* = 2.75$.

ANSWER:
P = *p*-value = *P*(*F* > 2.75 | 20, 20) $\Rightarrow$0.01 < **P** < 0.025

315. Determine the *p*-value that would be used to test the null hypothesis $H_o: \sigma_1^2/\sigma_2^2 = 1$ vs. the alternative hypothesis $H_a: \sigma_1^2/\sigma_2^2 \neq 1$, with $n_1 = 31$, $n_2 = 61$ and $F^* = 1.94$.

ANSWER:
P = *p*-value = 2 *P*(*F* > 1.94 | 30, 60) $\Rightarrow$2(0.01) < **P** <2(0.025) $\Rightarrow$ 0.02 < **P** < 0.05

316. Draw an approximate *F*-curve and determine the critical region and critical value(s) that would be used to test $H_o: \sigma_1^2 = \sigma_2^2$ vs. $H_a: \sigma_1^2 > \sigma_2^2$, with $n_1 = 10,\ n_2 = 16,$ and $\alpha = 0.05$.

ANSWER:

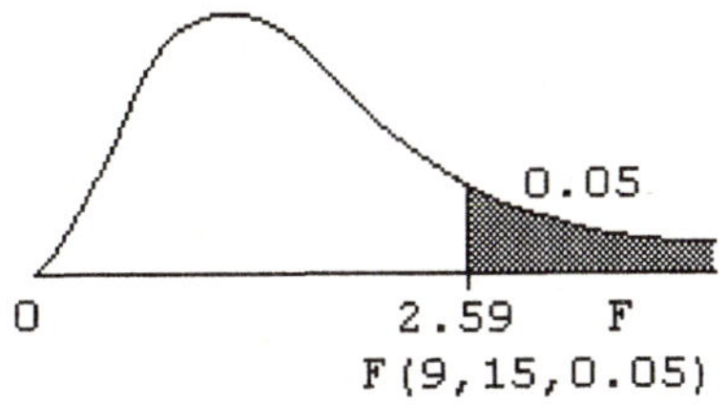

317. Draw an approximate *F*-curve and determine the critical region and critical value(s) that would be used to test $H_o: \sigma_1^2/\sigma_2^2 = 1$ vs. $H_a: \sigma_1^2/\sigma_2^2 \neq 1$, with $n_1 = 25,\ n_2 = 31,$ and $\alpha = 0.05$.

ANSWER:

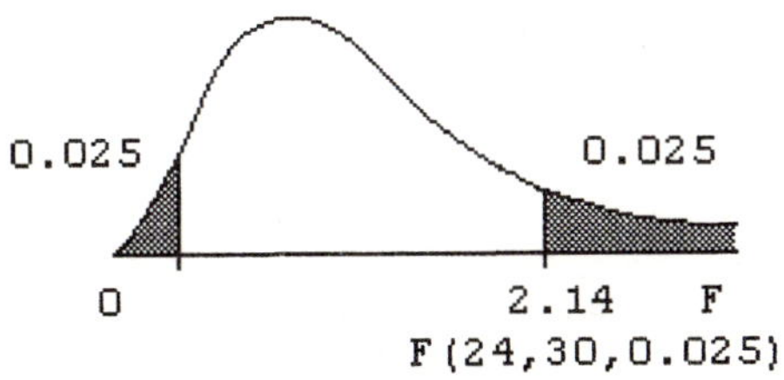

318. Draw an approximate *F*-curve and determine the critical region and critical value(s) that would be used to test $H_o:\sigma_1^2/\sigma_2^2=1$ vs. $H_a:\sigma_1^2/\sigma_2^2>1$, with $n_1=10$, $n_2=10$, and $\alpha=0.01$.

ANSWER:

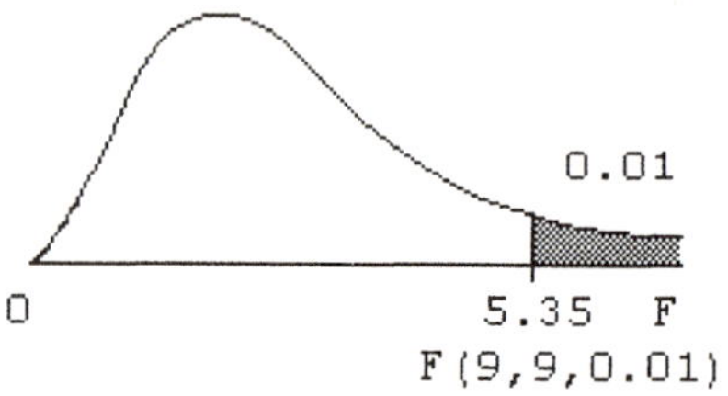

319. Draw an approximate *F*-curve and determine the critical region and critical value(s) that would be used to test $H_o:\sigma_1^2=\sigma_2^2$ vs. $H_a:\sigma_1^2<\sigma_2^2$, with $n_1=25$, $n_2=16$, and $\alpha=0.01$.

ANSWER:

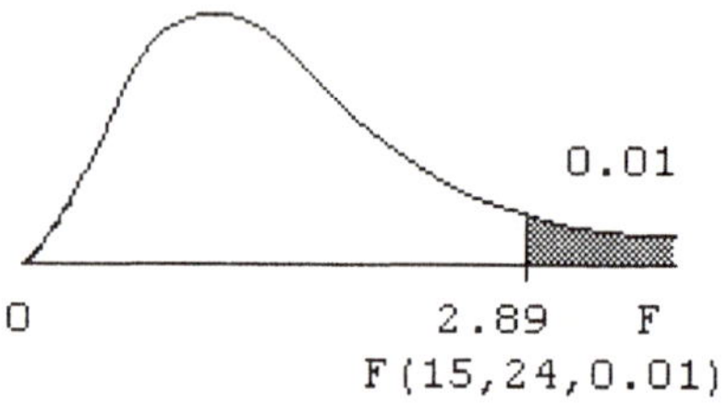

320. Use the table of critical values of the *F*-distribution to find *F*(9, 40, 0.01).

ANSWER:
2.89

321. Two independent samples of sizes 3 and 4, respectively, are drawn form a normally distributed population. Find the probability that the variance of the first sample is at least 16 times larger than the variance of the second sample.

ANSWER:
$P(s_1^2\geq 16s_2^2)=P(s_1^2/s_2^2\geq 16)=(F\geq 16\,|\,df=2,3)$ 0.025, since *F*(2, 3, 0.025) = 16.

<u>QUESTIONS 322 THROUGH 324 ARE BASED ON THE FOLLOWING INFORMATION:</u>
Consider testing $H_o:\sigma_1^2/\sigma_2^2=1\,(\leq)$ vs. $H_a:\sigma_1^2/\sigma_2^2>1$ given that $n_1=16, s_1=4.1, n_2=20, s_2=2.5$.

322. Calculate the value of the test statistics.

ANSWER:
$F^*=s_1^2/s_2^2=(4.1)^2/(2.5)^2$ = 2.6896 $\approx$ 2.69

323. Test the hypothesis at the 0.05 level of significance using the *p*-value approach.

ANSWER:
P = p-value = *P*(*F* > 1.46 | *df* =15, 19) $\Rightarrow$ 0.01 < **P** < 0.025
Since *p*-value < α = 0.05, we reject H_o.

324. Test the hypothesis at the 0.05 level of significance using the classical approach.

ANSWER:
The critical value is *F*(15,19,0.05) = 2.23.
Since F^* = 2.69 falls in the rejection region, we reject H_o.

325. Two independent samples, each of size 3, are drawn form a normally distributed population. Find the probability that one of the sample variances is at least 19 times larger than the other one.

ANSWER:
$P(s_1^2 \geq 19s_2^2 \text{ or } s_2^2 \geq 19s_1^2) = P(s_1^2/s_2^2 \geq 19) + P(s_2^2/s_1^2 \geq 19) = 2P(F \geq 19 \mid df = 2,2)$
= 2(0.05) = 0.10, since *F*(2, 2, 0.05) = 19

326. Two independent samples of sizes 3 and 5, respectively, are drawn form a normally distributed population. Find the probability that the variance of the first sample is at least 18 times larger than the variance of the second sample.

ANSWER:
$P(s_1^2 \geq 18s_2^2) = P(s_1^2/s_2^2 \geq 18) = (F \geq 18 \mid df = 2,4)$ = 0.01, since *F*(2, 4, 0.01) = 18

<u>QUESTIONS 327 THROUGH 330 ARE BASED ON THE FOLLOWING INFORMATION:</u>
The standard deviation of GRE scores for 37 female students was 23.9, and the standard deviation for 36 male students was 6.8. Assume that GRE scores are normally distributed for both groups. The director of a graduate program at one of the 15 Michigan public universities believes that the standard deviation of GRE scores for females is larger than the standard deviation of GRE scores for males.

327. State the appropriate null and alternative hypotheses for this situation.

ANSWER:
$H_o: \sigma_F/\sigma_M = 1\ (\leq)$ and $H_a: \sigma_F/\sigma_M > 1$

328. Calculate the value of the test statistic.

ANSWER:
$F^* = s_F^2/s_M^2 = (23.9)^2/(6.8)^2$ = 12.35

329. Test the hypothesis in question 327 at α = 0.01 using the *p*-value approach.

ANSWER:
P = *p*-value = *P*(*F* > 12.35 | *df* = 36, 35) $\Rightarrow$ **P** < 0.01
Since *p*-value < α = 0.01, we reject H_o. There is sufficient evidence to support the director's belief that that the standard deviation of GRE scores for females is larger than the standard deviation of GRE scores for males.

330. Test the hypothesis in question 327 at $\alpha = 0.05$ using the classical approach.

ANSWER:
The critical value is $F(36, 35, 0.01) = 2.24$.

Since $F^* = 12.35$ falls in the rejection region, we reject H_o. We reach the same conclusion as stated in question 329.

QUESTIONS 331 THROUGH 334 ARE BASED ON THE FOLLOWING INFORMATION:
A study was conducted to determine whether or not there was equal variability in male and female systolic blood pressures. Random samples of 16 men and 13 women were used to test the experimenter's claim that the variances were unequal. The data are given below:

Men:	116	118	116	110	118	112	122	108
	120	123	124	100	115	110	110	120
Women	104	100	110	121	106	122	98	
	102	114	100	120	118	125		

331. Use computer to calculate summary measures for the two samples.

ANSWER:

Summary stats for two samples			
		Men	Women
	Sample sizes	16	13
	Sample means	115.125	110.769
	Sample standard deviations	6.438	9.671
	Sample variances	41.450	93.526

332. State the null and alternative hypotheses.

ANSWER:
$H_o: \sigma_W^2 / \sigma_M^2 = 1$ vs. $H_a: \sigma_W^2 / \sigma_M^2 \neq 1$

333. Calculate the value of the test statistic.

ANSWER:
$F^* = s_W^2 / s_M^2 = 93.526 / 41.45 = 2.256$

334. Do the sample data support the experimenter's claim at the 0.05 level of significance? Use the classical approach.

ANSWER:
The critical values for this test are: left tail, $F(12, 15, 0.975)$ and right tail, $F(12, 15, 0.025)$. However, since we chose the sample with the larger variance for the numerator, the value of F^* is greater than one, and will be in the right-hand tail; therefore, only the right-hand critical value is needed. Since $F(12, 15, 0.025) = 2.96$ and $F^* = 2.256$, we fail to reject H_o. There is not sufficient evidence to support the experimenter's claim that the variances were unequal.

Chapter 11

APPLICATONS OF CHI-SQUARE

Sections 11.1 Through 11.3

True-False Questions

1. For the chi-square distribution, the mean equals the mode.
ANSWER: F

2. The value of χ^2 may be negative, zero, or positive.
ANSWER: F

3. The chi-square distribution is skewed to the right.
ANSWER: T

4. A hypothesis test involving a multinomial experiment is always a left-tail test.
ANSWER: F

5. When using the chi-square distribution in a hypothesis test for a multinomial experiment, the number of degrees of freedom is the number of cells.
ANSWER: F

6. In a multinomial experiment, $n = \sum_{all\,cells} O = \sum_{all\,cells} E$.
ANSWER: T

7. In a multinomial experiment, $\sum(O-E)$ must equal zero since $\sum O = \sum E = n$.
ANSWER: T

8. The expected frequency in a chi-square test of a multinomial experiment is found by multiplying the hypothesized probability of a cell by the number of pieces of data in the sample.
ANSWER: T

9. In the multinomial experiment we have (r -1) times (c -1) degrees of freedom, where r is the number of rows, and c is the number of columns.
ANSWER: F

10. A multinomial experiment consists of n identical independent trials.
ANSWER: T

11. A multinomial experiment arranges the data into a two-way classification such that the totals in one direction are predetermined.
ANSWER: F

12. The chi-square test statistic $\chi^2* = \sum (O-E)^2 / E$ has a distribution that is approximately normal.
ANSWER: F

13. The data used in a chi-square multinomial test are always enumerative in nature.
ANSWER: T

14. The shape of the chi-square distribution depends on the size of the sample.
ANSWER: F

15. The chi-square distribution is skewed to the left (negatively skewed).
ANSWER: F

16. The chi-square distribution can assume only positive values.
ANSWER: T

17. The critical value at the 0.05 level of significance for a chi-square multinomial test where there are six categories is 11.07.
ANSWER: T

18. If the value of the chi-square test statistic is less than the critical value, the null hypothesis must be rejected at a predetermined level of significance.
ANSWER: F

19. The chi-square multinomial test can be applied if there are equal or unequal expected frequencies.
ANSWER: T

20. A multinomial experiment, in general, differs from a binomial experiment in that each trial has three or four outcomes rather than two outcomes.
ANSWER: F

21. The chi-square distribution will be used to test hypotheses concerning enumerated data.
ANSWER: T

22. The middle 0.95 portion of the chi-square distribution with 9 degrees of freedom has table values of 3.33 and 16.9 respectively.
ANSWER: F

23. Suppose that we have k cells into which n observations have been sorted, where the observed frequencies in each cell are denoted by $O_1, O_2, ..., O_k$ and the expected or theoretical frequencies are denoted by $E_1, E_2, ..., E_k$. Then $\sum_{i=1}^{k} O_i$, $\sum_{i=1}^{k} E_i$, and n must be exactly the same value.
ANSWER: T

24. In hypothesis testing, the null hypothesis is always a statement about a population parameter.
ANSWER: F

25. $\sum(O - E)$ must always equal zero, where the symbols O and E refer to the observed and expected frequencies, respectively.
ANSWER: T

26. All multinomial experiments result in equal expected frequencies.
ANSWER: F

27. A multinomial experiment, where the outcome of each trial can be classified into one of two categories, is identical to the binomial experiment.
ANSWER: T

28. For a chi-square distributed random variable with 10 degrees of freedom and a level of significance of 0.025, the chi-square critical value is 20.5. If the computed value of the test statistics is 17.87, this will lead us to reject the null hypothesis.
ANSWER: F

Multiple-Choice Questions

29. When computing χ^2 we see that:

A) large values of χ^2 indicate agreement between the two sets of frequencies.
B) large values of χ^2 indicate disagreement between the two sets of frequencies.
C) χ^2 uses only continuous variables.
D) χ^2 uses both continuous and categorical variables.
ANSWER: B

30. If H_o: $P(A) = 0.15$, $P(B) = 0.25$, $P(C) = 0.35$, and $P(D) = 0.25$ is the null hypothesis in a hypothesis test for a multinomial experiment, what is the appropriate alternative hypothesis?

A) H_a: $P(A) = 0.25$, $P(B) = 0.25$, $P(C) = 0.25$, and $P(D) = 0.25$
B) H_a: $P(A) \neq 0.15$, $P(B) = 0.25$, $P(C) \neq 0.35$, and $P(D) = 0.25$
C) H_a: all the probabilities are distributed differently from those listed in H_o.
D) H_a: one of the probabilities is distributed differently from those listed in H_o.
ANSWER: D

Chapter 11

31. In a multinomial experiment with more than five cells and with $\alpha \leq 0.10$, which of the following could not be a critical value of χ^2?

A) 6.00
B) 10.00
C) 14.00
D) 18.00
ANSWER: A

32. In a chi-square test comparing observed to expected frequencies, we fail to reject the null hypothesis whenever the observed frequencies are

A) each approximately equal to their corresponding expected frequency.
B) significantly greater than the expected frequencies.
C) considerably smaller than the expected frequencies.
D) not equal.
ANSWER: A

33. If $\alpha > 0.50$, which of the following is a possible value of $\chi^2(16, \alpha)$?

A) 17.0
B) 18.0
C) 19.0
D) None of these is possible.
ANSWER: D

34. Which of the following is *not* a characteristic of a multinomial experiment?

A) There are *n* identical independent trials.
B) The sum of the observed frequencies is *n*.
C) There are *k* possible cells.
D) The expected frequency for cell *i* is $P_i + O_i$ where P_i is the probability for cell *i* and O_i is the observed frequency for cell *i*.
ANSWER: D

35. Which of the following statements is false?

A) In repeated sampling, the calculated value of the test statistic $\chi^{2*} = \sum_{all\ cells} (O-E)^2 / E$ will have a sampling distribution that can be approximated by the standard normal distribution when *n* is large.
B) The chi-square distributions, like the Student *t*-distributions, are a family of probability distributions, each one being identified by the parameter number of degrees of freedom, *df*.
C) A *categorical variable* is a variable that classifies or categorizes each individual into exactly one of several cells or classes; these cells or classes are all-inclusive and mutually exclusive.
D) None of the above.
ANSWER: A

36. Which of the following statements is false?

A) $\sum(O-E)$ must always equal zero, where *O* and *E* are the observed and expected frequencies, respectively.
B) In a multinomial experiment, *df* =*k*, where *k* is the number of cells.
C) Not all multinomial experiments result in equal expected frequencies.
D) None of the above.
ANSWER: B

37. In a chi-square test of multinomial parameters, suppose that a sample showed that the observed frequency O_i and expected frequency E_i were equal for each cell *i*. Then, the null hypothesis is

A) rejected at $\alpha = 0.05$ but is not rejected at $\alpha = 0.025$.
B) not rejected at $\alpha = 0.05$ but is rejected at $\alpha = 0.025$.
C) rejected at any level of significance α.
D) not rejected at any α level.
ANSWER: D

38. In a chi-square test of multinomial parameters, suppose that the value of the test statistic is 13.08 and the number of degrees of freedom is 6. At the 5% significance level, the null hypothesis is

A) rejected and *p*-value for the test is smaller than 0.05.
B) not rejected, and *p*-value for the test is greater than 0.05.
C) rejected, and *p*-value for the test is greater than 0.05.
D) not rejected, and *p*-value for the test is smaller than 0.05.
ANSWER: A

39. Of the values for a chi-square test statistic listed below, which one is likely to lead to rejecting the null hypothesis in a goodness-of-fit test?

A) 0.78
B) 2.02
C) 1.94
D) 45.1
ANSWER: D

40. If we use the χ^2 test of multinomial parameters to test for the differences among 5 proportions, the degrees of freedom are equal to:

A) 2.
B) 3.
C) 4.
D) 5.
ANSWER: C

Chapter 11

Short-Answer Questions

41. Explain what is mean by "a categorical variable."

ANSWER:
Categorical variable categorizes each individual into exactly one of several cells or classes, are all-inclusive and mutually exclusive.

42. Find $\chi^2(27, 0.10)$.

ANSWER:
36.7

43. One guideline to ensure a good approximation to the χ^2 distribution is that $E_i \geq 5$. If this is not possible, what would be a possible solution?

ANSWER:
Combine smaller cells

44. Find $\chi^2(15, 0.99)$.

ANSWER:
5.23

45. Complete the following statement: multinomial experiments will always use a ___________ critical region.

ANSWER:
positive

46. Find $\chi^2(11, 0.05)$.

ANSWER:
19.7

47. Briefly discuss the assumptions for using chi-square distribution to make inferences based on enumerative data.

ANSWER:
The sample information is obtained using a random sample drawn from a population in which each individual is classified according to the categorical variable(s) involved in the test.

48. What is meant by categorical variable?

ANSWER:
A *categorical variable* is a variable that classifies or categorizes each individual into exactly one of several cells or classes; these cells or classes are all-inclusive and mutually exclusive.

Applied and Computational Questions

49. Classes at a large university that meet on Monday, Wednesday, and Friday were sampled for student absence. Using the following results, state the null and alternative hypotheses to test the claim that absences occur on the three days with equal frequency

Day	Monday	Wednesday	Friday
Number of Students Absent	238	197	267

ANSWER:

$H_o: p_1 = 1/3,\ p_2 = 1/3,\ p_3 = 1/3$;

H_a: at least one of the probabilities in H_o is different from the others.

50. A water slide has five different runs. To determine if the runs are equally popular, a count of usage is kept over a period of one week. Using the following results, test for uniform usage at $\alpha = 0.05$ and give the critical region, χ^{2*} and your conclusion.

Run	Observed Number
1	400
2	500
3	450
4	500
5	150

ANSWER:

$H_o: p_1 = 0.20,\ p_2 = 0.20,\ p_3 = 0.20,\ p_4 = 0.20,\ p_5 = 0.20$;

H_a: at least one of the probabilities in H_o is different from the others.

Critical region: $\chi^2 \geq 9.49$; Value of the test statistic: $\chi^2{*} = 212.5$;
Conclusion: reject of uniform usage.

51. The following table gives theoretical distribution over four categories and the actual observed distribution. Why would you be reluctant to apply the chi-square analysis to determine the goodness of fit in this sample?

Category	Theoretical Percent	Observed Number
1	0.05	5
2	0.45	15
3	0.39	15
4	0.11	5

ANSWER:
The chi-square analysis should not be applied to determine the goodness of fit in this sample because the expected frequencies are not greater than 5 in two of the categories.

Chapter 11

QUESTIONS 52 AND 53 ARE BASED ON THE FOLLOWING INFORMATION:
Mars, Inc., the manufacturer of M&M candies, claims that the distribution of the different colors of candies in a bag of M&Ms (brown, red, yellow, green, orange, and blue) will appear in the ratio 3:2:2:1:1:1. In testing this claim, Mars, Inc. obtained frequencies of 38, 15, 33, 4, 6, and 4, respectively.

52. State the null and alternative hypotheses to test the claim to support this ratio.

ANSWER:

H_o: 3:2:2:1:1:1 is ratio of candies in bag;

H_a : 3:2:2:1:1:1 is not ratio of candies in bag

53. Find the computed value of χ^2. If α = 0.05, what decision would be made?

ANSWER:

	Color					
	Brown	Red	Yellow	Green	Orange	Blue
Expected %	30	20	20	10	10	10
Observed %	38	15	33	4	6	4

$\chi^2 *$ = 20.633, and the critical value is χ^2(5, 0.05) = 11.10. Reject H_o at α = 0.05, and conclude that Mars' claim is not correct.

QUESTIONS 54 THROUGH 57 ARE BASED ON THE FOLLOWING INFORMATION:
A research report gives the following seasonal distribution of colds. A researcher randomly selects 200 cases from a large clinic that have been diagnosed as a cold and observed the results shown in the table below. The researcher wishes to test that the clinic has the reported seasonal distribution at α = 0.05.

Season	*Report Percent*	*Observed frequency*
Winter	20	30
Spring	35	80
Summer	10	25
Fall	35	65

54. State the null and alternative hypotheses.

ANSWER:

$H_o : p_1 = 0.20,\ p_2 = 0.35,\ p_3 = 0.25,\ p_4 = 0.25$

H_a : at least one of the probabilities in H_o is different from the others.

55. Calculate the value of the test statistic.

ANSWER:

Test statistic: $\chi^2 *$ = 5.54

56. Determine the *p*-value.

ANSWER:

0.10 < *p*-value < 0.25

57. State the decision and conclusion.

ANSWER:
Since *p*-value > α , fail to reject H_o that the clinic has the reported seasonal distribution.

QUESTIONS 58 THROUGH 60 ARE BASED ON THE FOLLOWING INFORMATION:
Using a deck containing 52 cards and 4 suits, a gambler draws one card and noted whether a club, diamond, heart, or spade is drawn. The card is replaced and another one is drawn. This experiment is performed 100 times, and the results are shown in the table below. The gambler wishes to determine if the results indicate an equal number of clubs, diamonds, hearts, and spades in the deck?

Category	*Observed Number*
Club	25
Diamond	15
Heart	30
Spade	30

58. State the null and alternative hypotheses.

ANSWER:
$H_o: p_1 = 0.25,\ p_2 = 0.25,\ p_3 = 0.25,\ p_4 = 0.25$
H_a : at least one of the probabilities in H_o is different from the others.

59. Determine the critical region at α = .05 and calculate the value of the test statistic.

ANSWER:
Critical region: $\chi^2 \geq 7.82$;

60. State the decision and conclusion.

ANSWER:
χ^2 * = 6.00. Fail to reject H_o since $X^2* < X^2$. The data indicated an equal number of clubs, diamonds, hearts, and spades in the deck.

QUESTIONS 61 THROUGH 63 ARE BASED ON THE FOLLOWING INFORMATION:
If a fair coin is tossed three times, the number of heads to occur has a binomial distribution with the probability distribution given in the table below. A coin is tossed three times, with the experiment repeated 100 times. The observed frequencies are shown in the table. One wishes to determine if the coin is fair at α = 0.05.

Number of Heads *x*	*P*(*x*)	Observed frequency
0	0.125	20
1	0.375	25
2	0.375	35
3	0.125	20

61. State the null and alternative hypotheses.

ANSWER:
H_o: Coin is fair, (Each *p* = 0.50).

H_a: Coin is unfair, (At least one *p* is different from the other).

62. Determine the critical region and calculate the value of the test statistic.

ANSWER:
The critical region is $\chi^2 \geq 7.82$, and the test statistic is $\chi^2* = 13.3$.

63. State the decision and conclusion.

ANSWER:
Reject H_o at $\alpha = 0.05$ since $\chi^2* = 13.3 > \chi^2 = 7.82$. There is sufficient evidence to indicate that the coin is unfair.

64. Suppose we have a multinomial experiment with the cells shown below. What observed frequencies *a*, *b*, *c*, *d*, and *e* would result in $\chi^2 = 0$, if we were testing the hypothesis that I, II, III, IV, and V occur in the ratio 10:7:5:4:2 with a random sample of size 840?

I	II	III	IV	V
a	*b*	*c*	*d*	*e*

ANSWER:
a = 300, *b* = 210, *c* = 150, *d* = 120, and *c* = 60

QUESTIONS 65 THROUGH 67 ARE BASED ON THE FOLLOWING INFORMATION:
An instructor claims that final grades in his course occur in the ratio 1:3:5:2:1 for the grades of A, B, C, D, and F. A random sample of 240 of the students showed that 15 received a grade of A, 55 received a grade of B, 90 received a grade of C, 50 received a grade of D, and 30 received a grade of F. Find the computed value of χ^2. If $\alpha = 0.05$, what decision would be made?

65. State the null and alternative hypotheses.

ANSWER:
H_o: Final grades ratio is 1:3:5:2:1
H_a: Final grades ratio is not 1:3:5:2:1

66. Calculate the value of the test statistic and determine the critical region at $\alpha = 0.05$.

ANSWER:
Test statistic: $\chi^2* = 10.167$; Critical region: $\chi^2 \geq 9.49$

67. State the decision and conclusion.

ANSWER:
Reject H_o since $\chi^2* > \chi^2$. There is sufficient evidence to conclude that final grades ratio is not as claimed by the instructor.

68. In a multinomial experiment with three cells we are testing the claim that $p_1 = p_2 = p_3$ using α = 0.05. If the observed frequencies in the first two cells are 20 and 16, what are the possible observed frequencies in the third cell which would cause us to fail to reject the claim?

ANSWER:
The possible observed frequencies would be 8, 9, 10, . . . , 29, 30, 31.

69. At a large university five different professors teach the same course. Random samples of 50 students taking the course from each of the instructors were selected. The number of students earning satisfactory grades in the course (A, B, or C) and the number earning unsatisfactory grades in the course were determined. The number of satisfactory grades from each of the instructors were 35, 42, 30, 40, and 39. Does the sample evidence support the claim that satisfactory grades are given in the same proportion by all five instructors? Use α = 0.05. Find the computed value of χ^2 and state the decision.

ANSWER:
H_o: Satisfactory grades are given in the same proportion by all five instructors.

H_a: Satisfactory grades are not given in the same proportion by all five instructors.

$\chi^2 *$ = 9.53, critical region: χ^2 (4, 0.05) = 9.49. We barely reject the null hypothesis at α = 0.05. We conclude that the sample evidence does not support the claim that satisfactory grades are given in the same proportion by all five instructors.

70. State the null hypothesis, H_o, and the alternative hypothesis, H_a, that would be used to test the following statement: The five numbers: 10, 11, 12, 13, and 14, are equally likely.

ANSWER:
H_o: *P*(10) = *P*(11) = *P*(12) = *P*(13) = *P*(14) = 0.20

H_a: The numbers are not equally likely.

71. State the null hypothesis, H_o, and the alternative hypothesis, H_a, that would be used to test the following statement: The multiple-choice question with choices A, B, C, D, and E has a history of students selecting answers in the ratio of 2:3:2:1:2.

ANSWER:
H_o: *P*(A) = 0.20 , *P*(B) = 0.30, *P*(C) = 0.20, *P*(D) = 0.10, *P*(E) = 0.20

H_a: The probabilities are distributed differently than listed in H_o.

72. State the null hypothesis, H_o, and the alternative hypothesis, H_a, that would be used to test the following statement: The poll will show a distribution of 17%, 37%, 40%, and 6% for the possible ratings of excellent (E), good (G), fair (F), and poor (P) on a specific issue.

ANSWER:
H_o: *P*(E) = 0.17, *P*(G) = 0.37, *P*(F) = 0.40, *P*(P) = 0.06

H_a: The percentages are different than specified in H_o.

Chapter 11

QUESTIONS 73 THROUGH 77 ARE BASED ON THE FOLLOWING INFORMATION:
A manufacturer of floor polish conducted a consumer-preference experiment to determine which of five different floor polishes was the most appealing in appearance. A sample of 100 consumers viewed five patches of flooring that had each received one of the five polishes. Each consumer indicated the patch he or she preferred. The lighting and background were approximately the same for all patches. The results were as follows:

polish	A	B	C	D	E	Total
Frequency	30	17	14	21	18	100

73. State the null hypothesis for "no preference" in statistical terminology.

ANSWER:
H_o: P(A) = P(B) = P(C) = P(D) = P(E) = 0.20

74. What test statistic will be used in testing the null hypothesis in question 73?

ANSWER:
χ^2 test statistic

75. Calculate the value of the test statistic.

ANSWER:
The expected values are calculated as follows:
$E = np$ = 100(0.20) = 20, for all five cells

The observed and expected frequencies are shown in the table below:

polish	A	B	C	D	E	Total
Observed	30	17	14	21	18	100
Expected	20	20	20	20	20	100
$(O-E)^2/E$	5.00	0.45	1.80	0.05	0.20	7.50

$$\chi^{2*} = \sum_{allcells} [(O-E)^2/E] = 7.50.$$

76. Complete the hypothesis test at the 0.10 level of significance using the *p*-value approach and the classical approach.

ANSWER:
P = *p*-value = $P(\chi^2 > 7.50 \mid df = 4)$; Using the table of χ^2 distribution: 0.10 < **P** < 0.25. Since **P** > α = 0.10, fail to reject H_o, and conclude that the preferences of polish are not significantly different from equal proportions.

77. Complete the hypothesis test at the 0.10 level of significance using the classical approach.

ANSWER:

The critical region is $\chi^2(4,\ 0.10) \geq 7.78$. Since the test statistic χ^{2*} falls in the non-critical region, we fail to reject H_o at $\alpha = 0.10$, and conclude that the preferences of polish are not significantly different from equal proportions.

QUESTIONS 78 THROUGH 81 ARE BASED ON THE FOLLOWING INFORMATION:

Carter's supermarket carries four qualities of ground beef: A, B, C, and D, respectively. Customers are believed to purchase these four qualities with probabilities of 0.15, 0.30, 0.35, and 0.20, respectively, from the least to most expensive. A sample of 200 purchases resulted in sales of 18, 65, 77, and 40 of the respective qualities.

78. State the null and alternative hypotheses.

ANSWER:

H_o: P(A) = 0.15, P(B) = 0.30, P(C) = 0.35, P(D) = 0.20

H_a: The proportions are different than specified in H_o.

79. Calculate the value of the test statistic.

ANSWER:

The expected values are calculated according to the formula $E = np$. The observed and expected frequencies are shown in the table below:

Quality	A	B	C	D	Total
Observed	18	65	77	40	200
Expected	30	60	70	40	200
$(O-E)^2/E$	4.80	0.417	0.70	0.00	5.917

$\chi^{2*} = \sum[(O-E)^2/E] = 5.917$

80. Does this sample contradict the expected proportions at $\alpha = 0.05$? Solve using the *p*-value approach.

ANSWER:

P = *p*-value = $P(\chi^2 > 5.917 \mid df = 3)$; Using the table of χ^2 distribution: 0.10 < **P** < 0.25.

Since **P** > α = 0.05, fail to reject H_o, and conclude that the proportions of meat qualities bought at Carter's are not significantly different from the claimed proportions.

81. Does this sample contradict the expected proportions at $\alpha = 0.05$? Solve using the classical approach.

ANSWER:

The critical region is $\chi^2(3,\ 0.05) \geq 7.82$. Since the test statistic χ^{2*} falls in the noncritical region, we fail to reject H_o at $\alpha = 0.05$, and conclude that the proportions of meat qualities bought at Carters are not significantly different from the claimed proportions.

Chapter 11

QUESTIONS 82 THROUGH 85 ARE BASED ON THE FOLLOWING INFORMATION:
It is believed that about 40% of Americans own guns for hunting, 30% for protection, 18% for both hunting and protection, and 12% for other reasons. A survey in Detroit of 1000 individuals gave the following results.

Reasons for owning a gun	*Number Responding*
Hunting	370
Protection	300
Hunting and Protection	180
Other	150

Suppose you are interested in test the hypothesis that the distribution of reasons for owning a gun is the same in Detroit as it is nationally known.

82. State the null and alternative hypotheses.

ANSWER:

H_o: The proportions of reasons for owing a handgun are 0.40, 0.30, 0.18, 0.12.

H_a: The proportions are different than specified in H_o.

83. Calculate the value of the test statistic.

ANSWER:
The expected values are calculated according to the formula $E = np$ as follows:
The observed and expected frequencies are shown in the table below:

Handgun	Hunting	Protection	Hunting and protection	Other	Total
Observed	370	300	180	150	1000
Expected	400	300	180	120	1000
$(O-E)^2/E$	2.25	0.0	0.0	7.5	9.75

$$\chi^{2*} = \sum_{allcells} [(O-E)^2/E] = 9.75$$

84. Complete the hypotheses test at $\alpha = 0.05$ using the *p*-value approach.

ANSWER:

P = *p*-value = $P(\chi^2 > 9.75 \mid df = 3)$; Using the table of χ^2 distribution: 0.01 < **P** < 0.025. Since $P < \alpha = 0.05$, reject H_o, and conclude that the proportions for reasons for owning a handgun in Detroit are significantly different from those nationally at the 0.05 level of significance.

85. Complete the hypotheses test at $\alpha = 0.05$ using the classical approach.

ANSWER:

The critical region is: $\chi^2(3, 0.05) \geq 7.82$. Since the test statistic χ^{2*} falls in the critical region, therefore we reject H_o and conclude that the proportions for reasons for owning a handgun in Detroit are significantly different from those nationally at the 0.05 level of significance.

QUESTIONS 86 THROUGH 89 ARE BASED ON THE FOLLOWING INFORMATION:
A sample of 500 individuals are tested for their blood type: A, B, O, or AB, and the results are used to test the hypothesized distribution of blood types that 41% A, 9% B, 46% O, and 4% AB. The observed results were as follows:

Blood Type	A	B	O	AB
Number	190	50	245	15

A doctor wishes to determine if there is sufficient evidence to show that the stated distribution is incorrect.

86. State the null and alternative hypotheses.

ANSWER:
H_o: *P*(A) = 0.41, *P*(B) = 0.09, *P*(O) = 0.46, *P*(AB) = 0.04

H_a: The proportions are different than stated in H_o.

87. Calculate the value of the test statistic.

ANSWER:
The expected values are calculated according to the formula *E* = *np* as follows:
The observed and expected frequencies are shown in the table below:

	A	B	O	AB	Total
Observed	190	50	245	15	500
Expected	205	45	230	20	500
$(O-E)^2/E$	1.098	0.556	0.978	1.250	3.882

$$\chi^{2*} = \sum_{allcells} [(O-E)^2/E] = 3.882.$$

88. Complete the hypothesis test at the 0.05 level of significance using the *p*-value approach.

ANSWER:
P = *p*-value = $P(\chi^2 > 3.882 \mid df = 3)$; Using the table of χ^2 distribution: 0.25 < **P** < 0.50.
Since $P > \alpha = 0.05$, fail to reject H_o, and conclude that we do not have sufficient evidence to show that the hypothesized distribution of blood types is incorrect.

89. Complete the hypothesis test at the 0.05 level of significance using the classical approach.

ANSWER:
The critical region is $\chi^2 \geq 7.82$. Since the test statistic χ^{2*} falls in the noncritical region, we fail to reject H_o at $\alpha = 0.05$, and conclude that we do not have sufficient evidence to show that the hypothesized distribution of blood types is incorrect.

Chapter 11

QUESTIONS 90 THROUGH 93 ARE BASED ON THE FOLLOWING INFORMATION:
A biology professor claimed that the proportions of grades in his classes are the same. A sample of 100 students showed the following frequencies:

Grade	A	B	C	D	F
Frequency	18	20	28	23	11

90. State the null and alternative hypotheses to be tested.

ANSWER:

H_o : *P*(A) = *P*(B) = *P*(C) = *P*(D) = *P*(F) = 0.20

H_a : At least one proportion differs from its specified value.

91. Determine the rejection region at the 5% significance level.

ANSWER:

Reject H_o if $\chi^{2*} > \chi^2(4, 0.05) = 9.49$.

92. Compute the value of the test statistics.

ANSWER:

$\chi^{2*} = 7.90$

93. Do the data provide enough evidence to support the professor's claim?

ANSWER:

Since $\chi^{2*} = 7.90 < 9.49$, we fail to reject H_o. The data provide enough evidence to support the professor's claim.

QUESTIONS 94 THROUGH 97 ARE BASED ON THE FOLLOWING INFORMATION:
The mathematics department at a certain college in Texas claims that the grades in its introductory algebra course are distributed as follows: 10% A's, 20% B's, 40% C's, 20% D's, and 10% F's. In a poll of 400 randomly selected students who had completed this course, it was found that 45 had received A's, 95 B's, 150 C's, 60 D's, and 50 F's.

94. State the null and alternative hypotheses.

ANSWER:

H_o: The distribution of grades is 10% A's, 20% B's, 40% C's, 20% D's, 10% F's.

H_a: The distribution of grades is different than stated in H_o.

95. Calculate the value of the test statistic.

ANSWER:
The observed and expected frequencies are shown in the table below, where *E* = *np*.

	A	B	C	D	F	Total
Observed	45	95	150	60	50	400
Expected	40	80	160	80	40	400

$\chi^{2*} = \sum[(O-E)^2/E]$ = 0.625 + 2.813 + 0.625 + 5.0 + 2.5 = 11.563

96. Does this sample contradict the department's claim at the 0.05 level? Solve using the *p*-value approach.

ANSWER:
P = *p*-value = $P(\chi^2 > 11.563 \mid df = 4)$; Using the table of χ^2 distribution: 0.01< **P** < 0.025. Since **P**< α = 0.05, reject H_o. There is sufficient evidence at the 0.05 level of significance to show that the grade distribution is different than claimed. In other words, this sample contradicts the department's claim.

97. Does this sample contradict the department's claim at the 0.05 level? Solve using the classical approach.

ANSWER:
The critical region is $\chi^2 \geq 9.50$. Since the test statistic χ^{2*} falls in the critical region, we reject H_o at α = 0.05. There is sufficient evidence at the 0.05 level of significance to show that the grade distribution is different than claimed. In other words, this sample contradicts the department's claim.

98. Find the critical value $\chi^2(20, 0.01)$.

ANSWER:
37.6

99. Find the critical value $\chi^2(18, 0.025)$.

ANSWER:
21.6

100. Find the critical value $\chi^2(30, 0.10)$.

ANSWER:
40.3

101. Find the critical value $\chi^2(45, 0.025)$.

ANSWER:
It is approximately (59.3 + 71.4) / 2 = 65.35.

102. Find the critical value $\chi^2(10, 0.05)$.

ANSWER:
18.3

103. Find the critical value $\chi^2(12, 0.01)$.

ANSWER:
26.2

Chapter 11

104. Find the critical value $\chi^2(50, 0.975)$.

ANSWER:
32.4

105. State the null hypothesis, H_o, and the alternative hypothesis, H_a, that would be used to test the following statement: "The numbers, 1, 2, 3, and 4, are equally likely to be drawn."

ANSWER:
H_o: $P(1) = P(2) = P(3) = P(4) = 0.25$
H_a: The numbers are not equally likely.

106. State the null hypothesis, H_o, and the alternative hypothesis, H_a, that would be used to test the following statement: "That multiple-choice question with four possible answers, A, B, C, and D, has a history of students selecting answers in the ratio of 2:3:2:1, respectively."

ANSWER:
H_o: $P(A) = 2/8$, $P(B) = 3/8$, $P(C) = 2/8$, $P(D) = 1/8$
H_a: The possible answers A, B, C, and D are not equally likely.

107. Find the critical value $\chi^2(65, 0.90)$.

ANSWER:
It is approximately (46.5 + 55.3) / 2 = 50.9.

108. State the null hypothesis, H_o, and the alternative hypothesis, H_a, that would be used to test the following statement: "The poll will show a distribution of 7%, 15%, 38%, and 40% for the possible ratings of excellent (E), good (G), fair (F), and poor (P) on US foreign policy in the Middle East during George W. Bush administration."

ANSWER:
H_o: $P(E) = 0.07$, $P(G) = 0.15$, $P(F) = 0.38$, $P(P) = 0.40$
H_a: At least one percentage is different than specified in H_o.

109. Place bounds on the *p*-value for testing the null hypothesis $H_o: P(1) = P(2) = P(3) = P(4) = P(5) = 0.20$, given that the value of the test statistic $\chi^2{*} = 12.89$.

ANSWER:
$\mathbf{P} = p\text{-value} = P(\chi^2 > 12.89 \mid df = 4) \Rightarrow 0.01 < \mathbf{P} < 0.025$

110. Determine the critical value and critical region that would be used in the classical approach of a multinomial experiment to test the null hypothesis $H_o: P(1) = P(2) = P(3) = P(4) = P(5) = P(6) = 1/6$, with level of significance $\alpha = 0.05$.

ANSWER:
The critical value = $\chi^2(5, 0.05) = 11.1$ and the critical region is the right hand-tail area that is greater than 11.1. The null hypothesis H_o is rejected if the value of the test statistic $\chi^{2*} > 11.1112$.

111. Determine the critical value and critical region that would be used in the classical approach of a multinomial experiment to test the null hypothesis H_o : *P*(A) = 0.28, *P*(B) = 0.37, *P*(C) = 0.35, with $\alpha = 0.01$.

ANSWER:
The critical value = χ^2(2, 0.01) = 9.21 and the critical region is the right hand-tail area that is greater than 9.21. The null hypothesis H_o is rejected if the value of the test statistic χ^{2*} > 9.21.

112. In 2004, Brand *A* microwaves had 45% of the market, Brand *B* had 35%, and Brand *C* had 20%. This year the makers of brand *C* launched a heavy advertising campaign. A random sample of appliance stores shows that of 10,000 microwaves sold, 4350 were Brand *A*, 3450 were Brand *B*, and 2200 were Brand *C*. Has the market changed? Test at $\alpha = 0.01$.

ANSWER:
$H_o : p_1 = 0.45$, $p_2 = 0.35$, $p_3 = 0.20$
H_a : At least two proportions differ from their specified values.
The critical value is $\chi^2(2, 0.01)$ = 9.21, and the value of the test statistic: $\chi^{2*} = 25.714$. Therefore, we reject the null hypothesis. There is sufficient evidence to indicate that the market has changed since 2004.

113. Place bounds on the *p*-value for testing the null hypothesis H_o : *P*(A) = 0.25, *P*(B) = 0.30, *P*(C) = 0.35, *P*(D) = 0.10 given that the value of the test statistic χ^2 * = 8.95.

ANSWER:
P = *p*-value = *P*(χ^2 > 8.95 | *df* = 3) $\Rightarrow$ 0.025 < **P** < 0.05

QUESTIONS 114 THROUGH 121 ARE BASED ON THE FOLLOWING INFORMATION:
A certain type of flower seed will produce magenta, chartreuse, and ochre flowers in the ratio 6:3:1 (one flower per seed). A total of 150 seeds are planted and all germinate, yielding the following results:

Magenta	Chartreuse	Ochre
78	54	18

114. State the null and alternative hypotheses.

ANSWER:
H_o: *P*(Magenta) = 0.60, *P*(Chartreuse) = 0.30, *P*(Ochre) = 0.10
H_a : At least one of the proportions is different than specified in H_o.

115. If the null hypothesis is true, what is the expected number of magenta flowers?

ANSWER:
E(magenta) = $n \cdot p$ = 150 (0.60) = 90

116. If the null hypothesis is true, what is the expected number of chartreuse flowers?

ANSWER:
E(Chartreuse) = $n \cdot p$ = 150 (0.30) = 45

117. If the null hypothesis is true, what is the expected number of ochre flowers?

ANSWER:
E(Ochre) = $n \cdot p$ = 150 (0.10) = 15

118. How many degrees of freedom are associated with chi-square?

ANSWER:
$k - 1 = 3 - 1 = 2$

119. Calculate the value of the test statistics.

ANSWER:
$\chi^{2*} = 1.6 + 1.8 + 0.6 = 4.0$

120. Complete the hypothesis test at $\alpha = 0.10$, using the p-value approach.

ANSWER:
P = p-value = $P(\chi^2 > 4.0 \mid df = 2) \Rightarrow 0.10 <$ **P** < 0.25.
Since p-value > $\alpha = 0.10$, we fail to reject H_o. There is no significant evidence to suggest that this type of flower seed will not produce magenta, chartreuse, and ochre flowers in the ratio 6:3:1. In other words, the proportions of the three colors are not significantly different from the 6:3:1 ratio.

121. Compute the hypothesis test at $\alpha = 0.10$ using the classical approach.

ANSWER:
The critical value = $\chi^2(2, 0.10) = 4.61$. Since $\chi^{2*} = 4$ does not fall in the rejection region, we fail to reject H_o. We reach the same conclusion as stated in question 120.

QUESTIONS 122 THROUGH 125 ARE BASED ON THE FOLLOWING INFORMATION:
A large supermarket carries four types of fish. Customers are believed to purchase these four types with probabilities of 0.10, 0.30, 0.35, and 0.25, respectively, from the least to most expensive type. A sample of 400 purchases resulted in sales of 37, 130, 153, and 80 of the respective types.

122. State the null and alternative hypotheses.

ANSWER:
H_o: $p_1 = 0.10$, $p_2 = 0.30$, $p_3 = 0.35$, $p_4 = 0.25$
H_a: At least one of the proportions is different than specified in H_o.

123. Calculate the value of the test statistic.

ANSWER:
The expected values are calculated according to the formula $E = np$. The observed and expected frequencies are shown in the table below:

	Quality				
	Type 1	Type 2	Type 3	Type 4	Total
Observed (O)	37	130	153	80	400
Expected (E)	40	120	140	100	200

$\chi^{2*} = \sum[(O-E)^2/E]$ = 6.27

124. Does this sample contradict the expected proportions? Test at the 0.05 level of significance using the *p*-value approach.

ANSWER:
P = *p*-value = *P*(χ^2 > 6.27 | *df* = 3) $\Rightarrow$ 0.05 < **P** < 0.10. Since **P** > α = 0.05, we fail to reject H_o. We conclude that the proportions of fish qualities bought at Carter's are not significantly different from the claimed proportions.

125. Does this sample contradict the expected proportions? Test at the 0.05 level of significance using the classical approach.

ANSWER:
The critical value = χ^2(3, 0.05) = 7.82. Since χ^{2*} = 6.27 does not fall in the rejection region, we fail to reject H_o. We reach the same conclusion as stated in question 124.

QUESTIONS 126 THROUGH 128 ARE BASED ON THE FOLLOWING INFORMATION:
A program for generating random numbers on a computer is to be tested. The program is instructed to generate 150 single-digit integers between 0 and 9. The frequencies of the observed integers were as follows:

Integer	0	1	2	3	4	5	6	7	8	9
Frequency	16	12	11	10	15	15	12	17	21	21

The programmer has sufficient reason to believe that the integers are not being generated uniformly.

126. State the null and alternative hypotheses.

ANSWER:
H_o: *P*(0) = *P*(1) = *P*(2) = ………… = *P*(9) = 0.10
H_a: At least one proportion is different than specified in H_o.

127. Test the hypotheses in question 126 at α = 0.10 using the *p*-value approach.

ANSWER:
Each of the expected frequencies = (150)(0.10) =15. $\chi^{2*} = \sum(O-E)^2/E$ = 9.07
P = *p*-value = *P*(χ^2 > 9.07 | *df* = 9) $\Rightarrow$ 0.25 < **P** < 0.50. Since **P** > α = 0. 10, we fail to reject H_o. There is sufficient reason to support the programmer's belief that the integers are being generated uniformly.

128. Test the hypotheses in question 126 at $\alpha = 0.10$ using the classical approach.

ANSWER:
The critical value = $\chi^2(9,\ 0.10)$ = 14.7 Since $\chi^{2*} = 9.07$ does not fall in the rejection region, we fail to reject H_o. We reach the same conclusion as stated in question 127.

QUESTIONS 129 THROUGH 132 ARE BASED ON THE FOLLOWING INFORMATION:
Skittles Original Fruit bite size candies are multiple colored candies in a bag and you can "Taste the Rainbow" with their five colors and flavors: Green-Lime, Purple-Grape, Yellow-Lemon, Orange-Orange, and Red-Strawberry. Unlike some of the other multi-colored candies available, Skittles claims their 5 colors are equally likely. In an attempt to reject this claim, an 8-ounce bag of Skittles was purchased and colors counted.

Red	Orange	Yellow	Green	Purple
34	40	44	32	50

129. State the null and alternative hypotheses.

ANSWER:
H_o: *P*(R) = *P*(O) = *P*(Y) = *P*(G) = *P*(P) = 0.20
H_a: At least one proportion is different than specified in H_o.

130. Does this sample contradict Skittles' claim? Test at the .05 level of significance using the *p*-value approach.

ANSWER:
Each of the expected frequencies = (200)(0.20) = 40. $\chi^{2*} = \sum (O-E)^2 / E$ = 5.40
P = *p*-value = $P(\chi^2 > 5.4 \mid df = 4) \Rightarrow 0.10 < \mathbf{P} < 0.25$ [Almost 0.25 since $\chi^2(4,\ 5.39) = 0.25$] Since **P** > α = 0. 05, we fail to reject H_o. There is no sufficient evidence to contradict Skittles's claim to conclude that these 5 colors are not equally likely.

131. Does this sample contradict Skittles' claim? Test at the .05 level of significance using the classical approach.

ANSWER:
The critical value = $\chi^2(4,\ 0.05)$ = 9.49. Since $\chi^{2*} = 5.4$ does not fall in the rejection region, we fail to reject H_o. We reach the same conclusion as stated in question 130.

132. Suppose we purchase a 16-ounce bag and count the five colors. The results are shown below:

Red	Orange	Yellow	Green	Purple
68	80	88	64	100

Calculate the value of chi-square for these data. How is the new chi-square value related to the one found in question 130? What effect does this new value have on the test results? Explain.

ANSWER:
Each of the expected frequencies = (400)(0.20) = 80, and the new chi-square value χ^{2*} = 10.80. This value is exactly twice the value found in question 130. In this case, we reject H_o since χ^{2*} = 10.80 falls in the rejection region $\chi^2 > 9.49$. Now, we can say that there is sufficient evidence to contradict Skittles's claim. We may conclude that these 5 colors are not equally likely.

QUESTIONS 133 THROUGH 136 ARE BASED ON THE FOLLOWING INFORMATION:
When interbreeding two strains of roses. We expect the hybrid to appear in three genetic classes in the ratio 1:3:4. The results of an experiment yield 60 hybrids of the first type, 255 of the second type, and 285 of the third type.

133. State the null and alternative hypotheses.

ANSWER:
H_o: $p_1 = 0.125$, $p_2 = 0.375$, $p_3 = 0.500$
H_a: At least one of the proportions is different than specified in H_o.

134. Calculate the value of the test statistic.

ANSWER:
The expected values are calculated according to the formula *E* = *np*. The observed and expected frequencies are shown in the table below:

	Quality Type 1	Type 2	Type 3	Total
Observed (O)	60	255	285	600
Expected (E)	75	225	300	600

$$\chi^{2*} = \sum_{allcells} [(O-E)^2/E] = 3.0 + 4.0 + 0.75 = 7.75$$

135. Do we have sufficient evidence to reject the hypothesized genetic ratio at the 0.05 level of significance? Test using the *p*-value approach.

ANSWER:
P = *p*-value = $P(\chi^2 > 7.75 \mid df = 2) \Rightarrow 0.01 < \mathbf{P} < 0.025$. Since $\mathbf{P} < \alpha = 0.05$, we reject H_o There is sufficient evidence to reject the hypothesized genetic ratio at the 0.05 level of significance.

136. Do we have sufficient evidence to reject the hypothesized genetic ratio at the 0.05 level of significance? Test using the classical approach.

ANSWER:
The critical value = $\chi^2(2, 0.05) = 5.99$ Since $\chi^{2*} = 7.75$ does fall in the rejection region, we reject H_o. We reach the same conclusion as stated in question 135.

Chapter 11

QUESTIONS 137 THROUGH 140 ARE BASED ON THE FOLLOWING INFORMATION:

A national survey states that 67% of college students are under the age of 25, 21% are between the ages of 25 and 30, 8% are between 30 and 40, and 4% are over 40. A random sample of 250 students at Grand Rapids Community College yielded the following data:

Age	*Frequency*
Under 25	138
25 but under 30	62
30 but under 40	32
Over 40	18

137. State the null and alternative hypotheses to test whether the distribution of students' ages at Grand Rapids Community College agrees with the national survey.

ANSWER:

Let p_i be the proportion of students in age category *i*; where *i* = 1, 2, 3, and 4 for the four age groups as they appear on the frequency table above. The hypotheses to be tested are: $H_o : p_1 = 0.67, \quad p_2 = 0.21, \quad p_3 = 0.08, \quad p_4 = 0.04$

H_a : At least one p_i is different from the specified value.

138. Compute the value of the test statistic.

ANSWER:

The expected cell counts for each of the four age categories, computed by using the formula $E_i = np_i$, are 167.5, 52.5, 20, and 10, respectively. The chi-square test statistic can now be calculated as: $\chi^{2*} = \sum (O_i - E_i)^2 / E_i$ = 20.52.

139. Set up the appropriate rejection region for α = 0.05.

ANSWER:

With *df* = *k* – 1 = 3, and α = 0.05, we reject H_o when $\chi^{2*} > \chi^2(3, 0.05) = 7.82$

140. What is the appropriate conclusion?

ANSWER:

Since χ^{2*} = 20.52 > 7.82, reject H_o. We conclude the distribution of students' ages at Grand Rapids Community College does not agree with the national survey.

Section 11.4

True-False Questions

141. In the chi-square test of independence, data are classified according to two categorical variables.
ANSWER: T

142. In a contingency table, the sum of the observed frequencies in a given row equals the sum of the expected frequencies for the same row.
ANSWER: T

143. The chi-square test of independence is always a two-tailed test.
ANSWER: F

144. A contingency table is an arrangement of data into a two-way classification.
ANSWER: T

145. The number of degrees of freedom in the chi-square test of independence, where the contingency table has r rows and c columns, is determined by $df = r \cdot c$.
ANSWER: F

146. The chi-square test of homogeneity is used when the two categorical variables in the contingency table are controlled by the experimenter so that the row (or column) totals are predetermined.
ANSWER: F

147. The observed frequency of a cell should not be allowed to be smaller than 5 when a chi-square test of homogeneity is being conducted.
ANSWER: F

148. The charts for both the multinomial experiment and the contingency table must be set in such a way that each piece of data will fall into exactly one of the categories.
ANSWER: T

149. The null hypothesis being tested by a test of homogeneity is that the distribution of proportions is the same for each of the subpopulations.
ANSWER: T

150. For a contingency table, the expected frequency for a cell is determined by dividing the column total by the grand total.
ANSWER: F

151. The sum of the observed frequencies in a chi-square test of independence need not equal the sum of the expected frequencies.
ANSWER: F

152. Chi-square tests of independence are always lower-tailed because a perfect fit between observed and expected frequencies makes the test statistic χ^{2*} equal to zero.
ANSWER: F

153. The degrees of freedom associated with a chi-square test of independence where data are summarized in a contingency table with *r* rows and *c* columns equal the number of rows times the number of columns in the table minus two; that is, *rc* -2.
ANSWER: F

154. A chi-square test for independence is applied to a contingency table with 4 rows and 4 columns for two qualitative variables. The degrees of freedom for this test must be 9.
ANSWER: T

155. In a chi-square test of independence, if the value of the test statistic was $\chi^{2*} = 16.55$, and the critical value at $\alpha = 0.025$ was 14.5, then we must reject the null hypothesis at $\alpha = 0.05$.
ANSWER: T

156. A chi-square test for independence with 10 degrees of freedom results in a test statistic of 18.89. Using the chi-square table, the most accurate statement that can be made about the *p*-value for this test is that 0.025 < *p*-value < 0.05.
ANSWER: T

157. The chi-square test statistic for a contingency table with *r* rows and *c* columns can be negative if *r* is much smaller than *c*.
ANSWER: F

158. A chi-square test for independence is applied to a contingency table with 3 rows and 5 columns for two qualitative variables. The number of degrees of freedom for this test is 8.
ANSWER: T

159. In a chi-square test of independence with 6 degrees of freedom and a level of significance of 0.05, the critical value from the chi-square table is 12.6. The computed value of the test statistics is 10.97. This will lead us to reject the null hypothesis.
ANSWER: F

160. A chi-squared test for independence is applied to a contingency table with 3 rows and 5 columns for two qualitative variables. The degrees of freedom for this test must be 15.
ANSWER: F

Multiple-Choice Questions

161. Which of the following statements about contingency tables is correct?

A) In the test of independence, one set of marginal totals (either row totals or column totals) is known before the data are collected.
B) In the test for homogeneity, the null hypothesis says, "The distribution of proportions is the same in all subpopulations."
C) In the test of independence, the number of degrees of freedom is $r + c - 1$.
D) In the test for homogeneity, the number of degrees of freedom is $rc - 1$, where *r* is the number of rows and *c* is the number of columns in the contingency table.
ANSWER: B

162. A contingency table is set up based on type of treatment (A, B, or C) and results of condition (improved, worse, or no change). Given that $\chi^2 * = 12.3$, then the *p*-value results would be:

A) 0.005 < *p*-value < 0.01.
B) 0.01 < *p*-value < 0.025.
C) 0.025 < *p*-value < 0.05.
D) 0.05 < *p*-value < 0.10.
ANSWER: B

163. You have calculated the chi-square test statistic in a test of independence and determined that $\chi^2 * = -4.23$. Therefore, you will know that you

A) automatically reject H_o.
B) automatically fail to reject H_o.
C) Observed frequencies that were greater than the corresponding expected frequencies.
D) made a mistake in the calculation.
ANSWER: D

164. What is your conclusion for a chi-square test of independence with critical value of 17.34 and $\chi^2 * = 2.54$?

A) Reject the null hypothesis
B) Fail to reject the null hypothesis
C) Unable to reject or fail to reject the null hypothesis
D) None of the above
ANSWER: B

165. Which of the following statements is false?

A) A contingency table is an arrangement of data in a two-way classification. The data are sorted into cells, and the number of data in each cell is reported.
B) In the case of contingency tables, the number of degrees of freedom is exactly the same as the larger of the number of columns or rows in the table.
C) In general, the *r* x *c* contingency table (*r* is the number of rows; c is the number of columns) is used to test the independence of the row factor and the column factor.
D) None of the above.
ANSWER: B

166. Which of the following statements is false?

A) The actual testing procedure for independence and homogeneity with contingency tables is not the same
B) In a test of homogeneity, we are actually testing the null hypothesis: The distribution of proportions within the rows is the same for all rows.
C) In a test of homogeneity, the alternative hypothesis is stated as: The distribution of proportions within the rows is not the same for all rows; that is, at least one is different from the others.
D) All of the above.
ANSWER: A

Chapter 11

167. Which of the following statements is false?

A) The number of degrees of freedom in *r* x *c* contingency table is given by $df = (r-1)\cdot(c-1)$.
B) The *test of homogeneity* is used when one of the two variables in the contingency table is controlled by the experimenter so that the row or column totals are predetermined.
C) In general, the expected frequency at the intersection of the row *i* and the column *j* in an *r* x *c* contingency table is given by $E_{ij} = \text{row total} \times \text{column total} = R_i \cdot C_j$.
D) None of the above.

ANSWER: C

168. The number of degrees of freedom for a contingency table with 6 rows and 6 columns is

A) 36.
B) 25.
C) 12.
D) 6.

ANSWER: C

169. Consider a cell in a contingency table. Given the cell's row total of 80, the cell's column total of 60, and a sample size of 250, the cell's *expected frequency* is

A) 19.2.
B) 3.125.
C) 20.0.
D) 1.786.

ANSWER: A

170. A chi-square test of independence with 10 degrees of freedom results in a test statistic of 19.25. Using the chi-square table, the most accurate statement that can be made about the *p*-value for this test is that:

A) $p\text{-value} < 0.025$.
B) $0.025 < p\text{-value} < 0.05$.
C) $0.05 < p\text{-value} < 0.10$.
D) $0.10 < p\text{-value} < 0.20$.

ANSWER: B

171. The chi-square test of independence is based upon:

A) two qualitative variables.
B) two quantitative variables.
C) three or more qualitative variables.
D) three or more quantitative variables.

ANSWER: A

172. A chi-square test of independence is applied to a contingency table with 4 rows and 5 columns for two qualitative variables. The degrees of freedom for this test will be:

A) 20.
B) 16.
C) 15.
D) 12.

ANSWER: D

173. In a chi-square test of independence, the value of the test statistic was $X^2 = 9.572$, and the critical value at $\alpha = 0.025$ was 11.1433. Thus,

A) we fail to reject the null hypothesis at $\alpha = 0.025$.
B) we reject the null hypothesis at $\alpha = 0.025$.
C) we don't have enough evidence to accept or reject the null hypothesis at $\alpha = 0.025$.
D) we should decrease the level of significance in order to reject the null hypothesis.
ANSWER: A

Short-Answer Questions

174. Suppose we are interested in determining whether or not there is a particular difference in the preference for a particular product depending on the gender of the consumer. State the null and alternative hypotheses if we were to use a contingency table in the test.

ANSWER:
H_o: Gender and preference are independent.
H_a: Gender and preference are dependent.

175. In using a contingency table, what assumption allows us to compute the expected frequency for a cell as we do?

ANSWER:
The assumption of independence in the null hypothesis allows us to compute the expected frequency for a cell.

176. What must "*a*" and "*b*" equal in order that the chi-square value, $\chi^2 *$, be zero?

		Levels of A 1	2	Total
Levels of B	1	30	70	100
	2	*a*	*b*	200

ANSWER:
$a = 60$, $b = 140$

177. What hypothesis is being tested when a contingency table is used to perform a test of homogeneity?

ANSWER:
The proportions within the row are the same for all rows.

178. In performing a hypothesis test concerning a contingency table, discuss the implication of obtaining a computed value of χ^2 that is very close in value to zero.

ANSWER:
If $\chi^2 *$ is close to zero, then the observed frequencies are very close in value to the expected frequencies.

Chapter 11

179. What is a contingency table?

ANSWER:
A contingency table is an arrangement of data into a two-way classification system.

180. A contingency table has 4 rows and 5 columns. The computed value of χ^2 is given by $\chi^2 *$. Find bounds for the *p*-value.

ANSWER:
0.005 < *p*-value < 0.01

181. How does a test of homogeneity differ from a general contingency table problem?

ANSWER:
In test of homogeneity, experimenter controls 1 of the 2 variables so that row totals or the column totals are predetermined.

182. The "test of independence" and the "test of homogeneity" are completed identical fashion, using the contingency table to display and organize the calculations. Explain how these two hypothesis tests differ.

ANSWER:
The *test of independence* has one sample of data that is being cross-tabulated according to the categories of two separate variables; the *test of homogeneity* has multiple samples being compared side-by-side and together these samples form the entire sample used in the contingency table.

Applied and Computational Questions

<u>QUESTIONS 183 AND 184 ARE BASED ON THE FOLLOWING INFORMATION:</u>
A group of high school seniors was given both a math aptitude test as well as a computer aptitude test. They were then grouped into one of three math aptitude classes as well as one of three computer science aptitude classes as shown below. One wishes to test the null hypothesis that computer science aptitude is independent of mathematics aptitude at $\alpha = 0.05$.

		Computer Science Aptitude		
		Low	Medium	High
Math Aptitude	Low	40	25	10
	Medium	25	50	25
	High	20	40	15

183. Determine the critical region and calculate the value of the test statistic.

ANSWER:
The critical region is $\chi^2 \geq 9.49$, and the test statistic is $\chi^2 * = 18.571$.

184. State the decision and conclusion.

ANSWER:

Since $\chi^2* > \chi^2$, reject H_o. There is sufficient evidence to conclude the computer science aptitude is not independent of mathematics aptitude at α = 0.05.

185. If 400 people were classified as short or tall as well as leader or follower and height is independent of leader/follower classification, what numbers would you expect in the four cells?

	Leader	*Follower*	Total
Short			200
Tall			200
Total	100	300	

ANSWER:

	Leader	*Follower*
Short	50	150
Tall	50	150

QUESTIONS 186 AND 187 ARE BASED ON THE FOLLOWING INFORMATION:

Consider the following data regarding germination rates for treated and untreated seeds, test the null hypotheses that the germination rate is the same for the treated as the untreated seed, at α = 0.01.

	Germinated	*Not Germinated*
Treated	85	15
Untreated	120	30

186. Determine the critical region and calculate the value of the test statistic.

ANSWER:

Critical region is: $\chi^2 \geq 6.63$, Test statistic is: $\chi^2* = 1.016$.

187. State the decision and conclusion.

ANSWER:

Fail to reject the null hypothesis since $\chi^2* < \chi^2$. There is not sufficient evidence to conclude that germination rates do not differ for treated and untreated seeds.

QUESTIONS 188 THROUGH 191 ARE BASED ON THE FOLLOWING INFORMATION:

Veterans and non-veterans were surveyed concerning giving veteran preference in hiring for state government jobs. Suppose the results for the veteran preference were as follows:

	Yes	*No*
Veteran	800	200
Non-veteran	360	90

Chapter 11

188. What percent of the veterans favored giving veteran preference?

ANSWER:
80%

189. What percent of the non-veterans favored giving veteran preference?

ANSWER:
80%

190. What would these answers lead you to believe about the independence of veteran preference and veteran/non-veteran status?

ANSWER:
The two factors are independent

191. Find the value of the test statistic $\chi^2 *$.

ANSWER:
$\chi^2 * = 0$

QUESTIONS 192 THROUGH 194 ARE BASED ON THE FOLLOWING INFORMATION:
Consider the following 2×2 contingency table.

Preference	*Marital Status*		**Total**
	Single	Married	
Candidate A	40	30	70
Candidate B	20	30	50
Total	60	60	120

192. Let p_1 be the population proportion of singles who prefer Candidate A and p_2 be the population proportion of married who prefer Candidate A. Compute the test statistic, z^*, for testing $H_o : p_1 = p_2$ vs. $H_a : p_1 \neq p_2$.

ANSWER:
Value of the test statistic: $z^* = 1.852$

193. Compute the test statistic for testing that candidate preference is independent of marital status. That is, compute $\chi^2 *$.

ANSWER:
Value of the test statistic: $\chi^2 * = 3.429$

194. Show that $\chi^2 * = (z^*)^2$.

ANSWER:
$X^2 * = 3.429 = (1.852)^2 = (z^*)^2$.

195. Refer to the contingency table below with the given observed frequencies, what possible values of *a*, would cause us to fail to reject the claim that the row variable is independent of the column variable using α = 0.01?

18	20	16
14	30	*a*

ANSWER:
We fail to reject the claim if *a* were any one of the values 5, 6, 7, ..., 45, 46, or 47.

196. A study involving marijuana use and antisocial behavior resulted in the following data. Give the *p*-value for testing that the type of dominant antisocial behavior is independent of the level of marijuana use, and write your conclusion if α = 0.05.

Dominant Antisocial Behavior	*Level of Marijuana Use*		
	Light	Medium	Heavy
Insomnia	15	8	8
Aggressiveness	10	8	20
Transient Psychosis	8	12	7
None Apparent	15	10	6

ANSWER:
Value of the test statistic: $\chi^2 *$ = 13.995, 0.025 < *p*- value < 0.05. Since *p*-value < α, reject the null hypothesis that type of dominant antisocial behavior is independent of the level of marijuana use.

QUESTIONS 197 THROUGH 200 ARE BASED ON THE FOLLOWING INFORMATION:
The individuals in the following table have an eye irritation, or a nose irritation, or a throat irritation. They have only one of the three.

	Age (years)				
Type of Irritation	18-29	30-44	45-64	65 and over	Total
Eye	125	160	100	15	400
Nose	260	370	225	25	880
Throat	75	90	45	10	220
Total	460	620	370	50	1500

A physician wishes to determine if there is sufficient evidence to reject the hypothesis that the type of ENT irritation is independent of the age group.

197. State the null and alternative hypotheses.

ANSWER:
H_o: The type of ENT irritation is independent of age group.
H_a: The type of ENT irritation is not independent of age group.

Chapter 11

198. Calculate the value of the test statistic.

ANSWER:
The expected frequencies are shown in the table below:

	Age (years)			
Type of Irritation	18-29	30-44	45-64	65 and over
Eye	122.67	165.33	98.67	13.33
Nose	269.87	363.73	217.07	29.33
Throat	67.47	90.93	54.27	7.33

$$\chi^{2*} = \sum_{allcells} [(O-E)^2 / E]$$

$= 0.0443 + 0.1718 + 0.0179 + 0.2092 + 0.3610 + 0.1081 + 0.2897 + 0.6392 + 0.8404 + 0.0095 + 1.5834 + 0.9726$
$= 5.2471$

199. Solve using the *p*-value approach.

ANSWER:
P = *p*-value = $P(\chi^2 > 5.2471 \mid df = 6)$; Using the table of χ^2 distribution: 0.25 < **P** < 0.50. Since **P** > α = 0.05, we fail to reject H_o. There is not sufficient evidence to indicate that the type of ENT irritation is not independent of the age group at the 0.05 level of significance.

200. Solve using the classical approach.

ANSWER:
The critical region is $\chi^2 \geq 12.6$. Since the test statistic χ^{2*} falls in the noncritical region, we fail to reject H_o. We reach the same conclusion as stated in question 199.

QUESTIONS 201 THROUGH 204 ARE BASED ON THE FOLLOWING INFORMATION:
The manager of an assembly process wants to determine whether the number of defective parts manufactured depends on the day of the week the parts are produced. He collected the following information.

	Days of Week					
Quality of parts	Mon.	Tues.	Wed.	Thurs.	Fri.	Total
Nondefective	34	36	38	38	36	182
Defective	6	4	2	2	4	18
Total	40	40	40	40	400	200

201. State the null and alternative hypotheses.

ANSWER:
H_o: The number of defective parts is independent of the day of the week.

H_a: The number of defective parts is not independent of the day of the week.

202. Calculate the value of the test statistic.

ANSWER:
The expected frequencies are shown in the table below:

Day of Week	Mon.	Tues.	Wed.	Thurs.	Fri.
Nondefective	36.4	36.4	36.4	36.4	36.4
Defective	3.6	3.6	3.6	3.6	3.6

$$\chi^{2*} = \sum_{allcells} [(O-E)^2 / E]$$

= 0.1582 + 0.0044 + 0.0703 + 0.0703 + 0.0044 +
1.6000 + 0.0444 + 0.7111 + 0.7111 + 0.0444
= 3.4186.

203. Complete the hypothesis test using the *p*-value approach.

ANSWER:
P = *p*-value = $P(\chi^2 > 3.4186 \mid df = 4)$; Using the table of χ^2 distribution: 0.25 < **P** < 0.50. Since **P** > α = 0.05, we fail to reject H_o. There is not sufficient evidence to indicate that the number of defective parts is not independent of the day of the week on which they are produced.

204. Complete the hypothesis test using the classical approach.

ANSWER:
The critical region is $\chi^2 \geq 9.49$; χ^{2*} falls in the noncritical region, therefore we fail to reject H_o at the 0.05 level of significance. We reach the same conclusion as stated in question 203.

QUESTIONS 205 THROUGH 209 ARE BASED ON THE FOLLOWING INFORMATION:
Three professors are scheduled to teach an elementary statistics course next semester. A sample of previous grade distributions for these three professors is shown below.

	Professor		
Grades	#1	#2	#3
A	15	12	30
B	20	32	26
C	25	20	10
Other	20	26	24

The department head of statistics wishes to determine if there is there sufficient evidence to conclude that the distribution of grades is not the same for all three professors.

205. State the null and alternative hypotheses.

ANSWER:
H_o: The distribution of grades is the same for all professors.

H_a: The distribution of grades is not the same for all professors.

206. Calculate the value of the test statistic.

ANSWER:
The expected frequencies are shown in the table below:

	Professor		
Grades	#1	#2	#3
A	17.538	19.731	19.731
B	24.000	27.000	27.000
C	16.923	19.038	19.038
Other	21.538	24.231	24.231

$$\chi^{2*} = \sum_{allcells} [(O-E)^2 / E]$$

= 0.367 + 3.029 + 5.345 + 0.667 + 0.926 + 0.037 + 3.855 + 0.049 + 4.291 + 0.110 + 0.129 + 0.002
= 18.807

207. Complete the hypothesis test at the 0.01 level of significance using the *p*-value approach.

ANSWER:
P = *p*-value = $P(\chi^2 > 18.807 \mid df = 6)$; Using the table of χ^2 distribution: **P** < 0.005. Since **P** $< \alpha = 0.01$, we reject H_o. There is sufficient evidence to indicate that the distribution of grades is not the same for all professors, at the 0.0 level of significance.

208. Complete the hypothesis test at the 0.01 level of significance using the classical approach.

ANSWER:
The critical region is $\chi^2 \geq 16.8$. Since the test statistic χ^{2*} is in the critical region, we reject H_o. We reach the same conclusion as stated in question 207.

209. Which professor is the easiest grader? Explain, citing specific supporting evidence.

ANSWER:
Professor #3 gives A's in higher proportion and C's in lower proportions than expected if all graded the same. This can be supported by the value of chi-square that comes from those two cells.

QUESTIONS 210 THROUGH 213 ARE BASED ON THE FOLLOWING INFORMATION:
The table below reports the responses of 300 students selected from schools with low graduation rates to the question "Do tests required for graduation discourage some students from staying in school?"

	Urban	*Suburban*	*Rural*	*Total*
Yes	60	30	50	140
No	25	15	15	55
Unsure	45	25	35	105
Total	130	70	100	300

One wishes to determine if there is a relationship between a student's response and the school's location.

210. State the null and alternative hypotheses.

ANSWER:

H_o: The student's response and the school location are independent.

H_a: The student's response and the school location are not independent.

211. Calculate the value of the test statistic.

ANSWER:

The expected frequencies are shown in the table below:

	Urban	Suburban	Rural
Yes	60.667	32.667	46.667
No	23.833	12.833	18.833
Unsure	45.500	24.500	35.000

$$\chi^{2*} = \sum_{allcells} [(O-E)^2 / E]$$

= 0.007 + 0.218 + 0.238 + 0.057 + 0366 + 0.606 + 0.005 + 0.010 + 0.000 = 1.507.

212. Complete the hypothesis test at the 0.05 level of significance using the *p*-value approach.

ANSWER:

P = *p*-value = $P(\chi^2 > 1.507 \mid df = 4)$; Using the table of χ^2 distribution: 0.75 < **P** < 0.90. Since **P** > α = 0.05, we fail to reject H_o. There is not sufficient evidence to show that the student's response and the school location are not independent

213. Complete the hypothesis test at the 0.05 level of significance using the classical approach.

ANSWER:

The critical region is: $\chi^2 \geq 9.49$. Since the test statistic χ^{2*} falls in the noncritical region, we fail to reject H_o at the 0.05 level of significance. We reach the same conclusion as stated in question 212.

QUESTIONS 214 THROUGH 216 ARE BASED ON THE FOLLOWING INFORMATION:

Consider the following set of data.

	Response Yes	No	**Total**
Group 1	38	12	50
Group 2	35	15	50
Total	73	27	100

Chapter 11

214. Compute the value of the test statistic z^* that would be used to test the null hypothesis that $p_1 = p_2$ where p_1 and p_2 are the proportions of "yes" responses in the respective groups.

ANSWER:

$p_1 = x_1 / n_1 = 38 / 50 = 0.76$, and $p_2 = x_2 / n_2 = 35 / 50 = 0.70$

$p'_p = (x_1 + x_2)/(n_1 + n_2) = (38 + 35) / (50 + 50) = 0.73$, and $q'_p = 1 - p'_p = 1 - 0.73 = 0.27$.

The value of the test statistic is

$$z^* = \frac{p_1 - p_2}{\sqrt{p'_p q'_p[(1/n_1) + (1/n_2)]}} = \frac{0.76 - 0.70}{\sqrt{(0.73)(0.27)[(1/50) + (1/50)]}} = 0.6757$$

215. Compute the value of the test statistic χ^2* that would be used to test the hypothesis that "response is independent of group."

ANSWER:

The expected values are shown in the table below:

	Yes	No
Group 1	36.5	13.5
Group 2	36.5	13.5

$$\chi^{2*} = \sum_{allcells} [(O - E)^2 / E] = 0.0616 + 0.1667 + 0.0616 + 0.1667 = 0.4566$$

216. Show that $\chi^2* = (z*)^2$.

ANSWER:

$\chi^{2*} = 0.4566$ and $(z^*)^2 = (0.6757)^2 = 0.4566$, so they are equal.

217. State the null hypothesis H_o and the alternative hypothesis H_a that would be used to test the following statement: "In the recent Egyptian presidential election that was held September 7, 2005, the voters expressed preferences that were not independent of their party affiliations."

ANSWER:

H_o: Voters preference and voters party affiliation in Egypt are independent.

H_a: Voters preference and party affiliation in Egypt are not independent.

218. State the null hypothesis H_o and the alternative hypothesis H_a that would be used to test the following statement: "The distribution of opinions is the same for all five communities."

ANSWER:

H_o: The distribution is the same for all five communities.

H_a: The distribution is not the same for all five communities.

219. State the null hypothesis H_o and the alternative hypothesis H_a that would be used to test the following statement: "The proportion of *strongly agree* responses was the same for all categories surveyed."

ANSWER:
H_o: The proportion of *strongly agree* responses was the same in all categories sampled.
H_a: The proportion of *strongly agree* is not the same in all categories.

QUESTIONS 220 THROUGH 224 ARE BASED ON THE FOLLOWING INFORMATION:
The table below outlines the results of a survey conducted recently to collect information from Michigan high school students about their opinion on seatbelt usage. They were asked whether or not they rarely or never wear seatbelts when riding in someone else's car.

	Gender	
Seatbelt Usage	Female	Male
Rarely or never use seatbelt	284	442
Uses seatbelt	1660	1614

220. Suppose you wish to test the hypothesis that gender is independent of seatbelt usage, state the null and alternative hypotheses.

ANSWER:
H_o: Gender is independent of seatbelt usage.
H_a: Gender and seatbelt usage are not independent.

221. Calculate the table of expected frequencies.

ANSWER:

	Gender	
Seatbelt Usage	Female	Male
Rarely or never use seatbelt	352.84	373.16
Uses seatbelt	1591.16	1682.84

222. Calculate the value of the test statistic.

ANSWER:
$\chi^{2*} = \sum (O-E)^2 / E$ = 13.43 + 12.70 + 2.98 + 2.82 = 31.93

223. Using the classical approach at α = 0.05, does this sample present sufficient evidence to reject the null hypothesis that gender is independent of seatbelt usage?

ANSWER:
The critical value = χ^2(1, 0.05) = 3.84. Since $\chi^{2*} = 31.93$ falls in the rejection region, we reject H_o. There is sufficient evidence to indicate that seatbelt usage depends on gender.

224. Complete the hypothesis test at the 0.05 level of significance using the *p*-value approach.

ANSWER:
P = *p*-value = $P(\chi^2 > 31.93 \mid df = 1)$; Using the table of χ^2 distribution: **P** < 0.005.
Since **P** $< \alpha = 0.05$, we reject H_o. We reach the same conclusion as stated in question 223.

QUESTIONS 225 THROUGH 229 ARE BASED ON THE FOLLOWING INFORMATION:
A survey of randomly selected travelers who visited the restrooms in US 131 during their summer vacation in 2004 showed the following results:

	Quality of Restroom Facilities		
Gender of Respondent	Above Average	Average	Below Average
Female	56	48	16
Male	16	52	12

225. Suppose you wish to test the hypothesis that quality of responses is independent of the gender of the respondent, state the null and alternative hypotheses.

ANSWER:
H_o: Quality of responses is independent of the gender of the respondent.
H_a: Quality of responses is dependent of the gender of the respondent.

226. Calculate the table of expected frequencies.

ANSWER:

	Quality of Restroom Facilities		
Gender of Respondent	Above Average	Average	Below Average
Female	43.2	60.0	16.8
Male	28.8	40.0	11.2

227. Calculate the value of the test statistic.

ANSWER:
$\chi^{2*} = \sum (O-E)^2 / E$ = 3.79 + 2.40 + 0.04 + 5.69 + 3.60 + 0.06 = 15.58

228. Using the classical approach at α = 0.05, does this sample present sufficient evidence to reject the null hypothesis that quality of responses is independent of the gender of the respondent?

ANSWER:
The critical value = χ^2 (2, 0.05) = 5.99. Since χ^{2*} = 15.58 does fall in the rejection region, we reject H_o. There is sufficient evidence to indicate that quality of responses is dependent of the gender of the respondent.

229. Using the p-value approach at $\alpha = 0.05$, does this sample present sufficient evidence to reject the null hypothesis that quality of responses is independent of the gender of the respondent?

ANSWER:

P = p-value = $P(\chi^2 > 15.58 \mid df = 2)$; Using the table of χ^2 distribution: **P** < 0.005.

Since **P** < α = 0.05, we reject H_o. We reach the same conclusion as stated in question 223.

QUESTIONS 230 THROUGH 233 ARE BASED ON THE FOLLOWING INFORMATION:

Fear of darkness is a common emotion. The following data were obtained by asking 125 individuals in each age group whether they had serious fears of darkness.

	Age Group				
Fear of Darkness	Elementary	Jr. High	Sr. High	College	Adult
# Who Fear Darkness	44.2	44.2	44.2	44.2	44.2
# Who Do Not Fear Darkness	80.8	80.8	80.8	80.8	80.8

230. Suppose you wish to test the hypothesis that the same proportion of each age group has serious fears of darkness, state the null and alternative hypotheses.

ANSWER:

H_o: The proportion of individuals who has serious fears of darkness is the same in all five age group.

H_a: The proportion of individuals who has serious fears of darkness is not the same in all five age group.

231. Calculate the table of expected frequencies.

ANSWER:

	Age Group				
Fear of Darkness	Elementary	Jr. High	Sr. High	College	Adult
# Who Fear Darkness	52	45	30	23	71
# Who Do Not Fear Darkness	73	80	95	102	54

232. Calculate the value of the test statistic.

ANSWER:

$$\chi^{2*} = \sum (O-E)^2 / E$$

= 1.38 + 0.01 + 4.56 + 10.17 + 16.25 + 0.75 + 0.01 + 2.50 + 5.56 + 8.89

= 50.08

233. Using the classical approach at $\alpha = 0.01$, does this sample present sufficient evidence to reject the null hypothesis that the proportion of individuals who has serious fears of darkness is the same in all five age group?

ANSWER:

The critical value = $\chi^2(4, 0.01)$ = 13.3. Since $\chi^{2*} = 50.08$ falls in the rejection region, we reject H_o. There is sufficient evidence to indicate that the proportion of individuals who has serious fears of darkness is not the same in all five age group.

Chapter 11

234. A study of the purchase decisions of three stock portfolio managers, A, B, C, was conducted to compare the numbers of stock purchases that resulted in profits over a time period less than or equal to 1 year. One hundred randomly selected purchases were examined for each of the managers. Do the data provide evidence of differences among the rates of successful purchases for the three managers?

	Manager		
Portfolio	A	B	C
Profit	65	73	57
No Profit	35	27	43

ANSWER:
It is necessary to test a hypothesis of equivalence of the rates of successful purchases for three different managers, which is equivalent to a test of the equivalence of three binomial populations. The contingency table, including column and row totals and the estimated expected cell counts (in parentheses), follows.

	Manager			
	A	B	C	*Total*
Number Successes	65 (65)	73 (65)	57 (65)	195
Number of Failures	35 (35)	27 (35)	43 (35)	105
Total	100	100	100	300

The test statistic can be calculated as
$\chi^{2*} = \sum (O-E)^2 / E = 0.000 + 0.9846 + 0.9846 + 0.000 + 1.8286 + 1.8286 = 5.626.$
With $(r-1)(c-1) = 2$ *df*, the *p*-value is bounded between 0.05 and 0.10. Therefore, H_o is not rejected and the results are declared not significant. There is not enough information to conclude that the proportion of successful purchases will differ among the managers.

235. The personnel manager of a consumer product company asked a random sample of employees how they felt about the work they were doing. The table below gives a breakdown of their responses by gender. Do the data provide sufficient evidence to conclude that the level of job satisfaction is related to gender? Use $\alpha = 0.10$

	Response		
Gender	Very Interesting	Fairly Interesting	Not Interesting
Male	70	41	9
Female	35	34	11

ANSWER:

H_o : Job satisfaction and gender are independent

H_a : Job satisfaction and gender are dependent

The critical value is $\chi^2(2, 0.10) = 4.61$ and the value of the test statistic is $\chi^{2*} = 4.708$. Therefore, we reject the null hypothesis. There is sufficient evidence to conclude that job satisfaction is related to gender.

Chapter 12

ANALYSIS OF VARIANCE

Sections 12.1 through 12.3

True-False Questions

1. The ANOVA test assumes sampling from normal populations with equal variances.
ANSWER: T

2. In single-factor ANOVA, if the null hypothesis is rejected then all of the population means are declared to differ from one another.
ANSWER: F

3. We do not need to assume that the observations are independent to perform analysis of variance.
ANSWER: F

4. Experimental error is the name given to the variability that takes place among the replicates of an experiment as it is repeated under constant conditions.
ANSWER: T

5. The rejection of H_o in single-factor ANOVA indicates that you have identified the level(s) of the factor that is (are) different from the others.
ANSWER: F

6. To partition the sum of squares for the total in single-factor ANOVA is to separate the numerical value of SS(total) into two values, SS(factor) and SS(error), such that the sum of these two values is equal to SS(total).
ANSWER: T

7. In order to apply the *F*- test in ANOVA, the sample standard deviation from each factor level sample must be the same.
ANSWER: F

8. In single-factor ANOVA, a sum of squares is actually a measure of variance.
ANSWER: F

9. Independent samples were collected in order to test the effect a factor had on a variable of interest. The data is summarized in the ANOVA table shown below.

	df	*SS*
Factor	2	810
Error	8	720
Total	10	1530

The null hypothesis could be written as $H_o : \mu_1 = \mu_2 = \mu_3 = \mu_4$.

ANSWER: F

10. Fail to reject H_o in single-factor ANOVA is the desired decision when the means for the levels of the factor being tested are all different.
ANSWER: F

11. In single-factor ANOVA, the degrees of freedom for the factor are equal to the number of factor levels tested less one.
ANSWER: T

12. The measure of a specific level of a factor being tested in an ANOVA is the variance of the factor level.
ANSWER: F

13. Independent samples were collected in order to test the effect a factor had on a variable of interest. The data are summarized in the ANOVA table shown below.

	df	*SS*
Factor	2	84.5
Error	10	9.5
Total	12	94.0

The critical value of *F* at the 0.05 level of significance is 5.46.
ANSWER: F

14. In single-factor ANOVA, when the calculated value of the test statistic $F*$, is greater than the table value for *F*, the conclusion will be: "The factor being tested does have an effect on the variable."
ANSWER: T

15. In single-factor ANOVA, when the calculated value of the test statistic $F*$ is greater than the table value for *F*, then the decision will be: "Fail to reject H_o."

ANSWER: F

16. In single-factor ANOVA, if 10 is subtracted from every data value, then the calculated value of the test statistic $F*$ is also reduced by 10.
ANSWER: F

17. A possible interpretation of H_o in single-factor ANOVA is that "There is no difference between the mean values of the random variable at the various levels of the factor being tested"
ANSWER: T

18. A possible interpretation of H_o in single-factor ANOVA is that "There is no variance among the mean values of x for each of the different levels of the factor being tested"
ANSWER: T

19. A possible interpretation of H_a in single-factor ANOVA is that "The factor being tested has no effect on the random variable x."
ANSWER: F

20. In single-factor ANOVA, the sample size from each factor level must be the same in order to apply the F-test.
ANSWER: F

21. In single-factor ANOVA, we want to reject H_o and conclude that the factor has an effect on the variable when the amount of variance assigned to the factor is significantly larger than the variance assigned to error.
ANSWER: T

22. Independent samples were collected in order to test the effect a factor had on a variable of interest. The data are summarized in the ANOVA table shown below.

	df	*SS*
Factor	2	28.5
Error	12	125.3
Total	14	153.8

The null hypothesis could be written as $H_o : \mu_1 = \mu_2 = \mu_3$.
ANSWER: T

23. The F-distribution is symmetrical around the mean zero.
ANSWER: F

24. The F-distribution is based on two sets of degrees of freedom, one for the numerator, and the other for the denominator.
ANSWER: T

25. In single-factor ANOVA, if the computed value of F is F^* = 9.56, and the critical value is F = 6.39, we would conclude that all the population means are equal.
ANSWER: F

26. In single-factor ANOVA, the alternative hypothesis used in the F-test states that $\mu_1 = \mu_2 = \mu_3$.
ANSWER: F

27. One characteristic of the F-distribution is that the computed value of F can only range between −1.0 and +1.0, inclusive.
ANSWER: F

28. In single-factor ANOVA, if the computed value of F is F^* = 4.21, and the critical value is F = 8.89, we would fail to reject the null hypothesis.
ANSWER: T

29. The ANOVA technique simultaneously compares several populations to determine if their means are equal. This comparison is actually made by comparing the variances of the samples; hence the name "Analysis of Variance".
ANSWER: **T**

30. In single-factor ANOVA, df(total)= df(factor) + df(error).
ANSWER: **T**

31. In single-factor ANOVA, the calculated value of the test statistic F^* = has two types of degrees of freedom. The number of degrees of freedom for the numerator df_n = df(error), and the number of degrees of freedom for denominator df_d = df(factor).
ANSWER: **F**

Multiple-Choice Questions

32. When hypothesis testing involves more than two means, we use the ANOVA rather than the *t*-test. ANOVA stands for:

A) variation between the levels.
B) analysis of variances.
C) the estimation of the ratio of two population variances.
D) an optional variance analysis.
ANSWER: **B**

33. Which of the following is a correct interpretation of the null hypothesis for analysis of variance for one factor?

A) There is no difference between the mean values of the random variable at the various levels of the test factor.
B) The factor being tested had no effect on the random variable *x*.
C) There is no variance amongst the mean values of *x* for each of the different factor levels.
D) All of the above.
ANSWER: **D**

34. Given the following set of data, there are three degrees of freedom values. Identify the correct statement below.

		Replicates			
		A	B	C	D
Factor Levels	I	6	11	8	3
	II	9	10	11	10
	III	14	11	12	15

A) *df*(Factor) = 3
B) *df*(Error) = 9
C) *df*(Total) = 12
D) All of the above.
ANSWER: **A**

35. In single-factor ANOVA, when the calculated value of the test statistics $F*$ is greater than the table value for *F*, we will:

A) fail to reject H_o and conclude the factor being tested does have an effect on the variable.
B) fail to reject H_o and conclude the factor being tested does not have an effect on variable.
C) reject H_o and conclude the factor being tested does have an effect on variable.
D) reject H_o and conclude the factor being tested does not have an effect on variable.

ANSWER: C

36. Identify the correct statement about the analysis of variance technique.

A) The mean squares are measures of variance.
B) The "partitioning" of the variance occurs when the sum of squares for total is separated into two parts, *SS*(Factor) and *SS*(Error).
C) We reject the null hypothesis and conclude that the tested factor has an effect when the variance assigned to the factor is much larger than the variance assigned to error.
D) All of the above.

ANSWER: D

37. In single-factor ANOVA, if the test is conducted and the null hypothesis is rejected, what does this indicate?

A) All the population means are equal.
B) At least two of the population means are different.
C) The normal distribution should be used instead of the *F*-distribution to determine the critical value for the test.
D) None of the above.

ANSWER: B

38. What distribution does the *F*-distribution approach as the sample size increases?

A) Binomial
B) Normal
C) Student's *t* - distribution
D) Chi-square

ANSWER: B

39. ANOVA is used to compare two or more population:

A) variances.
B) proportions.
C) medians.
D) means.

ANSWER: D

40. Given the significance level α = 0.01, the critical *F*-value for the degrees of freedom, *d.f.* = (3, 8) is equal to

A) 7.59.
B) 27.5.
C) 5.42.
D) 4.07.
ANSWER: A

41. In a single-factor ANOVA, there are three treatments with sizes $n_1 = 5, n_2 = 6$ and $n_3 = 5$. Then the rejection region for this test at the 0.05 level of significance is

A) $F > 3.74$.
B) $F > 4.86$.
C) $F > 4.97$.
D) $F > 3.81$.
ANSWER: D

42. Given the significance level α = 0.025, the critical *F*-value for the degrees of freedom, *d.f.* = (3, 8) is

A) 7.59.
B) 27.5.
C) 5.42.
D) 4.07.
ANSWER: C

43. In a single-factor ANOVA test, the test statistic is F^* = 4.25. The rejection region is $F >$ 3.06 for the α = 0.05, $F >$ 3.8 for α = 0.025, and $F >$ 4.89 for α = 0.01. For this test, the approximate *p*-value is

A) greater than 0.05.
B) between 0.025 and 0.05.
C) between 0.01 and 0.025.
D) approximately 0.05.
ANSWER: C

44. A professor of statistics in Michigan State University wants to determine whether the average starting salaries among graduates of the 15 universities in Michigan are equal. A sample of 25 recent graduates from each university was randomly taken. The appropriate critical value for the ANOVA test is obtained from the *F*-distribution with numerator and denominator degrees of freedom, respectively, equal to:

A) 15 and 25
B) 14 and 360
C) 360 and 14
D) 25 and 15
ANSWER: B

45. Given the significance level α = 0.05, the critical F-value for the degrees of freedom, *d.f.* = (3, 8) is

A) 7.59.
B) 27.5.
C) 5.42.
D) 4.07.
ANSWER: D

46. The test statistic in the single-factor ANOVA equals the ratio

A) sum of squares for factor ÷ sum of squares for error.
B) sum of squares for error ÷ sum of squares for factor.
C) mean square for factor ÷ mean square for error.
D) mean square for error ÷ mean square for factor.
ANSWER: C

47. One-way ANOVA is performed on three independent samples with sizes $n_1 = 8$, $n_2 = 9$, and $n_3 = 10$. The critical value obtained from the F-table for this test at the 0.025 level of significance equals:

A) 3.69.
B) 4.32.
C) 3.72.
D) 5.61.
ANSWER: B

48. Which of the following statements is false regarding a single-factor ANOVA?

A) Between-sample variation and within-sample variation are compared in an ANOVA test.
B) The data values from repeated samplings are called replicates.
C) SS(total) = SS(factor) + SS(error)
D) None of the above
ANSWER: D

49. Which of the following statements is false regarding a single-factor ANOVA?

A) The factor degrees of freedom are 1 less than the number of levels (columns) for which the factor is tested; that is df(factor) = $c - 1$.
B) The error degrees of freedom are the sum of the degrees of freedom for all levels tested (columns in the data table). Since each column has k_i degrees of freedom; therefore, df(error) = $k_i + k_2 + k_3 + \cdots\cdots = \sum_i k_i = n$
C) The total degrees of freedom are 1 less than the total number of data; that is df(total) = $n - 1$.
D) None of the above
ANSWER: C

50. Which of the following statements is false regarding a single-factor ANOVA?

A) The mean square for the factor being tested, MS(factor), and the mean square for error, MS(error), are obtained by dividing the sum-of-squares value by the corresponding number of degrees of freedom; that is MS(factor)= SS(factor) / df(factor) and MS(error) = SS (error) / df(error).
B) MS(total) = MS(factor) + MS(error).
C) The calculated value of the test statistic, F^*, is found by dividing the MS(factor) by the MS(error).
D) None of the above

ANSWER: B

51. In a single-factor ANOVA, if the numerator and denominator degrees of freedom are 4 and 25, respectively, then the total number of observations must equal:

A) 24
B) 25
C) 29
D) 30

ANSWER: D

52. The number of degrees of freedom for the denominator in one-way ANOVA test involving 4 population means with 15 observations sampled from each population is:

A) 60
B) 19
C) 56
D) 45

ANSWER: C

53. A single-factor ANOVA is performed on three independent samples with $n_1 = 6$, $n_2 = 7$, and $n_3 = 8$. The critical value obtained from the *F*-table for this test at the 2.5% level of significance equals:

A) 3.55
B) 39.45
C) 4.56
D) 29.45

ANSWER: C

54. The *F*-statistic in one-way ANOVA represents the:

A) variation between the treatments plus the variation within the treatments.
B) variation within the treatments minus the variation between the treatments.
C) variation between the treatments divided by the variation within the treatments.
D) variation within the treatments divided by the variation between the treatments.

ANSWER: C

55. A single-factor ANOVA is applied to three independent samples having means 8, 11, and 16, respectively. If each observation in the third sample were increased by 20, the value of the *F*-statistics would:

A) increase
B) decrease
C) remain unchanged
D) increase by 20
ANSWER: **A**

Short-Answer Questions

56. In single-factor ANOVA, if *df*(Factor) = 3, what is the null hypothesis being tested?

ANSWER:
Since *df*(Factor)=3, then, number of levels of the factor = 4. The null hypothesis must be:
$H_o : \mu_1 = \mu_2 = \mu_3 = \mu_4$.

57. Explain how to determine *df*(Factor), *df*(Error), and *df*(Total) if *n* is the number of data in the total sample and *c* is the number of levels (columns) for which the factor is being tested.

ANSWER:
df(Factor)= $c - 1$; *df*(Error)=$n - c$; *df*(Total)=$n - 1$; Also, *df*(Total) = *df*(Factor) + *df*(Error)

58. When simultaneously comparing three or more population means, an efficient technique is called ________.

ANSWER:
ANOVA

59. In ANOVA, explain what is meant by "replicates.", and "levels of the tested factor."

ANSWER:
"Replicates" refers to data values from repeated sampling.
"Levels of the tested factor" refers to random samples at each level of the factor being tested.

60. What is meant by a single-factor ANOVA experiment?

ANSWER:
Experiment designed to determine the effect of one factor, at various levels, on a response variable.

61. In single-factor ANOVA, if MS(factor) is significantly larger than MS(error), what is your decision and conclusion?

ANSWER:
We reject H_o. There is sufficient evidence to conclude that the means for the factor levels being tested are not all the same.

62. In single-factor ANOVA, determine the values *A*, *B*, *C*, *D*, and *E* missing in the ANOVA table shown below:

	SS	*df*	*MS*	F^*
Factor	*A*	3	18	*E*
Error	*B*	15	*D*	
Total	162	*C*		

ANSWER:
$A = 54$, $B = 108$, $C = 18$, $D = 7.2$, $E = 2.5$

63. Briefly discuss the importance of Analysis of Variance (ANOVA).

AMSWER:
ANOVA is important simply because it is used to test a hypothesis about several population means. Specifically, The ANOVA techniques allow us to test the null hypothesis (all means are equal) against the alternative hypothesis (at least one mean value is different) with a specified level of significance α .

64. The single-factor ANOVA technique separated the variance among the sample data into two measures of variance. What are they? Briefly explain what does each one measure?

ANSWER:
(1) MS(factor), the measure of variance between the levels of the factor being tested, and
(2) MS(error), the measure of variance within the levels of the factor being tested.

65. In single-factor ANOVA, if MS(factor) is not significantly larger than MS(error), what is your decision and conclusion?

ANSWER:
We will not be able to reject H_o. There is not sufficient evidence to conclude that the means for the factor levels being tested are not all the same.

Applied and Computational Questions

QUESTIONS 66 THROUGH 70 ARE BASED ON THE FOLLOWING INFORMATION:
Consider the following table for a single-factor ANOVA.

	Factor levels		
Replicates	1	2	3
1	2	3	7
2	5	0	8
3	4	6	9

66. Find $x_{1,3}$.

ANSWER:
$x_{1,3} = 4$

67. Find $x_{3,2}$.

ANSWER:

$x_{3,2} = 8$

68. Find C_1.

ANSWER:

$C_1 = 11$

69. Find $\sum x$.

ANSWER:

$\sum x = 44$

70. Find $\sum (C_i)^2$.

ANSWER:

$\sum (C_i)^2 = 778$

71. The following ANOVA table shows results of independent samples collected to test the effect a factor had on a variable. Find the critical value for *F* at α = 0.05 and determine if H_o can be rejected.

	df	*SS*
Factor	2	810
Error	8	720
Total	10	1530

ANSWER:

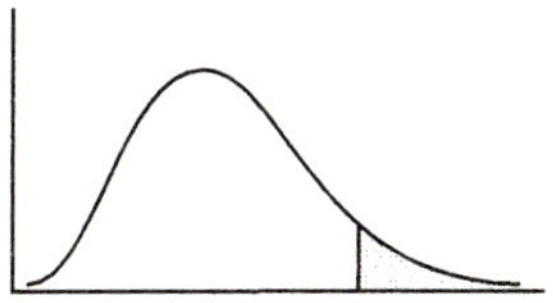

$F\,(2, 8, 0.05) = 4.46$

Since $F^* = (810/2) / (720/8) = 45$, and $F\,(2, 8, 0.05) = 4.46$, we reject the null hypothesis.

QUESTIONS 72 THROUGH 75 ARE BASED ON THE FOLLOWING INFORMATION:

Consider the set of data shown below:

		Replicates			
Factor Levels	*1*	120	100	100	80
	2	100	90	95	90
	3	80	80	85	89

72. Find SS(Factor), SS(Error), and SS(Total).

ANSWER:
SS(Factor) = 555, SS(Error) = 926, SS(Total) = 1481

73. Develop the ANOVA table.

ANSWER:

	SS	*df*	*MS*	*F**
Factor	555	2	277.5	2.697
Error	926	9	102.89	
Total	1481	11		

74. State the appropriate null and alternative hypotheses.

ANSWER:
$H_o : \mu_1 = \mu_2 = \mu_3$ vs. H_a :at least two of the population means are not the same.

75. Test the hypotheses in question 74 at α = 0.025.

ANSWER:
Since F^* = 2.697, and F (2, 9, 0.025) = 5.71, we fail to reject the null hypothesis.

QUESTIONS 76 THROUGH 78 ARE BASED ON THE FOLLOWING INFORMATION:
Independent samples were collected in order to test the effect a factor had on a variable. Consider the ANOVA table below.

	df	SS
Factor	2	810
Error	8	720
Total	10	1530

76. State the null and alternative hypotheses.

ANSWER:
$H_o : \mu_1 = \mu_2 = \mu_3$ vs. H_a :at least two of the population means are not the same.

77. Find the calculated value of F.

ANSWER:
F^* = 4.5

78. Test the hypotheses in question 76 at α = 0.01.

ANSWER:
Since F^* = 4.5, and F (2, 8, 0.01) = 8.65, we fail to reject the null hypothesis.

79. The following experimental results have the same factor level means. Compute F^* for both. What difference is in the two sets of results?

Experiment A–Factor Level			*Experiment B–Factor Level*		
1	2	3	1	2	3
28	35	42	20	32	29
32	37	40	40	42	39
30	39	35	30	37	49
$\bar{x}_1 = 30$	$\bar{x}_2 = 37$	$\bar{x}_3 = 39$	$\bar{x}_1 = 30$	$\bar{x}_2 = 37$	$\bar{x}_3 = 39$

ANSWER:

For *A*: F^* = 9.57. For *B*: F^* = 0.89. In *A* most of variability is between levels 1, 2, and 3; while in *B*, most of variability is within the three levels.

80. Complete the ANOVA table shown below by filling in the appropriate values for *A, B, C, D,* and *E.*

Source	*SS*	*df*	*MS*	F^*
Method	*A*	*B*	*D*	*E*
Error	137.5	*C*	12.5	
Total	175.5	13		

ANSWER:
A = 38.0, *B* = 2, *C* = 11, *D* = 19.0, *E* = 1.52

QUESTIONS 81 THROUGH 83 ARE BASED ON THE FOLLOWING INFORMATION:
Consider the following experiment that consists of only two factor levels.

Factor Levels

1	*2*
12.2	13.1
13.0	14.2
12.5	15.0
12.9	14.7

81. Develop the ANOVA table.

ANSWER:

	SS	*df*	*MS*	*F**
Factor	5.12	1	5.12	12.29
Error	2.50	6	0.4167	
Total	7.62	7		

82. Compute t^* for testing $H_o: \mu_1 = \mu_2$ versus $H_a: \mu_1 \neq \mu_2$.

ANSWER:
t^* = 3.506

83. Show that $(t^*)^2$ =. F^*

ANSWER:
$(t^*)^2$ = $(3.506)^2$ = 12.29 = F^*.

QUESTIONS 84 THROUGH 86 ARE BASED ON THE FOLLOWING INFORMATION:
Consider the data in the table below;

Treatments

1	2	*3*
2	6	7
3	6	6
2	9	8
		10

84. Construct the ANOVA table.

ANSWER:

	SS	*df*	*MS*	*F**
Treatment	55.48	2	27.74	12.592
Error	15.42	7	2.203	
Total	70.90	9		

85. State the null and alternative hypotheses.

ANSWER:
$H_o : \mu_1 = \mu_2 = \mu_3$ vs. H_a :at least two of the population means are not the same.

86. Test the hypotheses in question 85 at the 0.05 level of significance.

ANSWER:
Since F^* = 12.592, and the critical value is *F*(2, 7, 0.05) = 4.74, we reject the null hypothesis at α = 0.05, and conclude that at least two of the population means are not the same.

87. Place bounds on the *p*-value for the following situation: F^* = 4.21, df(Factor) = 3, and *df(*Error) = 10.

ANSWER:
$0.025 < P < 0.05$

88. Place bounds on the *p*-value for the following situation: F^* = 3.99, *df*(Factor) = 5, and *df*(Error) = 15.

ANSWER:
$0.01 < P < 0.025$

89. Suppose that an *F*-test has a *p*-value of 0.029. What is the interpretation of the situation if you had previously decided on a 0.05 level of significance?

ANSWER:
Reject the null hypothesis; since the *p*-value is less than the previously set value for α .

90. Determine the critical region(s) and critical value(s) that would be used to test $H_o : \mu_1 = \mu_2 = \mu_3 = \mu_4$ with $n = 18, \alpha = 0.05$. Sketch a graph to display the results.

ANSWER:

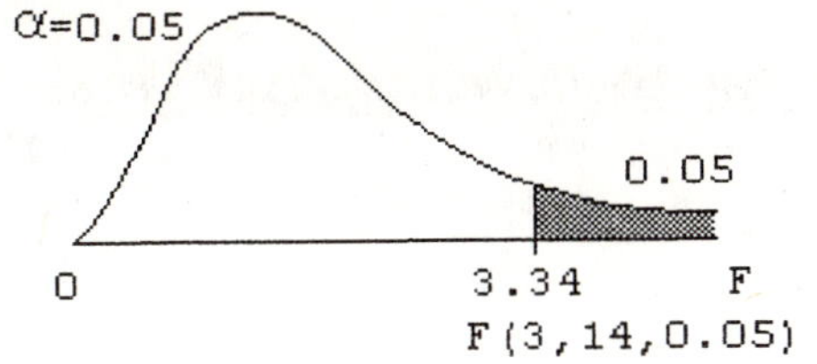

91. Determine the critical region(s) and critical value(s) that would be used to test $H_o : \mu_1 = \mu_2 = \mu_3 = \mu_4 = \mu_5$ with $n = 15, \alpha = 0.01$. Sketch a graph to display the results.

ANSWER:

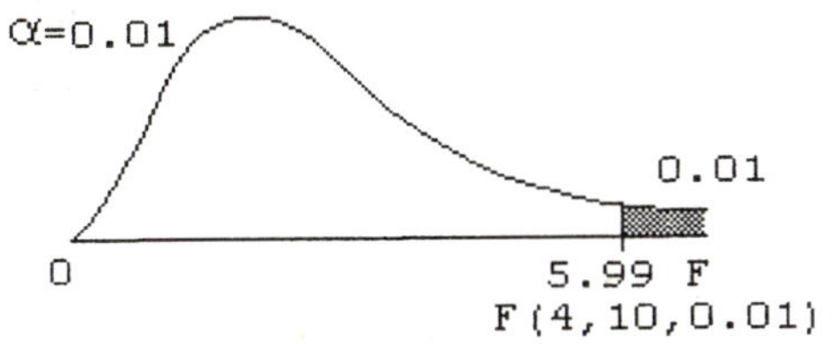

92. Determine the critical region(s) and critical value(s) that would be used to test $H_o : \mu_1 = \mu_2 = \mu_3$ with $n = 25, \alpha = 0.01$. Sketch a graph to display the results.

ANSWER:

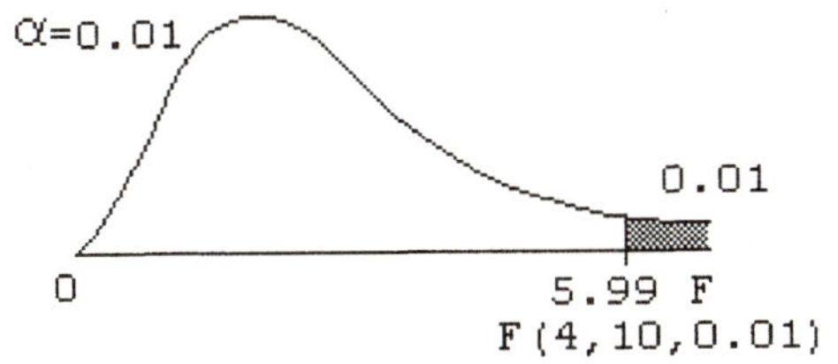

93. Suppose that an *F*-test has a *p*-value of 0.035. What is the interpretation of *p*-value = 0.035?

ANSWER:
0.035 of the probability distribution associated with *F* and a true null hypothesis is more extreme than F^*. That is, area under the curve and to the right of F^*.

94. Suppose that an *F*-test has a *p*-value of 0.073. What is the interpretation of the situation if you had previously decided on a 0.05 level if significance?

ANSWER:
Fail to reject the null hypothesis; since the *p*-value is greater than the set value for α.

Chapter 12

95. Each department at a large industrial plant is rated weekly. State the hypotheses used to test "the mean weekly ratings are the same in four departments."

ANSWER:
$H_o: \mu_1 = \mu_2 = \mu_3 = \mu_4$
H_a :Not all department mean weekly ratings are equal.

QUESTIONS 96 THROUGH 101 ARE BASED ON THE FOLLOWING INFORMATION:
Consider the following partial ANOVA table:

Source	*df*	*SS*	*MS*
Factor	3	*	*
Error	*	51.17	*
Total	20	93.44	

96. Find the 4 missing values, identified by *

ANSWER:

Source	*df*	*SS*	*MS*
Factor	3	42.27	14.09
Error	17	51.17	3.01
Total	20	93.44	

97. How many levels of the factor are being tested?

ANSWER:
df(factor) = 3 = c-1, where c is the number of levels of the factor $\Rightarrow$ c = 4 levels

98. Find the calculated value of the test statistic

ANSWER:
F^* = MS(factor) / MS(error) = 14.09 / 3.01 = 4.68.

99. State the null and alternative hypotheses

ANSWER:
$H_o: \mu_1 = \mu_2 = \mu_3 = \mu_4$
H_a :The means are not equal (that is, at least one mean is different)

100. Test the hypotheses in question 99 at the 0.05 level of significance using the *p*-value approach

ANSWER:
P = P(F > 4.66 | df_n = 3, df_d =17) $\Rightarrow$ 0.01 < **P** < 0.025
Since p-value > α = 0.05, we reject H_o.

101. Test the hypotheses in question 99 at the 0.05 level of significance using the classical approach.

ANSWER:
The critical value = F(3, 17, 0.05) = 3.20. Since F^* = 4.68 falls in the rejection region, we reject H_o.

QUESTIONS 102 THROUGH 106 ARE BASED ON THE FOLLOWING INFORMATION9:
In a one-factor ANOVA, assume there are "t" levels of the factor being tested, and the total number of observations is "N".

102. What are the degrees of freedom for SS(error)?

ANSWER:
$N - t$

103. What are the degrees of freedom for SS(factor)?

ANSWER:
$t - 1$

104. What are the degrees of freedom for SS(Total)?

ANSWER:
$N - 1$

105. The *F*-value is a ratio of two variance estimates. What variance is used as a denominator of the ratio?

ANSWER:
MS(error)

106. The *F*-value is a ratio of two variance estimates. What variance is used as a numerator of the ratio?

ANSWER:
MS(factor)

107. Fill in the blanks (identified by asterisks) in the following partial ANOVA table:

Source of Variation	*SS*	*df*	*MS*	*F*
Factor	*	*	195	*
Error	625	*	*	
Total	1600	25		

ANSWER:

Source of Variation	*SS*	*df*	*MS*	*F*
Factor	975	5	195	6.24
Error	625	20	31.25	
Total	1600	24		

QUESTIONS 108 THROUGH 110 ARE BASED ON THE FOLLOWING INFORMATION9:
In a single-factor ANOVA, 7 experimental units were assigned to the first level, 13 units to the second level, and 10 units to the third level. A partial ANOVA table for this experiment is shown below:

Source of Variation	*SS*	*df*	*MS*	*F*
Factor	*	*	*	1.50
Error	*	*	4	
Total	*	*		

108. Fill in the blanks (identified by asterisks) in the above ANOVA Table.

ANSWER:

Source of Variation	*SS*	*df*	*MS*	F^*
Treatments	12	2	6	1.50
Error	108	27	4	
Total	130	29		

109. State the null and alternative hypotheses.

ANSWER:

$H_o : \mu_1 = \mu_2 = \mu_3$

H_a : The population means are not equal (that is; at least one of the means is different)

110. Test at the 5% significance level to determine if differences exist among the three treatment means.

ANSWER:

Since the test statistics F^* = 1.50, and the critical region = *F*(2, 27, 0.05) $\approx$3.32, we fail to reject the null hypothesis. There is not sufficient evidence to conclude that the population means are not equal.

Sections 12.4

True-False Questions

111. The mathematical model for a particular problem is an equational statement showing the anticipated makeup of an individual piece of data.
ANSWER: T

112. Side-by-side dotplots are very useful in visualizing the within-sample variation, the between-sample variation, and the relationship between them.
ANSWER: T

113. In a single-factor ANOVA, the null hypothesis is that there is no difference between the levels of the factor being tested.
ANSWER: T

114. In a single-factor ANOVA, our goal is to investigate the effect that various levels of the factor being tested have on each other.
ANSWER: F

115. In a single-factor ANOVA, we must assume independence among all observations of the experiment.
ANSWER: T

116. In a single-factor ANOVA, we must assume that the effects due to chance and due to untested factors are *F*-distributed.
ANSWER: F

117. In a single-factor ANOVA, we must assume the variance caused by the effects due to chance is not the same as the variance caused by effects due to untested factors; otherwise the null hypothesis will always be rejected at any level of significance.
ANSWER: F

Multiple-Choice Questions

118. A Wal-Mart department store examined a sample of the 20 credit sales, and recorded the amounts charged for each of four types of credit cards as follows: 4 for American Express, 5 for Master Card, 6 for Visa, and 5 for Discover. What are the degrees of freedom for the *F* statistic?

A) 15 for the numerator, 4 for the denominator
B) 3 for the numerator, 16 for the denominator
C) 2 for the numerator, 17 for the denominator
D) 16 for the numerator, 4 for the denominator
ANSWER: B

Chapter 12

119. Five different fertilizers were applied to a field of tomato, in constructing the ANOVA table, how many degrees of freedom are there in the numerator?

A) 2
B) 3
C) 4
D) 5
ANSWER: **C**

120. One-way ANOVA is applied to three independent samples having means 12, 15, and 20, respectively. If each observation in the third sample were increased by 25, the value of the statistic would:

A) increase.
B) decrease.
C) remain unchanged.
D) increase by 25.
ANSWER: **A**

121. In a single-factor ANOVA, suppose that there are four levels of the factor being tested with $n_1 = 5$, $n_2 = 6$, $n_3 = 5$, and $n_4 = 4$. Then the rejection region for this test at the 5% level of significance is expressed as

A) $F^* > F(4, 20, 0.025)$
B) $F^* > F(4, 20, 0.05)$
C) $F^* > F(3, 16, 0.025)$
D) $F^* > F(3, 16, 0.05)$
ANSWER: **D**

122. In an ANOVA test, the test statistic is $F = 6.75$. The rejection region is $F > 3.97$ for the 5% level of significance, $F > 5.29$ for the 2.5% level, and $F > 7.46$ for the 1% level. For this test, the *p*-value is

A) greater than 0.05
B) between 0.025 and 0.05
C) between 0.01 and 0.025
D) approximately 0.05
ANSWER: **C**

123. In a single-factor analysis of variance, the null hypothesis of equal population means is rejected if:

A) MS (factor) is much smaller than MS (error)
B) MS (factor) is much larger than MS (error)
C) MS (factor) is equal to MS (error)
D) None of the above
ANSWER: **B**

124. Which of the following is not a required condition for one-way ANOVA?

A) The sample sizes must be equal.
B) The populations must all be normally distributed.
C) The population variances must be equal.
D) The samples for each treatment must be selected randomly and independently.
ANSWER: **A**

125. The distribution of the test statistic for analysis of variance is the:

A) normal distribution.
B) Student's *t*-distribution.
C) *F*-distribution.
D) chi-squared distribution.
ANSWER: **C**

Short-Answer Questions

126. In single-factor ANOVA, rejection of H_o implies that there is a difference between the levels. Discuss the problem that would follow.

ANSWER:
If we reject H_o, problem is to locate level or levels that are different. This may be main object of analysis.

127. In single-factor ANOVA, the null hypothesis is that there is no difference between the levels of the factor being tested. How would you interpret a "fail to reject H_o" decision?

ANSWER:
A "fail to reject H_o" decision must be interpreted as the conclusion that there is no evidence of difference due to the levels of the tested factor.

128. Why does *df*(Factor), the number of degrees of freedom associated with the factor, always appear first in the critical value notation *F*[*df*(factor), *df*(error), α]?

ANSWER:
df(factor) appears first in the critical number notation since MS(factor) is the numerator for the calculated value of the test statistic *F*.

129. For the single-factor ANOVA, the mathematical model formula $x_{c,k} = \mu + F_c + \varepsilon_{k(c)}$ is an expression of the composition of each piece of data entered in our data table. Interpret each term of this model.

ANSWER:
$x_{c,k}$ is the value of the variable at the kth replicate of level *c*.

μ is the mean value for all the data without respect to the test factor.

F_c, is the effect that the factor being tested has on the response variable at each different level of *c*.

$\varepsilon_{k(c)}$ is the experimental error that occurs among the *k* replicates in each of the *c* columns

130. In single-factor ANOVA, the null hypothesis is that there is no difference between the levels of the factor being tested. How would you interpret a "reject H_o" decision?

ANSWER:
A "reject H_o" decision implies that there is a difference between the levels. That is, at least one level is different from the others.

Applied and Computational Questions

QUESTIONS 131 THROUGH 137 ARE BASED ON THE FOLLOWING INFORMATION:
A study was designed to compare the fasting blood sugar readings for three groups of diabetic patients. One group used insulin to control their problem, one group used oral drugs, and one group used exercise and diet. The blood sugar readings for the three samples were as follows:

	Group	
Insulin	Oral Drug	Diet/Exercise
110	120	100
95	135	95
125	140	110
130	130	115
110	125	100

It is highly likely that $\mu_1 = \mu_2 = \mu_3$ is false.

131. Which group gave the largest sample mean?

ANSWER:
$\bar{x}_2 = 130$

132. Which group gave the smallest sample mean?

ANSWER:
$\bar{x}_3 = 104$

133. If asked to speculate on what two population means differ, what would your choice be?

ANSWER:
μ_2 and μ_3

134. Develop the ANOVA table for testing the claim of equal means.

ANSWER:

	SS	*df*	*MS*	*F**
Group	1720	2	860	8.00
Error	1290	12	107.5	
Total	3010	14		

135. Write the appropriate null and alternative hypotheses.

ANSWER:
$H_o : \mu_1 = \mu_2 = \mu_3$ vs. H_a : at least two of the population means are not the same.

136. Give the critical region for $\alpha = 0.05$,

ANSWER:
$F(2, 12, 0.05) = 3.89$,

137. Find a bound on the *p*-value, and write the conclusion.

ANSWER:
P = *p* - value < 0.01, we reject the null hypothesis at $\alpha = 0.05$, and conclude that at least two of the population means differ.

138. The coded values for the measure of elasticity in plastic, prepared by two different processes, for samples of six drawn randomly from each of the processes are shown below. Using the *F* test, at $\alpha = 0.05$, determine if the data presents sufficient evidence to indicate a difference in mean elasticity for the two processes.

Process A	6.1	7.1	7.8	6.9	7.6	8.2
Process B	9.1	8.2	8.6	6.9	7.5	7.9

ANSWER:

	SS	*df*	*MS*	*F**
Group	1.688	1	1.688	2.88
Error	5.862	10	0.5862	
Total	7.55	11		

Since $F^* = 2.88$, and critical region is $F \geq 4.96$, we fail to reject the null hypothesis. The data does not present sufficient evidence to indicate a difference in mean elasticity for the two processes.

QUESTIONS 139 THROUGH 141 ARE BASED ON THE FOLLOWING INFORMATION:
In order to better control inflation, the government suggested pay increases be limited to 8% or less. A member of the Inflation Fighters group compiled the following percent increases for three different industry groups.

Sales/Service	Produce Manufacturing	Research
4.5	6.0	5.0
5.0	5.5	5.5
6.2	5.5	6.0
5.5	7.1	5.8

139. Develop the ANOVA table.

ANSWER:

	SS	*df*	*MS*	*F**
Group	1.072	2	0.536	1.25
Error	3.855	9	0.428	
Total	4.927	11		

Chapter 12

140. Write the null and alternative hypotheses.

ANSWER:

$H_o: \mu_1 = \mu_2 = \mu_3$ vs. H_a :at least two of the population means are not the same.

141. Test for equal means at α = 0.05.

ANSWER:

Critical region: $F \geq 4.26$, F^* = 1.25, therefore we fail to reject H_o at α = 0.05, and conclude that the group means are identical.

QUESTIONS 142 THROUGH 144 ARE BASED ON THE FOLLOWING INFORMATION:
The table below shows the cars that ran out of gas during a one day period on the New York State Thruway for 4 observation periods.

Observation Period	*Number of Cars Running Out of Gas*			
Westbound, AM	37	34	38	36
Eastbound, AM	37	40	37	42
Westbound, PM	33	34	38	35
Eastbound, PM	41	36	40	39

142. Develop the ANOVA table.

ANSWER:

	SS	*df*	*MS*	*F**
Group	48.69	3	16.23	3.56
Error	54.75	12	4.5625	
Total	103.44	15		

143. State the null and alternative hypotheses.

ANSWER:

$H_o: \mu_1 = \mu_2 = \mu_3 = \mu_4$ vs. H_a :at least two of the population means are not the same.

144. At the 0.05 level of significance, does the data contradict the hypothesis that the mean number of cars running out of gas is the same in all four categories? Test at α = 0.05 using the classical approach..

ANSWER:

Critical region is: $F \geq 3.49$, F^* = 3.56, therefore we reject H_o at α = 0.05. The data contradicts the hypothesis that the mean number of cars running out of gas is the same in all four categories

QUESTIONS 145 THROUGH 147 ARE BASED ON THE FOLLOWING INFORMATION:
Four brands of gasoline were compared in an experiment. Sixteen small engines were used and the time of operation for one gallon of gasoline was measured. Four engines were randomly assigned to each brand.

	Brand		
A	B	C	D
25	30	32	35
30	30	34	30
30	35	36	35
32	34	30	38

145. Develop the ANOVA table.

ANSWER:

	SS	*df*	*MS*	*F**
Brand	58.50	3	19.50	2.33
Error	100.50	12	8.375	
Total	159.00	15		

146. State the null and alternative hypotheses.

ANSWER:

$H_o: \mu_1 = \mu_2 = \mu_3 = \mu_4$ vs. H_a :at least two of the population means are not the same.

147. Test for equal means at $\alpha = 0.01$ using the classical approach.

ANSWER:

Critical region is: $F \geq 5.95$, $F^* = 2.33$, therefore we fail to reject H_o at $\alpha = 0.01$. There is not sufficient reason to indicate that at least two population means of the four brands of gasoline are identical.

QUESTIONS 148 THROUGH 150 ARE BASED ON THE FOLLOWING INFORMATION:

A cookie salesman interested in increasing his sales volume arranged to have displays of his best selling cookie in three locations in a market as shown below.

Location	Volume Sold
by meat counter	24, 26, 25, 25, 30
by check out	26, 30, 35, 40, 45
in cookie section	24, 24, 32, 33, 43

148. Develop the ANOVA table.

ANSWER:

	SS	*df*	*MS*	*F**
Location	212.8	2	106.4	2.56
Error	499.6	12	41.633	
Total	712.4	14		

149. State the null and alternative hypotheses.

ANSWER:

$H_o: \mu_1 = \mu_2 = \mu_3$ vs. H_a :at least two of the population means are not the same.

150. Calculate p-value for testing equal means in the three locations based on the weekly sales volumes. What is your conclusion at $\alpha = 0.01$?

ANSWER:

p -value > 0.05; therefore we fail to reject H_o at $\alpha = 0.01$. Conclusion: The data provides sufficient evidence to conclude that the means in the three locations are equal.

QUESTIONS 151 THROUGH 155 ARE BASED ON THE FOLLOWING INFORMATION:
Four different statistical computer programs were tested for time (in seconds) required completing a particular task. The results are shown below.

	Seconds Required by Program			
Sample	A	B	C	D
1	18.4	16.6	22.4	31.4
2	17.6	16.9	21.5	30.1
3	19.6	17.0	22.6	33.4

151. How many treatments are there?

ANSWER:
4 treatments

152. At the 0.01 level, what is the critical value?

ANSWER:
Critical value = 7.59

153. What is the value of the test statistic?

ANSWER:

	SS	*df*	*MS*	*F**
Program	393.60	3	131.20	126.03
Error	8.33	8	1.041	
Total	401.93	11		

154. Write the appropriate null and alternative hypotheses, and your conclusion at the 0.01 level.

ANSWER:

$H_o: \mu_1 = \mu_2 = \mu_3 = \mu_4$ vs. H_a :at least two of the population means are not the same.

155. Use the classical approach to complete the hypothesis test.

ANSWER:
Since F^* = 126.03, and the critical value is F = 7.59, we reject the null hypothesis. There is sufficient evidence to indicate that the mean time (in seconds) required completing a particular task is different for at least two of the four statistical computer programs.

QUESTIONS 156 THROUGH 159 ARE BASED ON THE FOLLOWING INFORMATION:
A new worker was recently assigned to a crew of workers who perform a certain job. From the records of the number of units of work completed by each worker each day last month, a sample of size five was randomly selected for each of the two experienced workers and the new worker as shown in the table below.

	Workers		
	New	A	B
Units of Work (*replicates*)	10	13	12
	12	14	15
	11	12	11
	13	14	14
	10	15	15

156. State the null and alternative hypotheses.

ANSWER:

H_o: The mean values for workers are all equal.

H_a :The mean values for workers are not all equal.

157. Assume the data were randomly collected and are independent, and the effects due to chance and untested factors are normally distributed. Develop the ANOVA table.

ANSWER:

$n = 15, \quad C_1 = 56, \quad C_2 = 68, \quad C_3 = 67, \quad T = 191, \quad \sum x^2 = 2475$

Source	*df*	*SS*	*MS*	F^*
Work	2	17.733	8.867	4.22
Error	12	25.200	2.100	
Total	14	42.933		

158. At the 0.05 level of significance, does the evidence provide sufficient reason to reject the claim that there is no difference in the amount of work done by the three workers? Solve using the *p*-value approach.

ANSWER:

P = *p*-value = $P(F > 4.22 \mid df_n = 2, df_d = 12)$. Using the tables of *F*-distribution we get, 0.025 < **P** < 0.05. Since **P** $< \alpha$; reject H_o. There is sufficient evidence to indicate that there is significant difference between the workers with regards to mean amount of work produced.

159. At the 0.05 level of significance, does the evidence provide sufficient reason to reject the claim that there is no difference in the amount of work done by the three workers? Solve using the classical approach.

ANSWER:

The critical region is: $F \geq 3.89$. Since the value of the test statistic *F** falls in the critical region, we reject H_o. We reach the same conclusion as stated in question 158.

Chapter 12

QUESTIONS 160 THROUGH 163 ARE BASED ON THE FOLLOWING INFORMATION:
An experiment was designed to compare the lengths of time that four different drugs provided pain relief following heart surgery. The results (in hours) are shown in the following table.

Drug

A	B	C	D
9	7	7	5
7	7	9	5
5	5	9	3
3	5	9	
		11	

160. State the null and alternative hypotheses.

ANSWER:

H_o The mean amount of relief time is the same for all four drugs.

H_a : The mean amount of relief time is not the same for all four drugs.

161. Assume the data were randomly collected and are independent, and the effects due to chance and untested factors are normally distributed. Develop the ANOVA table.

ANSWER:

$n = 16, \quad C_A = 24, \quad C_B = 24, \quad C_C = 45, \quad C_D = 13, \quad T = 106, \quad \sum x^2 = 784$

Source	*df*	*SS*	*MS*	F^*
Drug	3	47.083	15.694	5.43
Error	12	34.667	2.889	
Total	15	81.75		

162. Is there enough evidence to reject the null hypothesis that there is no significant difference in the length of pain relief for the four drugs at $\alpha = 0.05$? Solve using the *p*-value approach.

ANSWER:

P = *p*-value = $P(F > 5.43 \mid df_n = 3, df_d = 12)$. Using the tables of *F*-distribution, we get 0.01 < **P** < 0.025. Since **P** < α ; reject H_o. There is sufficient evidence to indicate that there is a significant difference between the mean amount of relief time for these four drugs.

163. Is there enough evidence to reject the null hypothesis that there is no significant difference in the length of pain relief for the four drugs at $\alpha = 0.05$? Solve using the classical approach.

ANSWER:

The critical region is: $F \geq 3.49$. Since the test statistic *F** falls in the critical region, we reject H_o. We reach the same conclusion as stated in question 162.

QUESTIONS 164 THROUGH 167 ARE BASED ON THE FOLLOWING INFORMATION:
A certain vending company's soft-drink dispensing machines are supposed to serve eight ounces of beverage. Various machines were samples and the resulting amounts of dispensed drink were recorded, as shown in the following table.

	Machines				
	A	B	C	D	E
Amounts of Soft Drink Dispensed	7.0	9.8	7.6	9.7	9.6
	7.4	10.3	7.3	9.6	7.7
	7.3	9.9	7.1	9.4	8.5
	7.6		7.7		9.0

164. State the null and alternative hypotheses.

ANSWER:

H_o: The mean amounts dispensed by the machines are all equal.

H_a :The mean amounts dispensed by the machines are not all equal.

165. Assume the data were randomly collected and are independent, and the effects due to chance and untested factors are normally distributed. Develop the ANOVA table.

ANSWER:

$n = 18,\quad C_A = 29.3,\quad C_B = 30.0,\quad C_C = 29.7,\quad C_D = 28.7,\quad C_E = 34.8,\quad T = 152.5,$ and $\sum x^2 = 1315.01$

Source	*df*	*SS*	*MS*	F^*
Machine	4	20.454	5.1135	26.16
Error	13	2.542	0.1955	
Total	17	22.996		

166. Does this sample evidence provide sufficient reason to reject the null hypothesis that all five machines dispense the same average amount of soft drink? Solve using the *p*-value approach.

ANSWER:

P = *p*-value = $P(F > 26.16 \mid df_n = 4, df_d = 13)$ .Using the tables of *F*-distribution, we get **P** < 0.01. Since **P**< α ; we reject H_o. There is sufficient evidence to indicate that here is a significant difference between the machines with regards to mean amount of soft drink dispensed.

167. Does this sample evidence provide sufficient reason to reject the null hypothesis that all five machines dispense the same average amount of soft drink? Solve using the classical approach.

ANSWER:

The critical region is: $F \geq 5.21$. Since the test statistic *F** falls in the critical region, we reject H_o. We reach the same conclusion as stated in question 166.

QUESTIONS 168 THROUGH 171 ARE BASED ON THE FOLLOWING INFORMATION:
It is believed that the median family incomes for three counties in Michigan are as follows: Wexford $37,780, Osceola $32,135, and Macomb $39,630. The following data represent the family incomes (in thousands) for nine randomly selected individuals from each of the three counties.

Wexford	Osceola	Macomb
46.3	33.2	41.8
40.8	31.3	43.4
43.3	38.4	46.1
36.2	36.5	40.6
41.4	29.8	41.9
38.2	38.7	39.1
45.0	32.1	52.0
47.8	38.7	48.9
49.6	27.0	42.4

168. State the null and alternative hypotheses.

ANSWER:

H_o The mean family income is the same for all three counties.

H_a : The mean family income is not the same for at least two of the three counties.

169. Assume the data were randomly collected and are independent, and the effects due to chance and untested factors are normally distributed. Develop the ANOVA table.

ANSWER:

$n = 27,\ C_W = 388.6,\quad C_O = 305.7,\quad C_M = 396.2,\quad T = 1090.5,\quad \sum x^2 = 45050.19$

Source	*df*	*SS*	*MS*	F^*
Counties	2	560.016	280.008	15.06
Error	24	446.091	18.587	
Total	26	1006.107		

170. Is there sufficient evidence to conclude that the mean family income is the same for each of the three counties at the 0.05 level of significance? Solve using the *p*-value approach.

ANSWER:

P = *p*-value = $P(F > 15.06 \mid df_n = 2, df_d = 24)$. Using the tables of *F*-distribution, we get **P** < 0.01. Since **P** < α ; reject H_o. There is sufficient evidence to indicate that the mean family income is not the same at least two of the three counties.

171. Is there sufficient evidence to conclude that the mean family income is the same for each of the three counties at the 0.05 level of significance? Solve using the classical approach.

ANSWER:

The critical region is: $F \geq 3.40$. Since the test statistic *F** falls in the critical region, we reject H_o. We reach the same conclusion as stated in question 170.

QUESTIONS 172 THROUGH 175 ARE BASED ON THE FOLLOWING INFORMATION:
A consumer research organization is attempting to determine whether there is any difference in mpg for fully loaded 22-foot trucks leased from three companies: A, B, and C. Five of these trucks are rented from each company. Each truck is driven with the same weight cargo over the same 200 mile route and the mpg recorded. The results of the test are:

A	B	C
3.4	5.1	7.9
4.2	2.0	8.5
5.1	8.7	5.2
4.9	6.7	8.0
3.1	6.1	8.1

172. State the null and alternative hypotheses.

ANSWER:
$H_o: \mu_1 = \mu_2 = \mu_3$ (Average mpg is the same for all three rental companies).
H_a: Not all of the mean mpg is the same for the three companies.

173. Develop the ANOVA table.

ANSWER:

Source	*df*	*SS*	*MS*	F^*
Trucks	2	28.948	14.474	5.05
Error	12	34.392	2.866	
Total	14	63.340		

174. Is there any difference in mean mpg? Perform the appropriate test at $\alpha = 0.05$ using the critical approach.

ANSWER:
The critical value is: $F(2,12,0.05) = 3.89$. Since the value of the test statistic is $F^* = 5.05 > 3.89$, we reject H_o. There is sufficient evidence at the .05 level of significance to indicate that the consumer research organization does not find support for the equality of mean mpg for the three companies.

175. Is there any difference in mean mpg? Perform the appropriate test at $\alpha = 0.05$ using the *p*-value approach.

ANSWER:
P = *p*-value = $P(F > 5.06 \mid df_n = 2, df_d = 12)$. Using the tables of *F*-distribution, we have **P** < 0.01. Since **P** < α; we reject H_o. We reach the same conclusion as stated in question 174.

176. State the null hypothesis H_o and the alternative hypothesis H_a that would be used to test the following statement: "The mean scores are the same at all five levels of the experiment."

ANSWER:
$H_o: \mu_1 = \mu_2 = \mu_3 = \mu_4 = \mu_5$ vs. H_a: Not all mean scores are equal.

177. State the null hypothesis H_o and the alternative hypothesis H_a that would be used to test the following statement: "The test scores are the same at all three sections."

ANSWER:
$H_o: \mu_1 = \mu_2 = \mu_3$
H_a: Not all test mean scores are equal.

178. State the null hypothesis H_o and the alternative hypothesis H_a that would be used to test the following statement: "The three levels of the test factor do not significantly affect the data."

ANSWER:
$H_o: \mu_1 = \mu_2 = \mu_3$ (The test factor has no effect)
H_a: Not all test means are equal. (The test factor has an effect)

179. State the null hypothesis H_o and the alternative hypothesis H_a that would be used to test the following statement: "The four different methods of treatment do affect the variable."

ANSWER:
$H_o: \mu_1 = \mu_2 = \mu_3 = \mu_4$ (The different methods of treatment have no effect)
H_a: Not all test means are equal. (The different methods of treatment have an effect)

180. Place bounds on the p-value for the following situation: F* = 4.85, df(Factor) = 2, df(Error) = 10

ANSWER:
P = P(F > 4.85 | df_n = 2, df_d =12) $\Rightarrow$ 0.025 < **P** < 0.05

181. Place bounds on the p-value for the following situation: F* = 4.89, df(Factor) = 4, df(Error) = 15

ANSWER:
P = P(F > 4.89 | df_n = 4, df_d =15) $\Rightarrow$ **P** = 0.01

182. Sketch an approximate F-curve and use the classical approach to determine the critical region(s) and critical value(s) that would be used to test $H_o: \mu_1 = \mu_2 = \mu_3 = \mu_4$ with $n = 20,\ \alpha = 0.05$.

ANSWER:

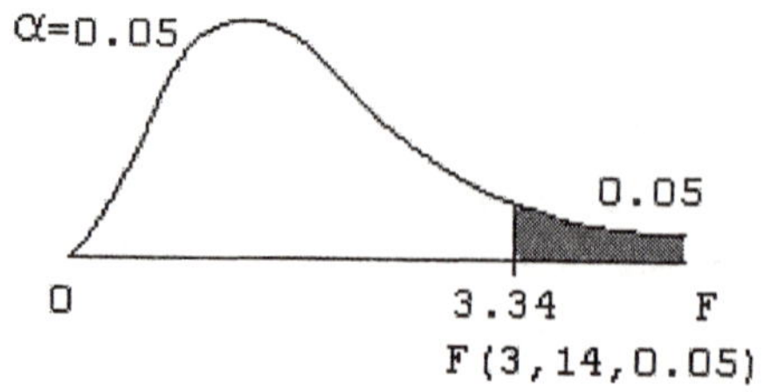

183. Place bounds on the *p*-value for the following situation: F* = 3.57, df(Factor) = 6, df(Error) = 21

ANSWER:
P = P(F > 3.57 | df_n = 6, df_d = 21) $\Rightarrow$ 0.01 < **P** < 0.025

184. Sketch an approximate *F*-curve and use the classical approach to determine the critical region(s) and critical value(s) that would be used to test the null hypothesis $H_o : \mu_1 = \mu_2 = \mu_3$ with $n = 25,\ \alpha = 0.05$

ANSWER:

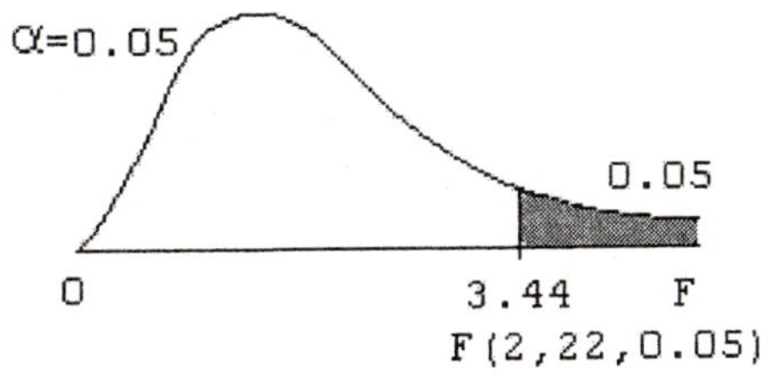

185. Sketch an approximate *F*-curve and use the classical approach to determine the critical region(s) and critical value(s) that would be used to test
$H_o : \mu_1 = \mu_2 = \mu_3 = \mu_4 = \mu_5$ with $n = 15,\ \alpha = 0.01$

ANSWER:

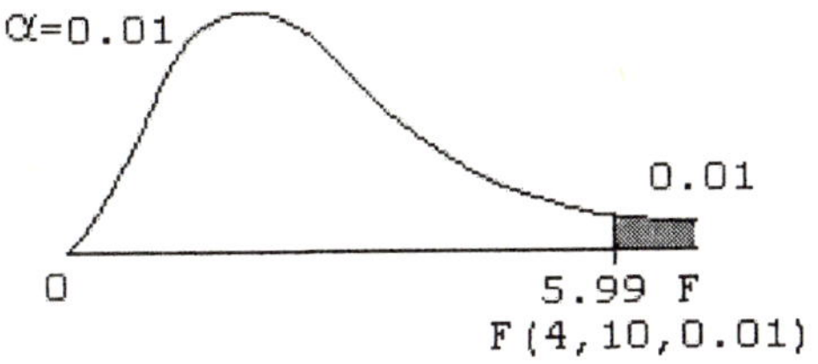

QUESTIONS 186 THROUGH 188 ARE BASED ON THE FOLLOWING INFORMATION:
Suppose that an *F*- test (as described in this chapter using the p-value approach) has a *p*-value of 0.039.

186. What is the interpretation of *p*-value = 0.039?

ANSWER:
p-value = 0.039 can be interpreted as 0.039 of the probability distribution associated with F and a true null hypothesis is more extreme than the value of the test statistic F^*. That is, area under the curve and to the right of F^*.

187. What is the interpretation of the situation if you had previously decided on a 0.05 level of significance?

ANSWER:
Reject the null hypothesis; since the *p*-value is smaller than the previously set value for α .

188. What is the interpretation of the situation if you had previously decided on a 0.025 level of significance?

ANSWER:
Fail to reject the null hypothesis; since the *p*-value is greater than the previously set value for α .

QUESTIONS 189 THROUGH 196 ARE BASED ON THE FOLLOWING INFORMATION:
The single-factor analysis of variance (ANOVA) is used to test a hypothesis about several population means. Assume that c is the number of levels (columns) for which the factor is tested, k_i is the number of replicates at each level tested, and n = $\sum k_i$ is the number of data in the total sample.

189. State the null hypothesis, in a general form, for the one-way ANOVA.

ANSWER:
H_o: The test factor has no effect on the mean at the tested levels.

190. State the alternative hypothesis, in a general form, for the one-way ANOVA.

ANSWER:
H_a: The test factor has does have an effect on the mean at the tested levels.

191. What must happen in order to "reject H_o" if using p-value approach?

ANSWER:
P = *p*-value = $P(F > F^*)$ must be $\leq \alpha$.

192. What must happen in order to "reject H_o" if using the classical approach?

ANSWER:
The calculated value of F; namely F^*, must fall in the critical region; that is, the variance between levels of the factor must be significantly larger than variance within the levels.

193. How would a decision of "reject H_o" be interpreted?

ANSWER:
The tested factor has a significant effect on the variable.

194. What must happen in order to "fail to reject H_o" If using the *p*-value approach?

ANSWER:
P = *p*-value = P($F > F^*$) must be > α .

195. What must happen in order to "fail to reject H_o" If using the classical approach?

ANSWER:
The calculated value of F; namely F^*, must fall in the non-critical region; that is, the variance between levels of the factor must not be significantly larger than variance within the levels.

196. How would a decision of "fail to reject H_o" be interpreted?

ANSWER:
The tested factor does not have a significant effect on the variable.

QUESTIONS 197 THROUGH 203 ARE BASED ON THE FOLLOWING INFORMATION:
Three new drugs are being tested for their effect on the number of days of hospitalization needed by the patient following surgery. There is a control group receiving a placebo and three treatment groups with each receiving one of three new drugs, all developed to promote recovery. The results of an analysis of variance used to analyze the data are shown here.

One-way ANOVA: Days versus Group

Source of Variation	*df*	*SS*	*MS*	F^*	*P-value*
Group	3	13.5	4.5	1.875	0.175
Error	16	38.4	2.4		
Total	19	51.9			

197. How many patients were there?

ANSWER:
df(Total) = *n* -1 = 19 ⇒ *n* = 20 patients

198. How do these results verify that there was one control and three test groups?

ANSWER:
df(Group) = c-1 = 3 ⇒ c = 4 groups

199. Using the SS values, verify the two mean square values.

ANSWER:
MS(Group) = SS(Group) / *df*(Group) = 13.5 / 3 = 4.5
MS(Error) = SS(Error) / *df*(Error) = 38.4 / 16 = 2.4

200. Using the MS values, verify the F-value

ANSWER:
F-value = MS(Group) / MS(Error) = 4.5 / 2.4 = 1.875

201. Verify the P-value

ANSWER:
Using Minitab Statistical Software, *p*-value = 0.175

Chapter 12

202. State the null and alternative hypotheses

ANSWER:

$H_o: \mu_1 = \mu_2 = \mu_3 = \mu_4$

H_a : The means are not all equal (that is, at least one mean is different)

203. State the decision and conclusion reached as a result of the analysis at the 0.05 level of significance

ANSWER:

Decision: Fail to reject H_o : since p-value = 0.175 > α = 0.05.

Conclusion: There is no evidence at the 0.05 level of significance of difference between the means due to the levels of the tested factor.

QUESTIONS 204 THROUGH 207 ARE BASED ON THE FOLLOWING INFORMATION:

A new operator was recently assigned to a crew of workers who perform a certain job. From the records of the number of units of work completed by each worker each day last month, a sample of size five was randomly selected for each of the three experienced workers and the new worker as shown in the table below. There is a reason to believe that there is no difference in the amount of work done by the three workers.

	Workers			
	New	A	B	C
Units of Work (replicates)	9	12	11	13
	11	13	14	12
	10	11	10	11
	12	13	13	12
	9	14	14	13

204. State the null and alternative hypotheses.

ANSWER:

H_o : The mean values for workers are all equal.

H_a : The mean values for workers are not all equal.

205. Use computer and statistical software to develop the ANOVA table.

ANSWER:

Source of Variation	*SS*	*df*	*MS*	*F*	*P-value*	*F crit*
Between Groups	18.55	3	6.183	3.533	0.0389	3.239
Within Groups	28.00	16	1.750			
Total	46.55	19				

206. Test the hypotheses in question 204 at the 0.05 level of significance using the *p*-value approach.

ANSWER:
Since p-value = 0.0389 < α = 0.05, we reject H_o. There is sufficient evidence to indicate that the mean values for workers are not all equal. In other words, there is significant difference between the workers with regards to mean amount of work produced.

207. Test the hypotheses in question 204 at the 0.05 level of significance using the classical approach.

ANSWER:
The critical value is 3.239. Since the value of the test statistic F^* = 3.533 falls in the rejection region, we reject H_o. We reach the same conclusion as stated ion question 206.

QUESTIONS 208 THROUGH 212 ARE BASED ON THE FOLLOWING INFORMATION:
A new all-purpose cleaner is being test-marketed by placing sales displays in three different locations within various supermarkets. The number of bottles sold from each location within each of the supermarkets tested is reported below.

Locations	I	42	37	46	40
	II	34	40	32	37
	III	47	50	52	54

Based on past experience, there is no sufficient evidence to doubt that the location of the sales display had no effect on the number of bottles sold.

208. State the null and alternative hypotheses.

ANSWER:
H_o : The location of the sales display had no effect on sales.
H_a : The location of the sales display did have an effect on sales.

209. Develop the ANOVA table by using a computer and statistical software.

ANSWER:

Source of Variation	*SS*	*df*	*MS*	*F*	*P-value*	*F crit*
Between Groups	460.667	2	230.333	19.511	0.0005	8.022
Within Groups	106.250	9	11.806			
Total	566.917	11				

210. Using the information obtained in question 209, state the decision and conclusion to the hypothesis test at the 0.01 level of significance using the *p*-value approach.

ANSWER:
Since *p*-value = 0.0005 < α = 0.01, we reject H_o. There is sufficient evidence to indicate that the location of the sales display had an effect on sales.

211. State the decision and conclusion to the hypothesis test at the 0.01 level of significance using the classical approach.

ANSWER:

The critical value is 8.022. Since the value of the test statistic $F^* = 19.511$ falls in the rejection region, we reject H_o. We reach the same conclusion as stated in question 210.

212. What is the practical interpretation of the *p*-value in this case? Explain.

ANSWER:

Since the *p*-value is very small (0.0005), it tells us the sample data is very unlikely to have occurred under the assumed conditions and a true null hypothesis. Therefore, the decision was to reject H_o.

QUESTIONS 213 THROUGH 217 ARE BASED ON THE FOLLOWING INFORMATION:

An experiment was designed to compare the lengths of time that four different drugs provided pain relief following brain surgery. The results (in hours) are shown in the following table. A doctor claims that there is no significant difference in the length of pain relief for the four drugs

Drug

A	B	C	D
10	14	12	14
10	16	12	12
8	16	10	10
	16	10	8
	18		

213. State the null and alternative hypotheses.

ANSWER:

H_o: The mean length of pain relief time is the same for all four drugs.

H_a: The mean length of pain relief time is not the same for all four drugs.

214. Develop the ANOVA table by using a computer and statistical software.

ANSWER:

Source of Variation	*SS*	*df*	*MS*	*F*	*P-value*	*F crit*
Between Groups	108.333	3	36.111	12.5	0.0005	3.490
Within Groups	34.667	12	2.889			
Total	143.000	15				

215. Is there enough evidence to reject the null hypothesis In question 213 at $\alpha = 0.05$? Use the *p*-value approach.

ANSWER:

Since p-value = 0.0005 < α = 0.05, we reject H_o. There is sufficient evidence to indicate that the mean length of pain relief time is not the same for all four drugs.

216. Is there enough evidence to reject the null hypothesis In question 213 at α = 0.05? Use the classical approach.

ANSWER:

The critical value is 3.49. Since the value of the test statistic F^*= 12.5 falls in the rejection region, we reject H_o. We reach the same conclusion as stated ion question 215.

217. What is the practical interpretation of the p-value in this case? Explain.

ANSWER:

Since the *p*-value is very small (0.0005), it tells us the sample data is very unlikely to have occurred under the assumed conditions and a true null hypothesis. Therefore, the decision was to reject H_o.

QUESTIONS 218 THROUGH 221 ARE BASED ON THE FOLLOWING INFORMATION:

To compare the effectiveness of three different methods of teaching reading, 27 children of equal reading aptitude were divided into three equal groups of 9 children each. Each group was instructed for a given period of time using one of the three methods. After completing the instruction period, all students were tested. The test results, shown in the following table, are used to determine if there is sufficient evidence that all three instruction-methods are equally effecting.

Methods of Teaching

	Method 1	Method 2	Method 3
Test Scores (replicates)	46	45	46
	45	51	52
	47	46	49
	45	56	51
	41	52	47
	44	52	49
	47	46	46
	50	48	49
	45	51	48

218. State the null and alternative hypotheses

ANSWER:

H_o: All three methods of instruction are equally effective, as measured by the mean test scores.

H_a: All three methods of instruction are not equally effective, as measured by the mean test scores.

219. Use Minitab or Excel to provide summary statistics table and ANOVA table for this data

ANSWER:

SUMMARY STATISTICS

Groups	Count	Sum	Average	Variance
Method 1	9	410	45.556	6.028
Method 2	9	447	49.667	13.250
Method 3	9	437	48.556	4.278

ANOVA TABLE

Source of Variation	SS	df	MS	F	P-value	F crit
Between Groups	81.407	2	40.704	5.184	0.01345	3.403
Within Groups	188.444	24	7.852			
Total	269.852	26				

220. Using the information in the computer printout in question 219, state the decision and the conclusion to the hypothesis test at $\alpha = 0.05$ using the *p*-value approach.

ANSWER:
Since p-value = 0.01345 < α = 0.05, we reject H_o. There is sufficient evidence to indicate that all three methods of instruction are not equally effective, as measured by the mean test scores.

221. Using the information in the computer printout in question 219, state the decision and the conclusion to the hypothesis test at α = 0.05 using the classical approach

ANSWER:
The critical value is 3.403. Since the value of the test statistic F^* = 5.184 falls in the rejection region, we reject H_o. We reach the same conclusion as stated ion question 220.

Chapter 13

Linear Correlation and Regression Analysis

Sections 13.1 through 13.3

True-False Questions

1. If x and y are highly correlated, then x is said to cause y to occur.
ANSWER: F

2. The variance of y about the line of best fit is the same as the variance of the error e where $e = y - \hat{y}$.
ANSWER: T

3. The covariance of x and y is defined by the equation: $covar(x, y) = \sum (x - \overline{x})(y - \overline{y}) / n$.
ANSWER: F

4. Generally speaking, the higher the correlation between x and y, the better will be the predictions which are made using the line of best fit provided the prediction is made for an x-value within the range of observed x-values.
ANSWER: T

5. The linear correlation coefficient is used to measure the strength of the linear relationship between two variables.
ANSWER: T

6. The coefficient of linear correlation is also commonly referred to as Pearson's product moment, r.
ANSWER: T

7. In general $\sum (x - \overline{x})(y - \overline{y}) = 0$ since it is always true that $\sum (x - \overline{x}) = 0$ and $\sum (y - \overline{y}) = 0$.
ANSWER: F

8. Correlation analysis attempts to find the equation of the line of best fit for two variables.
ANSWER: F

9. The coefficient of linear correlation is given by the equation $r = SS(xy)/\sqrt{SS(x)\cdot SS(y)}$
ANSWER: T

10. The linear correlation coefficient for the population is always a number between 0 and 1.
ANSWER: F

11. Covariance measures the strength of the linear relationship and is a standardized measure.
ANSWER: F

12. Analysis of linear dependency between two variables uses two measures: covariance and the coefficient of linear correlation.
ANSWER: T

13. Like the variance and standard deviation, the covariance of a single set of bivariate data is always positive.
ANSWER: F

14. Inferences about the linear correlation coefficient are about the pattern of behavior of the two variables involved and the usefulness of one variable in predicting the other.
ANSWER: T

15. The covariance of a single set of data is positive if the graph is dominated by points to the upper right and to the lower left of the centroid $(\bar{x}, \bar{y})$.
ANSWER: T

16. A confidence interval may be used to estimate the value of ρ, the linear correlation coefficient of the population. Usually this is accomplished by using the *t*-table with degrees of freedom equal to *n* -1.
ANSWER: F

17. The biggest disadvantage of covariance as a measure of linear dependency is that it does not have a standardized unit of measure.
ANSWER: T

18. Failure to reject the null hypothesis $H_o: \rho = 0$ is interpreted as meaning that a linear relationship between the two variables in the population has been shown.
ANSWER: F

Multiple-Choice Questions

19. The values below are suggested coefficients of correlation, *r*. The one that indicates the strongest negative relationship between the input variable *x* and the output variable *y* is:

A) -1.5.
B) -0.7.
C) 0.0.
D) 0.8.
ANSWER: B

20. The values below are suggested coefficients of correlation, *r*. The one that indicates the strongest positive relationship between the input variable *x* and the output variable *y* is:

A) 1.2.
B) 0.7.
C) 0.0.
D) 0.8.
ANSWER: **D**

21. An indication of no linear relationship between two variables would be:

A) a coefficient of correlation of +1.
B) a coefficient of correlation of -1.
C) a coefficient of correlation of 0.
D) a coefficient of correlation of -2.
ANSWER: **C**

22. In publishing the results of some research work, the following values of the correlation coefficient were listed. Which one would appear to be incorrect?

A) 1.05
B) 1.0
C) 0.95
D) -0.95
ANSWER: **A**

23. Which of the following statements is false?

A) The linear correlation coefficient *r* is a quantity that measures the strength of a linear relationship (dependency) between two variables.
B) Analysis of linear dependency between two variables uses two measures: covariance and the coefficient of linear correlation.
C) The covariance of *x* and *y* is defined as the sum of the products of the distances of all values of *x* and *y* from centroid $(\bar{x}, \bar{y})$.
D) None of the above
ANSWER: **C**

24. Which of the following formulas is false?

A) covar(*x*, *y*) = $\sum_{i=1}^{n}(x_i - \bar{x})(y_i - \bar{y})/(n-1)$.

B) $r = \sum_{i=1}^{n}(x_i - \bar{x})(y_i - \bar{y})/(s_x \cdot s_y)$

C) $r = SS(xy)/\sqrt{SS(x) \cdot SS(y)}$

D) None of the above
ANSWER: **B**

25. Which of the following statements is false regarding the covariance of a set of bivariate data?

A) It can be negative.
B) It can be positive.
C) It can be zero.
D) It is always zero since $\sum(x-\bar{x})$ and $\sum(y-\bar{y})$ are always zero and the covariance is defined as $\sum(x-\bar{x})(y-\bar{y})$ divided by $(n-1)$.

ANSWER: D

26. Which of the following statements is false?

A) The sign of the covariance is the opposite of the sign of the slope of the regression line.
B) The covariance of a single set of data is positive if the graph is dominated by points to the upper right and to the lower left of the centroid $(\bar{x}, \bar{y})$.
C) If the majority of the points are to the upper left and the lower right of the centroid $(\bar{x}, \bar{y})$, then the covariance is negative.
D) None of the above.

ANSWER: A

27. If the coefficient of linear correlation for a single set of bivariate data is 0.0698, while the standard deviation of x is 4.099 and the standard deviation of y is 2.098, then the covariance of x and y is

A) 0.205.
B) 0.300.
C) 0.286.
D) 0.146.

ANSWER: B

28. For a bivariate set of data, if SS(xy) =200, SS(x) = 350 and SS(y) = 125, then the Pearson's product moment is

A) 0.956.
B) 0.005.
C) 1.046.
D) None of the above.

ANSWER: A

29. If the coefficient of linear correlation and the covariance for a single set of bivariate data are 0.582 and 0.854, respectively, and the standard deviation of x is 1.625, then the standard deviation of y is

A) 0.681.
B) 0.526.
C) 1.107.
D) 0.903.

ANSWER: D

30. If the covariance for a single set of bivariate data is 0.75, while the standard deviation of *x* is 2.5 and the standard deviation of *y* is 3.2, then the coefficient of linear correlation is

A) 0.234.
B) 0.300.
C) 0.094.
D) 0.265.
ANSWER: C

31. Which of the following statements is false?

A) Inferences about the linear correlation coefficient are about the pattern of behavior of the two variables involved and the usefulness of one variable in predicting the other.
B) Significance of the linear correlation coefficient means that you have established a cause-and-effect relationship.
C) The linear correlation coefficient of the population is denoted by the Greek letter ρ.
D) None of the above.
ANSWER: B

32. Which of the following statements is false?

A) The biggest disadvantage of covariance as a measure of linear dependency is that it does not have a standardized unit of measure.
B) We must find some way to eliminate the effect of the spread of the data when we measure dependency using the covariance. One way to achieve this is to standardize the original *x* and *y* variables and compute the covariance of standardized variables x' and y'.
C) The coefficient of linear correlation standardizes the measure of dependency and allows us to compare the relative strengths of dependency of different sets of data.
D) None of the above.
ANSWER: D

33. Which of the following statements is false regarding the assumptions for inferences about the linear correlation coefficient?

A) The set of (*x*, *y*) ordered pairs forms a random sample.
B) The *y* values at each *x* have a normal distribution.
C) Inferences about the linear correlation coefficient use the *t*-distribution with $(n - 1)$ degrees of freedom.
D) None of the above.
ANSWER: C

34. Which of the following statements is false?

A) The test statistic used to test the null hypothesis $H_o: \rho = 0$ is the calculated value of *r* from the sample data.
B) When we perform a hypotheses test about ρ, the linear correlation coefficient for the population, the number of degrees of freedom for the *r* statistic is 2 less than the sample size; that is, $df = n - 2$.
C) Rejection of the null hypothesis $H_o: \rho = 0$ means that there is no evidence of a linear relationship between the two variables in the population.
D) None of the above
ANSWER: C

Chapter 13

Short-Answer Questions

35. Suppose you are given a particular set of data and found that $r = 2.5$. How would you interpret this result?

ANSWER:
You would have made a computation error since it is always true that $-1 \leq r \leq 1$.

36. If a scatter diagram for a bivariate data set results in a horizontal or vertical line, what value does *r* take on?

ANSWER:
r is undefined since $r = \text{covar}(x, y)/(s_x \cdot s_y)$ and either s_x or s_y would equal zero and division by zero is undefined.

37. Indicate whether a negative or a positive correlation coefficient would be expected in a study involving the following two indicated variables: As the dosage of Heparin is increased, the Partial Thronboplain time (PTT) increases.

ANSWER:
Positive

38. Indicate whether a negative or a positive correlation coefficient would be expected in a study involving the following two indicated variables: As atmospheric oxygen decreases, the Hemoglobin count in the blood increases.

ANSWER:
Negative

39. Indicate whether a negative or a positive correlation coefficient would be expected in a study involving the following two indicated variables: As the amount of aspirin increases, the platelet aggregation decreases.

ANSWER:
Negative

40. Indicate whether a negative or a positive correlation coefficient would be expected in a study involving the following two indicated variables: Increasing the dosage of Dopamine Hydrochloride tends to increase the blood pressure.

ANSWER:
Positive

41. Thirty-four students in an Algebra course were given a math competency test on the first day of class. Thirty-two students completed the course and their scores on a comprehensive final exam were recorded. The correlation coefficient between math competency scores and final exam scores was computed. Give the critical region for testing $H_o: \rho = 0(\leq)$ vs. $H_a: \rho > 0$ at $\alpha = 0.05$.

ANSWER:
Critical region: $r > 0.296$

42. What is a disadvantage of using the covariance as a measure of linear dependency?

ANSWER:
Spread of data is a strong factor in size of covariance. Covariance does not have a standardized unit of measure.

43. Indicate whether the symbol ρ is a parameter or statistic. Justify your answer.

ANSWER:
The symbol ρ is a parameter since it represents the population correlation coefficient.

44. Indicate whether the symbol r is a parameter or statistic. Justify your answer.

ANSWER:
The symbol r is a statistic since it represents the sample correlation coefficient.

45. What is the primary question we answer in linear correlation analysis?

ANSWER:
Are the two variables under study linearly related?

46. What is the best analysis to describe a linear relationship between two variables?

ANSWER:
Linear correlation

47. Describe why the method used to define the correlation coefficient is referred to as "a product moment."

ANSWER:
A "moment" is the distance from the mean, and the product of both the horizontal moment and the vertical moment is summed in calculating the correlation coefficient.

48. State the null hypothesis H_o, and the alternative hypothesis H_a, that would be used to test the following statement: "The linear correlation coefficient is positive".

ANSWER:
$H_o: \rho = 0\ (\leq)$ vs. $H_a: \rho > 0$

49. State the null hypothesis H_o, and the alternative hypothesis H_a, that would be used to test the following statement: "There is no linear correlation".

ANSWER:
$H_o: \rho = 0$ vs. $H_a: \rho \neq 0$

50. State the null hypothesis H_o, and the alternative hypothesis H_a, that would be used to test the following statement: "There is evidence of negative correlation".

ANSWER:
$H_o: \rho = 0\ (\geq)$ vs. $H_a: \rho < 0$

51. State the null hypothesis H_o, and the alternative hypothesis H_a, that would be used to test the following statement: "There is positive linear relationship".

ANSWER:
$H_o : \rho = 0\ (\leq)$ vs. $H_a : \rho > 0$

52. Does the value of the sample linear correlation coefficient, *r*, indicate that there is a linear dependency between the two variables in the population from which the sample was drawn? Briefly explain how to answer this question.

ANSWER:
To answer this question we can perform a hypothesis test. The null hypothesis is: The two variables are linearly unrelated ($\rho = 0$), where ρ is the linear correlation coefficient for the population. The alternative hypothesis may be either one-tailed or two-tailed. Most frequently it is two-tailed, $\rho \neq 0$. However, when we suspect that there is only a positive or only a negative correlation, we should use a one-tailed test. The alternative hypothesis of a one-tailed test is $\rho > 0$ or $\rho < 0$.

Applied and Computational Questions

53. Calculate the correlation coefficient for the following set of data. What property do the points exhibit when plotted on a scatter diagram?

x	1	3	0	2	4
y	17.5	12.5	20.0	15.0	10.0

ANSWER:
$r = -1$. All the points fall on a straight line having a negative slope.

QUESTIONS 54 AND 55 ARE BASED ON THE FOLLOWING INFORMATION:
The scores (*x*) on a computer science aptitude test range from 0 to 25, and the course grade (*y*) with possible values: 0.0, 1.0, 1.5, 2.0, 2.5, 3.0, 3.5, 4.0, were recorded for 20 students in an introductory computer science course as shown below.

x	18	10	15	20	18	13	20	16	12	22
y	2.5	1.5	3.0	3.5	2.5	2.0	2.5	3.0	2.0	4.0

x	5	8	16	20	11	14	20	16	15	24
y	0.0	1.0	1.5	3.0	2.0	2.5	4.0	4.0	2.5	3.5

54. Find the coefficient of linear correlation for this data.

ANSWER:
$r = 0.833$

55. Give the *p*-value if this data is used to test $H_o : \rho = 0(\leq)$ vs. $H_a : \rho > 0$ at $\alpha = 0.1$. What is your decision?

ANSWER:
P = *p*-value < 0.005. Since **P** < α, we reject H_o.

56. The following data represent the number of credit hours, x, and the cost for textbooks, y, for five students. Calculate SS(x), SS(y), SS(xy), and the coefficient of linear correlation.

x	15	12	18	9	12
y	110	86	105	65	94

ANSWER:
SS(x) = 46.8, SS(y) = 1262, SS(xy) = 213, r = 0.88

57. Considering the following set of bivariate data, find the value of k that would result in a coefficient of linear correlation equal to exactly +1.

x	2	5	7	8	11
y	5.9	12.2	16.4	18.5	k

ANSWER:
k = 24.8

58. A set of bivariate data has a Pearson's product moment equal to 0.55, and the standard deviation of x equals 15.5, and the standard deviation of y equals 14.0. Find the covariance of x and y.

ANSWER:
Covar (x, y) = $r \cdot s_x \cdot s_y$ = (0.55)(15.5)(14.0) = 119.35

59. Find the covariance of x and y and the centroid of the data shown in the table below

x	1.5	2.7	3.5	4.0	5.0
y	2.7	4.2	7.2	9.0	9.5

ANSWER:
Covariance (x, y) = 3.8, Centroid = $(\overline{x}, \overline{y})$ = (3.34, 6.52)

60. Two different scales were used to measure the weights of 20 different objects. Find a 95% confidence interval for ρ if r = 0.4.

ANSWER:
(–0.05 to 0.70)

QUESTIONS 61 AND 62 ARE BASED ON THE FOLLOWING INFORMATION:
Accurate methods to calculate tree heights are difficult and expensive and hard to find. An inexpensive but less accurate method uses aerial photographic methods to estimate tree heights.

Ground (x)	95	69	110	90	95	77	84	92
Aerial (y)	80	72	105	86	87	90	86	90

61. Use the given data to find a 95% confidence interval for ρ, the population correlation between heights obtained on the ground and heights determined from aerial photographs.

ANSWER:
r = 0.75, $0.10 < \rho < 0.95$

62. What would you need to do to estimate ρ more closely?

ANSWER:
Increase the sample size.

63. A sample of size 52 was used to test $H_o: \rho = 0$ vs. $H_a: \rho \neq 0$. Give a bound on the *p*-value if $r^* = 0.31$.

ANSWER:
$0.02 < p$ –value < 0.05

64. Compute the coefficient of linear correlation for the following set of bivariate data and find a 95% confidence interval for ρ.

x	3	3	4	5	8	8
y	5.8	1.2	9.7	8.5	7.3	13.7

ANSWER:
$r = 0.655$. A 95% confidence interval for ρ is (–0.30 to 0.93).

QUESTIONS 65 THROUGH 67 ARE BASED ON THE FOLLOWING INFORMATION:
Consider the following bivariate data

x	2.1	3.4	3.5	4.7	5.3	5.4
y	9.4	9.5	9.1	12.1	12.9	13.1

65. Find the critical value of *r*.

ANSWER:
$r = 0.882$.

66. Find the calculated value of *r*, namely r^*.

ANSWER:
$r^* = 0.912$

67. State the decision.

ANSWER:
Reject the null hypothesis since $r^* > r$.

68. A sample of size twenty was used to test $H_o: \rho = 0 (\geq)$ vs. $H_a: \rho < 0$. Give a bound on the *p*-value if $r^* = -0.48$.

ANSWER:
$0.01 < p$ –value < 0.025

69. A study was conducted to determine the relationship between actual areas of planted corn and estimates of those areas obtained from earth observation satellites as shown in the table below.

Actual Area	160	640	120	300	1100	110	600
Estimated Area	172	605	98	280	1050	105	590

Give the *p*-value for testing $H_o: \rho = 0$ vs. $H_a: \rho \neq 0$. What is your conclusion at α =0.05.

ANSWER:
r = 0.999, p –value < 0.01. We reject the null hypothesis at α =0.05, and conclude that $\rho \neq 0$.

QUESTIONS 70 THROUGH 75 ARE BASED ON THE FOLLOWING INFORMATION:
Consider the following bivariate set of data.

Point	A	B	C	D	E	F	G	H	I	J
x	2	2	4	4	6	6	8	8	10	10
y	2	3	3	4	4	5	5	6	6	7

70. Construct a scatter diagram of the data.

ANSWER:

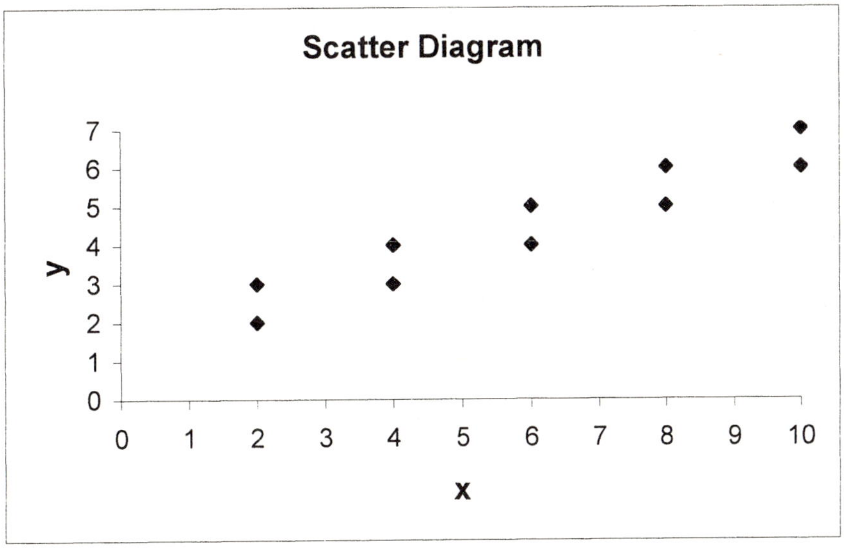

71. Calculate $\sum x$, $\sum y$, $\bar{x}$, $\bar{y}$, $\sum(x-\bar{x})(y-\bar{y})$, $\sum x^2$, $\sum xy$, and $\sum y^2$.

ANSWER:
$\sum x = 60$, $\sum y = 45$, $\bar{x} = 6$, $\bar{y} = 4.5$, $\sum(x-\bar{x})(y-\bar{y}) = 40$, $\sum x^2 = 440$, $\sum xy = 310$, $\sum y^2 = 225$.

72. Calculate the covariance.

ANSWER:
$\text{covar}\ (x, y) = [\sum (x - \bar{x})(y - \bar{y})]/(n-1) = 40 / 9 = 4.444$

73. Calculate s_x and s_y.

ANSWER:
$s_x = \sqrt{[\sum x^2 - (\sum x)^2 / n]/(n-1)} = \sqrt{[440 - (60^2/10)]/9} = \sqrt{8.889} = 2.981$
$s_y = \sqrt{[\sum y^2 - (\sum y)^2 / n]/(n-1)} = \sqrt{[225 - (45^2/10)]/9} = \sqrt{2.50} = 1.581$

74. Calculate *r* using the formula $r = \text{covar}\ (x, y)/[s_x \cdot s_y]$.

ANSWER:
$r = \text{covar}\ (x, y)/[s_x \cdot s_y] = 4.444 / [(2.981)(1.581)] = 0.943$

75. Calculate *r* using the formula: $r = SS(xy)/\sqrt{SS(x) \cdot SS(y)}$.

ANSWER:
$SS(y) = \sum x^2 - [(\sum x)^2 / n] = 440 - (60^2/10) = 80$
$SS(y) = \sum y^2 - [(\sum y)^2 / n] = 225 - (45^2/10) = 22.5$
$SS(xy) = \sum xy - [(\sum x)(\sum y)/n] = 310 - [(60)(45)/10] = 40$
$r = SS(xy)/\sqrt{SS(x) \cdot SS(y)} = 40/\sqrt{(80)(22.5)} = 0.943$

76. Use the confidence belts chart for the correlation coefficient to determine a 95% confidence interval for the true population linear correlation coefficient based on the following sample statistics: $n = 8$, $r = 0.20$.

ANSWER:
– 0.55 to 0.76

77. Use the confidence belts chart for the correlation coefficient to determine a 95% confidence interval for the true population linear correlation coefficient based on the following sample statistics: $n = 100$, $r = -0.40$.

ANSWER:
– 0.55 to – 0.22

78. Use the confidence belts chart for the correlation coefficient to determine a 95% confidence interval for the true population linear correlation coefficient based on the following sample statistics: $n = 25$, $r = +0.65$.

ANSWER:
0.34 to 0.82

79. Use the confidence belts chart for the correlation coefficient to determine a 95% confidence interval for the true population linear correlation coefficient based on the following sample statistics: n = 15, r = -0.23.

ANSWER:
– 0.65 to 0.31

QUESTIONS 80 AND 81 ARE BASED ON THE FOLLOWING INFORMATION:
The Test-Retest Method is one way of establishing the reliability of a test. The test is administered and at a later time, the same test is re-administered to the same individuals. The correlation coefficient is computed between the two sets of scores. The following test scores were obtained in a Test-Retest situation.

First Score	78	90	63	78	99	83	71	87	50	75
Second Score	74	92	54	77	96	80	74	82	55	72

The following summary statistics were given:
$n = 10,\ \sum x = 774,\ \sum y = 756,\ \sum x^2 = 61{,}662,\ \sum xy = 60{,}142,\ \sum y^2 = 58{,}810$

80. Calculate the linear correlation coefficient r.

ANSWER:
$$SS(x) = \sum x^2 - [(\sum x)^2 / n] = 61662 - (774^2/10) = 1754.4$$
$$SS(y) = \sum y^2 - [(\sum y)^2 / n] = 58810 - (756^2/10) = 1656.4$$
$$SS(xy) = \sum xy - [(\sum x)(\sum y)/n] = 60142 - [(774)(756)/10] = 1627.6$$
$$r = SS(xy)/\sqrt{SS(x)\cdot SS(y)} = 1627.6/\sqrt{(1754.4)(1656.4)} = 0.955$$

81. Set a 95% confidence interval for ρ.

ANSWER:
From the chart of confidence belts for the correlation coefficient we determine that the 95% confidence interval for ρ is 0.78 to 0.98.

82. State the null hypothesis H_o, and the alternative hypothesis H_a, that would be used to test the following statements: "The linear correlation coefficient is positive."

ANSWER:
$H_o: \rho = 0$ vs. $H_a: \rho > 0$

83. State the null hypothesis H_o, and the alternative hypothesis H_a, that would be used to test the following statements: "There is no linear correlation."

ANSWER:
$H_o: \rho = 0$ vs. $H_a: \rho \neq 0$

84. State the null hypothesis, H_o, and the alternative hypothesis H_a, that would be used to test the following statements: "There is evidence of negative correlation."

ANSWER:
$H_o: \rho = 0$ vs. $H_a: \rho < 0$

85. State the null hypothesis, H_o, and the alternative hypothesis H_a, that would be used to test the following statements: "There is positive linear relationship."

ANSWER:
$H_o: \rho = 0$ vs. $H_a: \rho > 0$

86. If a sample of size 20 has a linear correlation coefficient of –0.547, is there sufficient evidence to conclude that the linear correlation coefficient of the population is negative? Use $\alpha = 0.01$, and apply both the *p*-value approach and the classical approach.

ANSWER:
$H_o: \rho = 0$ vs. $H_a: \rho < 0$

Assume normality for *y* at each x. Since *n* = 20, then *df* = *n* –2 = 18. *r* = - 0.547, and the test statistic r^* = -0.547. α = 0.01.
The *p*-value approach: **P** = *P*(*r* < -0.547)
Using the table of "critical values of *r* when ρ = 0" we get 0.005 < **P** < 0.01. Since **P** < α; reject H_o.
The classical approach: Critical region: $r \leq$ -0.516.
r^* falls in the critical region, therefore we reject H_o. There is sufficient evidence to indicate at the 0.01 level of significance that the linear correlation coefficient of the population is negative.

87. Is a value of *r* = + 0.295 significant in trying to show that ρ is greater than zero for a sample of 52 data at the 0.05 level of significance? Use the *p*-value approach.

ANSWER:
$H_o: \rho = 0$ vs. $H_a: \rho > 0$; Assume normality for *y* at each x. Since *n* = 52, then df = *n* –2 = 50. *r* = 0.295 and the test statistic r^* = 0.295. α = 0.05.
P = *p*-value = *P*(*r* > 0.295). Using the table of "critical values of *r* when ρ = 0" we have 0.01 < **P** < 0.025. Since **P** < α; reject H_o. There is sufficient evidence to indicate at the 0.01 level of significance that the linear correlation coefficient of the population is positive.

88. Is a value of *r* = + 0.295 significant in trying to show that ρ is greater than zero for a sample of 52 data at the 0.05 level of significance? Use the classical approach.

ANSWER:
The test statistic r^* = 0.295, and the critical region is: $r \geq$ 0.231. Since r^* falls in the critical region, we reject H_o. There is sufficient evidence to indicate at the 0.05 level of significance that the correlation coefficient is positive.

QUESTIONS 89 THROUGH 91 ARE BASED ON THE FOLLOWING INFORMATION:
The population (in millions) and the violent crime rate (per 1000) were recorded for ten metropolitan areas in Illinois. The data are shown in the following table:

Population	10.1	1.4	2.2	7.1	4.5	0.4	0.4	0.3	0.3	0.5
Crime Rate	12.2	9.7	9.4	8.6	8.4	7.5	7.3	7.2	7.1	7.1

The following summary statistics are given:
$n = 10, \quad \sum x = 27.2, \quad \sum y = 84.5, \quad \sum x^2 = 180.22, \quad \sum xy = 270.1, \quad \sum y^2 = 738.01$

89. Calculate the linear correlation coefficient, *r*.

ANSWER:
$SS(x) = \sum x^2 - [(\sum x)^2 / n] = 180.22 - (27.2^2 / 10) = 106.236$
$SS(y) = \sum y^2 - [(\sum y)^2 / n] = 738.01 - (84.5^2 / 10) = 23.985$
$SS(xy) = \sum xy - [(\sum x)(\sum y) / n] = 270.1 - [(27.2)(84.5) / 10] = 40.26$
$r = SS(xy) / \sqrt{SS(x) \cdot SS(y)} = 40.26 / \sqrt{(106.236)(23.985)} = 0.798$

90. Do these data provide evidence to reject the null hypothesis that $\rho = 0$ in favor of the alternative $\rho \neq 0$ at $\alpha = 0.05$. Use the *p*-value approach.

ANSWER:
$H_o: \rho = 0$ vs. $H_a: \rho \neq 0$. Assume normality for *y* at each *x*. Since $n = 10$, then df = $n - 2 = 8$. $r = 0.798$ and the test statistic $r^* = 0.798$. $\alpha = 0.05$.
P = *p*-value = $P(r < -0.798) + P(r > 0.798) = 2\,P(r > 0.798)$. Using the table of "critical values of *r* when $\rho = 0$" we get **P** < 0.01. Since **P** < $\alpha = 0.05$; reject H_o.

91. Do these data provide evidence to reject the null hypothesis that $\rho = 0$ in favor of $\rho \neq 0$ at $\alpha = 0.05$. Use the classical approach.

ANSWER:
The critical regions are: $r \leq -0.632$ and $r \geq 0.632$. Since r^* falls in the critical region, we reject H_o. There is sufficient evidence to indicate at the 0.05 level of significance that the correlation coefficient is different from zero.

92. Consider a set of paired bivariate data (*x*, *y*). Describe the relationship of the ordered pairs that will cause $\sum[(x - \bar{x}) \cdot (y - \bar{y})]$ to be positive.

ANSWER:
The set of data will be predominantly ordered pairs which have coordinates such that both the *x* and *y* values are larger than $\bar{x}$ and $\bar{y}$, and both smaller than $\bar{x}$ and $\bar{y}$; this will result in the product $(x - \bar{x})(y - \bar{y})$ being positive. Graphically, the points will be mostly located in the upper right and the lower left of the four quarters of the graph formed by the vertical line $x = \bar{x}$ and the horizontal line $y = \bar{y}$.

93. Consider a set of paired bivariate data (x, y). Describe the relationship of the ordered pairs that will cause $\sum[(x-\bar{x})\cdot(y-\bar{y})]$ to be negative.

ANSWER:
The set of data will be predominantly ordered pairs which have coordinates such that either the x value is larger than $\bar{x}$ and y is smaller than $\bar{y}$, or x is smaller than $\bar{x}$ and y is larger than $\bar{y}$; this will result in the product (x-$\bar{x}$)(y-$\bar{y}$) being negative. Graphically, the points will be mostly located in the upper left and the lower right of the four quarters of the graph formed by the vertical line $x = \bar{x}$ and the horizontal line $y = \bar{y}$.

QUESTIONS 94 THROUGH 99 ARE BASED ON THE FOLLOWING INFORMATION:
The following set of 25 scores was randomly selected from Dr. Maas' inferential statistics class. Let x be the pre-final average and y the final examination score. (The final examination had a maximum of 100 points.)

Student	1	2	3	4	5	6	7	8	9	10	11	12	13
x	80	91	73	88	62	71	60	89	66	73	69	81	76
y	87	88	80	82	76	84	71	90	82	79	75	86	89

Student	14	15	16	17	18	19	20	21	22	23	24	25
x	78	83	76	91	76	99	99	64	86	63	95	97
y	85	89	85	94	78	95	98	72	94	81	90	98

The following summary statistics are given:

$$n = 25, \quad \sum x = 1986, \quad \sum y = 2128, \quad \sum x^2 = 161,246, \quad \sum xy = 170,971, \quad \sum y^2 = 182,522$$

94. Draw a scatter diagram for these data.

ANSWER:

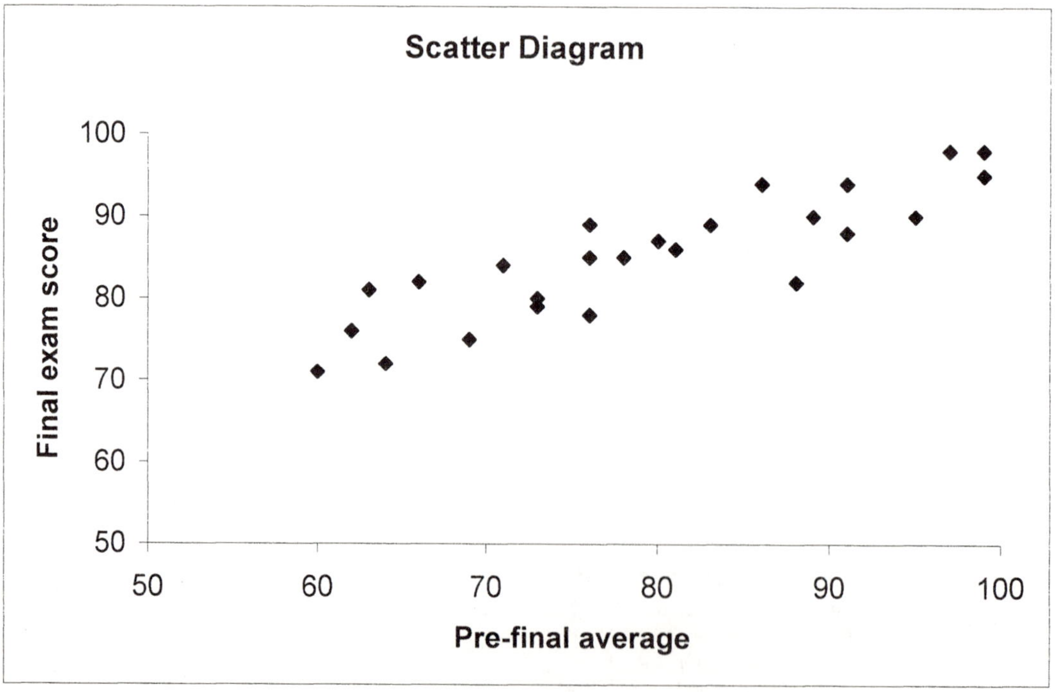

95. Calculate the equation of the line of best fit.

ANSWER:

$SS(x) = \sum x^2 - [(\sum x)^2 / n] = 161246 - (1986^2 / 25) = 3478.16$

$SS(xy) = \sum xy - [(\sum x)(\sum y) / n] = 170971 - [(1986)(2128) / 25] = 1922.68$

$b_1 = SS(xy) / SS(x) = 1922.68 / 3478.16 = 0.5528$

$b_0 = [\sum y - b_1 \sum x] / n = [2128 - (0.5528)(1986)] / 25 = 41.2057$

The equation of the line of best fit is: $\hat{y} = b_0 + b_1 x = 41.2057 + 0.5528x$

96 Draw the line of best fit on your graph.

ANSWER:

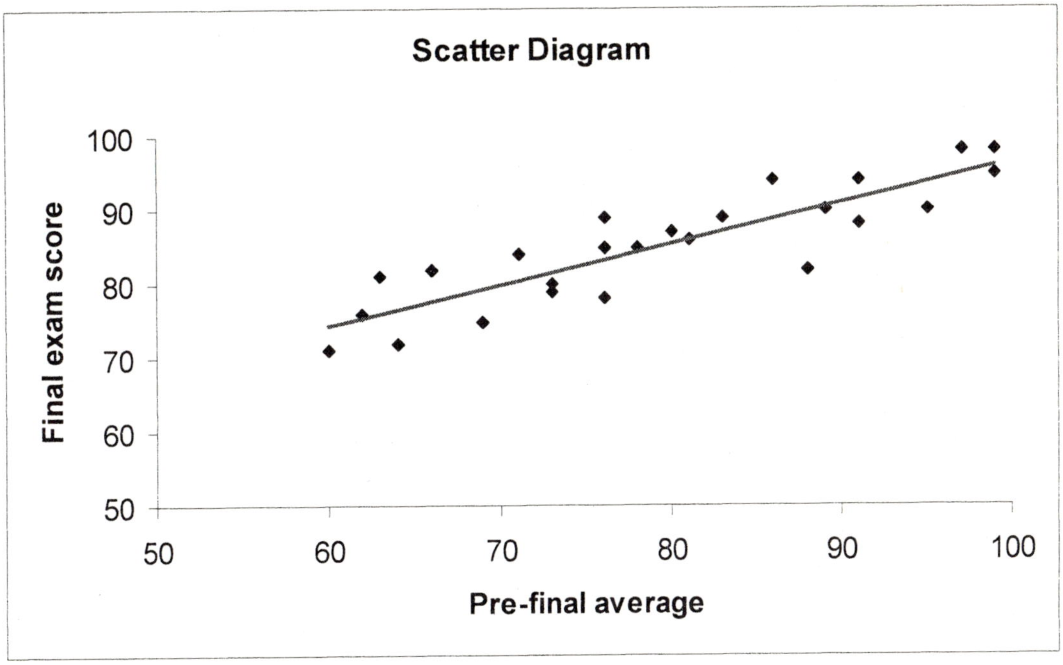

97. Calculate the linear correlation coefficient.

ANSWER:

$SS(y) = \sum y^2 - [(\sum y)^2 / n] = 182522 - (2128^2 / 25) = 1386.64$

$r = SS(xy) / \sqrt{SS(x) \cdot SS(y)} = 1922.68 / \sqrt{(3478.16)(1386.64)} = 0.875$

98. Test the significance of r at $\alpha = 0.10$ using the p-value approach and the classical approach.

ANSWER:

The linear correlation coefficient for the population is ρ.

$H_o : \rho = 0.0$ vs. $H_a : \rho \neq 0.0$

Assume normality for y at each x. Since n = 25, then $df = n - 2 = 23$. $\alpha = 0.10$, $r^* = 0.875$

The *p*-value approach:
P = 2*P*(r > 0.875). Using the table of "critical values of r when $\rho = 0$:" we get **P** < 0.01. Since **P** < α ; reject H_o.

The classical approach:
The critical regions are: $r \leq -0.34$, and $r \geq 0.34$. Since the test statistic r^* is in the critical region, we reject the null hypothesis. There is sufficient evidence to conclude that there is a correlation between the pre-final average and the final examination score.

99. Find the 95% confidence interval for the true value of ρ.

ANSWER:
The 0.95 interval for ρ (from the confidence belts chart for the correlation coefficient) is 0.70 to 0.92.

QUESTIONS 100 THROUGH 104 ARE BASED ON THE FOLLOWING INFORMATION:
The data below are for the number of unemployed persons (in millions) and the federal unemployment insurance payments (in billions of dollars) for the years 1978 – 1985. Some economists state that these two variables are positively related.

	Year							
	1978	1979	1980	1981	1982	1983	1984	1985
Federal Unemployment Insurance Payments	11.8	10.7	18.0	19.7	23.7	31.5	18.4	16.8
# Unemployed Persons	6.2	6.1	7.6	8.3	10.7	10.7	8.5	8.3

100. Assume that a simple linear regression model is appropriate for these data. Identify the dependent and independent variables.

ANSWER:
Dependent variable *Y*: Number of unemployed persons
Independent variable *X*: Federal unemployment insurance payments

101. Develop a scatter diagram for these data. What does the scatter diagram indicate about the relationship between these two variables?

ANSWER:

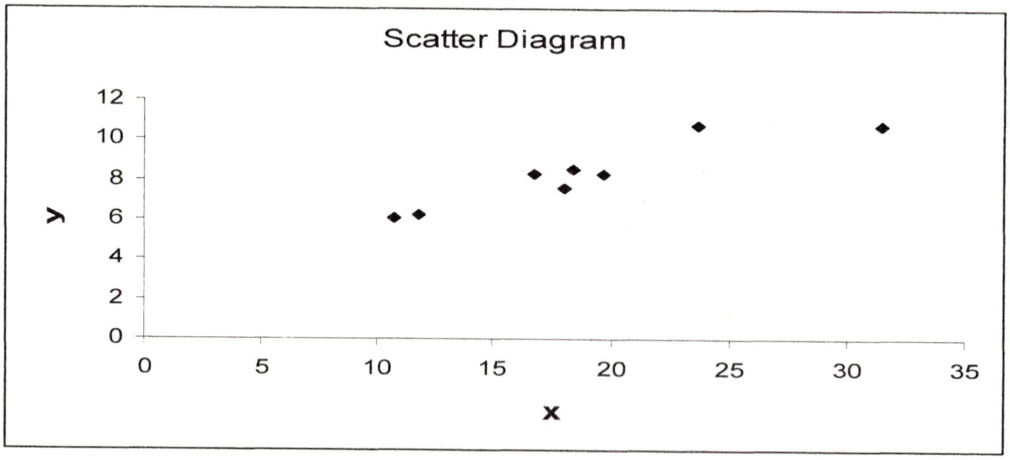

The number of unemployed persons and the federal unemployment insurance payments appear to be positively linearly related.

102. Use these data to develop an estimated regression equation.

ANSWER:
$\hat{y}$ = 3.664 + 0.2463*x*

103. Calculate the coefficient of correlation.

ANSWER:
r = 0.9327

104. Test the null hypothesis that the true population coefficient of correlation equals zero using the 0.05 significance level and the classical approach.

ANSWER:
$H_o : \rho = 0$ (There is no linear relationship) vs. $H_o : \rho \neq 0$ (A linear relationship exists)
n = 8, *df* = *n* – 2 = 6, α = 0.05
The rejection regions are: $r \leq -0.632$, and $r \geq 0.632$. Since the test statistic r^* = 0.9327 falls in the rejection region; we reject the null hypothesis at the 0.05 level of significance. There is sufficient evidence to indicate that a linear relationship exists between these two variables.

105. Consider a set of paired bivariate data (*x*, *y*). Describe the relationship of the ordered pairs that will cause $\sum[(x-\bar{x})\cdot(y-\bar{y})]$ to be near zero.

ANSWER:
The set of data will be ordered pairs, which have coordinates such that the product (x-$\bar{x}$)(y-$\bar{y}$) being distributed between positive, negative and zero so that the sum is near zero. Graphically, the points will be approximately evenly distributed between the four quarters of the graph formed by the vertical line *x* = $\bar{x}$ and the horizontal line *y* = $\bar{y}$.

QUESTIONS 106 THROUGH 108 ARE BASED ON THE FOLLOWING INFORMATION:
Consider the following set of bivariate data: (25,15), (35,55), (65,35), (85,25), (115,65) and (125,15).

106. Calculate the covariance.

ANSWER:
$$SS(xy) = \sum xy - \left[\left(\sum x\right)\left(\sum y\right)\right]/n = 16050 - (450)(210)/6 = 300.$$
Covar(x, y) = *SS*(*xy*) / (*n*-1) = 300 / 5 = 60.

107. Calculate the standard deviation of the six x-values and the standard deviation of the six *y*-values.

ANSWER:
$$s_x = \sqrt{\left[\sum x^2 - \left(\sum x\right)^2 / n\right]/(n-1)} = \sqrt{[42150-(450)^2/6]/5} = 40.988$$

$$s_y = \sqrt{\left[\sum y^2 - \left(\sum y\right)^2 / n\right]/(n-1)} = \sqrt{[9550-(210)^2/6]/5} = 20.976$$

Chapter 13

108. Calculate r, the coefficient of linear correlation.

ANSWER:
$r = \text{covar}(xy)/(s_x \cdot s_y)$ = 60 / [(40.988)(20.976)] = 0.0698

QUESTIONS 109 THROUGH 113 ARE BASED ON THE FOLLOWING INFORMATION:
Consider the following set of bivariate data:

x	42	49	46	57	64	58
y	19	23	18	25	30	29

109. Calculate $\sum x$, $\sum y$, $\sum x^2$, $\sum xy$, and $\sum y^2$.

ANSWER:

	x	y	x^2	xy	y^2
	42	19	1764	798	361
	49	23	2401	1127	529
	46	18	2116	828	324
	57	25	3249	1425	625
	64	30	4096	1920	900
	58	29	3364	1682	841
Sum	316	144	16990	7780	3580

$\sum x = 316,\ \sum y = 144,\ \sum x^2 = 16990,\ \sum xy = 7780,$ and $\sum y^2 = 3580$

110. Calculate SS(*x*), SS(*y*), SS(*xy*).

ANSWER:

$$SS(x) = \sum x^2 - \left(\sum x\right)^2 / n = 16990 - (316)^2/6 = 347.333$$

$$SS(y)\sum y^2 - \left(\sum y\right)^2 / n = 3580 - (144)^2/6 = 124.0$$

$$SS(xy) = \sum xy - \left[\left(\sum x\right)\left(\sum y\right)\right]/n = 7780 - (316)(144)/6 = 196.0$$

111. Calculate s_x and s_y.

ANSWER:

$$s_x = \sqrt{\left[\sum x^2 - \left(\sum x\right)^2 / n\right]/(n-1)} = \sqrt{[16990 - (316)^2/6]/5} = 8.335$$

$$s_y = \sqrt{\left[\sum y^2 - \left(\sum y\right)^2 / n\right]/(n-1)} = \sqrt{[3580 - (144)^2/6]/5} = 4.980$$

112. Calculate the covariance.

ANSWER:
Covar(x, y) = SS(*xy*) / (*n*-1) = 196 / 5 = 39.2

113. Calculate Pearson's product moment using two different ways.

ANSWER:

$r = SS(xy)/\sqrt{SS(x)\cdot SS(y)} = (196)/\sqrt{(347.333)(124)} = 0.944$, or

$r = \text{covar}(xy)/(s_x \cdot s_y)$ = 39.2 / [(8.335)(4.980)] = 0.944

QUESTIONS 114 THROUGH 119 ARE BASED ON THE FOLLOWING INFORMATION:

Consider the accompanying bivariate data.

x	2	3	3	4	5	6	7	8	8	9
y	8	8	9	6	7	4	5	2	3	3

114. Draw a scatter diagram for the data.

ANSWER:

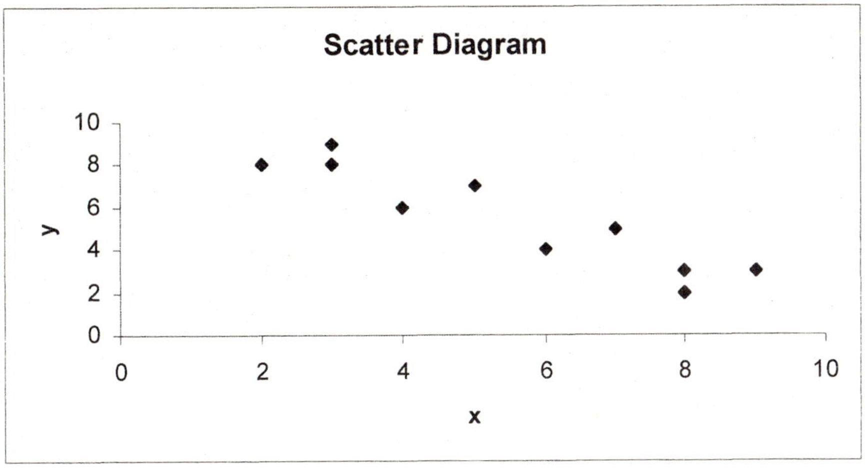

115. What does the scatter diagram tell you about the relationship between *x* and *y*?

ANSWER:
There is a strong negative linear relationship between the two variables.

116. Calculate the covariance.

ANSWER:

	x	y	$x-\bar{x}$	$y-\bar{y}$	$(x-\bar{x})(y-\bar{y})$
	2	8	-3.5	2.5	-8.75
	3	8	-2.5	2.5	-6.25
	3	9	-2.5	3.5	-8.75
	4	6	-1.5	0.5	-0.75
	5	7	-0.5	1.5	-0.75
	6	4	0.5	-1.5	-0.75
	7	5	1.5	-0.5	-0.75
	8	2	2.5	-3.5	-8.75
	8	3	2.5	-2.5	-6.25
	9	3	3.5	-2.5	-8.75
Sum	55	55	0	0	-50.5
Mean	5.5	5.5			

Therefore the covariance is: covar(x, y) = $\sum_{i=1}^{n}(x_i-\bar{x})(y_i-\bar{y})/(n-1)$
= (-50.5) / 9 = -5.611.

117. Calculate s_x and s_y.

ANSWER:

	x	y	xy	x^2	y^2
	2	8	16	4	64
	3	8	24	9	64
	3	9	27	9	81
	4	6	24	16	36
	5	7	35	25	49
	6	4	24	36	16
	7	5	35	49	25
	8	2	16	64	4
	8	3	24	64	9
	9	3	27	81	9
Sum	55	55	252	357	357

$$s_x=\sqrt{\left[\sum x^2-\left(\sum x\right)^2/n\right]/(n-1)}=\sqrt{[357-(55)^2/10]/9}=2.461$$

$$s_y=\sqrt{\left[\sum y^2-\left(\sum y\right)^2/n\right]/(n-1)}=\sqrt{[357-(55)^2/10]/9}=2.461$$

118. Use your answers to questions 116 and 117 to calculate the coefficient of linear correlation, *r*.

ANSWER:
$r=\text{covar}(x,y)/(s_x\cdot s_y)$ = (-5.611) / [(2.461)(2.461)] = -0.926

119. Use the formula $SS(xy)=\sum xy-\left[\left(\sum x\right)\left(\sum y\right)\right]/n$ to calculate Pearson's product moment.

ANSWER:

$$SS(xy)=\sum xy-\left[\left(\sum x\right)\left(\sum y\right)\right]/n=252-(55)(55)/10=-50.5$$

$$SS(x)=\sum x^2-\left(\sum x\right)^2/n=357-(55)^2/10=54.5$$

$$SS(y)\sum y^2-\left(\sum y\right)^2/n=357-(55)^2/10=54.5$$

$$r=SS(xy)/\sqrt{SS(x)\cdot SS(y)}=(-50.5)/\sqrt{(54.5)(54.5)}=-0.926$$

QUESTIONS 120 THROUGH 124 ARE BASED ON THE FOLLOWING INFORMATION:
The table of "Confidence Belts for the Correlation Coefficient $1-\alpha = 0.95$" available in your textbook is used to determine a 95% confidence interval for the true population linear correlation coefficient based on the sample statistics *n* and *r*.

120. Find the 95% confidence interval for ρ if $n = 8$, and $r = 0.20$.

ANSWER:
-0.55 to 0.77

121. Find the 95% confidence interval for ρ if $n = 100$, and $r = -0.40$.

ANSWER:
-0.55 to -0.22

122. Find the 95% confidence interval for ρ if $n = 25$, and $r = +0.65$.

ANSWER:
0.34 to 0.82

123. Find the 95% confidence interval for ρ if $n = 15$, and $r = -0.23$.

ANSWER:
-0.65 to 0.31

124. Find the 95% confidence interval for ρ if $n = 50$, and $r = 0.60$.

ANSWER:
0.40 TO 0.73

QUESTIONS 125 THROUGH 130 ARE BASED ON THE FOLLOWING INFORMATION:
The Test-Retest Method is one way of establishing the reliability of a test. The test is administered and then, at a later date, the same test is re-administered to the same individuals. The correlation coefficient is computed between the two sets of scores. The following test scores were obtained in a Test-Retest situation.

1st Score	48	73	61	85	99	69	81	76	76	88
2nd Score	54	71	53	81	95	73	79	76	73	91

125. Use computer to find the linear correlation coefficient, *r*.

ANSWER:

Table of correlations

	1st Score	2nd Score
1st Score	1.000	
2nd Score	0.955	1.000

The linear correlation coefficient $r = 0.955$.

126. Set a 95% confidence interval for ρ .

ANSWER:
The 95% confidence interval for ρ is read from the "Confidence Belts for the Correlation Coefficient $1-\alpha=0.95$" table available in your textbook. The values are 0.78 and 0.98.

127. State the null and alternative hypotheses for testing "The Test-Retest Method led to a reliable test".

ANSWER:
$H_o:\rho=0$ vs. $H_a:\rho\neq 0$

128. Test the hypotheses in question 127 at the 0.05 level of significance using the *p*-value approach.

ANSWER:
The value of the test statistic is r^* = 0.955.
P = *p*-value = *P*(*r* < -0.955) + *P*(*r* > 0.955) = 2 · *P*(*r* > 0.955) with *df* = *n* – 2 = 8.
We use the "Critical Values of *r* When ρ = 0" table available in your textbook to place bounds on the *p*-value. This implies that **P** < 0.01. Since *p*-value < α = 0.05, we reject H_o . There is sufficient evidence of a linear relationship between the two sets of scores.

129. Test the hypotheses in question 127 at the 0.05 level of significance using the classical approach.

ANSWER:
The critical value is found at the intersection of the *df* = 8 row and the two-tailed 0.05 column of the "Critical Values of r When ρ = 0" table available in your textbook. The value is 0.632. Since this is a two-tailed test; we have two critical values: $\pm$0.632. But r^* = 0.955 > 0.632, so we reject H_o . We reach the same conclusion as stated in question 128.

130. Can you use the 95% confidence interval for ρ in question 126 for testing the hypotheses in question 127? Explain in detail.

ANSWER:
Since the hypothesized value $\rho=0$ is not included in the 95% confidence interval (0.78, 0.098) for ρ , we reject H_o . We reach the same conclusion as stated in question 128.

131. Place bounds on the *p*-value resulting from a sample with *n* = 15 and *r* = 0.525, if H_a is two-tailed.

ANSWER:
We use the "Critical Values of *r* When ρ = 0" table available in your textbook to place bounds on the *p*-value. Since *df* = *n* – 2 = 13, then 0.02 < **P** =*p*-value < 0.05.

132. Place bounds on the *p*-value resulting from a sample with $n = 20$ and $r = 0.405$, If H_a is one-tailed.

ANSWER:
We use the "Critical Values of *r* When ρ = 0" table available in your textbook to place bounds on the *p*-value. Since $df = n - 2 = 18$, then $0.05 < 2\mathbf{P} = 2$ *p*-value $< 0.10 \Rightarrow 0.025 < \mathbf{P} < 0.05$.

133. Determine the bounds on the *p*-value that would be used in testing $H_o: \rho = 0$ vs. $H_a: \rho \neq 0$, using the *p*-value approach with $n = 15$, and $r = 0.552$.

ANSWER:
We use the "Critical Values of *r* When ρ = 0" table available in your textbook to place bounds on the *p*-value. Since $df = n - 2 = 13$, then $0.02 < \mathbf{P} =$ *p*-value < 0.05.

134. Determine the bounds on the *p*-value that would be used in testing $H_o: \rho = 0$ vs. $H_a: \rho > 0$ using the *p*-value approach with $n = 8$, and $r = 0.772$.

ANSWER:
We use the "Critical Values of *r* When ρ = 0" table available in your textbook to place bounds on the *p*-value. Since $df = n - 2 = 6$, then $0.02 < 2\mathbf{P} = 2$ *p*-value < 0.05. This implies that $0.01 < \mathbf{P} < 0.025$.

135. Determine the bounds on the *p*-value that would be used in testing $H_o: \rho = 0$ vs. $H_a: \rho < 0$ using the *p*-value approach with $n = 22$, and $r = -0.396$.

ANSWER:
We use the "Critical Values of *r* When ρ = 0" table available in your textbook to place bounds on the *p*-value. Since $df = n - 2 = 20$, then $0.05 < 2\mathbf{P} = 2$ *p*-value $< 0.10 \Rightarrow 0.025 < \mathbf{P} < 0.05$.

136. What are the critical values of *r* for α = 0.05 and $n = 27$ if H_a is two-tailed?

ANSWER:
The critical value is found at the intersection of the *df* = 25 row and the two-tailed 0.05 column of the "Critical Values of r When ρ = 0" table available in your textbook. The value is 0.381. Since this is a two-tailed test; we have two critical values: ± 0.381.

137. What are the critical values of *r* for $\alpha = 0.05$ = 0.05 and $n = 42$ If H_a is one-tailed?

ANSWER:
The critical value is found at the intersection of the *df* = 40 row and the two-tailed 0.10 column of the "Critical Values of *r* When ρ = 0" table available in your textbook. The value is 0.257. Since this is a one-tailed; the value is -0.257, if left tail critical region; and 0.257, if right tail.

138. Determine the critical values that would be used in testing $H_o: \rho = 0$ vs. $H_a: \rho \neq 0$ using the classical approach with n = 21, α = 0.05.

ANSWER:
The critical value is found at the intersection of the df = 19 row and the two-tailed 0.05 column of the "Critical Values of r When ρ = 0" table available in your textbook. The value is 0.433. Since this is a two-tailed test; we have two critical values: ± 0.433.

139. Determine the critical values that would be used in testing $H_o: \rho = 0$ vs. $H_a: \rho < 0$ using the classical approach with n = 16, α = 0.05.

ANSWER:

The critical value is found at the intersection of the df = 14 row and the two-tailed 0.10 column of the "Critical Values of r When ρ = 0" table available in your textbook. Since this is a one-tailed with critical region at the left tail; the value is -0.426 as shown in the graph

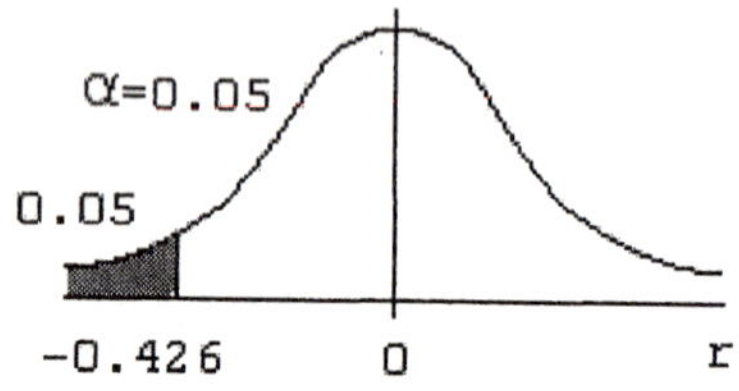

140. Determine the critical values that would be used in testing $H_o: \rho = 0$ vs. $H_a: \rho > 0$ using the classical approach with n = 52, α = 0.01.

ANSWER:
The critical value is found at the intersection of the df = 50 row and the two-tailed 0.02 column of the "Critical Values of r When ρ = 0" table available in your textbook. Since this is a one-tailed with critical region at the right tail; the value is 0.322.

141. If a sample of size 20 has a linear correlation coefficient of 0.467, is there sufficient evidence to conclude that the linear correlation coefficient of the population is positive? Use the p-value approach at α = 0.01.

ANSWER:
$H_o: \rho = 0$ vs. $H_a: \rho > 0$
We use the "Critical Values of r When ρ = 0" table available in your textbook to place bounds on the p-value. Since $df = n - 2 = 18$, then 0.02 < 2**P** = 2 p-value < 0.05 $\Rightarrow$ 0.01 < **P** < 0.025. Since p-value $> \alpha$, we fail to reject H_o. There is not sufficient evidence to conclude that the linear correlation coefficient of the population is positive.

142. A sample of 20 pieces of bivariate data has a linear correlation coefficient of r = 0.489. Does this provide sufficient evidence to reject the null hypothesis that ρ = 0 in favor of a two-sided alternative? Use the classical approach at α = 0.10.

ANSWER:
$H_o: \rho = 0$ vs. $H_a: \rho \neq 0$

The critical value is found at the intersection of the *df* = 18 row and the two-tailed 0.10 column of the "Critical Values of r When ρ = 0" table available in your textbook. The value is 0.378. Since this is a two-tailed test; we have two critical values: ± 0.378 as shown below.

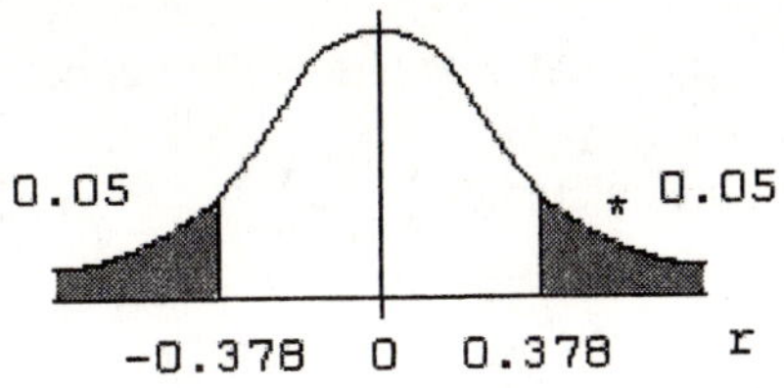

Since r^* = 0.489 falls in the rejection region, we reject H_o. There is sufficient evidence to conclude that the linear correlation coefficient of the population is not zero.

143. If a sample of size 14 has a linear correlation coefficient of -0.517, is there significant reason to conclude that the linear correlation coefficient of the population is negative? Use the *p*-value approach at α = 0.05.

ANSWER:
$H_o: \rho = 0$ vs. $H_a: \rho < 0$
We use the "Critical Values of *r* When ρ = 0" table available in your textbook to place bounds on the *p*-value. Since $df = n - 2 = 12$, then 0.05 < 2**P** = 2 *p*-value < 0.10 $\Rightarrow$ 0.025 < **P** < 0.05. Since *p*-value < α, we reject H_o. There is sufficient evidence to conclude that the linear correlation coefficient of the population is negative

QUESTIONS 144 THROUGH 151 ARE BASED ON THE FOLLOWING INFORMATION:
The population (in millions) and the violent crime rate (per 1000) were recorded for ten metropolitan areas. The data are shown in the following table:

Population	9	2	5.6	1	6.2	10.2	7.2	2.8	3.2	4.6
Crime Rate	10.5	5.5	9.3	4	8	12.1	8.5	6.2	6.9	8.3

144. Construct a scatter diagram for the data.

ANSWER:

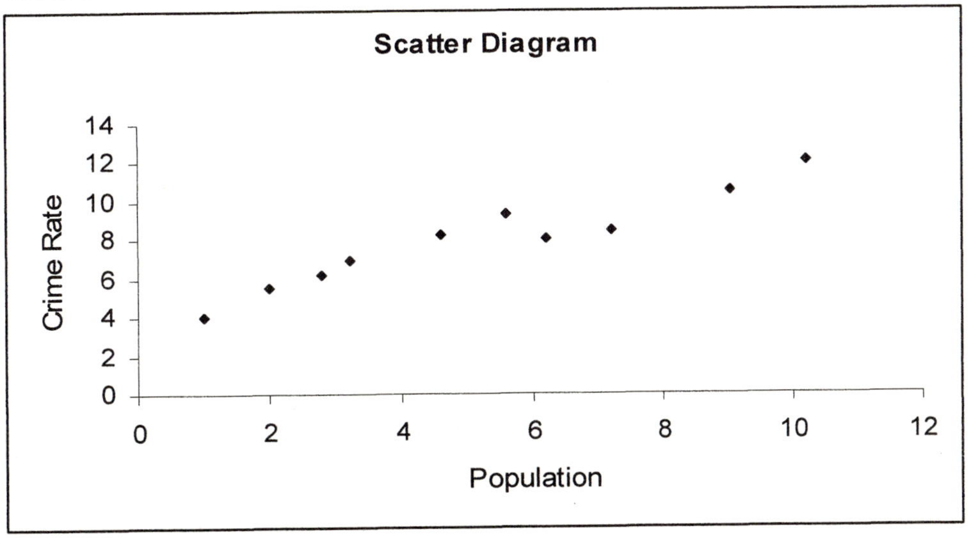

145. What does the scatter diagram in question 144 tell you about the relationship between the two variables?

ANSWER:
There is a positive linear relationship between population size and violent crime rate.

146. Use a computer to form the extensions table and calculate $\sum x$, $\sum y$, $\sum xy$, $\sum x^2$ and $\sum y^2$.

ANSWER:
$\sum x = 51.8$, $\sum y = 79.3$, $\sum xy = 473.42$, $\sum x^2 = 350.92$, and $\sum y^2 = 680.59$

147. Find SS(x), SS(y), and SS(xy).

ANSWER:

$$SS(xy) = \sum xy - \left[\left(\sum x\right)\left(\sum y\right)\right]/n = 473.42 - (51.8)(79.3)/10 = 62.646$$

$$SS(x) = \sum x^2 - \left(\sum x\right)^2/n = 350.92 - (51.8)^2/10 = 82.596$$

$$SS(y) = \sum y^2 - \left(\sum y\right)^2/n = 680.59 - (79.3)^2/10 = 51.741$$

148. Calculate the coefficient of linear correlation, r.

ANSWER:

$$r = SS(xy)/\sqrt{SS(x)\cdot SS(y)} = (62.646)/\sqrt{(82.596)(51.741)} = 0.958$$

149. Use computer to verify the value of r in question 148.

ANSWER:

Table of correlations

	Population	Crime Rate
Population	1.000	
Crime Rate	0.958	1.000

150. Do these data provide evidence to reject the null hypothesis that $\rho = 0$ in favor of $\rho \neq 0$ at $\alpha = 0.05$? Use the p-value approach.

ANSWER:
$H_o: \rho = 0$ vs. $H_a: \rho \neq 0$
We use the "Critical Values of r When $\rho = 0$" table available in your textbook to place bounds on the p-value. Since $df = n - 2 = 8$, then **P** = p-value < 0.01. Since p-value < α = 0.05, we reject H_o. There is sufficient evidence to conclude that the linear correlation coefficient of the population is not zero.

151. Do these data provide evidence to reject the null hypothesis that $\rho = 0$ in favor of $\rho \neq 0$ at $\alpha = 0.05$? Use the classical approach.

ANSWER:
$H_o : \rho = 0$ vs. $H_a : \rho \neq 0$
The critical value is found at the intersection of the *df* = 8 row and the two-tailed 0.05 column of the "Critical Values of r When ρ = 0" table available in your textbook. The value is 0.632. Since this is a two-tailed test; we have two critical values: ± 0.632. Since r^* = 0.958 > 0.632 falls in the rejection region, we reject H_o. We reach the same conclusion stated in question 150.

Section 13.4

True-False Questions

152. The random variable, *e* (also known as the residual), is positive when the predicted value $\hat{y}$ is greater than the observed value of *y*, and is negative when $\hat{y}$ is less than *y*.
ANSWER: F

153. The slope β_1 of the regression line of the population can be estimated by means of a confidence interval that is determined by the formula $b_1 \pm z(\alpha/2) \cdot s_{b_1}$.
ANSWER: F

154. We test the hypothesis $H_o: \beta_1 = 0$ to determine whether the equation for the line of best fit is of any real value in predicting the output variable *y*.
ANSWER: T

155. The hypothesis $H_o: \beta_1 = 0$ is tested using the Student's *t*-distribution with $df = n - 1$.
ANSWER: F

156. The line of best fit always passes through the centroid $(\bar{x}, \bar{y})$.
ANSWER: T

157. In regression analysis, the error term must be normally distributed if references are to be made.
ANSWER: T

158. The value of the input variable *x* must be randomly selected to achieve valid regression results.
ANSWER: F

159. The output variable *y* must be normally distributed about the regression line for each value of the input variable *x*.
ANSWER: T

160. The sum of squares for error is the name given to the numerator portion of the formula used to calculate the variance of *y* about the regression line.
ANSWER: T

161. The line of best fit results from an analysis of two (or more) related quantitative variables.
ANSWER: T

162. The line of best fit, provided one exists, will best predict the value of the dependent, or output, variable from a value of the independent, or input, variable.
ANSWER: T

Multiple-Choice Questions

163. What does b_1 represent in the regression equation?

A) Level of correlation
B) y – intercept
C) Slope of the line
D) Dependent variable
ANSWER: C

164. What is the linear model used to explain relationship between two variables in a population?

A) $y = b_0 + b_1 x + e$
B) $y = \alpha + \beta x$
C) $y = \beta + bx$
D) $y = \beta_0 + \beta_1 x + \varepsilon$
ANSWER: D

165. The regression analysis is used to determine

A) the strength of a relationship between two variables.
B) what is the relationship between two variables.
C) a cause-and-effect situation.
D) the value of r.
ANSWER: B

166. If all of the values of an independent variable x are equal, then regressing a dependent variable y on x will result in a correlation coefficient, r of:

A) -1.0.
B) 0.0.
C) 1.0.
D) 1.2.
ANSWER: B

167. The vertical spread of the data points about the regression line is measured by:

A) the correlation coefficient.
B) the standard error of the estimate.
C) the y-intercept.
D) the slope of the regression line.
ANSWER: B

168. In a regression problem the following pairs of (x, y) are given: (3, 2), (3, 1), (3, 0), (3, -1) and (3, -2). That indicates that the correlation coefficient if

A) 2.
B) 1.
C) 0.
D) -1.
ANSWER: C

Chapter 13

169. A regression analysis between sales (y in \$1000) and advertising (x in \$100) resulted in the following least squares line: $\hat{y} = 75 + 5x$. This implies that if advertising is \$800, then the predicted amount of sales (in dollars) is:

A) \$79,000.
B) \$75,040.
C) \$115,000.
D) \$4,075.
ANSWER: C

170. A regression analysis between weight (y in pounds) and height (x in inches) resulted in the following least squares line: $\hat{y} = 130 + 5x$. This implies that if the height is increased by 1 inch, the weight, on average, is expected to:

A) increase by 1 pound.
B) increase by 5 pounds.
C) decrease by 5 pounds.
D) decrease by 1 pound.
ANSWER: B

171. Which of the following statements is false?

A) When there is no relationship between the variables, a horizontal line of best fit will result.
B) A horizontal line has a slope of zero, which implies that the value of the input variable has no effect on the output variable.
C) The linear model used to explain the behavior of linear bivariate data in the population is $\hat{y} = \beta_0 + \beta_1 x + \varepsilon$, where β_0 is the y-intercept, β_1 is the slope, and ε (lowercase Greek letter "epsilon") is the random experimental error in the observed value of y at a given value of x.
D) None of the above.
ANSWER: D

172. Which of the following statements is false?

A) The equation of the line of best fit takes the form $\hat{y} = b_0 - b_1 x$.
B) When the line of best fit is plotted, it shows us a pictorial representation of the line.
C) When the line of best fit is plotted, It tells us whether or not there really is a linear relationship between the two variables.
D) When the line of best fit is plotted, it tells us the quantitative (equation) relationship between the two variables.
ANSWER: A

173. Which of the following formulas represent the sum of squares for error (SSE)?

A) $\sum (y - \hat{y})^2$
B) $\sum (y - b_0 - b_1 x)^2$
C) $\sum y^2 - b_0 \left(\sum y\right) - b_1 \left(\sum xy\right)$
D) All of the above
ANSWER: D

174. Which of the following statements is false?

A) The sum of the errors (residuals) for all values of *y* for a given value of *x* is exactly zero.
B) The variance of the error e (also known as the residual) is estimated by the formula $s_e^2 = \sum(y-\hat{y})^2/(n-1)$ where *n* – 2 is the number of degrees of freedom.
C) The variance of y about the line of best fit is the same as the variance of the error *e*. Recall that *e* = *y* – $\hat{y}$.
D) None of the above.

ANSWER: **B**

Short-Answer Questions

175. Suppose 20 bivariate observations produced *SSE* = 8.82, find s_e^2.

ANSWER:
s_e^2 = *SSE* / (*n* – 2) = 8.82 / 18 = 0.49

176. Indicate whether the symbol β_1 is a parameter of statistic.

ANSWER:
Parameter

177. Indicate whether the symbol b_1 is a parameter of statistic.

ANSWER:
Statistic

178. What are the primary questions we answer in linear regression analysis?

ANSWER:
What is linear relationship between these two variables?

179. If you know the value of *r* is very close to zero, what value would you anticipate for b_1? Explain.

ANSWER:
The value of b_1 would be close to zero also. The formulas used to calculate *r* and b_1 have the same numerator; namely, SS(*xy*).

180. Describe why the method used to find the line of best fit is referred to as "the method of least squares".

ANSWER:
The vertical distance from a potential line of best fit to the data point is measured by $(y-\hat{y})$. The line of best fit is defined to be the line that results in the smallest possible total when the squared values of $(y-\hat{y})$ are totaled. Thus "the method of least squares".

181. Comment on the statement "The two coefficients for the line of best fit have the same sign." as sometimes true, always true, or never true. Explain your response if your answer is "sometimes true" or "never true"

ANSWER:
Sometimes true. The two coefficients (slope and *y*-intercept) measure two completely different concepts. Their signs are unrelated.

Applied and Computational Questions

QUESTIONS 182 THROUGH 184 ARE BASED ON THE FOLLOWING INFORMATION:

The following summary data are given: $n = 5$, $\sum x = 13$, $\sum y = 246$, $\sum x^2 = 51$, $\sum y^2 = 12{,}946$, and $\sum xy = 760$.

182. Find the equation of the line of best fit.

ANSWER:
$\hat{y} = 31 + 7x$

183. Show that $s_e = 0$.

ANSWER:
$s_e^2 = [\sum y^2 - b_0 \sum y - b_1 \sum xy]/(n-2)$ = [12,946-(31)(246)-(7)(760)] / 3 = 0 / 3 = 0.
Hence, $s_e = 0$.

184. What do you know about this set of bivariate data?

ANSWER:
The data must fall exactly on a straight line.

QUESTIONS 185 AND 186 ARE BASED ON THE FOLLOWING INFORMATION:

The following summary data are given: $n = 10$, $\sum x = 39$, $\sum y = 35.1$, $\sum x^2 = 193$, $\sum y^2 = 130.05$, and $\sum xy = 152.7$.

185. Find the equation of the line of best fit.

ANSWER:
$\hat{y} = 2 + 0.387x$

186. Find s_e.

ANSWER:
$s_e^2 = [\sum y^2 - b_0 \sum y - b_1 \sum xy]/(n-2)$ = [130.05-(2)(35.1)-(0.387)(152.7)] / 8 = 0.0944
Hence, $s_e = 0.307$.

QUESTIONS 187 AND 188 ARE BASED ON THE FOLLOWING INFORMATION:
Consider the following set of bivariate data:

x	1.0	0.0	3.0	2.0	6.0
y	4.0	1.5	9.0	6.5	16.5

187. Find s_e.

ANSWER:
$s_e = 0$

188. Based on the value of s_e, what do you know about this bivariate data?

ANSWER:
The data must fall exactly on a straight line.

QUESTIONS 189 THROUGH 193 ARE BASED ON THE FOLLOWING INFORMATION:
The following data show the number of hours (x) studied for a final exam, and the score (y) received on the exam for a random sample of 15 students.

x	3	4	4	5	5	6	6	7	7	7	8	8	8	9	9
y	53	59	74	58	77	78	86	68	90	83	79	97	100	89	96

189. Draw a scatter diagram of the data.

ANSWER:

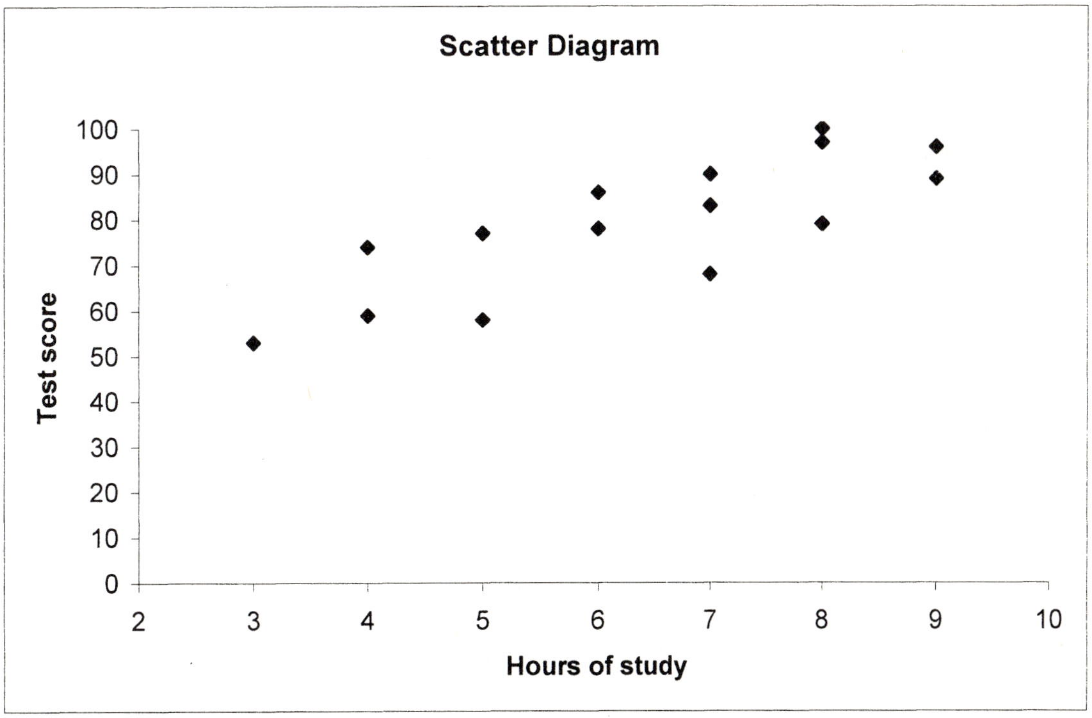

Chapter 13

190. Find the equation of the line of best fit.

ANSWER:
Summary of data:

$n = 15,\ \sum x = 96,\ \sum y = 1187,\ \sum x^2 = 664,\ \sum xy = 7910,\ \sum y^2 = 96939$

$SS(x) = \sum x^2 - [(\sum x)^2 / n] = 664 - (96^2 / 15) = 49.6$

$SS(xy) = \sum xy - [(\sum x)(\sum y)/n] = 7910 - [(96)(1187) / 15] = 313.2$

$b_1 = SS(xy)/SS(x) = 313.2 / 49.6 = 6.315$

$b_0 = [\sum y - b_1 \sum x]/n = [1187 - (6.315)(96)] / 15 = 38.717$

The equation of the line of best fit is: $\hat{y} = b_0 + b_1 x = 38.717 + 6.315x$

191. Find the ordinates $\hat{y}$ that correspond to x = 3, 4, 5, 6, 7, 8, and 9.

ANSWER:
If x = 3, then $\hat{y}$ = 57.662
If x = 4, then $\hat{y}$ = 63.977
If x = 5, then $\hat{y}$ = 70.292
If x = 6, then $\hat{y}$ = 76.607
If x = 7, then $\hat{y}$ = 82.922
If x = 8, then $\hat{y}$ = 89.237
If x = 9, then $\hat{y}$ = 38.717 + 6.315 (9) = 95.552

192. Find the five values of *e* that are associated with the points where *x* = 4 and *x* = 7.

ANSWER:

x	4	4	7	7	7
y	59	74	68	90	83
$\hat{y}$	63.977	63.977	82.922	82.922	82.922
e	-4.977	10.023	-14.922	7.078	0.078

193. Find the variance s_e^2 of all the points about the line of best fit.

ANSWER:

$$s_e^2 = \frac{\sum y^2 - b_0 \sum y - b_1 \sum xy}{n-2} = [96939 - (38.717)(1187) - (6.315)(7910)] / 13$$

$= 79.252$

194. Using the following bivariate data, calculate the standard error of estimate.

x	2	3	3	5	7	7	8
y	5	6	9	4	2	2	0

ANSWER:
1.66

195. Find the equation of the line of best fit for the data shown below. Then, find the variance error by evaluating $\sum(y-\hat{y})^2/(n-2)$.

x	0	1	3	4
y	4	4	10	12

ANSWER:
The equation of the line of best fit is $\hat{y} = 3.1 + 2.2x$, and the variance error is $s_e^2 = 1.3$

QUESTIONS 196 THROUGH 204 ARE BASED ON THE FOLLOWING INFORMATION:
The average number of client contacts per month, x, and the sales volume, y (in $1000), were recorded of each of 10 salespeople.

x	22	16	50	48	57	14	25	52	18	52
y	35	30	100	85	135	20	35	95	35	115

196. Draw a scatter diagram of the data

ANSWER:

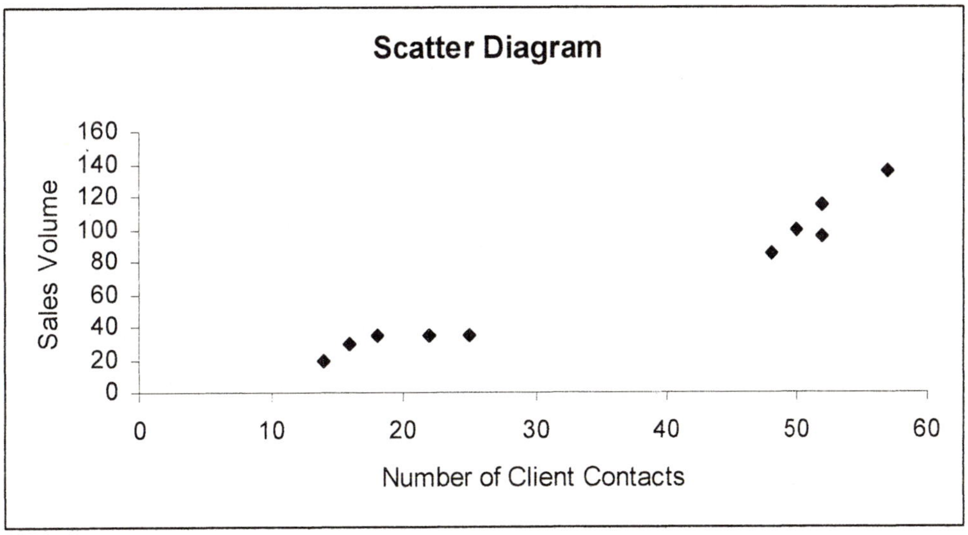

197. Does the scatter diagram suggest a linear relationship between *x* and *y*?

ANSWER:
The scatter diagram suggests a linear relationship between *x* and *y*.

Chapter 13

198. Calculate $\sum x$, $\sum y$, $\sum x^2$, $\sum y^2$, and $\sum xy$.

ANSWER:

	x	y	xy	x^2	y^2
	22	35	770	484	1225
	16	30	480	256	900
	50	100	5000	2500	10000
	48	85	4080	2304	7225
	57	135	7695	3249	18225
	14	20	280	196	400
	25	35	875	625	1225
	52	95	4940	2704	9025
	18	35	630	324	1225
	52	115	5980	2704	13225
Sum	354	685	30730	15346	62675

$\sum x = 354$, $\sum y = 685$, $\sum x^2 = 15346$, $\sum y^2 = 62675$, and $\sum xy = 30730$

199. Calculate SS(*x*) and SS(*xy*).

ANSWER:

$$SS(xy) = \sum xy - \left[\left(\sum x\right)\left(\sum y\right)\right]/n = 30730 - [(354)(685)]/10 = 6481$$

$$SS(x) = \sum x^2 - \left(\sum x\right)^2/n = 15346 - (354)^2/10 = 2814.4$$

200. Calculate the slope and *y*-intercept for the line of best fit.

ANSWER:
The slope b_1 = SS(*xy*) / SS(*x*) = 6481 / 2814.4 = 2.3028.
The *y*-intercept $b_0 = [\sum y - (b_1 \cdot \sum x)]/n$ = [685 – (2.3028)(354)] / 10 = -13.0191.

201. What is the equation of the line of best fit?

ANSWER:
$\hat{y} = -13.0191 + 2.3028x$

202. Predict the sales volume for a salesperson who contacted 50 clients.

ANSWER:
$\hat{y}$ = -13.0191 + 2.3028 (50) = 102.1209 (in $1000) or $102,120.9

203. Calculate the sum of squares for error.

ANSWER:
SSE = $\sum y^2 - (b_0)(\sum y) - (b_1)(\sum xy)$
= 62675 – (-13.0191)(685)-(2.3028)(30730) = 828.0395

204. Determine the variance of *y* about the line of best fit.

ANSWER:
The variance of y about the line of best fit is the same as the variance of the error *e*.
s_e^2 = SSE / (n – 2) - 828.0395 / 8 = 103.5049.

QUESTIONS 205 THROUGH 212 ARE BASED ON THE FOLLOWING INFORMATION:
The price (in $) and the carat weight of a diamond are its two most known characteristics. In order to understand the role carat weight has in determining the price of a diamond, the carat weight and price of 20 loose round diamonds, all of color D and clarity VS1, were obtained recently as shown below.

Carat Weight	Price	Carat Weight	Price
0.56	2789	0.59	2841
0.60	2517	0.65	2853
0.53	2645	0.51	2024
0.57	2367	0.54	2609
0.53	2673	0.51	2603
0.61	3029	0.51	2159
0.53	2701	0.57	2398
0.51	2549	0.52	2061
0.67	2959	0.56	2328
0.54	2276	0.51	2047

205. Draw a scatter diagram of the data: carat weight (x) and price (y).

ANSWER:

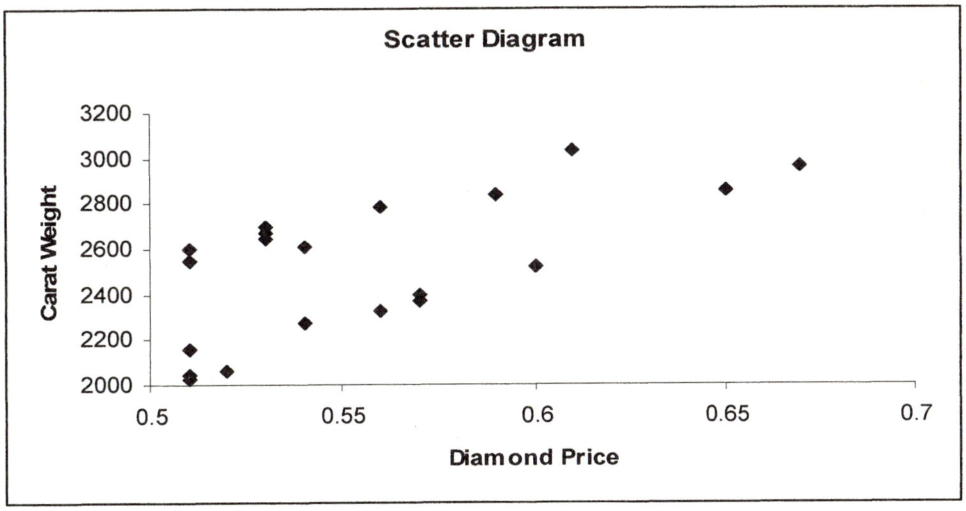

206. Does the data suggest a linear relationship for the domain 0.50 to 0.66 carats?? Discuss your findings in question 205.

ANSWER:
There is a linear pattern to the data, however the data falls into two groups forming two parallel linear patterns, one forming the top and the other forming the bottom of the total pattern.

207. Diamonds smaller than 0.50 carats and diamonds larger than 0.66 carats may not fit the linear pattern demonstrated by this data. Explain.

ANSWER:
Since we only have data in this weight range, we cannot predict with confidence outside this range. Smaller values than 0.50 carats and larger values than 0.66 carats decrease and increase, respectively, exponentially.

208. Use computer to find the equation for the line of best fit.

ANSWER:

Regression coefficients		
		Coefficient
	Constant	187.2943
	Carat Weight	4198.0319

The equation for the line of best fit is $\hat{y}$ = 187.2943 + 4198.0319*x*

209. According to the results obtained in question 208, what would be a typical price for a 0.50 carat loose diamond of this quality?

ANSWER:
$\hat{y}$ = 187.2943 + 4198.0319 (0.50) = $2,286.3

210. On the average, by how much does the price increase for each extra 0.01 carat in weight? Within what interval of *x*-values would you expect this to be true?

ANSWER:
The price, on the average, increases by $41.98 for each extra 0.01 carat in weight. We would expect this to be true for *x*-values within the interval 0.50 to 0.66 carats.

211. Use computer to find the variance of *y* about the regression line.

ANSWER:

Summary measures		
	Multiple R	0.6517
	R-Square	0.4247
	StErr of Est	237.9051

The variance of *y* about the regression line = $s_e^2 = (237.9051)^2 = 56,598.84$

212. Graph and display the line of best fit on the scatter diagram. What characteristics in the scatter diagram support the large value obtained in question 211?

ANSWER:

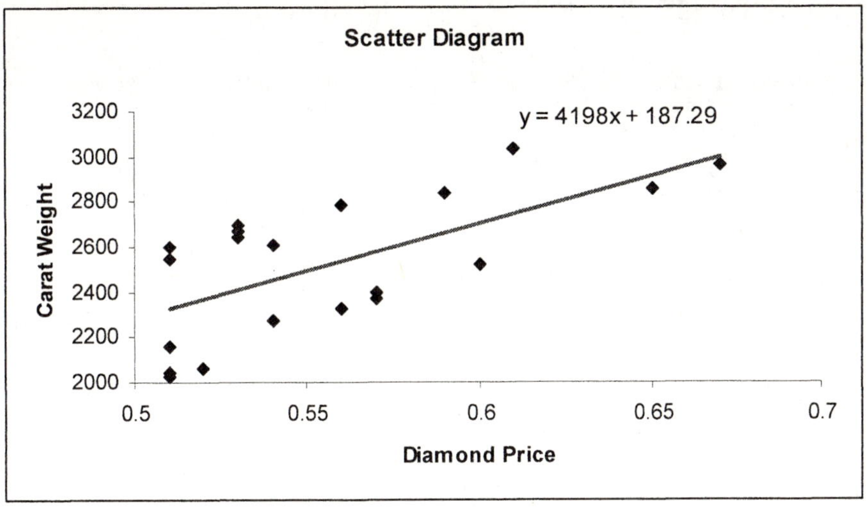

The scatter diagram shows a sizeable amount of vertical distance between the top and bottom points along the line of best fit.

Sections 13.5 through 13.7

True-False Questions

213. There are $n-1$ degrees of freedom involved with the inferences about the regression line.
ANSWER: F

214. The best point estimate, or prediction, for both $\mu_{y|x_0}$ and y_{x_0} is $\hat{y}$.
ANSWER: T

215. The conference interval for $\mu_{y|x_0}$ and the prediction interval for y_{x_0} are constructed in a similar fashion.
ANSWER: T

216. The symbol $\mu_{y|x_0}$ refers to the mean of the population *y*-values at a given value of *x*, while y_{x_0} refers to the individual *y*-value selected at random that will occur at a given value of *x*.
ANSWER: T

217. The *standard error of* regression (slope) is σ_{b_1} and is estimated by s_{b_1}; the estimate of the variance of the error about the regression line.
ANSWER: T

218. The best point estimate, or prediction for both μ_{y/x_0} and y_{x_0}, is the actual value of *y*.
ANSWER: F

219. The prediction interval for an individual value of *y* is wider than the confidence interval for the mean value of *y*; both calculated at the same value x_0.
ANSWER: T

220. The confidence interval for an individual value of *y* is wider than the prediction interval for the mean value of *y*; both calculated at the same value x_0.
ANSWER: F

Multiple-Choice Questions

221. In a simple linear regression problem, which of the following table values would be appropriate for a 95% confidence interval for the mean of *y* for a given value of *x* if the sample size is 10?

A) 1.86
B) 1.81
C) 2.31
D) 2.36
ANSWER: C

222. In a simple linear regression problem including eight observations, which of the following table values would be appropriate for a 90% prediction interval of the value of a single randomly selected *y*?

A) 1.40
B) 1.86
C) 1.44
D) 1.94
ANSWER: D

223. Which of the following statements is false?

A) The linear correlation coefficient can be used to indicate the usefulness of *x* as a predictor of *y* in the case where the linear model is appropriate.
B) The concepts of linear correlation and regression are quite similar because both measure the same characteristics.
C) It is possible to have data that yield a strong linear correlation coefficient and have the wrong model.
D) Regression analysis should be used to answer questions about the relationship between two variables.
ANSWER: B

224. Which of the following statements is false regarding the assumptions for Inferences about linear regression?

A) The set of (*x, y*) ordered pairs forms a random sample.
B) The *y* values at each *x* have a normal distribution
C) Since the population standard deviation is unknown and replaced with the sample standard deviation, the normal distribution will be used.
D) None of the above
ANSWER: C

225. Which of the following statements is false?

A) The slope β_1 of the regression line of the population can be estimated by means of a confidence interval. The confidence interval is determined by $b \pm z(\alpha/2)$.
B) The null hypothesis $H_o: \beta_1 = 0$ will be tested using the Student's *t*-distribution with ($n - 2$) degrees of freedom.
C) The test statistics t^* found by using the formula $t^* = (b_1 - \beta_1)/s_{b_1}$ is used for testing $H_o: \beta_1 = 0$
D) None of the above
ANSWER: A

226. Which of the following statements is false?

A) The best point estimate, or prediction for both μ_{y/x_0} and y_{x_0}, is $\hat{y}$. This is the *y* value obtained when an *x* value is substituted into the equation of the line of best fit.
B) The sampling distribution of $\hat{y}$ is the Student's *t*-distribution with $df = n - 2$.
C) The prediction interval for an individual value of *y* is wider than the confidence interval for the mean value of *y*; both calculated at the same value x_0.
D) None of the above
ANSWER: B

227. Which of the following statements is false?

A) Regression only measures movement between *x* and *y*; it never prove causation.
B) The regression equation is meaningful only in the domain of the *x* variable studied. Estimation outside this domain is extremely dangerous; it requires that we know or assume that the relationship between *x* and *y* remains the same outside the domain of the sample data.
C) The regression equation is meaningful only in the domain of the *y* variable studied. Estimation outside this domain is extremely dangerous; it requires that we know or assume that the relationship between *x* and *y* remains the same outside the domain of the sample data.
D) None of the above

ANSWER: C

Short-Answer Questions

228. State the null hypothesis H_o, and the alternative hypothesis H_a, that would be used to test the statement: "There is evidence that the slope of the line of best fit is negative".

ANSWER:
$H_o : \beta_1 = 0\ (\geq)$ vs. $H_a : \beta_1 < 0$

229. Determine the *p*-value for testing $H_a : \beta_1 < 0$, with $n = 50$, $b_1 = -1.20$, $s_{b_1} = 0.80$.

ANSWER:
$t^* = b_1 / s_{b_1} = -1.20 / 0.80 = -1.50$
P = *p*-value = $P(t < -1.50 \mid df = 48) = 0.07$

230. State the null hypothesis H_o, and the alternative hypothesis H_a, that would be used to test the statement: "The slope for the line of best fit is greater than 1.0".

ANSWER:
$H_o : \beta_1 = 1\ (\leq)$ vs. $H_a : \beta_1 > 1$

231. Determine the *p*-value for testing $H_a : \beta_1 > 0$, with $n = 20$, $t^* = 2.8$.

ANSWER:
P = *p*-value = $P(t > 2.8 \mid df = 18) = 0.006$

232. State the null hypothesis H_o, and the alternative hypothesis H_a, that would be used to test the statement: "There is no significant relationship between the *x* and *y* variables".

ANSWER:
$H_o : \beta_1 = 0$ vs. $H_a : \beta_1 \neq 0$

233. Determine the critical value(s) and rejection region(s) that would be used with the classical approach in testing $H_o: \beta_1 = 0$ vs. $H_a: \beta_1 > 0$, with n = 30 and α = 0.025.

ANSWER:
Critical value = t (28, 0.025) = 2.05.
Rejection region: Reject H_o if $t^* \geq 2.05$.

234. Determine the p-value for testing $H_a: \beta_1 \neq 0$, with df = 12, b_1 = 0.20, and s_{b_1} = 0.125

ANSWER:
$t^* = b_1 / s_{b_1}$ = 0.20 / 0.125 = 1.60
P = p-value = 2 · $P(t > 1.6 \mid df = 12)$ = 2 (0.068) = 0.136

235. Determine the critical value(s) and rejection region(s) that would be used with the classical approach in testing $H_o: \beta_1 = 0$ vs. $H_a: \beta_1 \neq 0$, given that n = 18 and α = 0.10.

ANSWER:
Critical values = $\pm t(16, 0.05) = \pm 1.75$.
Rejection region: Reject H_o if $t^* \leq -1.75$ or $t^* \geq 1.75$.

Applied and Computational Questions

QUESTIONS 236 THROUGH 238 ARE BASED ON THE FOLLOWING INFORMATION:
The scores (x) on a computer science aptitude test range from 0 to 25, and the course grade (y) with possible values: 0.0, 1.0, 1.5, 2.0, 2.5, 3.0, 3.5, 4.0, were recorded for 20 students in an introductory computer science course as shown below.

x	18	10	15	20	18	13	20	16	12	22
y	2.5	1.5	3.0	3.5	2.5	2.0	2.5	3.0	2.0	4.0

x	5	8	16	20	11	14	20	16	15	24
y	0.0	1.0	1.5	3.0	2.0	2.5	4.0	4.0	2.5	3.5

236. Test the null hypothesis $H_o: \beta_1 = 0 (\leq)$ vs. $H_a: \beta_1 > 0$ by giving the critical region for α = 0.05, the value of the test statistic t^*, and the conclusion.

ANSWER:
Critical region: $t \geq 1.73$, $t^* = 6.39$; Reject H_o and conclude that $\beta_1 > 0$.

237. Find the equation of the line of best fit, and construct a 95% confidence interval for the mean course grade for students who score 15 on the computer science aptitude test.

ANSWER:
The equation of the line of best fit is $\hat{y}$ = -0.285+0.18x.
(2.13 to 2.69) is the 95% confidence interval for the mean course grade for students who score 15 on the computer science aptitude test.

238. Construct a 95% prediction interval for the course grade for a student who scores 15 on the computer science aptitude test.

ANSWER:
(1.13 to 3.69)

239. Determine the *p*-value for testing $H_a : \beta_1 < 0$, given that *n* = 14 and t^* = –1.5.

ANSWER:
0.080

240. Determine the *p*-value for testing $H_a : \beta_1 > 0$, given that *n* = 20 and t^* = 2.0.

ANSWER:
0.030

241. Determine the *p*-value for testing $H_a : \beta_1 \neq 0$, given that *n* = 10 and t^* = 2.4.

ANSWER:
0.044

242. Find the equation of the line of best fit for the data below, and estimate the value of *y* when *x* is 6.

x	2	3	3	5	7	7	8
y	5	6	9	4	2	2	0

ANSWER:
The equation of the line of best fit is $\hat{y} = 9.4412 - 1.088x$

The estimate the value of *y* when *x* is 6 is $\hat{y}_{x=6}$ = 2.9132

QUESTIONS 243 THROUGH 245 ARE BASED ON THE FOLLOWING INFORMATION:
Ten experimental plots were utilized to investigate the relationship between the amount of fertilizer per plot and the yield of potatoes (in pounds) per plot. It is known that s_e^2 = 0.0879, SS(*x*) = 7.6, and the equation of the line of best fit is $\hat{y} = 6.63 + 0.51x$.

243. Find $s_{b_1}^2$.

ANSWER:
$s_{b_1}^2 = s_e^2$ / SS(*x*) = 0.879 / 7.6 = 0.1157

244. Test $H_o : \beta_1 = 0 (\leq)$ vs. $H_a : \beta_1 > 0$ at α = 0.05 using the classical approach.

ANSWER:
$t^* = (b_1 - \beta_1) / s_{b_1} = (0.51 - 0.0) / \sqrt{0.1157}$ = 1.499

Since *t**=1.499, and the critical region is *t* > 1.86, we fail to reject H_o at α = 0.05. There is no sufficient evidence to indicate that β_1 > 0.

245. Set a 95% confidence interval for β_1.

ANSWER:
(0.26 to 0.76)

QUESTIONS 246 AND 247 ARE BASED ON THE FOLLOWING INFORMATION:
The following data were collected on eight insulin dependent diabetics. The variable *x* is the average of thirty fasting blood sugar readings taken over the past month, and *y* is the hemoglobin A1C reading obtained at the end of the month in which the blood sugar determinations were made. The data are shown in the table below:

x	120	145	210	105	108	150	160	115
y	6.8	7.2	9.2	5.5	8.5	6.5	7.9	6.2

246. Test the hypothesis $H_o : \beta_1 = 0$ vs. $H_a : \beta_1 > 0$ at the 0.05 level of significance. Report t^*, the *p*-value for the test, and your conclusion.

ANSWER:
$t^* = 7.2$, p –value < 0.005. We reject H_o at α = 0.05. There is sufficient evidence to indicate that the slope of the line of best fit in the population is greater than zero.

247. Find a 95% confidence interval for the mean of *y* when *x* = 140.

ANSWER:
(6.6 to 7.3)

248. Use the following bivariate data to set a 95% confidence interval on β_1.

x	1	5	6	6	7	9
y	8.4	10.1	11.9	13.1	14.5	16.9

ANSWER:
(0.52 to 1.63)

QUESTIONS 249 THROUGH 251 ARE BASED ON THE FOLLOWING INFORMATION:
Waist measurements, *x*, and weights, *y*, were obtained for eighteen males under 30 years of age. The results were as follows:

x	33	33	30	34	34	40	35	35	32	38	34	32
y	160	187	156	179	187	230	197	196	173	201	174	163

x	35	32	32	34	36	30
y	163	167	151	195	227	155

249. Set a 99% confidence interval for β_1.

ANSWER:
(4.0 to 11.4)

250. Set a 95% confidence interval on the mean weight for all those males with 34-inch waist measurements.

ANSWER:
(175.84 to 189.06)

251. Set a 95% confidence interval on the weight of a given adult male with a 34-inch waist.

ANSWER:
(153.69 to 211.22)

QUESTIONS 252 AND 253 ARE BASED ON THE FOLLOWING INFORMATION:
Varying amounts of fertilizer were used on ten different plots and the yield of corn in bushels per plot was measured for each plot. Let *x* represents the amount of fertilizer and *y* represents the yield of corn. A summary of the results are as follows: $\hat{y}$ = 6.63 + 0.51*x* and s_e = 0.3. The *x*-values ranged from 2.0 to 4.5 and $\bar{x}$ = 3.2 and *SS*(*x*) = 7.6.

252. Construct a 95% confidence interval for the mean yield for all plots that have 3.0 units of fertilizer added.

ANSWER:
(7.94 to 8.38)

253. Construct a 95% confidence interval for the yield of an individual plot to which have 3.0 units of fertilizer added.

ANSWER:
(7.43 to 8.89)

QUESTIONS 254 AND 255 ARE BASED ON THE FOLLOWING INFORMATION:
Consider the following bivariate data:

x	12	14	16	20	23	46	48	50	50	55
y	14	24	30	28	30	80	90	85	110	120

254. Construct a 95% confidence interval for the mean of the population *y* - values when *x* = 30.
ANSWER:
(46.7 to 60.6)

255. Construct a 95% prediction interval for an individual *y*-value when *x* = 30.

ANSWER:
(31.1 to 76.2)

256. State the null hypothesis H_o, and the alternative hypothesis H_a, that would be used to test the following statement: "The slope for the line of best fit is positive."

ANSWER:
$H_o : \beta_1 = 0$ vs. $H_a : \beta_1 > 0$

257. State the null hypothesis H_o, and the alternative hypothesis H_a, that would be used to test the following statement: "There is no regression."

ANSWER:
$H_o: \beta_1 = 0$ vs. $H_a: \beta_1 \neq 0$

258. State the null hypothesis H_o, and the alternative hypothesis H_a, that would be used to test the following statement: "There is evidence of negative regression."

ANSWER:
$H_o: \beta_1 = 0$ vs. $H_a: \beta_1 < 0$

259. State the null hypothesis H_o, and the alternative hypothesis H_a, that would be used to test the following statement: "There is evidence of positive regression."

ANSWER:
$H_o: \beta_1 = 0$ vs. $H_a: \beta_1 > 0$

260. Determine the *p*-value for testing $H_a: \beta_1 > 0$, with $n = 20$, and $t^* = 2.2$.

ANSWER:
To determine the *p*- value for the test of the slope of the regression line, the table of probability values for Student's *t*-distribution is used with *df* = *n* – 2, and the value of test statistic is $t^* = (b_1 - \beta_1)/s_{b_1}$. **P** = *P*(*t* > 2.20 | df = 18) = 0.021.

261. Determine the *p*-value for testing $H_a: \beta_1 \neq 0$, with $n = 14$, $b_1 = 0.21$, and $s_{b_1} = 0.07$.

ANSWER:
To determine the *p*- value for the test of the slope of the regression line, the table of probability values for Student's *t*-distribution is used with df = *n* – 2, and the value of test statistic is $t^* = (b_1 - \beta_1)/s_{b_1}$. **P** = 2 *P*(*t* > 3.0 | df = 12) = 2(0.006) = 0.012.

262. Determine the *p*-value for testing $H_a: \beta_1 < 0$, with $n = 27$, $b_1 = -1.20$, and $s_{b_1} = 0.75$

ANSWER:
To determine the *p*- value for the test of the slope of the regression line, the table of probability values for Student's *t*-distribution is used with df = *n* – 2, and the value of test statistic is $t^* = (b_1 - \beta_1)/s_{b_1}$. **P** = *P*(*t* < - 1.6 | *df* = 25) = *P*(*t* >1.6 | *df* = 25) = 0.061.

QUESTIONS 263 THROUGH 269 ARE BASED ON THE FOLLOWING INFORMATION:
A sample of ten students were asked by their statistics professor for the distance (rounded to nearest mile) and the time (rounded to nearest minute) required to commute to college daily. The data collected are shown in the following table.

Distance	2	4	5	6	7	7	8	9	10	12
Time	6	13	15	20	18	23	20	25	28	30

Chapter 13

The following summary values are given:

$n = 10,\ \sum x = 70,\ \sum y = 198,\ \sum x^2 = 568,\ \sum xy = 1571,\ \sum y^2 = 4392$

263. Draw a scatter diagram of these data.

ANSWER:

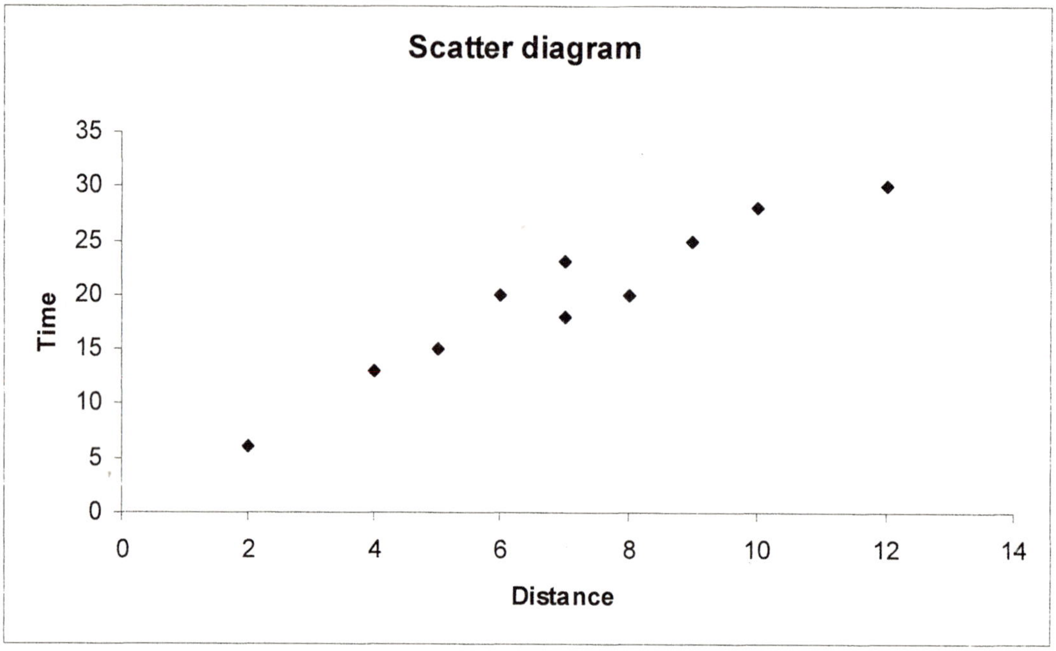

264. Find the equation that describes the regression line for these data.

ANSWER:

$SS(x) = \sum x^2 - [(\sum x)^2 / n] = 568 - (70^2/10) = 78$

$SS(xy) = \sum xy - [(\sum x)(\sum y)/n] = 1571 - [(70)(198)/10] = 185$

$b_1 = SS(xy)/SS(x) = 185/78 = 2.372$

$b_0 = [\sum y - b_1 \sum x]/n = [198 - (2.372)(70)]/10 = 3.196$

The equation of the line of best fit is: $\hat{y} = b_0 + b_1 x = 3.196 + 2.372x$.

265. Give a point estimate for the mean time required to commute four miles.

ANSWER:

When $x = 4$, $\hat{y} = 3.196 + 2.372\,(4) = 12.684$. Then, the point estimate for $\mu_{y|x=4} = 12.684$

266. Does the value of b_1 show sufficient strength to conclude that β_1 is greater than zero at the $\alpha = 0.05$ level? Test using the classical approach.

ANSWER:

β_1 is the slope of the line of best fit for the population of distances and their corresponding times required for students to commute to college.

$H_o: \beta_1 = 0$ vs. $H_a: \beta_1 > 0$. Assume normality for *y* at each *x*. Since *n* = 10, then df = *n* – 2 = 8. b_1 = 2.372 and α = 0.05.

$$s_e^2 = \frac{\sum y^2 - b_0 \sum y - b_1 \sum xy}{n-2} = [4392 - (3.196)(198) - (2.372)(1571)] / 8 = 4.0975$$

$$s_{b_1} = \sqrt{s_e^2 / SS(x)} = \sqrt{4.0975/78} = 0.2292$$

$$t^* = (b_1 - \beta_1)/s_{b_1} = (2.372 - 0) / 0.2292 = 10.349$$

The critical region is t $\geq$ 1.86. Since the test statistic t^* falls in the critical region; we reject H_o. There is sufficient evidence at the 0.05 level of significance to indicate that the slope is significantly greater than zero.

267. Find the 98% confidence interval for the estimation of β_1.

ANSWER:

$$b_1 \pm t \cdot s_{b_1} = 2.372 \pm (2.90)(0.2292) = 2.372 \pm 0.665.$$

The 98% confidence interval for β_1 is 1.707 to 3.037.

268. Give a 90% confidence interval for the mean travel time required to commute four miles.

ANSWER:

$\mu_{y|x=4}$ = 12.684 is the mean travel required to commute four miles. Normality assumed for *y* at each *x*. $n = 10,\ x_0 = 4,\ \overline{x} = \sum x/n = 7.0,\ s_e = \sqrt{4.0975} = 2.0242,\ \hat{y} = 12.684$

Since α /2 = 0.05, df = 8; then *t*(8, 0.05) = 1.86, and

$$E = t(n-2, \alpha/2) \cdot s_e \cdot \sqrt{(1/n) + [(x_0 - \overline{x})^2 / SS(x)}$$

$$= (1.86)(2.0242)\sqrt{(1/10) + [(4-7)^2/78]}$$

$$= (1.86)(2.0242)(0.4641) = 1.747$$

Hence $\hat{y} \pm E = 12.684 \pm 1.747$, and the 90% confidence interval for $\mu_{y|x=4}$ is 10.937 to 14.431.

Chapter 13

269. Give a 90% prediction interval for the travel time required for one person to commute four miles.

ANSWER:

$y_{x=4}$ is the travel time required for one person to commute four miles.

$n = 10, \quad x_0 = 4, \quad \bar{x} = \sum x/n = 7.0, \quad s_e = \sqrt{4.0975} = 2.0242, \quad \hat{y} = 12.684$

Since $\alpha/2 = 0.05$, df = 8; then $t(8, 0.05) = 1.86$, and

$$E = t(n-2, \alpha/2) \cdot s_e \cdot \sqrt{1 + (1/n) + [(x_0 - \bar{x})^2 / SS(x)}$$

$$= (1.86)(2.0242)\sqrt{1 + (1/10) + [(4-7)^2 / 78]}$$

$= (1.86)(2.0242)(1.1024) = 4.151$

Hence $\hat{y} \pm E = 12.684 \pm 4.151$, and the 90% prediction interval for $y_{x=4}$ is 8.533 to 16.835.

QUESTIONS 270 THROUGH 274 ARE BASED ON THE FOLLOWING INFORMATION:

People not only live longer today but they also are living independently longer, even an individual may become temporarily dependent at some age. The table shown below includes two variables: people's age at which they became dependent (*x*) and the number of independent years they had remaining (*y*).

x	65	66	67	68	70	72	74	76	78	80	83	85
y	11.1	10.0	10.4	9.3	8.2	6.8	6.8	4.4	5.4	2.5	2.7	0.9

The following summary values are given:

$n = 12, \ \sum x = 878, \ \sum y = 82.3, \ \sum x^2 = 64742, \ \sum xy = 5772.2, \ \sum y^2 = 695.87$

270. Draw a scatter diagram.

ANSWER:

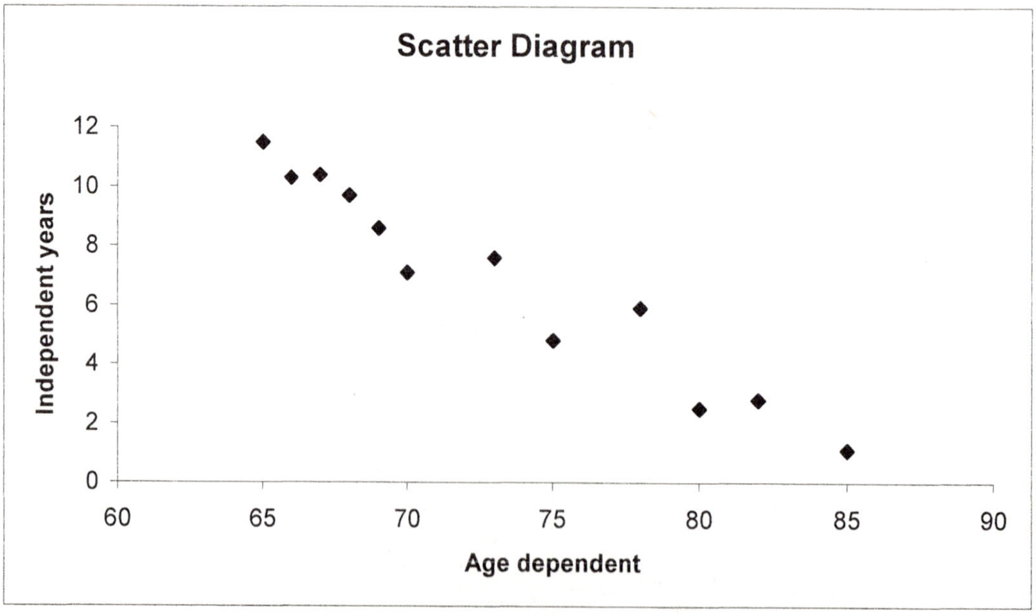

271. Find the equation for the line of best fit.

ANSWER:

$SS(x) = \sum x^2 - [(\sum x)^2 / n] = 64742 - (878^2 / 12) = 501.6667$

$SS(xy) = \sum xy - [(\sum x)(\sum y) / n] = 5772.2 - [(878)(82.3) / 12] = -249.4167$

$b_1 = SS(xy) / SS(x) = -249.4167 / 501.6667 = -0.497$

$b_0 = [\sum y - b_1 \sum x] / n = [82.3 - (-0.497)(878)] / 12 = 43.222$

The equation of the line of best fit is: $\hat{y} = b_0 + b_1 x = 43.222 - 0.497x$

272. Draw the line of best fit on the scatter diagram.

ANSWER:

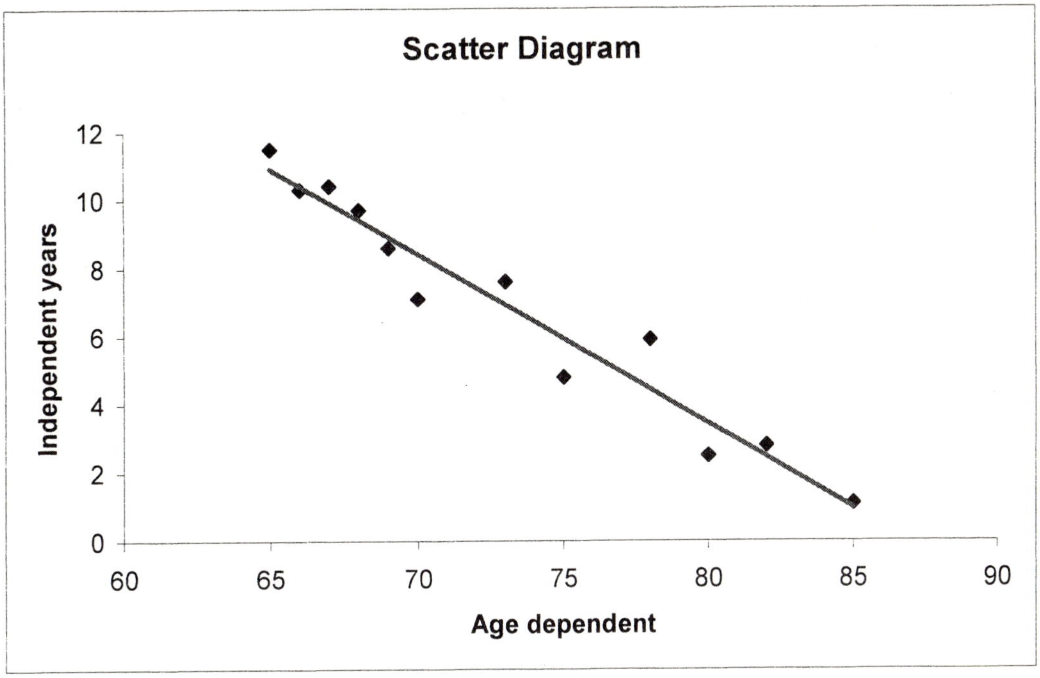

273. For a person who becomes dependent at age 80, how many years of independent living can be expected to remain? Find the answer two different ways; use the equation for the line of best fit found in question 271 and use the line on the scatter diagram in question 272.

ANSWER:
When $x = 80$, $\hat{y} = 43.222 - 0.497(80) = 3.462$. Reading from the graph in question 272, $\hat{y} \approx 3.5$ when $x = 80$.

274. Construct a 99% prediction interval for the number of years of independent living remaining for a person who becomes dependent at age 80.

ANSWER:

$y_{x=80}$ is the number of years of independent living remaining for a person who becomes dependent at age 80. $n = 12, \quad x_0 = 80, \quad \bar{x} = \sum x/n = 73.17,$

$s_e^2 = \dfrac{\sum y^2 - b_0 \sum y - b_1 \sum xy}{n-2}$ = [695.87-(43.222)(82.3)-(-0.497)(5772.2)]/10 = 0.74828

$s_e = \sqrt{0.74828} = 0.865$; Since $\alpha/2$ = 0.005, df = 10; then $t(10, 0.005)$ = 3.17, and

$E = t(n-2, \alpha/2) \cdot s_e \cdot \sqrt{1 + (1/n) + [(x_0 - \bar{x})^2 / SS(x)}$

$= (3.17)(0.865)\sqrt{1 + (1/12) + [(80 - 73.17)^2 / 501.6667]}$ = 2.974

But $\hat{y} = 3.462$, then $\hat{y} \pm E = 3.462 \pm 2.974$, and the 99% prediction interval for $y_{x=80}$ is 0.488 to 6.436.

QUESTIONS 275 THROUGH 279 ARE BASED ON THE FOLLOWING INFORMATION:

The following set of 25 scores was randomly selected from Dr. Maas' inferential statistics class. Let *x* be the pre-final average and *y* the final examination score. (The final examination had a maximum of 100 points.)

Student	1	2	3	4	5	6	7	8	9	10	11	12	13
x	80	91	73	88	62	71	60	89	66	73	69	81	76
y	87	88	80	82	76	84	71	90	82	79	75	86	89

Student	14	15	16	17	18	19	20	21	22	23	24	25
x	78	83	76	91	76	99	99	64	86	63	95	97
y	85	89	85	94	78	95	98	72	94	81	90	98

275. Given that $\sum y = 2128, \sum y^2 = 182.522,$ and $\sum xy = 170.971$, find the standard deviation of the *y*-values about the regression line $\hat{y} = 41.2057 + 0.5528x$,

ANSWER:

$$s_e^2 = \frac{\sum y^2 - b_0 \sum y - b_1 \sum xy}{n-2}$$

= [182.522 – (41.2066)(2128) – (0.5528)(170971] / 23 = 13.982

$s_e = \sqrt{13.982} = 3.793$

276. Calculate a 95% confidence interval for the true value of the slope given that SS(*x*) = 3478.16.

ANSWER:

Since $s_{b_1} = \sqrt{s_e^2 / SS(x)} = \sqrt{13.982/3478.16}$ = 0.0634, then

$b_1 \pm t(n-2) \cdot s_{b_1} = 0.5528 \pm 2.07(0.0634) = 0.5528 \pm 0.1312$.

Hence, the 95% confidence interval for β_1 is 0.4216 to 0.684

277. Test the significance of the slope at $\alpha = 0.05$ using the p-value approach and the classical approach.

ANSWER:

β_1 is the slope of the line of best fit for the population of pre-final averages and final exam. $H_o: \beta_1 = 0$ vs. $H_a: \beta_1 > 0$ (Note: the alternative hypothesis can be either one-tailed or two-tailed. Since the slope is positive, a one-tail test is appropriate). Assume normality for *y* at each *x*. $n = 25$, $df = n - 2 = 23$, $b_1 = 0.5528$, $s_{b_1} = 0.0634$

The test statistic $t^* = (b_1 - \beta_1)/s_{b_1} = (0.5528 - 0) / 0.0634 = 8.719$

The *p*-value approach:

P = $P(t > 8.719 \mid df = 23)$. Using the table of critical values of Student's *t*-distribution, we get **P** < 0.005. Since **P** < α = 0.05; reject H_0. There is sufficient evidence to indicate at the 0.05 level of significance that the slope is significantly greater than zero.

The classical approach:

The critical region is $t \geq 1.71$. Since the test statistic t^* falls in the critical region; we reject H_o. We reach the same conclusion as stated above in the *p*-value approach.

278. Estimate the mean final-exam grade that all students with an 85 pre-final average will obtain (95% confidence interval).

ANSWER:

$\mu_{y|x=85}$ is the mean final exam grade that all students with an 85 pre-final average will obtain. Normality assumed for *y* at each *x*.

$n = 25, \quad x_0 = 85, \quad \bar{x} = \sum x/n = 79.44, \quad s_e = 3.793, \quad \hat{y} = 88.19$

Since $\alpha/2 = 0.025$, $df = 23$; then t(23, 0.025) = 2.07, and

$$E = t(n-2, \alpha/2) \cdot s_e \cdot \sqrt{(1/n) + [(x_0 - \bar{x})^2 / SS(x)]}$$

$$= (2.07)(3.793)\sqrt{(1/25) + [(85 - 79.44)^2 / 3478.16]}$$

$$= (2.07)(3.793)(0.2211) = 1.736$$

Hence, $\hat{y} \pm E = 88.19 \pm 1.736$, and the 95% confidence interval for $\mu_{y|x=85}$ is 86.454 to 89.926.

279. Using the 95% prediction interval, predict the score that Terri will receive on her final, knowing that her pre-final average is 80.

ANSWER:

$y_{x=78}$ is the final exam score that Terri will receive on her final, knowing that her pre-final average is 80. $n = 25, \quad x_0 = 80, \quad \bar{x} = \sum x/n = 79.44, \quad s_e = 3.793, \quad \hat{y} = 85.43$

Since $\alpha/2 = 0.025$, df = 23; then *t*(23, 0.025) = 2.07, and

$$E = t(n-2, \alpha/2) \cdot s_e \cdot \sqrt{1 + (1/n) + [(x_0 - \bar{x})^2 / SS(x)}$$

$$= (2.07)(3.793)\sqrt{1 + (1/25) + [(80 - 79.44)^2 / 3478.16]}$$

$$= (2.07)(3.793)(1.0198) = 8.01$$

Hence, $\hat{y} \pm E = 85.43 \pm 8.01$, and the 95% prediction interval for Terri is 77.42 to 93.44.

280. The data below are for the number of unemployed persons (in millions) and the federal unemployment insurance payments (in billions of dollars) for the years 1978 – 1985. Some economists state that these two variables are positively related.

	Year							
	1978	1979	1980	1981	1982	1983	1984	1985
Federal Unemployment Insurance Payments	11.8	10.7	18.0	19.7	23.7	31.5	18.4	16.8
# Unemployed Persons	6.2	6.1	7.6	8.3	10.7	10.7	8.5	8.3

Use the classical approach at the 0.05 level of significance and a computer to test the null hypothesis that the population slope is zero.

ANSWER:

$H_o : \beta_1 = 0$ (There is no linear relationship) vs. $H_a : \beta_1 \neq 0$ (A linear relationship exists)

The test statistic is $t^* = 6.335$; and the critical regions are: $t \leq -2.45$ or $t \geq 2.45$. Therefore we reject the null hypothesis. There is sufficient evidence at $\alpha = 0.05$ to indicate that a linear relationship exists between the two variables.

281. Sketch a t-curve to determine the critical value(s) and rejection regions that would be used with the classical approach in testing $H_o : \beta_1 = 0$ vs. $H_a : \beta_1 < 0$, with $n = 16$, $\alpha = 0.05$

ANSWER:

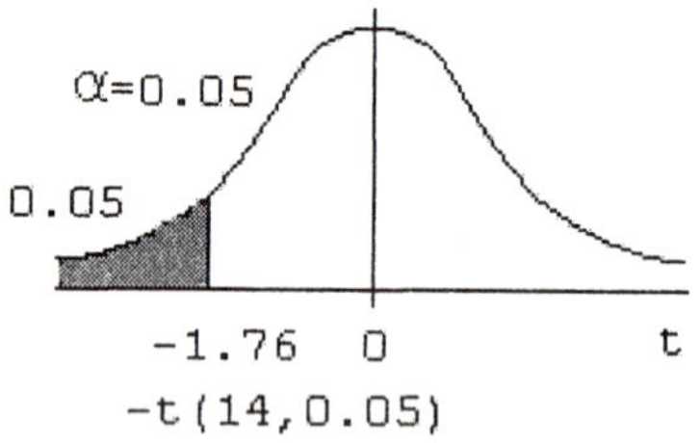

QUESTIONS 282 THROUGH 285 ARE BASED ON THE FOLLOWING INFORMATION:

A company manager is interested in the relationship between x = number of years that an employee has been with the company and y = the employee's annual salary (in thousands of dollars). The following MINITAB output is from a regression analysis for predicting y from x for n = 15 data points.

Predictor	Coef	StDev	*t*-ratio	*p*
Constant	16.8221	0.3887	43.28	0.000
X	0.64983	0.02617	24.83	0.000

s = 0.8081 R-sq = 97.9% R-sq(adj) = 97.8%

282. What is the equation of the line of best fit?

ANSWER:

$\hat{y} = 16.8221 + 0.64983x$

283. What are the estimates of the slope and y - intercept?

ANSWER:
Slope = b_1 = 0.64983, and y - intercept = b_0 = 16.8221

284. Interpret the estimated slope and y- intercept in question 283

ANSWER:
The slope b_1: For each additional year an employee is with this company, his or her salary increases, on average, by $650.
The y-intercept b_0: An employee just starting a job with this company has a starting salary of $16,820.

285. Does a linear relationship exist between x any y? Test using α = 0.05.

ANSWER:
$H_o : \beta = 0$ vs. $H_a : \beta \neq 0$
Since p-value = 0.0 $< \alpha$, reject H_o. There is sufficient evidence to indicate that a linear relationship does exist between x and y.

QUESTIONS 286 THROUGH 294 ARE BASED ON THE FOLLOWING INFORMATION:
An experiment was conducted to study the effect of a new drug in lowering the heart rate in adults. The data collected are shown in the following table.

Drug Dose in mg. (x)	1.75	2.50	0.50	2.00	2.75	2.25	0.75	1.25	1.50	1.00
Heart Rate Reduction (y)	13	17	9	19	20	19	6	11	14	14

286. Draw a scatter diagram of the data.

ANSWER:

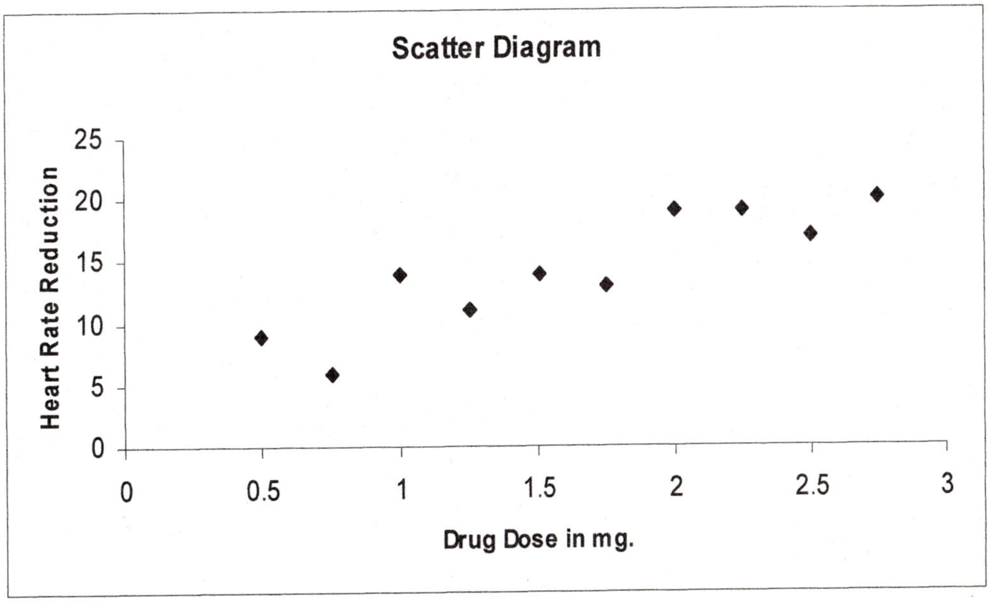

287. Does the scatter diagram suggest a linear relationship between drug dose and heart rate reduction?

ANSWER:
The scatter diagram suggests a positive linear relationship between drug dose and heart rate reduction.

288. Use computer to determine the equation of the line of best fit.

ANSWER:

Regression coefficients		
		Coefficient
	Constant	5.3758
	Drug Dose in mg.	5.4303

The equation of the line of best fit is $\hat{y}$ = 5.3758 + 5.4303*x*.

289. What is the estimated or predicted heart rate reduction for a dose of 2.00 mg?

ANSWER:
$\hat{y}$ = 5.3758 + 5.4303(2) = 16.24

290. Calculate the error sum of squares and SS(*x*).

ANSWER:
SSE = $\sum y^2 - (b_0)(\sum y) - (b_1)(\sum xy)$ = 2210 – (5.3758)(142) – (5.4303)(258.75) = 41.546

$$SS(x) = \sum x^2 - \left(\sum x\right)^2 / n = 31.5625 - (16.25)^2 / 10 = 5.1563$$

291. Find the 95% confidence interval for the mean heart-rate reduction for a dose of 2.00 mg.

ANSWER:
$s_e = \sqrt{SSE/(n-2)} = \sqrt{41.546/8} = 2.2789$ and $t(n-2, \alpha/2)$ = t(8,0.025) = 2.31.

$$E = t(n-2, \alpha/2) \cdot s_e \cdot \sqrt{\frac{1}{n} + \frac{(x_0 - \overline{x})^2}{SS(x)}} = (2.31)(2.2789)\sqrt{\frac{1}{10} + \frac{(2-1.625)^2}{5.1563}} = 1.88$$

The lower and upper confidence limits for the mean heart-rate reduction when x = 2 are,
LCL = $\hat{y}$ - E = 16.24 – 1.88 = 14.36
UCL = $\hat{y}$ + E = 16.24 + 1.88 = 18.12.

292. Find the 95% prediction interval for the heart-rate reduction expected for an individual receiving a dose of 2.00 mg.

ANSWER:

$$E = t(n-2, \alpha/2) \cdot s_e \cdot \sqrt{1 + \frac{1}{n} + \frac{(x_0 - \overline{x})^2}{SS(x)}} = (2.31)(2.2789)\sqrt{1 + \frac{1}{10} + \frac{(2-1.625)^2}{5.1563}} = 5.59$$

$\hat{y} \pm E = 16.24 \pm 5.59$
Thus, 10.65 to 21.83 is the 95% prediction interval for y when x = 2.

293. Use computer software to verify the confidence and prediction intervals found in questions 291 and 292.

ANSWER:
Using Minitab, we obtain the following results:

```
Predicted Values for New Observations

New Obs       Fit    SE Fit         95% CI               95% PI
   1       16.236     0.813   (14.361, 18.111) (10.657, 21.816)


Values of Predictors for New Observations

             Drug Dose in mg.
New Obs
   1             2.00
```

294. Comment on the widths of the two intervals formed in questions 291 and 292.

ANSWER:
The width of the 95% confidence interval = 18.12 – 14.36 = 3.76
The width of the 95% prediction interval = 21.83 – 10.65 = 11.18
It is always the case that the prediction interval for an individual value of *y* is wider than the confidence interval for the mean value of *y*; both calculated at the same value x_0 (2 in our case).

<u>QUESTIONS 295 THROUGH 303 ARE BASED ON THE FOLLOWING INFORMATION:</u>
Some people believe that the height (y) in inches and shoe size (x) are related according to the equation $y = 2x + 50$. A random sample of 30 college students' heights and shoe sizes was taken to test this relationship. The data are shown below:

Shoe Sizes	13	10	9	13	10	8	8.5	9	11	7	10	12	8.5	12	10
Height	75	72	66	73	72	67	69	67	71	62	66	70	66	71	70

Shoe Sizes	8.5	7.5	12	8.5	8.5	9.5	13	13	6	13	8.5	7.5	12	7.5	6.5
Height	66	64	73	67	66	67	74	74	65	77	67	63	69	64	62

295. Construct a scatter diagram of the data, and comment on the visual linear relationship.

ANSWER:
The linear relationship between shoe size and height seems appropriate.

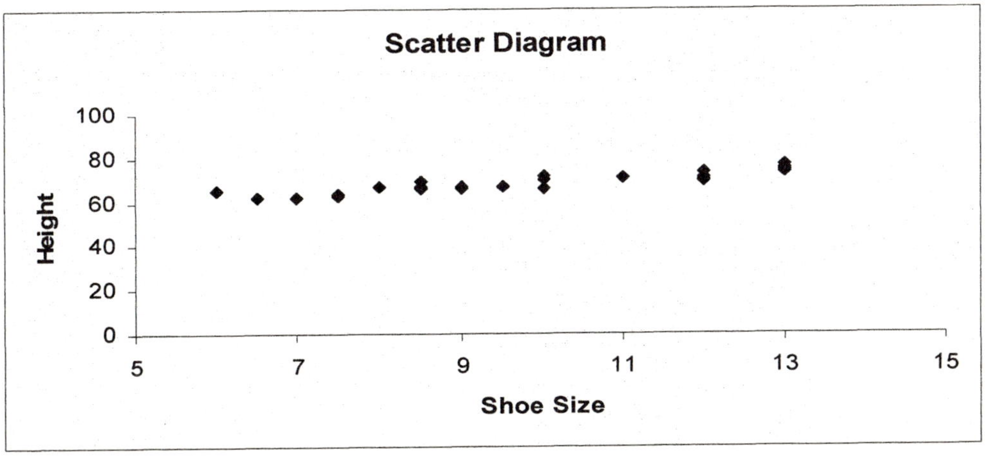

296. Use computer to calculate the correlation coefficient, *r*.

ANSWER:

Table of correlations

	Shoe Size	Height
Shoe Size	1.000	
Height	0.902	1.000

297. Is the population correlation coefficient significant? Test the appropriate hypotheses at the 0.05 level of significance?

ANSWER:

$H_o: \rho = 0$ vs. $H_a: \rho \neq 0$

The value of the test statistic r^* = 0.902
The critical values are found in the "Critical Values of *r* When ρ = 0" table, at the intersection of the *df* = 8 row and the two-tailed 0.05 column of the table. These values are ± 0.632. Since r^* = 0.902 > 0.632, we reject H_o. There is sufficient evidence to indicate that there is a linear dependency between the shoe size and the height of a person in the population from which this sample was drawn.

298. Use computer to calculate the line of best fit.

ANSWER:

Regression coefficients

	Coefficient
Constant	52.1932
Shoe Size	1.6725

The line of best fit is $\hat{y}$ = 52.1932 + 1.6725*x*

299. Compare the slope and *y*-intercept in question 298 to the slope and intercept of the equation *y* = 2*x* + 50. List similarities and differences.

ANSWER:
The results are shown in the table below:

	Original equation: *y* = 2*x* + 50	Line of best fit: $\hat{y}$ = 52.1932 + 1.6725*x*
Slope	2.0	1.6725
y - intercept	50.0	52.1932

While the slope of the original equation is slightly larger than the slope of the line of best fit (2.0 vs. 1.6725), its y-intercept is slightly smaller (50 vs. 52.1932).

300. Using the line of best fit found in question 298, estimate height for a student with a size 10 shoe. Compare results.

ANSWER:

Line of best fit: $\hat{y}$ = 52.1932 + 1.6725 (10) = 68.9182 $\approx$ 69 inches

Original equation: $\hat{y}$ = 2(10) + 50 = 70 inches

The line of best fit provides an excellent estimate compared to the original equation.

301. Use computer to construct the 95% confidence interval for the mean height of all college students with a size 10 shoe using the equation formed in question 298. Is your estimate using $y = 2x + 50$ for a size 10 included in this interval?

ANSWER:

```
Predicted Values for New Observations

New Obs        Fit      SE Fit          95% CI
   1         68.918      0.325     (68.253, 69.584)

Values of Predictors for New Observations

New Obs        Shoe Size
   1             10.0
```

The height for a student with a size 10 shoe is estimated using the equation $y = 2x + 50$ to be 70 inches, which is not included in this confidence interval.

302. Construct the 95% prediction interval for the individual heights of all college students with a size 10 shoe using the equation formed in question 298.

ANSWER:

```
Predicted Values for New Observations

New Obs        Fit      SE Fit          95% PI
   1         68.918      0.325     (65.238, 72.598)

Values of Predictors for New Observations

New Obs        Shoe Size
  1              10.0
```

303. Comment on the widths of the two intervals formed in questions 301 and 302. Explain.

ANSWER:

The width of the 95% confidence interval = 69.584 – 68.253 = 1.331

The width of the 95% prediction interval = 72.598 – 65.238 = 4.346

It is always the case that the prediction interval for an individual value of y is wider than the confidence interval for the mean value of y; both calculated at the same value x_0 (2 in our case).

Chapter 13

304. Comment on the statement "The correlation coefficient has the same sign as the slope of the least squares line fitted to the same data." as sometimes true, always true, or never true. Explain your response if your answer is "sometimes true" or "never true" .

ANSWER:
Always true

305. Explain why a 95% confidence interval for the mean value of y at a particular x is much narrower than a 95% prediction interval for an individual y-value at the same value of x.

ANSWER:
According to Central Limit Theorem, the standard error for $\bar{x}$'s is much smaller than the standard deviation for individual *x*'s. Thus the confidence interval for the mean value of *y* will be narrower than the prediction interval for an individual y-value at the same value of x.

QUESTIONS 306 THROUGH 310 ARE BASED ON THE FOLLOWING INFORMATION:
Consider the following set of bivariate date:

x	5	7	9	11	13
y	10	12	14	16	18

306. Calculate $\sum x$, $\sum y$, $\sum x^2$, $\sum y^2$, and $\sum xy$.

ANSWER:

	x	y	xy	x^2	y^2
	5	10	50	25	100
	7	12	84	49	144
	9	14	126	81	196
	11	16	176	121	256
	13	18	234	169	324
Sum	45	70	670	445	1020

$\sum x = 45$, $\sum y = 70$, $\sum x^2 = 445$, $\sum y^2 = 1020$, and $\sum xy = 670$

307. Calculate SS(*x*) , SS(*y*), and SS(xy).

ANSWER:

$$SS(x) = \sum x^2 - \left(\sum x\right)^2 / n = 445 - (45)^2 / 5 = 40$$

$$SS(y) = \sum y^2 - \left(\sum y\right)^2 / n = 1020 - (70)^2 / 5 = 40$$

$$SS(xy) = \sum xy - \left[\left(\sum x\right)\left(\sum y\right)\right] / n = 670 - [(45)(70)] / 5 = 40$$

308. Calculate the slope for the line of best fit.

ANSWER:
The slope b_1 = SS(*xy*) / SS(*x*) = 40 / 40 = 1.0.

309. Calculate the sample correlation coefficient, r.

ANSWER:
$r = SS(xy)/\sqrt{SS(x)\cdot SS(y)} = (40)/\sqrt{(40)(40)} = 1.0$

310. The sample correlation coefficient, *r*, is related to the slope of the line of best fit, b_1, by the equation $r = b_1 \cdot \sqrt{SS(x)/SS(y)}$. Verify this equation using this data set.

ANSWER:
$b_1 \cdot \sqrt{SS(x)/SS(y)} = 1 \cdot \sqrt{40/40} = 1 = r$

QUESTIONS 311 THROUGH 315 ARE BASED ON THE FOLLOWING INFORMATION:
A scientist is studying the relationship between wind velocity (*x*) and DC output of a windmill (*y*). The following MINITAB output is from a regression analysis for predicting *y* from *x*.

Predictor	Coef	StDev	*t*-ratio	*p*
Constant	-0.1346	0.1803	-0.75	0.470
X	0.28996	0.03050	9.51	0.000

s = 0.2435 R-sq = 88.3% R-sq(adj) = 87.3%

Analysis of Variance

Source	DF	SS	MS	*F*	*p*
Regression	1	5.3606	5.3606	90.40	0.000
Error	12	0.7116	0.0593		
Total	13	6.0721			

311. What is the equation of the line of best fit?

ANSWER:
$\hat{y}$ = -0.1346 + 0.28996*x*

312. Predict the DC output for a wind velocity of 22 mph.

ANSWER:
$\hat{y}$ = -0.1346 + 0.28996 (22) = 6.2445

313. What is the value of the residual sum of squares?

ANSWER:
Residual sum of squares = Error sum of squares = 0.7116

Chapter 13

314. One of the assumptions about the random error ε in the regression model is that the values of ε have a common variance equal to σ^2. What is the best estimator of σ?

ANSWER:

$s = \sqrt{\text{MSE}} = \sqrt{0.0593} = 0.2435$

315. Does a linear relationship exist between *x* any *y*? Test using $\alpha = 0.05$.

ANSWER:

$H_o: \beta = 0$ vs. $H_a: \beta \neq 0$ Since *p*-value = 0.0 $< \alpha$, reject H_o. There is sufficient evidence to indicate that a linear relationship does exist between *x* and *y*.

<u>QUESTIONS 316 THROUGH 320 ARE BASED ON THE FOLLOWING INFORMATION:</u>

A scientist is studying the relationship between *x* = inches of annual rainfall and *y* = inches of shoreline erosion. One study reported the following data. Use the following MINITAB output to answer the questions below.

x	30	25	90	60	50	35	75	110	45	80
y	0.3	0.2	5.0	3.0	2.0	0.5	4.0	6.0	1.5	4.0

The regression equation is *y* = - 1.7359 + 0.0731 *x*.

Predictor	Coef	StDev	*t*-ratio	*p*
Constant	-1.7359	0.1882	-9.22	0.000
x	0.073099	0.002867	25.50	0.000

s = 0.2416 R-sq = 98.8% R-sq(adj) = 98.6%

Analysis of Variance

Source	DF	SS	MS	*F*	*p*
Regression	1	37.938	37.938	650.14	0.000
Error	8	0.467	0.058		
Total	9	38.405			

316. Construct a scatter diagram of the data, display the estimated regression line on the graph, and comment on the visual linear relationship.

ANSWER:

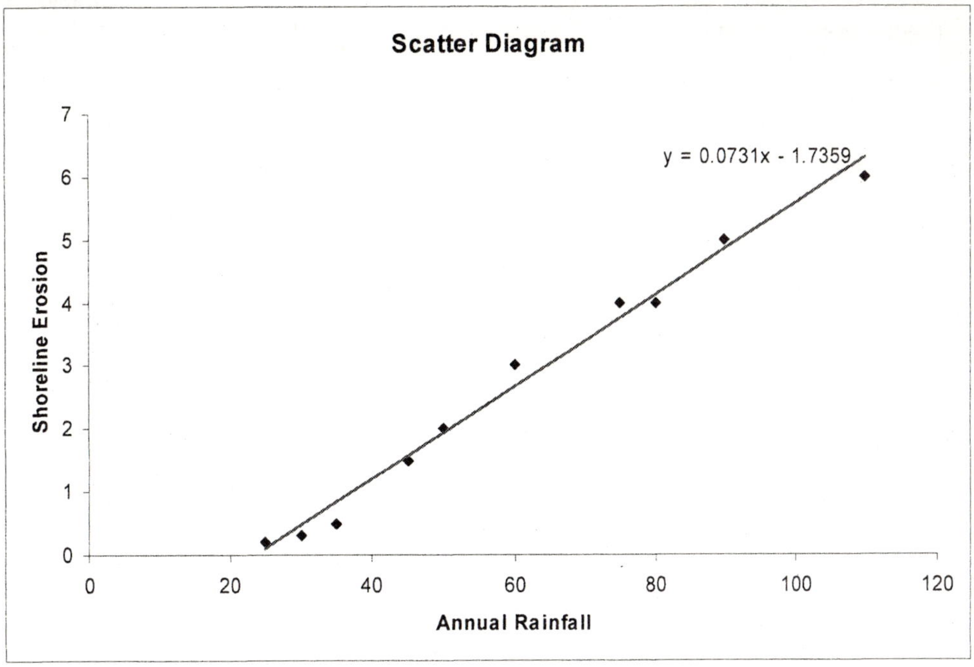

A linear relationship between inches of rainfall and inches of shoreline erosion seems appropriate.

317. Specify the *y*-intercept and the slope of the estimated regression line?

ANSWER:
The estimated regression line is $\hat{y}$ = -1.7359 + 0.0731*x*.. Its y-intercept b_0 = -1.7359 and slope b_1 = 0.0731.

318. Interpret the estimated slope of the regression line in question 317.

ANSWER:
The slope b_1 = 0.0731. This means that for each additional inch of annual rainfall, the shoreline erodes, on average, by 0.0731 inch,

319. If we wish to test the usefulness of the simple linear regression model for predicting shoreline erosion from a given amount rainfall, what are the appropriate null and alternative hypotheses?

ANSWER:
$H_o : \beta = 0$ vs. $H_a : \beta \neq 0$

320. Test the hypotheses in question 319 at α = 0.05 using the *p*-value approach.

ANSWER:
Since *p*-value = 0.0 < α , reject H_o . There is sufficient evidence to indicate that a linear relationship does exist between *x* and *y*. That is, the simple linear regression model is useful for predicting erosion from a given amount of rainfall

Chapter 14

Elements of Nonparametric Statistics

Sections 14.1 through 14.3

True-False Questions

1. One of the advantages that the nonparametric tests have is the necessity for less restrictive assumptions.
ANSWER: T

2. If a tie occurs in a set of ranked data, the data that form the tie are removed from the set.
ANSWER: F

3. The confidence level of a statistical hypothesis test is measured by $1-\beta$.
ANSWER: F

4. The efficiency of a nonparametric test is the probability that a false null hypothesis is rejected.
ANSWER: F

5. Distribution-free or nonparametric methods provide test statistics for an unspecified distribution.
ANSWER: T

6. When choosing between parametric and nonparametric tests, we are interested primarily in the control of error, the relative power of the test, and efficiency.
ANSWER: T

7. The Runs test is the nonparametric counterpart of the parametric *t*-test for two dependent means.
ANSWER: F

8. The power of a test $1 - \beta$, is the probability that we reject the null hypothesis when we should have rejected it. If two tests with the same level of significance α are equal candidates for use, then the one with the greater power is the one you would want to choose.
ANSWER: T

9. The nonparametric methods, or distribution-free methods as they are also known, do not depend on the distribution of the population being sampled, but they depend on the distribution of the sample itself.
ANSWER: F

10. While nonparametric methods require few assumptions about the parent population, they are generally harder to apply than their parametric counterparts.
ANSWER: F

11. Nonparametric methods can be used in situations where parametric methods cannot be used.
ANSWER: T

12. Suppose that you set the levels of risk you can tolerate for Type I and Type II error at α and β, respectively, and then you are able to determine the sample size it would take to meet your specified challenge. The test that required the larger sample size would seem to have the edge, since it would be more efficient.
ANSWER: F

13. When we compare two or more tests, they must be equally qualified for use. That is, each test has a set of assumptions that must be satisfied before it can be applied.
ANSWER: T

14. The nonparametric methods are also known as distribution-free methods.
ANSWER: T

15. If two tests with the same level of significance α are equal candidates for use, then the one with the smaller probability of Type I error is the one you would want to choose.
ANSWER: F

16. Efficiency is the ratio of the sample size of the best parametric test to the sample size of the best nonparametric test when compared under a fixed set of risk values.
ANSWER: T

Multiple-Choice Questions

17. Which one of the following statements is correct in describing the nonparametric tests?

A) The nonparametric methods require us to make assumptions about the distribution of the population from which the measurements come.
B) The underlying probability theory for the nonparametric method is often a binomial distribution.
C) The nonparametric methods generally use difficult to calculate test statistics.
D) The z is used as a test statistic for most nonparametric tests because we need to assume that the variable is normally distributed.

ANSWER: B

18. Nonparametric methods depend on:

A) the distribution of the population being sampled.
B) more confining restrictions than their parametric counterparts.
C) many assumptions about the parent population.
D) None of the above is correct.
ANSWER: D

19. Which one of the following statements is incorrect about the comparison of parametric and nonparametric statistical methods?

A) The efficiency of a nonparametric test is the ratio of the sample size of the best parametric test to the sample size of the nonparametric test.
B) If a set of sample data is such that it can be analyzed by using either a parametric or a nonparametric method, the parametric method is the better choice.
C) There are nonparametric methods for which there is no parametric counterpart.
D) Some nonparametric methods use a z-test statistic. When this occurs, the sample data is from a normal population.
ANSWER: D

20. When trying to control the risk of error and two tests are equal candidates, we should select the one with

A) no error.
B) lowest efficiency.
C) greatest power.
D) most calculations.
ANSWER: C

21. Which one of the following is *not* a nonparametric test?

A) Chi-square test of independence
B) The Mann-Whitney *U* test
C) The sign test
D) The Runs test
ANSWER: A

22. Which of the following statements is true?

A) Nonparametric methods can be applied to a wider variety of problems because they have more rigid requirements than parametric methods.
B) Unlike parametric methods, nonparametric methods cannot be applied to nominal data that lack numeric values.
C) Nonparametric methods can be applied to a wider variety of problems because they have less rigid requirements than parametric methods.
D) None of the above is true.
ANSWER: C

23. Nonparametric tests can be appropriate when:

A) one or more of the assumptions underlying a particular parametric statistical test has been violated.
B) the sample size is very large.
C) the underlying population can be assumed to be normally distributed.
D) all assumptions for a particular parametric statistical test have been met.
ANSWER: A

24. When choosing between parametric and nonparametric tests, we are interested primarily in

A) the control of error.
B) the relative power of the test.
C) efficiency.
D) All of the above.
ANSWER: D

25. Which of the following tests would be an example of nonparametric method?

A) The sign test
B) The Mann-Whitney *U* test
C) The runs test
D) All of the above.
ANSWER: D

26. Which of the following statements is false?

A) Nonparametric methods provide test statistics for an unspecified distribution.
B) When choosing between parametric and nonparametric tests, we are interested primarily in the control of error, the relative power of the test, and efficiency.
C) For a statistical procedure to be parametric, either we assume that the parent population is at least approximately normally distributed or we rely on the central limit theorem to give us a normal approximation.
D) None of the above.
ANSWER: D

27. The power of a test is the probability that we

A) reject the null hypothesis when it is false.
B) reject the null hypothesis when it is true.
C) fail to reject the null hypothesis when it is false.
D) fail to reject the null hypothesis when it is true.
ANSWER: A

28. Which of the following statements is true regarding efficiency?

A) It is the ratio of the sample size of the best nonparametric test to the sample size of the best parametric test when compared under a fixed set of risk values.
B) It is the ratio of the sample size of the best parametric test to the sample size of the best nonparametric test when compared under a fixed set of risk values.
C) It is the sum of the sample size of the best nonparametric test and the sample size of the best nonparametric test when compared under a fixed set of risk values.
D) It is the difference of the sample size of the best parametric test and the sample size of the best nonparametric test when compared under a fixed set of risk values.

ANSWER: **B**

29. Which of the following statements is false?

A) The risk associated with a Type I error is controlled directly by the level of significance α .
B) *P*(Type I error) = α
C) *P*(Type II error) = β .
D) It is α , not β , that we must control.

ANSWER: **D**

Applied and Computational Questions

30. The efficiency rating for the sign test is approximately 0.63. What does this mean?

ANSWER:
This means that a sample of size 63 with a parametric test will do the same job as a sample of size 100 will do with the sign test.

31. The power and the efficiency of a test cannot be used alone to determine the choice of a test. Explain in detail.

ANSWER:
Sometimes you will be forced to use a certain test because of the data you are given. When there is a decision to be made, the final decision rests in a trade-off of three factors: (1) the power of the test, (2) the efficiency of the test, and (3) the data (and the number of data) available.

32. Briefly discuss the reasons for the recent popularity of nonparametric statistics.

ANSWER:
a) Nonparametric methods require few assumptions about the parent population.
b) Nonparametric methods are generally easier to apply than their parametric counterparts.
c) Nonparametric methods are relatively easy to understand.
d) Nonparametric methods can be used in situations where the normality assumptions cannot be made.
e) Nonparametric methods are generally only slightly less efficient than their parametric counterparts.

33. Explain why nonparametric methods are also called distribution-free methods.

ANSWER:
Nonparametric methods do not depend on the distribution of the population being sampled. This is why they are called distribution-free methods.

34. What two factors influence our decision as to the "best" test?

ANSWER:
The two factors are the ability to control the risk of errors and the sample size required.

35. The efficiency of a particular nonparametric test is 0.82. For a fixed set of risk values, the sample size of the best nonparametric test is 50. Find the sample size of the parametric test.

ANSWER:
The sample size of the parametric test is 41.

Section 14.4

True-False Questions

36. The sign test is a versatile and an exceptionally easy-to-apply nonparametric method that uses only plus and minus signs.
ANSWER: T

37. The sign test can be used when the null hypothesis to be tested concerns the value of the population median.
ANSWER: T

38. The sign test is always a two-tailed test.
ANSWER: F

39. The sign test may be applied to a hypothesis test dealing with the median difference between independent data that result from two independent samples.
ANSWER: F

40. Two dependent means can be compared nonparametrically by using the sign test.
ANSWER: T

41. The sign test is a possible alternative to the Student's t - test for one mean value.
ANSWER: T

42. The sign test is a possible replacement for the F-test.
ANSWER: F

43. The sign test can be used to test the randomness of a set of data.
ANSWER: F

44. The sign test can be used in a hypothesis test concerning the median difference (paired difference) for two dependent samples.
ANSWER: T

45. In the sign test, if the observed value of the less frequent sign is larger than the critical value k displayed in the "Critical Values of the Sign Test" table available in your textbook, we reject H_o.
ANSWER: F

46. The sign test is the nonparametric alternative to the t-test used for one mean.
ANSWER: T

47. The sign test may be either one- or two-tailed test.
ANSWER: T

48. The sign test is always a one-tailed test.
ANSWER: F

49. The sign test can be applied to obtain a single-sample confidence interval for the unknown population mean μ.
ANSWER: F

50. The sign test can be applied to obtain a single-sample confidence interval for the unknown population median *M*.
ANSWER: T

Multiple-Choice Questions

51. The sign test is a nonparametric procedure for testing whether two populations have identical

A) means
B) medians
C) variance
D) Interquartile ranges
ANSWER: B

52. Which one of the following is a *dis*advantage of the sign test?

A) Tied pairs are not considered in the analysis.
B) Only the signs of the differences and not the actual values are used in the analysis.
C) Its inability to cope with small samples.
D) None of the above.
ANSWER: B

53. Which of the following statements is false regarding the sign test?

A) It is a versatile and exceptionally easy-to-apply nonparametric method that uses only plus and minus signs.
B) It can be used to construct confidence interval for the median of one population
C) It can be used in a hypothesis test concerning the value of the variance for one population.
D) It can be used in a hypothesis test concerning the median difference (paired difference) for two dependent samples.
ANSWER: C

54. Which of the following statements is false?

A) In the sign test, reject the null hypothesis whenever the number of the less frequent sign is extremely small.
B) If the number of the less frequent sign is less than or equal to the critical value *k* in the "Critical Values of the Sign Test" table available in your textbook, we will reject H_o.
C) If the observed value of the less frequent sign is larger than the critical value *k* in the "Critical Values of the Sign Test" table available in your textbook, we will fail to reject H_o.
D) In the sign test, reject the null hypothesis whenever the number of the less frequent sign is extremely large.
ANSWER: D

55. Which of the following statements is false regarding the sign test?

A) It is the nonparametric alternative to the t-test used for one mean
B) It is the nonparametric alternative to the z-test used for one proportion
C) It is the nonparametric alternative to the t-test used for the difference between two dependent means.
D) All of the above.

ANSWER: **B**

56. Which of the following statements is not always true regarding the sign test?

A) It can be used when the null hypothesis to be tested concerns the value of the population median.
B) It may be either one- or two-tailed test.
C) It uses only the plus and minus signs; therefore, the zeros are discarded and the usable sample size is adjusted accordingly.
D) Its test statistic is the number of the (+) signs; that is, $n(+)$.

ANSWER: **D**

57. Which of the following statements is false?

A) The sign test may be carried out by means of a normal approximation using the standard normal variable z.
B) The normal approximation to the sign test will be used if the "Critical Values of the Sign Test" table available in your textbook does not show the particular levels of significance desired or if n is large.
C) The sign test may be the easiest test procedure of all nonparametric tests to use.
D) None of the above.

ANSWER: **D**

Short-Answer Questions

58. What non-parametric test can be used in place of either the one mean t - test or the two dependent means t - test?

ANSWER:
The sign test

59. Explain how the sign test is based on the binomial distribution and is often approximated by the normal distribution?

ANSWER:
The sign test is a binomial experiment of n trials (the n data observations) with two outcomes for each data [(+) or (–)], and p = (+) = 0.5. The variable x is the number of the least frequent sign.

60. Why does the sign test use a null hypothesis about the median instead of the mean like a t - test uses?

ANSWER:
The median is the middle value such that 50% of the distribution is larger in value and 50% is smaller in value.

Chapter 14

61. A restaurant has collected data on which of two seating arrangements (*A* and *B*) its customers prefer. In a sign test to determine which one seating arrangement is significantly preferred, the null hypothesis would be: (a) $M = 0$, (b) $M = 0.5$, (c) $p = 0$, or (d) $p = 0.5$. Explain your choice.

ANSWER:
The right choice is (d); $p = P(+) = P$(prefer seating arrangement (*A*) = 0.5.

62. State the null hypothesis H_o, and the alternative hypothesis H_a, that would be used to test the following statement: "There is no change in weight from weight-in until after three weeks of the aerobic exercises".

ANSWER:
$H_o : P(+\text{ gain}) = 0.5 \text{ vs. } H_a : P(+\text{ gain}) \neq 0.5$

63. Briefly discuss the assumptions for inferences about the population median using the sign test.

ANSWER:
(a) The *n* random observations that form the sample are selected independently.
(b) The population is continuous in the vicinity of the median *M*.

64. State the null hypothesis H_o, and the alternative hypothesis H_a, that would be used to test the following statement: "The median tax rate is 5%".

ANSWER:
$H_o : \text{Median tax rate} = 0.05 \text{ vs. } H_a : \text{Median tax rate} \neq 0.05$

65. Briefly discuss the assumptions for inferences about the median of paired differences using the sign test.

ANSWER:
(a) The paired data are selected independently.
(b) The variables are ordinal or numerical.

66. State the null hypothesis H_o, and the alternative hypothesis H_a, that would be used to test the following statement: "The median length of vacation time taken by university administrators is less than 21 days per academic year."

ANSWER:
$H_o : \text{Median} = 21\ (\geq) \text{ vs. } H_a : \text{Median} < 21$

67. What advantages do nonparametric statistics have over parametric methods?

ANSWER:
The nonparametric statistics do not require assumptions about the distribution of the variable.

68. Explain why a nonparametric test is not as sensitive to an extreme datum as a parametric test might be.

ANSWER:
The extreme value in a set of data can have a sizeable effect on the mean and standard deviation in the parametric methods. The nonparametric methods typically use rank numbers. The extreme value with ranks is either 1 or *n*, and neither changes if the value is more extreme.

Applied and Computational Questions

69. A computer center claims that the median downtime for its large mainframe computer is 45 minutes. A random sample of 30 downtimes for this computer revealed that 17 exceeded 45 minutes, 3 equaled 45 minutes, and 10 were less than 45 minutes. Give the critical region, test statistic and conclusion for testing H_o: $M = 45$ vs. H_a: $M \neq 45$ at α = 0.05.

ANSWER:
Critical region: $x \leq 7$
Test statistic: $x = 10$
Conclusion: Fail to reject the null hypothesis

QUESTIONS 70 THROUGH 72 ARE BASED ON THE FOLLOWING INFORMATION:
A blood bank claims that the median usage for red blood cells in a liver transplant is 15 units. A random sample of 34 transplants revealed that 16 exceeded 15 units, 4 equaled 15 units, and 14 were less than 15 units.

70. State the null and alternative hypotheses.

ANSWER:
$H_o: M = 15$ vs. $H_a: M \neq 15$

71. If testing the claim, what would be the test statistic, critical region, and conclusion at α = 0.05?

ANSWER:
Test statistic: $x^* = n(-) = 14$,
Critical region for α = 0.05 and $n = n(-) + n(+) = 30$ is $x \leq 9$
Conclusion: Unable to reject the null hypothesis

72. Estimate the population median with a 95% confidence interval.

ANSWER:
$x_{10} < M < x_{24}$

73. A marketing company conducted a taste preference test for a new brand of peanut butter. Customers were asked to compare crunchy versus creamy. Seventy chose crunchy over creamy, twenty-five chose creamy over crunchy, and five said they were equally good. Give the critical region, x, and conclusion for testing that there is a difference in preference using α = 0.05.

ANSWER:

Critical region: $x \le 37$; and test statistics is $x^* = 25$.
Conclusion: Chunky is preferred over creamy.

74. A tire manufacturer claims that the median mileage for their Eagle tire is 40,000 miles. A consumer agency wishes to test H_o: M = 40,000 vs. H_a: M < 40,000 at α = 0.05. When the agency tested 100 such tires, sixty-five gave less than 40,000 miles of wear and thirty-five gave more than 40,000. A minus sign was recorded if a tire gave less than 40,000 miles and a plus was recorded if it gave 40,000 or more. Give the critical region, z^*, and the conclusion.

ANSWER:

Critical region: $z < -1.65$, and test statistic is $z^* = -29$.
Conclusion: Reject the manufacturer's claim

75. The median test score for a computer science exam is 15.5. Use the following sample of test scores at a given university to test that the median score at the university is different from the median obtained at several university across the country. Test at α = 0.05.

17	13	20	12	14	16	16	18	12	19
16	10	10	20	14	8	19	19	16	12
16	12	21	17	16	17	14	16	12	9

ANSWER:

Critical region: $x \le 9$, and test statistic is $x^* = 13$.
Conclusion: Sample evidence does not indicate any difference from the national norm.

<u>QUESTIONS 76 AND 77 ARE BASED ON THE FOLLOWING INFORMATION:</u>
The table of "critical values of the sign test" indicates that for n = 10 and a two-tailed test, the critical region for the sign test is $x \le 1$ if $\alpha = 0.05$.

76. Verify this by finding $P(x \le 1) + P(x \ge 9)$ for a binomial distribution with n =10 and p = 0.5.

ANSWER:
$P[(x \le 1) + (x \ge 9)] = 0.001 + 0.010 + 0.010 + 0.001 = 0.022$

77. Further verify it by finding $P(x \le 2) + P(x \ge 8)$.

ANSWER:
$P[(x \le 2) + P(x \ge 8)] = 0.001 + 0.010 + 0.044 + 0.044 + 0.010 + 0.001 = 0.11$

QUESTIONS 78 THROUGH 80 ARE BASED ON THE FOLLOWING INFORMATION:
A reading speed and comprehension test was given to a random sample of 10 individuals before and after a reading course. The scores are given below:

Before	82	76	94	62	70	81	90	68	77	80
After	78	82	94	69	75	79	91	65	85	90

The claim being tested is that scores will improve after the course. Use α = 0.05.

78. State the null and alternative hypotheses.

ANSWER:
$H_o : M = 0\ (\leq), H_a : M > 0$

79. Give the critical region, and the value of the test statistic.

ANSWER:
Critical region: $x \leq 1$, and test statistics is $x^* = n(-) = 3$.

80. State the decision, and conclusion.

ANSWER:
Fail to reject H_o, and conclude that the scores did not improve after the course.

81. The Computer Anxiety Index (CAIN) was administered to one hundred and fifty students in a statistics course that utilizes statistical packages. The CAIN was given at the beginning and the end of the course. Ninety showed a reduction in computer anxiety, ten showed no change, and fifty showed an increase. Test the null hypothesis of no change versus the hypothesis that anxiety was reduced at α = 0.05. Use the sign test and give the critical region, z^*, and conclusion.

ANSWER:
Critical region: $z \leq -1.65$, and test statistic is $z^* = -3.30$
Conclusion: Reject the null hypothesis and conclude that Computer Anxiety was reduced.

82. Estimate the population median with a 0.95 confidence interval for a given set of 100 pieces of ordered data: $x_1, x_2, \ldots, x_{100}$.

ANSWER:
(x_{40} to x_{61})

83. The following sample data represents the percent of yearly income spent on entertainment for 16 residents selected from various care centers. Find a 95% confidence interval on the median percent spent on entertainment for all such residents.

12.1	13.3	13.2	12.1	12.2	12.5	12.5	12.5
13.1	13.0	12.6	12.6	13.0	12.7	12.8	12.8

ANSWER:
$12.5\% < M < 13.0\%$

84. The following diastolic blood pressures were obtained from 40 females who were over 60 years in age. Find a 95% confidence interval for the median diastolic reading for the population of females who are over 60.

72	72	74	74	75	75	75	77	77	78
79	79	80	80	80	81	81	84	84	85
85	86	86	86	87	88	90	90	90	90
90	94	94	94	96	96	98	100	104	106

ANSWER:
80 to 90

85. The ages (rounded to the nearest year) for a sample of 30 students in the evening school at a particular college are shown below:

25	30	32	41	27	34	47	31	28	24	23
40	37	35	40	29	25	23	22	30	48	21
21	34	35	32	28	50	25	32			

Find a 90% confidence interval on the median age of all students in the evening school.

ANSWER:
(28, 34)

QUESTIONS 86 THROUGH 88 ARE BASED ON THE FOLLOWING INFORMATION:
The following daily highs were recorded in the city of Chicago on 20 randomly selected December days.

32	21	25	25	31	27	22	44	39	18
49	32	34	36	38	40	30	28	36	38

86. Use the sign test to determine the 95% confidence interval for the median daily high temperature in Chicago during December.

ANSWER:
Ranked data:

18	21	22	25	25	27	28	30	31	32
32	34	36	36	38	38	39	40	44	49

For n = 20 and 1 - α = 0.95, the critical value from the table of critical values of the sign test is k = 5. Then, $x_{k+1} = x_6 = 27$ and $s_{n-k} = x_{15} = 38$. Hence, 27 to 38 is the 95% confidence interval for the unknown population median M.

87. Use the sign test to test at $\alpha = 0.05$ the hypothesis that the median daily high temperature in the city of Chicago during the month of December is 40 degrees, using the p-value approach.

ANSWER:

$H_o : \text{Median} = 40$ vs. $H_a : \text{Median} \neq 40$

Assume random sample. Temperature is a continuous variable.
Let $x = n$ (least frequent sign), and + = temperature above 40.

Now n (+) = 2, n (0) = 1, and n (-) = 17. Hence, $x = n$ (+) = 2.
The usable sample size is $n = n$ (+) + n (-) = 2 + 17 = 19.
The p-value approach: $\boldsymbol{P} = 2P(x \leq 2 \mid n = 19)$;
Using the table of critical values of the sign test: $\boldsymbol{P} \approx 0.01$. Since $\boldsymbol{P} < \alpha$; reject H_o.

88. Use the sign test to test at $\alpha = 0.05$ the hypothesis that the median daily high temperature in the city of Chicago during the month of December is 40 degrees, using the classical approach.

ANSWER:
The classical approach: Critical region: n (least freq sign)≤ 4; $x^* = 2$ is in the critical region; therefore we reject H_o at $\alpha = 0.05$, and conclude that the median temperature in the city of Chicago during the month of December is significantly different from 40.

QUESTIONS 89 THROUGH 92 ARE BASED ON THE FOLLOWING INFORMATION:
A blind taste test was used to determine people's preference for the taste of the "classic" cola and "new" cola. The results showed that 710 preferred the new, 650 preferred the old, and 300 had no preference.

89. A blind taster wishes to test if the preference for the taste of the new cola significantly greater than one-half, state the null and alternative hypothesis.

ANSWER:
H_o : There is no preference; $p = P(\text{prefer}) = 0.5$.

H_a : There is a preference for the new, $p > 0.5$.

90. Calculate the appropriate value of the test statistic.

ANSWER:
x is binomial random variable and approximately normal. Let + = prefer new; then $n(+)$ = 710, (0) = 300, and $n(-)$ = 650; The usable sample size is $n = n$ (+) + n (-) = 710 + 650 = 1360; Since $x = n(+)$ = 710; then, $x' = x - 0.50$ = 709.5. The value of the test statistic is: $z^* = [x' - (n/2)]/(\sqrt{n}/2)$ = [709.5 – (1360/2)] / ($\sqrt{1360}/2$) = 1.60.

91. Test the hypothesis in question 89 at the 0.01 level of significance using the p-value approach.

ANSWER:
The p-value approach: **P** = P(z > 1.60) = 0.5000 – 0.4452 = 0.0548. Since **P** > α = .01; fail to reject H_o.

92. Test the hypothesis in question 89 at the 0.01 level of significance using the classical approach.

ANSWER:
The classical approach: Critical region: $z \geq 2.33$; The test statistic is not in the critical region; therefore we fail to reject H_o. The evidence does not allow us to conclude that there is a significant preference for the new cola.

Chapter 14

QUESTIONS 93 THROUGH 100 ARE BASED ON THE FOLLOWING INFORMATION:
A sample of 32 students received the following grades in an organic chemistry examination.

44	45	51	49	53	57	54	45	54	53	48
45	35	48	46	59	58	50	48	54	63	47
60	60	50	31	44	45	57	51	50	35	

93. If you wish to determine whether this sample show that the median score for the exam differs from 50. What are the null and alternative hypotheses?

ANSWER:
Using the Sign Test (one median):
H_o : Median score on exam = 50 vs. H_a : Median score on exam $\neq 50$

94. Calculate the appropriate value of the test statistic for testing the hypothesis in question 93.

ANSWER:
Assume sample is random. Exam score is continuous random variable.
$x = n$ (least frequent sign)
Let + = above 50, then n (+) = 14, n (0) = 3, and n (-) = 15.
The usable sample size is $n = n$ (+) + n (-) = + = 29.
The value of the test statistic is: $x = n$ (+) = 14.

95. Test the hypothesis in question 93 at $\alpha = 0.05$ using the p-value approach.

ANSWER:
P = p-value = $2P(x \leq 14 \mid n = 29)$. Using the table of critical values of the sign test, we get **P** > 0.25. Since **P** > α ; fail to reject H_o. The sample evidence is not sufficient to justify the claim that median score is different from 50.

96. Test the hypothesis in question 93 at $\alpha = 0.05$ using the classical approach.

ANSWER:
The critical region: n (least freq sign) $\leq$ 8. The test statistic is not in the critical region; therefore we fail to reject H_o. We reach the same conclusion as stated in question 95.

97. If you wish to determine whether this sample shows that the median score for the exam is less than 50, what are the null and alternative hypotheses?

ANSWER:
H_o: Median score on exam = 50 ($\geq$) vs. H_a : Median score on exam < 50.

98. Calculate the appropriate value of the test statistic for testing the hypothesis in question 97.

ANSWER:
Assume sample is random. Exam score is continuous random variable.
$x = n$ (least frequent sign)
Let + = above 50 or equal to 50, then $n(+) = 17$, $n(-) = 15$, and $n = 32$.
The value of the test statistic is: $x = n\,(-) = 15$.

99. Test the hypothesis in question 97 at α = 0.05 using the *p*-value approach.

ANSWER:
P = *p*-value = $P(x \leq 15 \mid n = 32)$. Using the table of critical values of the sign test we get **P** > 0.0125. Since **P** > α, we fail to reject H_o. The sample evidence is not sufficient to justify the claim that the median score is less than 50.

100. Test the hypothesis in question 97 at α = 0.05 using the classical approach.

ANSWER:
The critical region is: *n* (least freq sign) $\leq$ 10; Since the test statistic is not in the critical region; we fail to reject H_o. We reach the same conclusion as stated in question 99.

101. Suppose that we have 15 pieces of data in ascending order ($x_1, x_2, x_3, \ldots\ldots, x_{15}$). Explain how to form a 90% confidence interval for the population median *M*.

ANSWER:
The "Critical Values of the Sign Test" available in your textbook shows a critical value of 3 (*k* = 3) for *n* = 15 and α = 0.10 for a hypothesis test. This means that we drop the last three values on each end (x_1, x_2, and x_3 on the left; x_{13}, x_{14}, and x_{15} on the right). The confidence interval is bounded by x_4 and x_{12}, inclusively. That is, the 90% confidence interval is x_4 to x_{12} and is expressed as: x_4 to x_{12}, 90% confidence interval for *M*.

102. Ten randomly selected college students were each asked how many hours of television they watched last week. The results are: 22, 6, 30, 24, 15, 28, 20, 34, 50, and 31. Determine the 90% confidence interval estimate for the median number of hours of television watched per week by college students.

ANSWER:

Ranked data: 6 [15] 20 22 24 28 30 31 [34] 50.

For *n* = 10 and α = 0.10, the "Critical Values of the Sign Test" available in your textbook implies that *k* = 1. Then $x_{k+1} = x_2 = 15$ and $x_{n-k} = x_9 = 34$. Therefore the 90% confidence interval for the median is 15 to 34.

103. State the null hypothesis H_o, and the alternative hypothesis H_a, that would be used to test the following statement: "The median value is at least 40."

ANSWER:
H_o : Median = 40 ($\geq$) vs. H_a : Median < 40

104. State the null hypothesis, H_o, and the alternative hypothesis, H_a, that would be used to test the following statement: "People prefer the taste of the French bread made of wheat".

ANSWER:
$H_o : P(\text{prefer wheat}) = 0.5\ (\leq)$ vs. $H_a : (\text{prefer wheat}) < 0.5$

105. Using the classical approach and the sign test, determine the critical value that would be used to test H_o: $P(+) = 0.5$ vs. H_a: $P(+) \neq 0.5$, with $n = 20$, $\alpha = 0.05$.

ANSWER:

$x = n(\text{least frequent sign}) \leq 5$

106. Using the classical approach and the sign test, determine the critical value that would be used to test H_o: $P(+) = 0.5$ vs. H_a: $P(+) > 0.5$, with $n = 80$, $\alpha = 0.025$.

ANSWER:
$x = n(-) \leq 30$

107. Using the classical approach and the sign test, determine the critical value that would be used to test H_o: $P(+) = 0.5$ vs. H_a: $P(+) \neq 0.5$ with $n = 175$, $\alpha = 0.10$.

ANSWER:
$z \leq -1.645$ or $z \geq +1.645$

108. Using the classical approach and the sign test, determine the critical value that would be used to test H_o: $P(+) = 0.5$ vs. H_a: $P(+) \neq 0.5$ with $n = 100$, $\alpha = 0.05$.

ANSWER:

$x = n(\text{least frequent sign}) \leq 39$

109. Using the classical approach and the sign test, determine the critical value that would be used to test H_o: $P(+) = 0.5$ vs. H_a: $P(+) < 0.5$, with $n = 40$, $\alpha = 0.05$.

ANSWER:
$x = n(+) \leq 14$

110. Using the classical approach and the sign test, determine the critical value that would be used to test H_o: $P(+) = 0.5$ vs. H_a: $P(+) \neq 0.5$ with $n = 160$, $\alpha = 0.05$.

ANSWER:
$z \leq -1.96$ or $z \geq +1.96$

111. Using the critical values of the sign test available in your textbook, determine the *p*-value that would be used to test H_o: Median = 25 vs. H_a: Median $\neq 25$, given that $n(+) = 34$, $n(0) = 0$, $n(-) = 56$, and $\alpha = 0.05$. State your decision.

ANSWER:
$x = n(\text{least frequent sign}) = n(+0) = 34$, and $n = n(+) + n(-) = 90$
P = *p*-value = $2P(x \leq 34 \mid n = 90) \Rightarrow 0.01 < \mathbf{P} < 0.05$. Since *p*-value $< \alpha = 0.05$, we reject H_o.

112. Using the critical values of the sign test available in your textbook, determine the p-value that would be used to test H_o: Median = 24 ($\geq$) vs. H_a: Median < 24, given that $n(+) = 17$, $n(0) = 2$, $n(-) = 28$, and $\alpha = 0.05$. State your decision.

ANSWER:
$x = n$(least frequent sign)= $n(+0) = 17$, and $n = n(+) + n(-) = 45$
P = p-value = $P(x \leq 17 \mid n = 45) \Rightarrow 0.05 <$ **P** < 0.125. Since p-value $> \alpha = 0.05$, we fail to reject H_o.

QUESTIONS 113 THROUGH 116 ARE BASED ON THE FOLLOWING INFORMATION:
A recent study reported that the mean salary of a full professor in US academic institutions was $84,175. The following table lists the average salary for a random sample of 20 institutions in Virginia.

87300	58200	58700	57700	74400
62700	79000	70500	55900	82500
55000	84100	67500	96600	63500
50200	48200	82700	90200	61300

113. State the null and alternative hypotheses in testing the claim that the median salary of full professors in Virginia is lower than the mean for the whole country.

ANSWER:
H_o: Median = 84,175 ($\geq$) vs. H_a: Median < 84,175

114. Calculate the value of the test statistic.

ANSWER:
$n(+) = 3$, $n(0) = 0$, $n(-) = 17$.
The value of the test statistic $x = n$(least frequent sign) = 3.

115. Test the hypotheses in question 113 at $\alpha = 0.05$ using the p-value approach.

ANSWER:
P = p-value = $P(x \leq 3 \mid n = 20) \Rightarrow$ **P** = 0.01 / 2 = 0.005. Since p-value < $\alpha = 0.05$, we reject H_o. There is sufficient evidence to indicate that the median salary of full professors in Virginia is lower than the mean for the whole country.

116. Test the hypotheses in question 113 at $\alpha = 0.05$ using the classical approach.

ANSWER:
Critical region: n(least freq sign) = $x \leq 5$. Therefore, the test statistic is in the critical region, and H_o is rejected. We reach the same conclusion as stated in question 115.

117. A researcher compared baseline values for antithrobin III with antithrobin II values 7 days after a bone marrow transplant for 50 patients. The differences were found to be nonsignificant. Suppose 19 of the differences were positive and 31 were negative. The null hypothesis is that the median difference is zero, and the alternative hypothesis is that the median difference is not zero. Use the 0.05 level of significance. Complete the test and carefully state your conclusion.

ANSWER:
H_o : Median = 0 vs. H_a : Median $\neq 0$
Since $n(+) = 19$, $n(0) = 0$, $n(-) = 31$, then $n = n(+) + n(-) = 50$, and the value of the test statistic is $x = n$(least frequent sign)= $n(+0) = 19$.
P = p-value = $2P(x \leq 19 \mid n = 50) \Rightarrow 0.10 <$ **P** < 0.25. Since p-value > $\alpha = 0.05$, we fail to reject H_o. There is no sufficient evidence to indicate that the median difference is different from zero.

<u>**QUESTIONS 118 THROUGH 121 ARE BASED ON THE FOLLOWING INFORMATION:**</u>
A taste test was conducted with a regular beef pizza. Each of 130 individuals was given two pieces of pizza, one with a whole-wheat crust and the other with a white crust. Each person was then asked whether she or he preferred whole-wheat or white crust. The results were: 64 preferred whole-wheat to white crust, 52 preferred white to whole-wheat crust, and 14 had no preference.

118. A blind taster wishes to test if the whole-wheat crust is preferred to white crust, state the null and alternative hypothesis.

ANSWER:
H_o : There is no preference; $p = P$(prefer whole-wheat crust) = 0.5.
H_a : There is a preference for the whole-wheat crust; $p > 0.5$.

119. Calculate the appropriate value of the test statistic.

ANSWER:
x is binomial random variable and approximately normal. Let + = prefer whole-wheat crust; then $n(+) = 64$, $n(0) = 14$, and $n(-) = 52$; The usable sample size is $n = n(+) + n(-)$ $= 64 + 52 = 116$; Since $x = n(+) = 64$; then $x' = x - 0.50 = 63.5$. The value of the test statistic is: $z^* = [x' - (n/2)]/(\sqrt{n}/2) = (63.5 - 58)/(\sqrt{116}/2) = 1.02$.

120. Test the hypothesis in question 118 at the 0.05 level of significance using the p-value approach.

ANSWER:
P = p-value = $P(z > 1.02) = 0.5000 - 0.3461 = 0.1539$. Since **P** > $\alpha = 0.05$, we fail to reject H_o. There is no sufficient evidence to indicate that whole-wheat crust is preferred to white crust.

121. Test the hypothesis in question 118 at the 0.05 level of significance using the classical approach.

ANSWER:

The critical region is $z \geq 1.645$. Since the test statistic z^* = 1.02 does not fall in the critical region; we fail to reject H_o. We reach the same conclusion as stated in question 120.

Section 14.5

True-False Questions

122. The Mann-Whitney *U* Test is used to compare two dependent sample means.
ANSWER: F

123. The Mann-Whitney *U* test is a nonparametric alternative for the *t*- test for the difference between two independent means.
ANSWER: T

124. The calculation of the Mann-Whitney test statistic *U* is a two-step procedure. We first determine the sum of the ranks for each of the two samples. Then, using the two sums of ranks, we calculate a *U* score. The larger U score is the test statistic.
ANSWER: F

125. The Mann-Whitney test may be carried out by means of a normal approximation using the standard normal variable *z* whenever the two sample sizes n_1 and n_2 are both greater than 10.
ANSWER: T

126. The sum of U_a and U_b in the Mann-Whitney *U* test will always be equal to the product of the two sample sizes n_a and n_b.
ANSWER: T

127. The sum of U_a and U_b in the Mann-Whitney *U* test will always be equal to the sum of the two sample sizes n_a and n_b.
ANSWER: F

Multiple-Choice Questions

128. Which of the following statements is false regarding the Mann-Whitney *U* test?

A) It is a nonparametric alternative for the *t*-test for the difference between two dependent (matched pairs) means.
B) It is often used in situations in which two independent random samples are drawn from the same population of subjects but different "treatments" are used on each test.
C) One of the assumptions of the test is that the random variables are ordinal or numerical.
D) None of the above.
ANSWER: A

129. Which of the following statements is true in order to use the standard normal distribution to approximate the distribution of Mann-Whitney statistic *U*?

A) Whenever n_a and n_b are both greater than 5.
B) Whenever n_a and n_b are both greater than 10.
C) Whenever n_a and n_b are both smaller than 10.
D) Whenever n_a and n_b are both smaller than 5.

ANSWER: B

130. Which of the following equations is true regarding U_a and U_b and the two sample sizes n_a and n_b in the Mann-Whitney *U* test?

A) $U_a + U_b = n_a + n_b$
B) $U_a / U_b = n_a / n_b$
C) $U_a + U_b = n_a \cdot n_b$
D) $U_a / U_b = n_a - n_b$

ANSWER: C

Short-Answer Questions

131. Let n_1 and n_2 be the sample sizes in the Mann-Whitney *U* test. Let R_a and R_b be the two rank sums. Give a formula involving n_1 and n_2 which will always give the sum R_a + R_b.

ANSWER:
$R_a + R_b = (n_1 + n_2) \cdot (n_1 + n_2 + 1)/2$
We use the fact that if *n* is a positive integer, $1 + 2 + \cdots + n = n(n + 1) / 2$.

132. Suppose the Mann-Whitney *U* test were used to test a two-tailed alternative hypothesis at $\alpha = 0.05$. If two independent samples (each of size 40) were used, what partitions in the sum of ranks is necessary in order to reject the null hypothesis?

ANSWER:
Split of 1416 and 1823 or split that is more extreme than this.

133. Briefly discuss the assumptions for inferences about two populations using the Mann-Whitney *U* test.

ANSWER:
(a) The two independent random samples are independent within each sample as well as between samples.
(b) The random variables are ordinal or numerical.

Chapter 14

134. What parametric test procedure is equivalent to Mann-Whitney *U* test?

ANSWER:
The *t*-test for the difference between two independent means is the parametric test procedure that is equivalent to Mann-Whitney *U* test.

135. State the null hypothesis H_o, and the alternative hypothesis, H_a, that would be used to test the following statement :"The cholesterol level for group A is lower than for group B".

ANSWER:
H_o : The average cholesterol level is the same for both groups A and B.
H_a : The average cholesterol level for group A is lower than for group B.

136. State the null hypothesis H_o, and the alternative hypothesis, H_a, that would be used to test the following statement :"The average test score is not the same for both male and female groups".

ANSWER:
H_o : The average test score is the same for both male and female groups.
H_a : The average value is not the same for male and female groups.

137. Briefly discuss the calculation of the Mann-Whitney test statistic *U*.

ANSWER:
The calculation of the Mann-Whitney test statistic *U* is a two-step procedure:
(a) We first determine the sum of the ranks for each of the two samples A and B. Then, using the two sums of ranks, we calculate via a pair of formulas a *U* score for each sample, U_a and U_b respectively.
(b) The test statistic is the smaller of U_a and U_b.

138. What characteristic of the data used in a parametric test is not part of the data when using the Mann-Whitney *U* test?

ANSWER:
The actual size of the data is not used, only its rank.

139. State the null hypothesis H_o, and the alternative hypothesis H_a, that would be used to test the following statement: "There is a difference in the value of the variable between the two professional groups of people"

ANSWER:
H_o : The average value is the same for both professional groups.
H_a : The average value is different for the professional groups.

140. State the null hypothesis H_o, and the alternative hypothesis, H_a, that would be used to test the following statement: "The blood pressure for group A is higher than for group B".

ANSWER:
H_o : The average blood pressure is the same for both groups A and B.
H_a : The average blood pressure for group A is higher than for group B.

Applied and Computational Questions

141. The cholesterol readings for individuals under 40 was compared with cholesterol readings for individuals who are 40 or over. The claim was that those who were 40 or over would have higher readings. Use the Mann-Whitney U test to test the hypothesis. Give the critical region for $\alpha = 0.05$, the test statistic, and the conclusion.

Age	*Cholesterol Readings*					
Less than 40	180	185	190	195	198	
40 or over	187	192	196	200	205	210

ANSWER:

Critical region is: $U \leq 5$, and the test statistic is: $U^* = 6$.
Conclusion: Unable to reject the null.

142. The yield of two different varieties of citrus trees is being compared. Twenty-five similar plots are available on an experimental farm. Variety 1 was planted on 13 randomly selected plots and variety 2 was planted on the remaining plots. Two years later the yields from the 25 trees is recorded. The yields were jointly ranked and it was found that the sum of ranks for variety 1 was 208.5 and the sum of ranks for variety 2 was 116.5. Test the null hypothesis that the two varieties give equal yields versus the alternative that they do not, at $\alpha = 0.05$. Give the critical region, U^*, and the conclusion.

ANSWER:

Since the critical region is: $U \leq 41$, and the test statistic is $U^* = 38.5$, we reject the null hypothesis.
Conclusion: The two varieties don't give equal yields.

143. A new product was tested in 10 stores. Five stores were randomly selected and the item was placed at the check-out stand. In the other stores, the item was located in a section containing similar items. It is of interest to test for no difference in sales versus greater sales occurring at the checkout location. Using the following data, give the critical region for $\alpha = 0.05$, U^*, and your conclusion.

Check-out location: 40, 42, 45, 45, 50
Similar items location: 32, 38, 40, 43, 45

ANSWER:

Since the critical region is: $U \leq 4$, and the test statistic is $U^* = 5.5$, few ail to reject the null hypothesis.
Conclusion: No difference in sales; unable to conclude that location affects sales.

144. Two populations were compared by selecting samples of size 20 from each. The Mann-Whitney U test was selected to test a two-tailed alternative at $\alpha = 0.05$. If $R_a = 380$ and $R_b = 440$, find z^* and give the p-value for the test.

ANSWER:

$z^* = -0.81$, and p - value = 0.418.

Chapter 14

145. A driver kept track of her gasoline mileage using full tanks of two different brands of gasoline. The gasoline consumption in miles per gallon for the two brands is shown below:

Brand A	17	16	18	21	19	20	
Brand B	15	18	19	17	20	21	22

Test the claim that the two brands of gasoline result in the same gasoline consumption. Use the Mann-Whitney *U* Test with α = 0.05. Give the critical region, *U**, and the conclusion.

ANSWER:

Since the critical region is $U \le 6$, and the test statistic is $U^* = 11.5$; we fail to reject the null hypothesis.
Conclusion: The two brands result in the same gasoline consumption.

146. In order to compare two fertilizers, one is applied to thirteen plots randomly selected from 25 available plots and the other is applied to the remaining twelve plots. The yields for the two fertilizers are shown below:

Fertilizer A	36	25	36	27	38	29	40	29	30	30	34	34	34
Fertilizer B	36	18	35	20	32	20	24	26	26	30	30	27	

Test for differences in yields at α = 0.05. Give the critical region, the value of *U**, and your conclusion.

ANSWER:

Critical region: $U \le 41$, $U^* = 38.5$; reject the null hypothesis.
Conclusion: There is a difference in yields; Fertilizer A yields more than Fertilizer B.

147. The time required to assemble a product part was determined for 25 males and 25 females. Since the data indicated non-normality of the times for the two sexes, the Mann-Whitney *U* test was selected to determine whether the assembly times differed for males and females. Find the sum of ranks for males, R_m, and the sum of ranks for females, R_f, which would be the strongest evidence possible supporting the hypothesis that females were faster in assembling the product part.

ANSWER:

$R_m = 950$, and $R_f = 325$

QUESTIONS 148 THROUGH 151 ARE BASED ON THE FOLLOWING INFORMATION:
A study involving 14 adults in the age group 40–45 years, gave the following weight values (in pounds).

Men	185	195	170	210	180	160		
Women	170	200	185	200	195	160	150	205

148. If you wish to test the research hypothesis that the weigh values differ for the two groups, state the null and alternative hypotheses.

ANSWER:

H_o: Weight values are the same for both groups of boys and girls.

H_a : Weight values are not the same for both groups of boys and girls.

149. Calculate the appropriate value of the test statistic.

ANSWER:

Ranked Data	Rank	Source
150	1	G
160	2.5	G
160	2.5	B
170	4.5	G
170	4.5	B
180	6	B
185	7.5	G
185	7.5	B
195	9.5	G
195	9.5	B
200	11.5	G
200	11.5	G
205	13	G
210	14	B

$n_b = 6, \quad n_g = 8$

R_b = 2.5 + 4.5 + 6.0 + 7.5 + 9.5 + 14 = 44

R_g = 1.0 + 2.5 + 4.5 + 7.5 + 9.5 + 11.5 + 11.5 + 13.0 = 61

Then, $U_b = n_b \cdot n_g + [n_g(n_g+1)/2] - R_g$ = (6)(8) + [(8)(9)/2] – 61 = 23

$U_g = n_b \cdot n_g + [n_b(n_b+1)/2] - R_b$ = (6)(8) + [(6)(7)/2} – 44 = 25.

The value of the test statistic is U^* = min(U_b, U_g) = min (23, 25) = 23.

150. Test the hypothesis in question 148 at the 0.05 level of significance using the *p*-value approach.

ANSWER:

P = *p*-value = 2*P*($U \leq 23$ | n_b = 6, n_g = 8).Using the table of critical values of *U* in the Mann-Whitney test, we get **P** > 0.10. Since **P** $> \alpha = 0.05$; fail to reject H_o.The evidence does not allow us to conclude that there is a significant difference between the boys and girls weight values.

151. Test the hypothesis in question 148 at the 0.05 level of significance using the classical approach.

ANSWER:
The critical region is $U \le 8$. Since the test statistic does not fall in the critical region; we fail to reject H_o. We reach the same conclusion as stated in question 150.

QUESTIONS 152 THROUGH 156 ARE BASED ON THE FOLLOWING INFORMATION:
Commercial airlines are often evaluated on the basis of two major performance categories: on-time arrivals and baggage handling. United Airlines received the following competitive ratings (the lower the better) on each of these dimensions over a 13-month period:

Month	On-Time Arrival	Baggage Handling
Aug.	8	5
Sept.	9	8
Oct.	9	6
Nov.	10	7
Dec.	7	5
Jan.	5	7
Feb.	7	8
Mar.	5	5
Apr.	8	6
May	5	6
June	3	2
July	3	4
Aug.	3	2

152. Convert the table to a table of ranks of the on-time arrivals (A) and baggage handling (B) for United.

ANSWER:

Rating	Rank	Dimension	Rating	Rank	Dimension
2	1.5	B	6	14	B
2	1.5	B	6	14	B
3	4	A	7	17.5	B
3	4	A	7	17.5	A
3	4	A	7	17.5	B
4	6	B	7	17.5	A
5	9.5	B	8	21.5	A
5	9.5	B	8	21.5	B
5	9.5	A	8	21.5	B
5	9.5	A	8	21.5	A
5	9.5	B	9	24.5	A
5	9.5	A	9	24.5	A
6	14	B	10	26	A

153. If you wish to test the hypothesis that baggage handling obtained higher ratings than on-time arrivals during the period, state the null and alternative hypotheses.

ANSWER:

H_o : Baggage handling scores are not higher than on-time arrivals.

H_a : Baggage handling scores are higher than on-time arrivals.

154. Calculate the value of the Mann-Whitney test statistic U^*.

ANSWER:

$n_a = 13,\ n_b = 13$;

$R_a = 4 + 4 + 4 + 9.5 + 9.5 + 9.5 + 17.5 + 17.5 + 21.5 + 21.5 + 24.5 + 24.5 + 26 = 193.5$

$R_b = 1.5 + 1.5 + 6 + 9.5 + 9.5 + 9.5 + 14 + 14 + 14 + 17.5 + 17.5 + 21.5 + 21.5 = 157.5$; Then,

$U_a = n_a \cdot n_b + [(n_b)(n_b + 1)/2] - R_b$ = (13)(13) + [(13)(14)/2] – 157.5 = 102.5

$U_b = n_a \cdot n_b + [(n_a)(n_a + 1)/2] - R_a$ = (13)(13) + [(13)(14)/2] – 193.5 = 66.5;

The value of the test statistic is U^* = min(U_a, U_b) = min (102.5, 66.5) = 66.5.

155. Test the hypothesis in question 153 at the 0.05 level of significance using the *p*-value approach.

ANSWER:

P = *p*-value $= P(U < 66.5)$. Using the table of critical values of *U* in the Mann-Whitney test, we get: **P** > 0.05. Since **P** $> \alpha = .05$, we fail to reject H_o. There is no sufficient evidence to indicate that United baggage handling did obtain higher ratings than on-time arrivals at the 0.05 level of significance.

156. Test the hypothesis in question 153 at the 0.05 level of significance using the classical approach.

ANSWER:

The critical region is $U \leq 51$; The test statistic is not in the critical region; therefore we fail to reject H_o. We reach the same conclusion as stated in question 155.

QUESTIONS 157 THROUGH 160 ARE BASED ON THE FOLLOWING INFORMATION:

Twenty students were randomly divided into two equal groups. Group 1 was taught a biology course using a standard lecture approach. Group 2 was taught using a computer-assisted approach. The test scores on a comprehensive final exam were as follows:

Group 1	80	88	65	94	82	97	93	95	60	75
Group 2	82	97	95	90	77	64	70	97	95	84

157. In you wish to test the claim that a computer-assisted approach produces higher achievement (as measured by final exam scores) in biology courses than does a lecture approach, what are the null and alternative hypotheses.

ANSWER:
Using the Mann-Whitney *U* Test (independent samples):

H_o: No effect due to teaching approach used.

H_a : The computer assisted instruction approach produced higher achievement.

158. Calculate the value of the appropriate test statistic.

ANSWER:

Ranked Data	Rank	Group
60	1	1
64	2	2
65	3	1
70	4	2
75	5	1
77	6	2
80	7	1
82	8.5	1
82	8.5	2
84	10	2
88	11	1
90	12	2
93	13	1
94	14	1
95	16	1
95	16	2
95	16	2
97	19	1
97	19	2
97	19	2

$n_1 = 10,\quad n_2 = 10,\; R_1 = 97.5,\quad R_2 = 112.5$

$U_1 = n_1 \cdot n_2 + [(n_2)(n_2 + 1)/2] - R_2$ = (10)(10) + [(10)(11)/2] – 112.5 = 42.5

$U_2 = n_2 \cdot n_1 + [(n_1)(n_1 + 1)/2] - R_1$ = (10)(10) + [(10)(11)/2] – 97.5 = 57.5

The value of the test statistic is *U** = min (U_1, U_2) = 42.5.

159. Test the hypothesis in question 157 at the .05 level of significance using the *p*-value approach.

ANSWER:
P = *p*-value = $P(U < 42.5)$. Using the table of critical values of *U* in the Mann-Whitney test, we get **P** > 0.05. Since **P**$> \alpha = 0.05$, we fail to reject H_o. The evidence does not allow us to conclude that the computer assisted instruction approach produced higher achievement scores.

160. Test the hypothesis in question 157 at the .05 level of significance using the classical approach.

ANSWER:
The critical region is $U \le 27$. Since the test statistic does not fall in the critical region; we fail to reject H_o. We reach the same conclusion as stated in question 159.

161. In a Mann-Whitney U test, suppose all 10 data values of sample "A" come before the smallest of the 10 data values in sample "B" when they are ranked together. Calculate the value *U** of the test statistic.

ANSWER:
R_a = 55 and R_b = 155. Therefore,

$$U_a = n_a \cdot n_b + \frac{n_b(n_b+1)}{2} - R_b = (10)(10) + \frac{(10)(10+1)}{2} - 155 = 0 \text{, and}$$

$$U_b = n_a \cdot n_b + \frac{n_a(n_a+1)}{2} - R_a = (10)(10) + \frac{(10)(10+1)}{2} - 55 = 100$$

Hence, *U** = smaller (U_a and U_b) = 0.

162. In a Mann-Whitney U test, suppose each sample has 8 data values and that both samples were perfectly matched; that is, a score in each sample is identical to one in the other sample. Calculate the value *U** of the test statistic.

ANSWER:
R_a = R_b = 68. Therefore, $U_a = U_b = (8)(8) + \frac{(8)(8+1)}{2} - 68 = 32$

Hence, *U** = smaller (U_a and U_b) = 32.

163. Determine the *p*-value when testing H_o: Average score for group A = Average score for group B vs. H_a: Average score for group A > Average score for group B , given that n_a = 15, n_b = 15 and *U* = 80.

ANSWER:
P = *p*-value > 0.05

164. Determine the *p*-value when testing H_o: Average weight for group A = Average weight for group B vs. H_a: Average weight for group A ≠ Average weight for group B, given that n_a = 9, n_b = 10, and *U* = 22.

ANSWER:
P = *p*-value = 2 $P(U \le 22) \Rightarrow 0.05 <$ **P** < 0.10

165. Determine the *p*-value when testing H_o: The average height is the same for both groups A and B. vs. H_a: Group A average heights are less than those for group B, given that with n_a = 50, n_b = 45, and *z* = -2.18.

ANSWER:
P = *p*-value = $P(z < -2.18) = 0.5000 - 0.4854 = 0.0146$

166. Use the classical method to determine the critical region that would be used to test H_o: Average(A) = Average(B) vs. H_a: Average(A) > Average (B) for an experiment involving two independent samples, given that n_a = 12, n_b = 20 and α = 0.05.

ANSWER:
Critical region is: $U \leq 77$

167. Use the classical method to determine the critical region that would be used to test H_o: The average score is the same for both groups A and B vs. H_a: Group A average scores are less than those for group B, for an experiment involving two independent samples, given that n_a = 78, n_b = 45, and α = 0.05.

ANSWER:
Critical region is: $z \leq -1.645$

QUESTIONS 168 THROUGH 174 ARE BASED ON THE FOLLOWING INFORMATION:
Pulse rates were recorded for 16 men and 13 women. The results are shown below:

Males	62	74	59	65	71	65	73	61	66	81	56	73	57	57	75	66
Females	82	57	69	55	75	63	79	67	77	107	75	69	96			

Assume a doctor wishes to test the hypothesis that the distribution of pulse rates differs for men and women.

168. State the null and alternative hypotheses.

ANSWER:
H_o: Average pulse rates are the same for both males and females.
H_a: Average pulse rates are not the same for males and females.

169. Identify the test statistic to be used in testing the hypotheses in question 168.

ANSWER:
The Mann-Whitney U statistic.

170. Calculate the value of the test statistic.

ANSWER:
Let "a" = Males and "b" = Females.
$n_a = 16,\ n_b = 13,\ R_a = 199,$ and $R_b = 236$. Therefore,

$$U_a = n_a \cdot n_b + \frac{n_b(n_b+1)}{2} - R_b = (16)(13) + \frac{(13)(14)}{2} - 236 = 63,$$ and

$$U_b = n_a \cdot n_b + \frac{n_a(n_a+1)}{2} - R_a = (16)(13) + \frac{(16)(17)}{2} - 199 = 145.$$

Hence, U^* = smaller (U_a and U_b) = 63.

171. Test the hypotheses in question 168 at the 0.05 level of significance using the *p*-value approach.

ANSWER:
Using the table of "critical values of U in the Mann-Whitney test", available in your textbook, we get **P** = *p*-value = $P(U \leq 63) \Rightarrow 0.05 <$ **P** < 0.10.
Since **P** > α = 0.05, we fail to reject H_o. There is no significant evidence to indicate that the average pulse rates are not the same for males and females.

172. Test the hypotheses in question 168 at the 0.05 level of significance using the classical approach.

ANSWER:
The critical region is $U \leq 59$. Since $U^* = 63$ does not fall in the rejection region, we fail to reject H_o. We reach the same conclusion as stated in question 171.

173. Approximate the distribution of the test statistic identified in question 169 using the normal distribution, and calculate the value of the standardized test statistic z^*.

ANSWER:
$\mu_U = (n_a \cdot n_b)/2 = (16)(13)/2 = 104$
$\sigma_U = \sqrt{n_a \cdot n_b \cdot (n_a + n_b + 1)/12} = \sqrt{(16)(13)(30)/12} = 22.804$
Then, $z^* = (U^* - \mu_U)/\sigma_U = (63 - 104)/22.804 = -1.80$

174. Calculate the *p*-value associated with the test statistic z^* in question 173 and use it to test the hypotheses in question 168 at the 0.05 level of significance.

ANSWER:
P = *p*-value = 2 $P(z < -1.80)$ = 2(0.5000- 0.4641) = 0.0718
Since **P** > α = 0.05, we fail to reject H_o. We reach the same conclusion as stated in question 171.

QUESTIONS 175 THROUGH 179 ARE BASED ON THE FOLLOWING INFORMATION:
A study involving 8 obese boys and 8 obese girls gave the following total-cholesterol values.

Obese Boys	186	199	165	205	175	177	210	195
Obese Girls	168	192	171	188	193	155	145	200

Suppose you wish to test the hypothesis that the total-cholesterol values differ for the two groups.

175. State the null and alternative hypotheses.

ANSWER:
H_o : Total cholesterol values are the same for both obese boys and girls.
H_a : Total cholesterol values are not the same for obese boys and girls.

176. Identify the test statistic to be used in testing the hypotheses in question 175.

ANSWER:
The Mann-Whitney *U* statistic

177. Calculate the value of the test statistic.

ANSWER:
Let "a" = Boys and "b" = Girls.
$n_a = 8,\ n_b = 8,\ R_a = 80,$ and $R_b = 56$. Therefore,

$$U_a = n_a \cdot n_b + \frac{n_b(n_b+1)}{2} - R_b = (8)(8) + \frac{(8)(9)}{2} - 56 = 44, \text{ and}$$

$$U_b = n_a \cdot n_b + \frac{n_a(n_a+1)}{2} - R_a = (8)(8) + \frac{(8)(9)}{2} - 80 = 20$$

Hence, U^* = smaller (U_a and U_b) = 20.

178. Test the hypotheses in question 175 at the 0.05 level of significance using the *p*-value approach.

ANSWER:
Using the table of "critical values of *U* in the Mann-Whitney test", available in your textbook, we get **P** = *p*-value = $P(U \leq 20) \Rightarrow$ **P** > 0.10.
Since **P** > α = 0.05, we fail to reject H_o. There is no significant evidence to indicate that total cholesterol values are not the same for obese boys and girls.

179. Test the hypotheses in question 175 at the 0.05 level of significance using the classical approach.

ANSWER:
The critical region is $U \leq 13$. Since U^* = 20 does not fall in the rejection region, we fail to reject H_o. We reach the same conclusion as stated in question 178.

Section 14.6

True-False Questions

180. The Runs test is most frequently used to test the randomness of or lack of randomness of data.
ANSWER: T

181 The Runs test is generally a one-tailed test.
ANSWER: F

182. The hypothesis test about randomness in the Runs test will be rejected when there are too few or two many runs.
ANSWER: T

183. The Runs test may be carried out by means of a normal approximation using the standard normal *z* variable whenever the two sample sizes n_1 and n_2 are both smaller than 20 or when the level of significance α is other than 0.025.
ANSWER: F

184. The Runs test is a nonparametric alternative to the difference between two independent means.
ANSWER: F

185. To complete the hypothesis test about randomness, when n_1 and n_2 are larger than 20 or when α is other than 0.05, we will use z, the standard normal random variable.
ANSWER: T

Multiple-Choice Questions

186. The following data were collected for a test of randomness:

2.2	3.3	3.3	3.5	3.6	3.7	3.8
3.8	3.8	3.9	4.1	4.2	4.5	

The rank assigned to the three observations of value 3.8 is:

A) 6.
B) 7.
C) 8.
D) 9.
ANSWER: C

Chapter 14

187. The following data were collected to determine whether the data points form a random sequence with regard to being above or below the median value.

13	13	14	15	17	18	18	19
20	21	22	24	24	24	24	27

The rank assigned to the four observations of value 24 is:

A) 12.0.
B) 12.5.
C) 13.0.
D) 13.5.
ANSWER: **D**

188. Which of the following statements is false regarding the runs test?

A) It is used most frequently to test the randomness of data (or lack of randomness).
B) Its test statistic is V, the number of runs observed.
C) It is generally a two-tailed test.
D) None of the above.
ANSWER: **D**

189. Which of the following statements is false when testing for randomness?

A) The Runs test is used.
B) It is the null hypothesis that states random, thereby making the "fail to reject" decision the desired outcome.
C) It is the alternative hypothesis that states random, thereby making the "reject" decision the desired outcome.
D) None of the above.
ANSWER: **C**

Short-Answer Questions

190. What is the assumption for inferences about randomness using the Runs test?

ANSWER:
The assumption is that each sample data can be classified into one of two categories (e.g. male or female)

191. In testing the randomness of data using the Runs test, when do we reject the null hypothesis?

ANSWER:
We will reject the hypothesis when there are too few runs because this indicated that the data are "separated" according to the two properties (e.g., "above" or "below" the median value.) We will also reject the null hypothesis when there are too many runs because that indicates that the data alternate between the two properties too often to be random.

Applied and Computational Questions

QUESTIONS 192 AND 193 ARE BASED ON THE FOLLOWING INFORMATION:
State the null hypothesis and the alternative hypotheses that would be used to test the following statements:

192. A die is tossed and a sequence of numbers (1, 2, 3, 4, 5, or 6) is recorded. The sequence of an odd digit or an even digit is not random.

ANSWER:
H_o: Odd and even numbers occurred in random order.

H_a: Odd and even numbers did not occur in random order.

193. Cars passing a toll booth were classified as either foreign or domestic. The sequence of foreign or domestic is not random.

ANSWER:
H_o: Order of passing toll booth by foreign or domestic was random.

H_a: Order of passing toll booth by foreign or domestic was not random.

194. The values in a set of data were replaced by the symbol "A" if the number was above the median and by the symbol "B" if the number was below the median. The following sequence was obtained.

A	B	B	A	B	A	B	B	A	A	A
B	B	A	A	B	B	A	B	B	A	

Give the critical region using α = 0.05, the computed test statistic and the conclusion for testing that the sequence is random versus it is not.

ANSWER:
Critical region: $V \leq 6$ or $V \geq 17$, and test statistic is $V^* = 13$; fail to reject the null hypothesis.
Conclusion: The sequence is random.

195. The outcomes for 25 rolls of a die were as follows:

2	5	6	1	1	2	2	3	4	6
4	3	1	2	5	1	2	3	6	5
4	6	4	3	1					

Give the critical region for α = 0.05, the value of the test statistic V^*, and the conclusion for determining whether or not this sequence of odd and even numbers is random.

ANSWER:
$n_1 = 12$, $n_2 = 13$; Critical regions are $V \leq 8$ or $V \geq 19$, Test statistic $V^* = 16$; fail to reject the null hypothesis of randomness.
Conclusion: The sequence is random.

Chapter 14

196. Thirty cars which enter a vehicle inspection station are monitored and it is noted whether they pass (P) or fail (F) the inspection. The following sequence was recorded.

P, P, P, P, P, F, F, F, P, P, P, P, P, P, P, F, F, F, F, P, P, P, P, P, P, F, F, P, P, F

Test at α = 0.05 to determine if the sequence is random. Give the critical region, V^*, and the conclusion.

ANSWER:
n_p = 20, n_f =10; Critical regions are: $V \leq 9$ or $V \geq 20$, Test statistic is $V^* = 8$; reject the null hypothesis of randomness. Conclusion: The number of runs is unusually small.

197. The following are the number of defective pieces turned out by a machine during 20 consecutive shifts:

10	12	13	12	15	16	17	17	18	10
10	14	17	17	17	12	12	16	16	16.

Test the null hypothesis that the numbers in the sample form a random sequence with respect to the two properties "above" and "below" the median value, versus the alternative hypothesis that the sequence is not random. Use α = 0.05. Give the critical region, V^*, and the conclusion.

ANSWER:
Critical regions are: $V \leq 6$ or $V \geq 16$, and test statistic is $V^* = 6$; reject the null hypothesis of randomness.
Conclusion: The number of runs is unusually small.

198. A product part is routinely selected from a production line and classified as either defective or non-defective. For the last 100 selected parts, 90 have been non-defective and 10 have been defective. If the defectives and non-defectives are displayed as a sequence of N's and D's, how many runs would you expect to see if their occurrences are random?

ANSWER:
$V = 19$, the mean number of runs for random occurrences.

199 A sequence of inspected items consists of 15 defectives and 105 non-defectives. If there are 11 runs in the sequence, find z^* and give the *p*-value for testing the alternative hypothesis that sequence is not random.

ANSWER:
$z^* = -6.89$, *p*-value is practically zero.

200. A sequence consists of 30 zeros and 20 ones. How many runs would there need to be to reject H_o: the sequence of 0's and 1's is random in favor of H_a: the sequence is non-random and there are more runs than should occur. Use α = 0.05.

ANSWER:
31 or more runs

201. A sequence consists of 15 items above the median (*a*) and 15 items below the median (*b*). We are testing at α = 0.05 the following hypotheses:

H_o: The sequence of *a*'s and *b*'s is random.

H_a: The sequence of *a*'s and *b*'s is not random.

How many runs would there need to be in order to fail to reject H_o?

ANSWER:
In order to fail to reject H_o, the number of runs must be 9, 10, 11, ..., 19, 20, or 21.

QUESTIONS 202 THROUGH 205 ARE BASED ON THE FOLLOWING INFORMATION:
A student was asked to perform an experiment that involved tossing a coin 25 times. After each toss, the student recorded the results as H (heads) or T (tails), as shown below:

T H T H T   H T H T H   H T H T H   T H T H H   T T H H T

202. If you wish to test student's claim that the results reported are random, state the null and hypotheses.

ANSWER:

H_o: The results are randomly ordered.

H_a: The results are not randomly ordered.

203. Calculate the value of the Runs test statistic.

ANSWER:
Since $n(H) = 13, n(T) = 12$, and there are 21 runs, then, the value of the test statistic is: $V^* = 21$.

204. Test the hypothesis in question 202 at the 5% level of significance using the *p*-value approach.

ANSWER:
Using the table of critical values for total number of runs (*V*); **P** < 0.05. Since **P** $< \alpha$; reject H_o.

205. Test the hypothesis in question 202 at the 5% level of significance using the classical approach.

ANSWER:
The critical regions are: $V \le 8$ or $V \ge 19$; The test statistic falls in the critical region; therefore we reject H_o There is sufficient evidence to indicate that the results are not randomly ordered.

QUESTIONS 206 THROUGH 209 ARE BASED ON THE FOLLOWING INFORMATION:
The following data were collected in an attempt to show that the number of minutes the city bus is late is steadily growing larger. The data are in order of occurrence.

Minutes: 7 2 4 10 11 11 3 6 6 7 13 4 8 9 10 5 6 9 12 15

206. If you wish to determine whether these data show sufficient lack of randomness to support the claim, what would be the null and alternative hypotheses?

ANSWER:

H_o : Random order of increase and decrease in value from previous value.

H_a : Lack of randomness (a trend, an increase in wait time).

207. Calculate the appropriate value of the test statistic.

ANSWER:
n (decreases) = 4, *n* (increases) = 13, and the value of the test statistic is $V^* = 8$.

208. Test the hypothesis in question 206 at the 0.05 level of significance using the *p*-value approach.

ANSWER:
Using the table of critical values for total number of runs (*V*), we get **P** > 0.05. Since **P** > α = 0.05, we fail to reject the null hypothesis. There is no sufficient evidence to conclude that there is an increase in wait time.

209. Test the hypothesis in question 206 at the 0.05 level of significance using the classical approach.

ANSWER:
The critical regions are $V \leq 3$ or $V \geq 10$; The test statistic is not in the critical region; therefore we fail to reject H_o. We reach the same conclusion as stated in question 208.

210. State the null hypothesis H_o, and the alternative hypothesis H_a that would be used to test the following statement: "The data did not occur in a random order about the median".

ANSWER:

H_o : The data did occur in a random order about the median.

H_a : The data did not occur in a random order about the median.

211. State the null hypothesis H_o, and the alternative hypothesis, H_a that would be used to test the following statement: "The sequence of head and tail is not random".

ANSWER:

H_o : Sequence of head/tail is in random order.

H_a : Sequence of head/tail is not in random order.

212. State the null hypothesis H_o, and the alternative hypothesis H_a that would be used to test the following statement: "The gender of students entering a college library was recorded; the entry is not random in order."

ANSWER:
H_o : The order of entry a college library by gender was random.
H_a : The order of entry a college library by gender was not random.

213. What aspect of randomness will be tested using the Runs test?

ANSWER:
The Runs test will test the order, or sequence, of occurrence for the numbers generated.

214. Most gambling rules are written using the phrase "99% confidence level" instead of "0.01 level of significance" as hypothesis tests typically use. Explain why this seems appropriate.

ANSWER:
When testing for randomness, it is the null hypothesis that states random, thereby making the "fail to reject" decision the desired outcome. The probability associated with that result is $1 - \alpha$, not the level of significance, and $1 - \alpha$ is known as the level of confidence.

215. Determine the *p*-value that would be used to complete the hypothesis test for the following Runs test:

H_o : The sequence of gender of students coming into the university gym was random
H_a: The sequence was not random; with *n*(*A*) = 10, *n*(*B*) = 12, and $V^* = 9$.

ANSWER:
We use the "critical values for total number of runs V " table, available in your textbook, to place bounds on the *p*-value. In this case, larger of *n*(*A*) and *n*(*B*) = 12, and smaller of *n*(*A*) and *n*(*B*) = 10. Then, **P** = *p*-value = $2 \cdot P(V \geq 9 \mid n(\mathrm{B}) = 12 \text{ and } n(\mathrm{A}) = 10) \Rightarrow$ **P** < 0.05.

216. Determine the *p*-value that would be used to complete the hypothesis test for the following runs test:

H_o : The new home cost prices collected occurred in random order above and below the median.
H_a: The new home cost prices did not occur in random order with $z = 1.52$.

ANSWER:
P = *p*-value = $2 \cdot P(z > 1.52) = 2\,(0.5000 - 0.4537) = 0.0926$

217. Determine the critical regions that would be used to complete the hypothesis test for the following Runs test using the classical approach.

H_o : The results collected occurred in random order above and below the median.
H_a: The results were not random; with n(A) = 10, n(B) = 16, and α = 0.05.

ANSWER:
Critical regions: $V \leq 8$ or $V \geq 19$

218. Determine the critical values that would be used to complete these hypothesis tests for the following runs tests using the classical approach.

H_o : The two properties alternated randomly.
H_a: The two properties didn't occur in random fashion; with $n(A)$ = 85, $n(B)$ = 55, and α = 0.05.

ANSWER:
The critical regions are: $z \leq -1.96$ or $z \geq 1.96$

219. My youngest daughter, Jessica, did not feel she was playing a game with a fair coin. She felt that if the coin was fair, the tossing of the coin should result in a random order of head and tail output. She performed her experiment 20 times. After each toss, Jessica recorded the results. The following data were reported (H= head, T= tail).

H T H H H   H T T H H   H T T H T   T T H H T

Use the runs test at the 0.05 level of significance to test Jessica's claim that the results reported are random. Use the p-value approach.

ANSWER:
H_o :The heads / tails sequence is random.
H_a: The heads / tails sequence is not of random order; n(H) = 11, n(T) = 9, and V^* = 10.
We use the table of "critical values for total number of runs V", available in your textbook, to place bounds on the p-value. In this case, larger of n(H) and n(T) = 11, and smaller of n(H) and n(T) = 9. Then, **P** = p-value = $2 \cdot P(V \geq 10 \mid n(\text{H}) = 11 \text{ and } n(\text{T}) = 9) \Rightarrow$ **P** > 0.05.
Since **P >** α = 0.05, we fail to reject H_o. There is no sufficient evidence to indicate that the heads / tails sequence is not of random order. In other words, we must support Jessica's claim that the results reported are random.

<u>QUESTIONS 220 THROUGH 223 ARE BASED ON THE FOLLOWING INFORMATION:</u>
The office of human resources at Western Michigan University recorded the gender of the last 35 individuals hired (M = male, F = female) as shown below:

F M M M M   M M F M F   M M F M M   M M F F F
M F F M M   M F F M F   F F M F F
Suppose you wish to determine whether this sequence is random.

220. State the null and alternative hypotheses.

ANSWER:
H_o :The male / female sequence is random.
H_a: The male / female sequence is not of random order.

221. At the α = 0.05 level of significance, are we correct in concluding that this sequence is not random? Test using the *p*-value approach.

ANSWER:

n(*M*) = 19 , *n*(*F*) = 16, and V^* = 17

We use the table of "critical values for total number of runs V ", available in your textbook, to place bounds on the *p*-value. In this case, larger of *n*(*M*) and *n*(*F*) = 19, and smaller of n(M) and n(F) = 16. Then, **P** = *p*-value = $2 \cdot P(V \geq 17 \mid n(M) = 19 \text{ and } n(F) = 16)$ imply that **P** > 0.05. Since **P** > α = 0.05, we fail to reject H_o. There is no sufficient evidence to indicate that the male / female sequence is not of random order.

222. At the α = 0.05 level of significance, are we correct in concluding that this sequence is not random? Test using the classical approach.

ANSWER:

Critical regions: $V \leq 12$ or $V \geq 25$

Since V^* = 17 does not fall in the rejection region, we fail to reject H_o. We reach the same conclusion as stated in question 221.

223. Use a computer to verify the above results in questions 221 and 222.

ANSWER:

Runs Test Results for GENDER		
	Number of obs	35
	Number above cutoff	19
	Number below cutoff	16
	Number of runs	17
	E(R)	18.371
	Stdev(R)	2.892
	Z-value	-0.474
	p-value (2-tailed)	0.635

QUESTIONS 224 THROUGH 230 ARE BASED ON THE FOLLOWING INFORMATION:

The number of absences recorded at a large lecture that met at 6 PM Tuesdays and Thursdays last winter semester were (in order of occurrence) as shown below:

5	17	6	10	18	13	16	20	14	17
11	14	10	6	8	13	15	4	5	5
6	12	7	18	25	6	7	5	10	19
6	8								

224. Use computer to determine the median number of absences.

ANSWER:

Summary measures for selected variables			
		Absences	
	Count	32.000	
	Median	10.000	

225. State the null and alternative hypotheses that can use in testing whether these data show randomness about the median value found in question 224.

ANSWER:

H_o : The numbers in the sample form a random sequence about the median value.

H_a :The sequence is not random.

226. Test the hypotheses in question 225 at α = 0.05 using the *p*-value approach.

ANSWER:

Runs Test Results for Absences		
	Number of obs	32
	Number above cutoff	18
	Number below cutoff	14
	Number of runs	13

We use the table of "critical values for total number of runs V ", available in your textbook, to place bounds on the *p*-value. In this case, larger of n_1 and n_2 = 18, and smaller of n_1 and n_2 =14, and V^* = 13. Then,

P = *p*-value = $2 \cdot P(V \geq 13 \mid n = 18 \text{ and } 14) \Rightarrow$ **P** > 0.05. Since **P >** α **=** 0.05, we fail to reject H_o. There is no sufficient evidence to indicate that the sequence is not of random order.

227. Test the hypotheses in question 225 at α = 0.05 using the classical approach.

ANSWER:

Critical regions: $V \leq 10$ or $V \geq 23$

Since V^* = 13 does not fall in the rejection region, we fail to reject H_o. We reach the same conclusion as stated in question 226.

228. Are the assumptions of using the normal approximation to complete the hypothesis test about randomness met in this situation? Discuss.

ANSWER:

The assumptions of using the normal approximation to complete the hypothesis test about randomness are: n_1 and n_2 are both larger than 20 or when the level of significance $\alpha \neq 0.05$. Therefore, these assumptions are not met in this particular situation.

229. Regardless of your answer to question 228, use computer and the normal approximation to test the hypotheses in question 225 at $\alpha = 0.05$.

ANSWER:

Normal Approximation to the Runs Test		
	E(R)	16.750
	Stdev(R)	2.738
	Z-value	-1.370
	p-value (2-tailed)	0.171

Since *p*-value = 0.171 > $\alpha = 0.05$, we fail to reject H_o. We reach the same conclusion as stated in question 226.

230. Did you reach the same conclusion in questions 226, 227, and 228?

ANSWER:
Yes; in the three questions, we failed to reject the null hypothesis at $\alpha = 0.05$.

Section 14.7

True-False Questions

231. The Spearman Rank Correlation coefficient and the Pearson Product Moment always give the same value.
ANSWER: F

232. The Spearman Rank Correlation coefficient, r_s, is a nonparametric alternative to the Pearson Product Moment, *r*.
ANSWER: T

233. The Spearman Rank Correlation coefficient, r_s, is determined by the equation $r_s = 6(\sum d)^2 /[n(n^2 - 1)]$, where *d* is the difference in the paired rankings, and *n* is the number of pairs of data.
ANSWER: F

234. The value of Spearman Rank Correlation coefficient, r_s, will range from 0 to 1.
ANSWER: F

235. Spearman's rank correlation coefficient is an alternative to using the linear correlation coefficient.
ANSWER: T

236. Charles Spearman developed the rank correlation coefficient in the early 1900's. It is a parametric alternative to the linear correlation coefficient (Pearson's product moment *r*).
ANSWER: F

237. The alternative hypothesis may be either two-tailed, there is correlation, or one-tailed if we anticipate either positive or negative correlation.
ANSWER: T

238. When there are only a few ties in either set of the order pairs of rankings, the value of the Spearman rank correlation coefficient (r_s) is exactly equal to the value of the Pearson product moment correlation coefficient (r).
ANSWER: F

239. The value of Spearman rank correlation coefficient, *r*, rages from -1 to + 1 and is used in much the same manner as Pearson's linear correlation coefficient *r* is used.
ANSWER: T

Multiple-Choice Questions

240. The rank correlation coefficient is used when one is:

A) correlating rankings of individual values for two variables.
B) correlating quantitative data.
C) analyzing data which are assumed to be linearly related.
D) not interested in drawing inferences from the study.

ANSWER: A

241. Which of the following statements is false?

A) The Spearman rank coefficient can be calculated by using Pearson's product moment formula with data rankings substituted for quantitative *x* and *y* values.
B) The null hypothesis that we will be testing by using the Spearman rank correlation coefficient r_s is: There is a correlation between the two rankings.
C) When the Spearman rank correlation test is used in cases where ties occur in either set of the ordered pairs of rankings, assign each tied observation the mean of the ranks that would have been assigned had there been no ties as is the case for the Mann-Whitney *U* test.
D) None of the above.

ANSWER: B

242. Which of the following statements is false?

A) The Spearman rank correlation test of significance will result in a failure to reject the null hypothesis when *r*, is close to zero.
B) The Spearman rank correlation test of significance will result in a rejection of the null hypothesis when *r*, is found to be close to + 1 or -1.
C) One of the assumptions for inferences about rank correlation is that the variables are nominal.
D) None of the above.

ANSWER: C

Short-Answer Questions

243. Briefly discuss the assumptions for Inferences about Rank Correlation.

ANSWER:
(a) The *n* ordered pairs of data form a random sample.
(b) The variables are ordinal or numerical.

244. State the null hypothesis, H_o, and the alternative hypothesis, H_a, that would be used to test the following statement: "There is no relationship between the two rankings".

ANSWER:

H_o : There is a no relationship between the two rankings.

H_a : There is a relationship between the two rankings

245. State the null hypothesis, H_o, and the alternative hypothesis, H_a, that would be used to test the following statement: "There is a positive correlation between the two variables"

ANSWER:
H_o : The is no correlation between the two variables.
H_a : There is positive correlation between the two variables.

246. State the null hypothesis, H_o, and the alternative hypothesis, H_a, that would be used to test the following statement: "Age has a decreasing effect on monetary value"

ANSWER:
H_o : Age has no effect on monetary value.
H_a : Age has a decreasing effect on monetary value.

247. State the null hypothesis, H_o, and the alternative hypothesis, H_a, that would be used to test the following statement: "The two variables are unrelated"

ANSWER:
H_o : The two variables are unrelated.
H_a : The two variables are related.

Applied and Computational Questions

248. Determine the critical value that would be used to test the hypotheses H_o: No correlation versus H_a: Negatively correlated for a multinomial experiment, with n = 14 and α = 0.05.

ANSWER:
The critical value is $r_s = -0.457$.

249. The hourly workers as well as the managers at a large manufacturing firm were asked to rank seven aspects related to working conditions at the firm. The overall rankings were obtained for both groups and the results were as follows:

	Aspect						
Ranking	1	2	3	4	5	6	7
Managers Ranking	4	3	1	7	2	5	6
Employee Ranking	6	2	3	5	1	4	7

Find the Spearman correlation coefficient.

ANSWER:
$r_s = 0.714$

QUESTIONS 250 THROUGH 252 ARE BASED ON THE FOLLOWING INFORMATION:
Consider the following bivariate pairs of ranks.

Rank of X	1	2	4	3	7	8	6	5
Rank of Y	2	3	1	4	6	8	5	7

250. Calculate the Pearson's product moment (r).

ANSWER:
0.786

251. Calculate the Spearman rank correlation coefficient (r_s).

ANSWER:
0.786

252. Compare your answers to questions 250 and 251. What did you notice?

ANSWER:
The Pearson's product moment (r) and the Spearman rank correlation coefficient (r_s) have the same value.

253. Consider the following data, which has the shape of a parabola. Find both the Pearson's product moment (r) and the Spearman rank correlation coefficient(r_s).

x	1	2	4	6	10	13	15	20	25	30
y	1	4	16	36	100	169	225	400	625	900

ANSWER:
$r = 0.963$, and $r_s = 1$.

QUESTIONS 254 AND 255 ARE BASED ON THE FOLLOWING INFORMATION:
The following data have several ties.

x	2	2	2	4	4	4	4	6	6	6
y	4	3	7	10	10	10	14	18	16	16

254. Rank both variables and then apply the Pearson product moment to the ranks to find the Spearman rank correlation coefficient.

ANSWER:
0.9585

255. Use the formula $r_s = 1 - \dfrac{6\sum(d)^2}{n(n^2-1)}$ to find the Spearman rank correlation coefficient.

ANSWER:
0.9606

Chapter 14

256. The weights and gestational ages of seven very low birth weight infants were recorded. The results were as follows:

Weight (Grams)	700	800	1050	725	990	1025	700
Age	25	28	30	27	32	31	27

Use the Spearman rank correlation coefficient (r_s) to test for positive correlation between the two variables at $\alpha = 0.01$.

ANSWER:

Since the critical region is $r_s \geq 0.893$; and test statistic is $r_s^* = 0.827$; we fail to reject the null hypothesis.

Conclusion: There is no sufficient evidence to conclude that there is a positive correlation between weight and age.

257. The ages for seven mated pairs of California gulls (in years) were recorded with the following results:

Males:	4	14	10	5	4	8	9
Females	3	12	10	7	4	6	5

Use the Spearman rank correlation coefficient to test the alternative hypothesis that the ages are positively correlated at $\alpha = 0.05$.

ANSWER:

Since the critical region is $r_s \geq 0.714$, and the test statistic is $r_s^* = 0.847$; we reject the null hypothesis.

Conclusion: There is sufficient evidence to conclude that the ages are positively related.

258. Consider the following bivariate data:

x	1	2	3	4
y	2	9	11	k

For what values of k will the Spearman rank correlation coefficient (r_s) = 1?

ANSWER:

$r_s = 1$ if k is any value greater than 11.

259. Determine the test criteria that would be used to test H_a: Variable B decreases as A increases given $n = 18$ and $\alpha = 0.01$ in a Spearman rank correlation experiment.

ANSWER:

Reject H_o if $r_s^* < -0.564$.

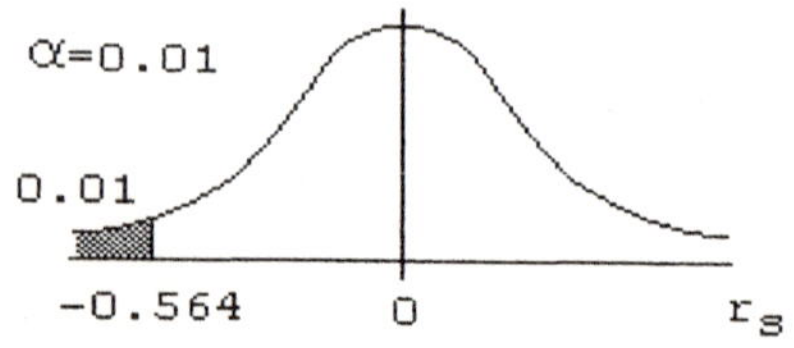

QUESTIONS 260 THROUGH 263 ARE BASED ON THE FOLLOWING INFORMATION:
Do foods high in fiber tend to have more sodium? The following table was obtained by selecting 11 soups from a list published in a health magazine. The soups were measured on the basis of both sodium content and fiber:

Soup	A	B	C	D	E	F	G	H	I	J	K
Sodium	490	840	520	470	500	590	430	300	460	440	400
Fiber	13	1	2	6	4	8	3	5	11	7	10

260. Rank the soups in ascending order based on the basis of their sodium content and on their fiber content, and show your results in a table.

ANSWER:

Soup	Sodium Rank	Fiber Rank	d	d^2
A	5	1	4	16
B	1	11	-10	100
C	3	10	-7	49
D	6	6	0	0
E	4	8	-4	16
F	2	4	-2	4
G	9	9	0	0
H	11	7	4	16
I	7	2	5	25
J	8	5	3	9
K	10	3	7	49

261. Compute the Spearman rank order correlation coefficient for the two sets of rankings.

ANSWER:

$$r_s = 1 - \frac{6\sum (d)^2}{n(n^2-1)} = 1 - \frac{6(284)}{11(120)} = 1 - 1.291 = -0.291$$

262. Does higher sodium content accompany foods that are higher in fiber? Test the null hypothesis that there is no relationship between the fiber and sodium content of the soups versus the alternative that there is a relationship between them at α = 0.05, using the *p*-value approach.

ANSWER:
$H_o : \rho_s = 0$ vs. $H_a : \rho_s > 0$ and the test statistic is $r_s{*} = -0.291$.
Using the table of critical values of Spearman's rank correlation coefficient we get: **P** > 0.10. Since **P** $> \alpha = 0.05$, we fail to reject H_o. There is not sufficient evidence presented by these data to enable us to conclude that there is any relationship between sodium content of soups and their fiber content.

263. Test the hypothesis stated in question 262 at α = 0.05 using the classical approach.

ANSWER:
The critical region is $r_s \geq 0.618$. Since the test statistic r_s * does not fall in the critical region; we fail to reject H_o. We reach the same conclusion as stated in question 262.

264. Determine the *p*-value that would be used to test " H_o: No relationship between the two variables vs. H_a: There is a positive relationship, " for the Spearman rank correlation experiment with *n* = 20 and r_s = 0.51.

ANSWER:
P = *p*-value = *P*(r_s ≥ 0.51 for *n* = 20) ⇒ 0.01 < **P** < 0.025

265. Determine the *p*-value that would be used to test " H_o: No correlation vs. H_a: There is a relationship," for the Spearman rank correlation experiment with *n* = 25, and r_s = 0.35.

ANSWER:
P = *p*-value = 2 · *P*(r_s ≥ 0.35 for *n* = 25) ⇒ 0.05 < **P** < 0.10

266. Determine the p-value that would be used to test " H_o: Variable A has no effect on Variable B vs. H_a: Variable B decreases as A increases" for the Spearman rank correlation experiment with *n* = 15, and r_s = 0.66.

ANSWER:
P = *p*-value = *P*(r_s ≥ 0.66 for *n* = 15) ⇒ 0.005 < **P** < 0.01

267. Determine the p-value that would be used to test " H_o: No correlation vs. H_a: There is a relationship," for the Spearman rank correlation experiment with *n* = 12, and r_s = 0.44.

ANSWER:
P = p-value = 2 · *P*(r_s ≥ 0.44 for *n* = 12) ⇒ **P** > 0.10

268. Determine the critical region(s) that would be used to test " H_o: No relationship between the two variables. vs. H_a: There is a relationship," for the Spearman rank correlation experiment with *n* = 15 and α = 0.05.

ANSWER:
The critical regions are r_s ≤ -0.525 or r_s ≥ 0.525

269. Determine the critical region(s) that would be used to test " H_o: No correlation vs. H_a: Positively correlated," for the Spearman rank correlation experiment with *n* = 24 and α = 0.05.

ANSWER:
Critical region: r_s ≥ 0.343

270. Determine the critical region(s) that would be used to test " H_o: Variable A has no effect on Variable B vs. H_a: Variable B decreases as A increases, "for the Spearman rank correlation experiment with n = 19 and α = 0.01.

ANSWER:
The critical region is $r_s \leq -0.549$

QUESTIONS 271 THROUGH 275 ARE BASED ON THE FOLLOWING INFORMATION:
The following data were collected on 12 business students who graduated from an MBA program, where U = Undergraduate GPA, and G = Graduate GPA at Graduation.

U	3.5	3.1	2.7	3.7	2.5	3.3	3.0	2.9	3.8	3.2	3.6	3.1
G	3.4	3.2	3.0	3.6	3.1	3.4	3.0	3.4	3.7	3.8	3.7	3.0

271. Rank the undergraduate GPA and the graduate GPA for the 12 students, and present your results in a table.

ANSWER:

Rankings

U	9	5.5	2	11	1	8	4	3	12	7	10	5.5
G	7	5	2	9	4	7	2	7	10.5	12	10.5	2

272. Compute the Spearman rank order correlation coefficient for the two sets of rankings.

ANSWER:
Let d = U – G.

Rankings

U	9	5.5	2	11	1	8	4	3	12	7	10	5.5
G	7	5	2	9	4	7	2	7	10.5	12	10.5	2
d_i	2	0.5	0	2	-3	1	2	-4	1.5	-5	-0.5	3.5
d_i^2	4	0.25	0	4	9	1	4	16	2.25	25	0.25	12.25

$$r_s = 1 - \frac{6 \cdot \sum d_i^2}{n(n^2-1)} = 1 - \frac{6(78)}{12(12^2-1)} = 1 - 0.2727 = 0.7273$$

273. State the appropriate null and alternative hypotheses in testing that a positive correlation exists between undergraduate GPA and GPA at graduation from a graduate business program.

ANSWER:
$H_o: \rho = 0\ (\leq)$ vs. $H_a: \rho > 0$

274. Test the hypotheses in question 273 at the 0.05 level of significance using the *p*-value approach.

ANSWER:

P = *p*-value = *P*($r_s \geq 0.727$ for n = 12) $\Rightarrow$ 0.005 < **P** < 0.01

Since **P** < α = 0.05, we reject H_o. There is sufficient evidence to indicate that a positive correlation exists between undergraduate GPA and GPA at graduation from a graduate business program (MBA).

275. Test the hypotheses in question 273 at the 0.05 level of significance using the classical approach.

ANSWER:

The critical region is $r_s \geq$ 0.0.497. Since the test statistic r_s^* = 0.7273 falls in the rejection region, we reject H_o. We reach the same conclusion as stated in question 274.